Chemical Modelling

Volume 12

Specialist Periodical Reports Online

Application for Free Access to Electronic Chapters

Customers purchasing a print volume are now entitled to free site-wide access to the electronic version of that title. (For the definition of a site, please consult **www.rsc.org/subagree**)

To apply for free access, please contact **ejournals@rsc.org** with proof of purchase (a scanned copy is acceptable) containing details of:

- **Title of book**
- **Volume number**
- **ISBN**
- **Address of your organisation**

By providing these details you will be agreeing to the terms of the Electronic Information Licence Agreement at **www.rsc.org/subagree**

ROYAL SOCIETY OF **CHEMISTRY**

A Specialist Periodical Report

Chemical Modelling

Volume 12

Editors
J.-O. Joswig, *Technische Universität Dresden, Dresden, Germany*
M. Springborg, *University of Saarland, Saarbrücken, Germany*

Authors
Alexey I. Baranov, *Max Planck Institute for Chemical Physics of Solids and TU Dresden, Germany*
Yong Cao, *Fudan University, China*
Michele Cascella, *University of Oslo, Norway*
Hélio A. Duarte, *Universidade Federal de Minas Gerais, Brazil*
Luciana Guimarães, *Universidade Federal de São João del Rei, Brazil*
Yi-Fan Han, *East China University of Science and Technology, China*
Xiaoming Huang, *Dalian University of Technology, China*
Miroslav Kohout, *Max Planck Institute for Chemical Physics of Solids, Germany*
Maicon P. Lourenço, *Universidade Federal de Minas Gerais, Brazil*
Doreen Mollenhauer, *Justus-Liebig-Universität Gießen, Germany*
Artem R. Oganov, *SUNY Stony Brook, NY, USA; Moscow Institute of Physics and Technology, Russia; Northwestern Polytechnical Univerisity, Xi'an, China*
Robert Ponec, *Institute of Chemical Process Fundamentals, Czech Republic*
Linwei Sai, *Hohai University, China*
Mathieu Salanne, *Sorbonne Universités, France*
Ruili Shi, *Dalian University of Technology, China*
Yan Su, *Dalian University of Technology, China*
Lingli Tang, *Dalian Nationalities University, China*
Pengfei Tian, *East China University of Science and Technology, China*
Jean Christophe Tremblay, *Freie Universität Berlin, Germany*
Stefano Vanni, *Université Nice Sophia Antipolis and CNRS, France*
Jing Xu, *East China University of Science and Technology, China*

Xin-Chao Xu, *East China University of Science and Technology, China*
Qingfeng Zeng, *Northwestern Polytechnical Univerisity, Xi'an, China*
Jijun Zhao, *University of Technology, China*
Xiangfeng Zhou, *Nankai Univeristy, Tianjin, China*
Qiang Zhu, *SUNY Stony Brook, NY, USA*

ROYAL SOCIETY OF CHEMISTRY

THE QUEEN'S AWARDS FOR ENTERPRISE: INTERNATIONAL TRADE 2013

Print ISBN: 978-1-78262-115-7
PDF eISBN: 978-1-78262-270-3
ISSN: 0584-8555
DOI: 10.1039/9781782622703

A catalogue record for this book is available from the British Library

Published by The Royal Society of Chemistry,
Thomas Graham House, Science Park, Milton Road,
Cambridge CB4 0WF, UK

Registered Charity Number 207890

For further information see our web site at www.rsc.org

Printed in the United Kingdom by CPI Group (UK) Ltd, Croydon,
CR0 4YY, UK

Preface

Michael Springborg[a] and Jan-Ole Joswig[b]
DOI: 10.1039/9781782622703-FP007

In this volume, the 12th volume of the *Specialist Periodical Reports* on *Chemical Modelling*, we, the editorial team, have continued to present reviews on attractive and current active areas of chemical modelling. As in the past years, we continue the annual appearance of the series and the wide range of different topics.

Volume 12 covers various topics from very accurate quantum-chemical modelling of small systems and reactions to multi-scale approaches for the descriptions of very large systems. For the first, one chapter focuses on a special catalytic reaction: the Fischer-Tropsch process that converts carbon monoxide and hydrogen gas into hydrocarbons. This process was developed 90 years ago in 1925. An example for the latter is an interesting contribution on bridging the modelling scales by developing accurate coarse-graining approaches for large-scale simulations. Additionally, a number of chapters deal with interesting nano-scale materials that are currently in the focus of the research community, such as imogolite nanotubes, gold clusters and nanoparticles and their interaction with nitrogen- and phosphine-ligands, as well as clusters in general, and materials for supercapacitors. More fundamental chapters review the development of structure prediction, delocalized chemical bonds in solids and vibrational quantum dynamics at metallic surfaces.

An exciting aspect about editing such a book series is that through the editing process we ourselves increase our knowledge about topics that are not directly related to our own daily research. We very much hope that also you will feel this excitement when reading Volume 12.

For the reader who is interested in further aspects of chemical modelling we mention that this field, of course, is not only of interest to theoreticians but also used in addressing problems in experimental studies. Thus, also other *Specialist Periodical Reports* on other topics contain contributions devoted to chemical modelling. This is, *e.g.*, the case for vol. 27 of the *Catalysis* series and for vol. 24 of the *Electron Paramagnetic Resonance* series.

[a]*Saarbrücken, Germany. E-mail: m.springborg@mx.uni-saarland.de*
[b]*Dresden, Germany. E-mail: jan-ole.joswig@chemie.tu-dresden.de*

CONTENTS

Cover

The icosahedral 'golden fullerene' WAu_{12} reproduced by permission of Pekka Pyykkö, Chemistry Department, University of Helsinki, Finland.

Toward accurate coarse-graining approaches for protein and membrane simulations

Michele Cascella*[a] and Stefano Vanni[b]
DOI: 10.1039/9781782622703-00001

1 Introduction

From the smallest biological molecules to complex living organisms, the organisation of the living matter follows highly hierarchical organisation. As depicted in Fig. 1, starting from the Ångström dimensionality, we first encounter atoms and molecules, then oligomers and polymer, like short RNAs or single-domain globular proteins; at larger scales, macromolecular assemblies give rise to cellular organelles and cells, that in turns, in superior organisms, form tissues, organs, and finally the whole body. Likewise, different biological phenomena occur at different size- and time-scales, and therefore can be understood by employing methods of investigation at the most pertinent level of resolution.[1]

Since the beginning of the informatics revolution in the 50's of the past century, major effort has been put in developing reliable mathematical and physical computational models of complex systems at different resolutions. In bottom-up approaches, the aim is to establish computational models based on fundamental physical principles that are able to predict the behaviour of the system of interest (Fig. 2). At the most fundamental level, quantum mechanics approaches[2] can be used to treat relatively small-sized molecular systems (up to $\sim 10^3$ atoms, and for times of the order of $10^{-12,-10}$ s).

Quantum mechanical calculations can nowadays reach up even to millions of atoms for static calculations,[3,4] also in this case depending on the degree of approximation with respect to the exact theoretical formulation (and according to the complexity of the system of interest).[5–13]

For larger systems and longer times ($\sim 10^6$ atoms, and for routinely times of $10^{-8,-7}$ s up to $10^{-6,-3}$ s) molecular models employing explicit representation of atoms (all-atom models, AA hereafter) interacting through parameterised mechanical effective potentials are the most commonly used approach.[14–17] Such potentials can be trained on both accurate quantum-mechanical calculations and on large experimental data set[18–22] and they can reliably reproduce molecular processes involving non-covalent intermolecular interactions or conformational changes.[23]

[a]*Department of Chemistry, and Centre for Theoretical and Computational Chemistry (CTCC), University of Oslo, Sem Sælands vei 26, 0371 Oslo, Norway. E-mail: michele.cascella@kjemi.uio.no*
[b]*Institut de Pharmacologie Moléculaire et Cellulaire, Universiteé Nice Sophia Antipolis and CNRS, 06560 Valbonne, France. E-mail: vanni@ipmc.cnrs.fr*

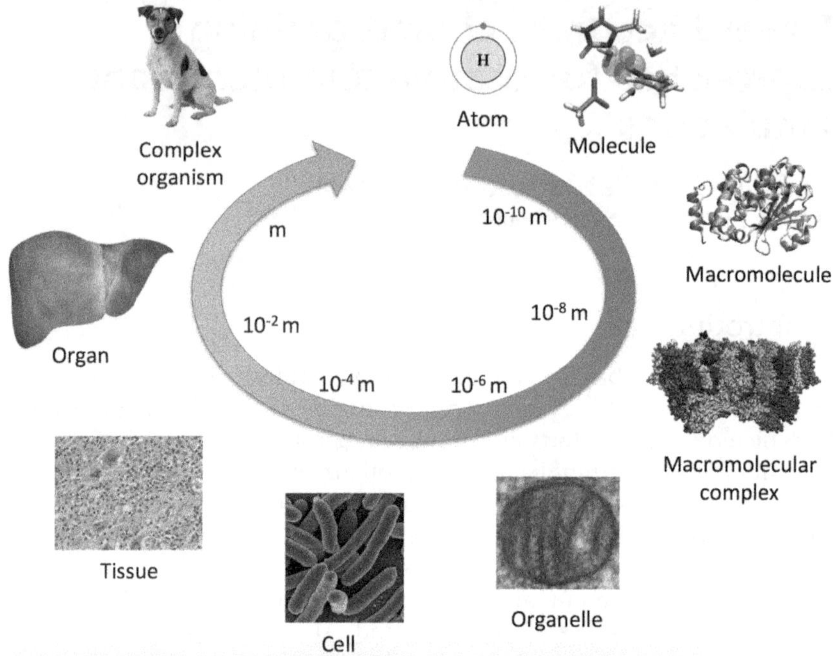

Fig. 1 Schematic representation of multi-scale assembly of constituents of matter from the atomic scale to the macroscopic one.

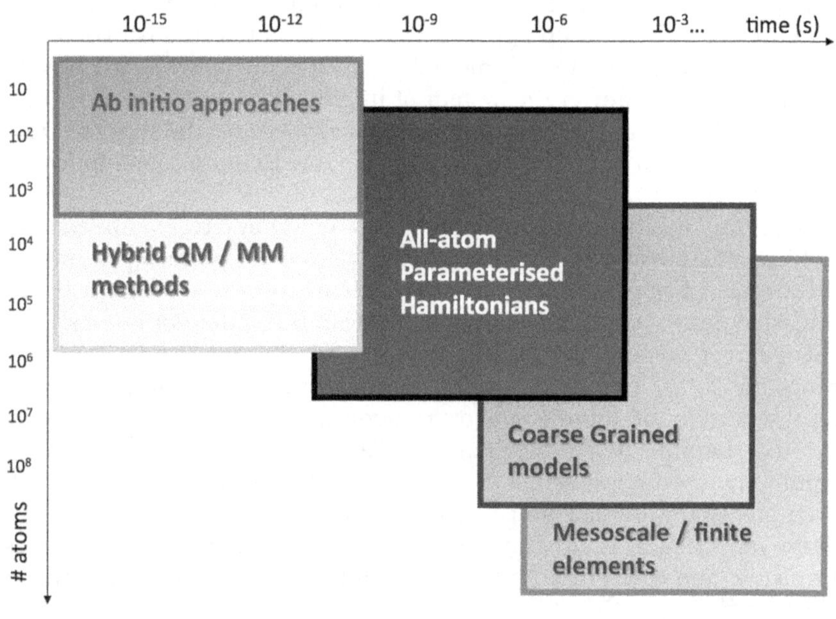

Fig. 2 Accessible time- and size-scales for different computational methods in routine studies using state-of-the-art implementations and architectures.

Combination of quantum mechanical and classical methods in a hierarchical structure is often used as a way of treating those biochemical phenomena that require quantum mechanical treatment while keeping a direct coupling with the environment.[24] Historically, the first multi-scale model proposed dates back to 1976 by Arieh Warshel and Michael Levitt,[25] where the idea of embedding quantum mechanical treatment of a chemically relevant portion of a biological system (like the active site of an enzyme) into a parameterised description of the environment was proposed. In the past decades, a large family of hybrid quantum mechanics/classical mechanics (QM/MM) methods stemmed from this seminal work, and have established what is today recognised as the standard common practice to treat quantum mechanical phenomena in biological systems.[26–36] For this fundamental theoretical work Profs. Warshel and Levitt, together with Prof. Martin Karplus, were awarded the Nobel Prize in Chemistry in 2013.

Even though atomistic simulations can now deal with systems as large as tens of millions of atoms, and for simulation times beyond the millisecond,[37,38] several biological processes involving large macromolecular complexes require description at time and sizes that go beyond even such dimensionalities. In order to overcome these bottlenecks, in the past decades several groups have been working on the development of reliable Coarse-Grained (CG) models.[39–56] Similarly to AA, effective CG potentials can be derived from higher-resolution AA simulations, or from direct match with specific experimental properties of interest. In such approaches the detailed atomic resolution is lost; nonetheless, some information on the topological structure of the molecular assembly is retained, as described in Fig. 3. These models can efficiently represent molecular systems composed by several millions of atoms, for effective times that can reach the second scale; therefore, they are in principle well-adapted to investigate the structure and dynamics of large macromolecular assemblies and multi-phase systems. The large number of reviews published on the subject in recent years highlights the strong interest by the scientific community in this topic (for example: ref. 1, 16, 50, 53, 57–67).

Treatment of very large systems, and for very long times, opens up a completely different view on the understanding of biological systems and phenomena. In fact, it is often the case that the complexity of such systems is irreducible to the fundamental properties of individual or relatively few molecules, but it requires the treatment of a large number of particles. Moreover, biochemical/biophysical processes are often not simply driven by simple thermodynamic equilibrium, but several kinetic effects, like for example diffusional barriers, may play a fundamental role.

As a pivotal example of the power of coarse-graining approaches in modern computational investigations, a very recent study by Marrink and co-workers was able to investigate the lipid composition, dynamics and diffusion in a realistic model of the plasma membrane.[68] Biological membranes are extremely complex environments formed by several lipophilic/amphiphilic compounds, which can behave in very different manners according to their specific composition. Building reliable models of such environments necessarily implies the use of large model systems to respect the relative concentration of the different species.

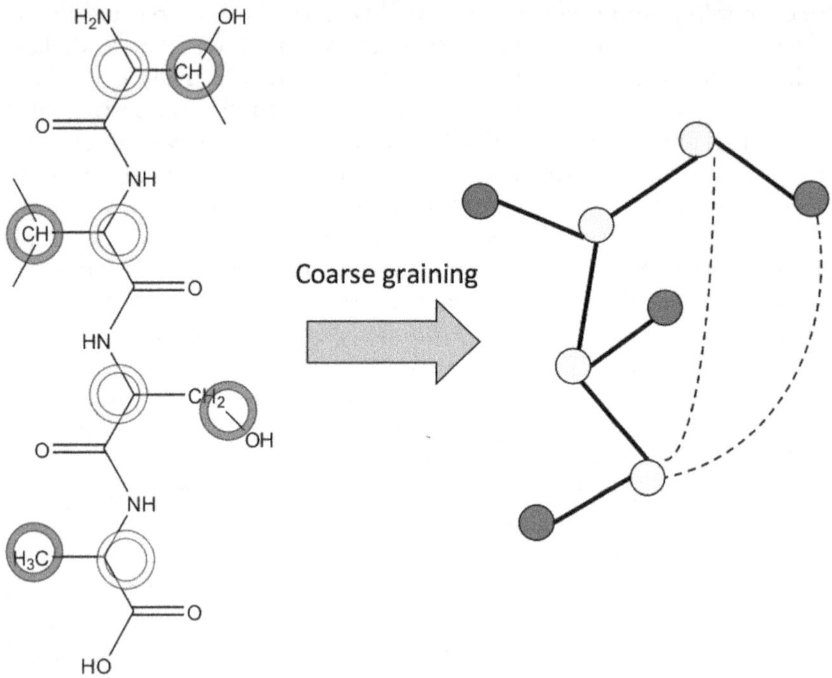

Fig. 3 The coarse graining approach. A molecular system (left) is mapped into a pseudo-molecular one (right), where each body is representative of a neighbouring group of atoms.

Moreover, properties such as lipid lateral diffusion, lipid flip/flop, membrane elasticity and surface tension can be heavily biased in MD simulations by too small boundaries. The study in ref. 68 reported a model made of over 60 different lipid types, in a stoichiometric ratio compatible to that experimentally determined using lipidomics approaches, for a total of ~20 000 lipid molecules, and simulated for 40 microseconds using coarse-grained potentials.

The major complication present in any multi-scale modelling approach of biological systems is associated to the fact that phenomena characteristic of a certain size/time scale may influence, directly or indirectly, properties that are intrinsic of a different scale. For example, the network of molecular interactions and molecular recognition patterns at interfaces directly influences the dynamics of large macromolecular complexes; on the other hand, the *in vivo* efficiency of an enzyme in a cell does not solely depend on its catalytic activity, but also on the accessibility to the substrates within the highly crowded cytoplasmic environment.

To date, the computational community struggles in the effort of building up, on one hand, more and more reliable and general models at the different resolution scales; on the other hand, it is becoming increasingly urgent to develop methods that combine and integrate information from multiple resolutions, in order to improve the predictive power of the implemented models.[48,49,57,69,70]

2 Coarse-grained modelling: basic ideas

The fundamental concepts related to coarse-graining root deeply into developments of statistical mechanics and the study of phase-transitions near the critical point. These studies evidenced how in such regions of the phase diagram the behaviour of the single particles become less relevant, while collective phenomena dominate the global behaviour of the system.[71–73] Although not necessarily near critical points, soft-matter systems, and proteins in particular, are usually characterised by a rather complex phase diagram, and at room conditions they often lay in marginally stable regions near several phase transition crossings.[74] This suggests that several of the physical properties of a polymer may be understood employing a description that is at a coarser resolution of the atomic one (for example: ref. 75–79).

Restricting our discussion to biological systems, the first CG model for proteins was proposed again by Levitt and Warshel.[80] The original model made use of single centroids to describe individual amino acids, and of an elastic network to reproduce the folded state structure and predict the folding kinetics. Through the last 40 years, several coarse grained models have been developed to study polymers, multiphase systems, proteins and nucleic acids, as previously referenced.

A coarse graining operation implies the mapping of a finely grained system formed by N particles and described by a Hamiltonian $\mathrm{H}_N(\boldsymbol{\alpha})$, defined by a set of parameters $\boldsymbol{\alpha}$, onto a second system composed by $M<N$ particles and responding to a new Hamiltonian $\mathrm{H}_M(\boldsymbol{\beta})$, which will depend on another set of parameters $\boldsymbol{\beta}$.

The mapping operation requires a transformation connecting the N and M particles of the two systems. Moreover, the mapping Hamiltonian $\mathrm{H}_M(\boldsymbol{\beta})$ must be such that the partition function Z_M:

$$Z_M = \int \mathrm{e}^{-\beta \mathrm{H}_M} \mathrm{d}\Gamma_M \tag{1}$$

is distributed according to the statistical distribution of the starting system.

In other words:

$$Z_M = \int \Xi(\boldsymbol{r}_N, \boldsymbol{p}_N \rightarrow \boldsymbol{r}_M, \boldsymbol{p}_M) \mathrm{e}^{-\beta \mathrm{H}_N} \mathrm{d}\Gamma_N \tag{2}$$

where $\Xi(\boldsymbol{r}_N, \boldsymbol{p}_N \rightarrow \boldsymbol{r}_M, \boldsymbol{p}_M)$ is the transformation mapping any conformation of the finely grained system into the coarse grained one.

Such transformation is evidently non-trivial for the following reasons:

(i) **The mapping transformation is ill defined.** In fact, there is no unique mathematical way of defining the mapping from one fine description into a coarse one. For example, a variable number of bodies may be used, yielding different levels of coarsening, or the same number of bodies may be used to address different structural parameters as degrees of freedom of the coarse grained Hamiltonian.

(ii) **The functional form of the coarse grained Hamiltonian is ill defined.** In fact, there is no universal transformation that defines an analytical potential function for any coarse grained representation. The choice of the functional form for the potential energy is typically made according to the properties of interest that the model should address.

(iii) **There is no consensus on how to parameterise the coarse grained Hamiltonian.** This is a consequence of the absence of a well-defined functional form for the potential energy. The interaction among coarse-grained bodies can be built to match as rigorously as possible the fine-grained system, or to fit chosen experimental data sets, or to model a specific property. In any case, this leads to potentials that have either limited transferability or reduced reliability.

Ultimately, the challenge in coarse-graining procedures lies in the definition of those finely grained degrees of freedom that will be disregarded in the coarse representation. It is evident that such choice is strictly dependent on the properties of interest that the coarse grained model should investigate. In fact, not all phenomena depend to the same extent on the same degrees of freedom or structural/dynamical parameters. Therefore, it is debatable whether the hunt for a universal coarse-grained model is by itself well placed, and that different models addressing multiple properties should instead be the right approach to coarse graining.

3 Protein representations

3.1 Atomistic *vs.* coarse-grained modelling

The study of the proteome constitutes one of the most fascinating challenges of molecular and cellular biology. Investigations may address different topics, comprising, among others, the relationship between the amino acid sequence, the structure and the function of individual proteins, protein folding, protein/protein interaction networks, dynamical effects on protein function (like in motor proteins), or interference of protein function by ligand binding.

Different phenomena associated to proteins can span the most different time and size scales. In fact, single protein filaments can have very different lengths (from few tens of amino acids like in small rubredoxins,[81] to hundred thousands like in titin[82]) and processes can be as fast as femtoseconds (like in early photo-activation of the visual signal in rhodopsin)[83,84] down to several hours, for example in weakly catalysed processes.[85] It is therefore unrealistic to imagine the establishment of a universal computational model able to describe so different phenomena at so diverse scales.

In the course of the decades, different computational protocols tackling modelling of proteins at different resolutions have been established. At the most accurate level, we find today quantum mechanics methods (mostly based on Density Functional Theory), used in combination with embedding methods.[24] The applicability of such methods is typically restricted to study those biochemical phenomena that strictly require a quantum-mechanical treatment. The most prominent examples include the study of enzymatic activity, photo-activation, or biological electron transfer and redox systems.[24] Even though the constant increase of computational power has allowed in recent years the appearance of the first quantum-mechanical studies on whole small proteins, even with explicit quantum mechanical treatment of the solvent,[4] it is unlikely that such approaches become feasible on a routinely basis to study biological phenomena occurring at larger time and size dimensionalities.

To date, the standard method to study structural and dynamical properties of proteins that provides the best compromise between accuracy and computational feasibility is classical molecular dynamics using parameterised potentials.[86] Within this approach, molecular systems are described starting from their atomic constituents. Atoms are represented as massive point objects; molecular structures are held together making use of mechanical stretching, bending and torsional effective potentials mimicking the binding effect on nuclei by the electronic cloud. The effect of bond polarization and van der Waals forces driving the interaction between non-bonded atoms is typically taken into account by associating an individual point electrostatic charge to any atomic centre, and by defining Lennard-Jones potentials between pairs of atoms. Overall, the general formula of an AA potential takes the following analytical form:

$$V^{\mathrm{AA}} = \sum_{\mathrm{bonds}} \frac{1}{2} k_b (r - r_0)^2 + \sum_{\mathrm{angles}} \frac{1}{2} k_a (\vartheta - \vartheta_0)^2 + \sum_{\mathrm{dihedral}} k_t [1 + \cos(n\varphi + \delta)] + \sum_{i<j} \frac{q_i q_j}{R_{ij}} + \sum_{i<j} \left[\frac{A_{ij}}{R_{ij}^{12}} - \frac{B_{ij}}{R_{ij}^{6}} \right]. \tag{3}$$

where the first three terms represent the stretching, bending and torsional energies, the set of (q_i) parameters define the Coulomb charges associated to each atom present in the system, and the (A_{ij}, B_{ij}) constants are the Lennard-Jones parameters associated to each i, j couple of atoms.

Parameterisation of the several constants defining the function V^{AA} can be achieved in multiple ways, but in general is determined by combining fitting over a series of experimental and high-level quantum mechanical data. In the course of the last decades, several sets of standardised parameters, generally named "force fields", have been developed.[18–22,87–89] These force fields are nowadays quite reliable in predicting molecular properties of most protein systems. In particular, several folding simulations have verified that the global free energy landscape of protein conformations can be reliably reproduced using such potentials.[90–92] Force fields are continuously re-parameterised to improve their reliability and transferability; therefore, we advise the reading of specialised literature and reviews to have a clear view on the performance of the different force fields for the calculation of different properties.[23]

To date, world-leading groups in molecular simulations have pushed molecular dynamics studies of proteins to reach time-scales on the millisecond range, and sizes as big as 10^7 atoms. These technical progresses allowed to investigate the folding of several globular proteins,[37,91,92] to elucidate signal transduction in G protein coupled receptors,[93,94] and to achieve structural refinement of low-resolution cryo-EM images of the HIV-1 capsid.[38]

Nonetheless, these dimensionalities are, on one hand, not accessible to the broad computational scientific community, and at the same time, they are not sufficient to cover the scales pertinent to large *in vivo* macromolecular complexes.

To date, several research groups are actively working to establish simplified CG potentials capable of reliably representing large macromolecular protein aggregates and the associated slow phenomena, and several reviews available in the literature provide detailed historical retrospective on the field of CG modelling.[1,16,50,53,57–67] Here we focus on the present challenges and on what are the currently debated possibilities of pushing the field forward.

3.2 Elastic network models

As explained in the former paragraphs, the CG procedure includes two steps: the definition of a simplified coarse system mapping the more complex and finer initial one, and the definition of a potential energy function for the coarse system capable of mapping the properties of interest of the finely-grained one.

The first model for proteins containing these ingredients dates to one of the fundamental seeding papers of the field of biomolecular simulations by Levitt and Warshel.[80] In their paper, they presented a simplified model for protein folding based on the concept of elastic networks. In these approaches, the structure of a folded protein (known from the experiment) is reduced to few centroids, usually one or two per amino acid, placed at the C_α (and C_β) position. The distances between consecutive C_α in the polymer chain are constrained to the experimental C_α–C_α length over a peptide bond. All C_α's interact with all other C_α's not contiguous in the sequence through an (an-)harmonic potential centred at the experimental distance of the C_α–C_α couple:

$$V^{\mathrm{EN}} = \sum_{i,j>i+1} V\left(\left|\boldsymbol{r}_{ij} - \boldsymbol{r}_{0_{ij}}\right|\right) \tag{4}$$

These methods accept intrinsically the native folded structure as a global minimum of their energy, and therefore can be studied to explore in a fast and efficient way the folding landscape. In particular, folded structures lay in a well-defined global minimum of the free energy of the protein. Therefore, small fluctuations of the energy should follow a quadratic law. This can be easily implemented in elastic networks, by imposition of a strictly harmonic analytical form of the potential V in eqn (4).

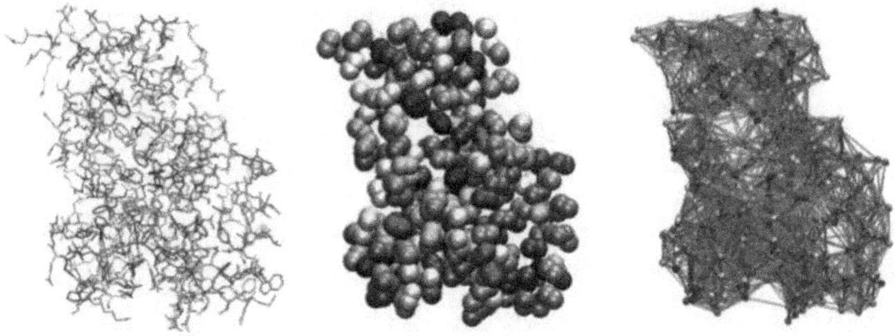

Fig. 4 The elastic network model approach. A folded protein (left) is reduced to a minimal set of beads representative of its amino acid sequence (centre). Beads are connected to each other (right) through a model pair-potential with global minimum at each pair's distance in the native structure.

Several studies in the past years have shown how such apparently simple approaches are indeed capable of representing well thermally associated local deformations of protein scaffolds, as well as how they can be used to reproduce principal component analysis coming from more complex all-atom simulations of the same targets.[95–105]

In 2004, based on previous studies by Bahar and coworkers,[95–97] Micheletti *et al.* worked on an extension of the model of elastic network where one single bead is introduced to mimic the presence of the side-chain (Fig. 4).[101] This minimal complication of the model is sufficient to significantly increase the accuracy of the predicted fluctuations.[102] In this so-called β-*Gaussian model*, a protein is reduced to a set of beads representing the C_α and the C_β of the various amino acids, and its total energy when in a trial configuration Γ takes the analytical form:

$$H(\Gamma) = H_B(\Gamma) + H_{\alpha\alpha}(\Gamma) + H_{\alpha\beta}(\Gamma) + H_{\beta\beta}(\Gamma) \tag{5}$$

Explicitly, the terms in the Hamiltonian are:

$$\begin{aligned} H_B(\Gamma) &= K\sum_i V\left(d_{i,i+1}^{C_\alpha C_\alpha}\right) \\ H_{\alpha\alpha}(\Gamma) &= \sum_{i<j} \Delta_{ij} V\left(d_{ij}^{C_\alpha C_\alpha}\right) \\ H_{\alpha\beta}(\Gamma) &= \sum_{ij} \Delta_{ij} V\left(d_{ij}^{C_\alpha C_\beta}\right) \\ H_{\beta\beta}(\Gamma) &= \sum_{i<j} \Delta_{ij} V\left(d_{ij}^{C_\beta C_\beta}\right) \end{aligned} \tag{6}$$

where Δ_{ij} represents the contact matrix between beads i,j in a predefined reference structure and d_{ij}^{XY} the separation of two beads in the configuration Γ. In the condition that the reference structure is the folded state of the protein, and thus sitting in the global minimum of the free energy landscape, the potentials $V\left(d_{ij}^{XY}\right)$ can be approximated by Taylor expansion truncated to its quadratic term:

$$V(d_{ij}^{XY}) \approx V(\boldsymbol{r}_{ij}) + \frac{k}{2}\sum_{m,n}\frac{r_{ij}^{m}r_{ij}^{n}}{r_{ij}^{2}}x_i^m x_j^n \tag{7}$$

where the distance d_{ij}^{XY} is expressed in terms of its equilibrium value $\boldsymbol{r}_{ij}$ and the instantaneous displacement $\boldsymbol{x}_{ij}$:

$$\boldsymbol{d}_{ij}^{XY} = \boldsymbol{r}_{ij} + \boldsymbol{x}_{ij} \tag{8}$$

and (m,n) are indexes that run over the three Cartesian coordinates. Furthermore, the coordinates of the C_β are embedded into the topological structure of the C_α, according the reconstruction scheme by Park and Levitt:[106]

$$\boldsymbol{r}_{C_\beta}^{i} = \boldsymbol{r}_{C_\alpha}^{i} + \ell\frac{2\boldsymbol{r}_{C_\alpha}^{i} - \boldsymbol{r}_{C_\alpha}^{i+1} - \boldsymbol{r}_{C_\alpha}^{i-1}}{\left|2\boldsymbol{r}_{C_\alpha}^{i} - \boldsymbol{r}_{C_\alpha}^{i+1} - \boldsymbol{r}_{C_\alpha}^{i-1}\right|} \tag{9}$$

where $\ell = 3$ Å. In this way, the coordinates of the C_α remain as the only degrees of freedom of the system, and thus the Hamiltonian of the system can be written simply as:

$$H = \frac{1}{2}k\sum_{ij,nm} x_{i,m}\mathcal{M}_{ij,mn}x_{j,n} \tag{10}$$

where $x_{i,m}$ is the deviation of the ith C_α along the m axis, and $\mathcal{M}$ is a $3N \times 3N$ symmetric matrix. The eigenvalues and the eigenvectors of $\mathcal{M}$ determine the elastic response of the system. Recently, such models have been also proposed as means of identification of evolutionary patterns in protein sequences.[107]

Despite their simplicity, elastic network models are based on very solid and intuitive mathematical-physical concepts (like polynomial expansion of the potential energy function from the global minimum), and this has constituted one of the reasons for their success in the literature in the past years. Nonetheless, at the same time, their simplicity restricts their applicability to a relatively small set of studies.

In particular, elastic network models reproduce the local shape of the folding free energy landscape of a protein by an artificial effective potential that does not contain any information on the physical means by which such shape is obtained. Moreover, they intrinsically bias the global minimum to a selected conformation, and thus they are not capable to capture the protein conformational changes and exploration of multiple structural minima that may be required for exploitation of the protein function (for example, in motor proteins, transporters *etc.*). Finally, these models do not contain any information on the external potential of the proteins; therefore, they cannot be in principle used to explore in detail interactions of proteins with the environment (*i.e.*, with membranes, other proteins or ligands).

3.3 Force-field based approaches

Establishment of CG protocols for proteins having broader applicability than elastic network models requires the definition of potential energy functions containing terms that are based on more general assumptions. The Holy Grail of the present research would be the definition of a universal form for such terms that would guarantee transferability over multiple systems keeping reasonable levels of reliability over several properties of interest.

In the next paragraph, we will discuss the general ideas behind force-field based coarse-graining, and will provide few examples of approaches present in the literature that could be used to push forward the field.

3.3.1 CG molecular Hamiltonian – simple concepts. The coarse-grained mapping of a molecular system can be done at different levels of resolution. We concentrate here on models at a relatively fine level of mapping that keep resemblance of the physicochemical nature of the polymer chain. Within such logic, models incorporate at least one body per amino acid (usually centred at the C_α position), but often they contain also one or multiple bodies to represent the side chains or other components of the backbone.

At this level of mapping, the coarse-grain systems still resemble the topology of the original molecular systems. It is therefore tempting to hypothesise that the CG potential should also resemble the all-atom one:

$$V^{CG} = V_{\text{bonded}} + V_{\text{nonbonded}} \tag{11}$$

The expression in eqn (11) is on purpose written in a very general form, because each term present there may or may not be expressed precisely as those of an AA force field (eqn (3)).

The most straightforward modelling, in fact, keeps all terms in eqn (11) analytically identical to atomistic models. Within this model, the Hamiltonian takes the form:

$$V^{CG}_{\text{mol}} = V_{\text{bond}} + V_{\text{angle}} + V_{\text{dihedral}} + V_{\text{nonbonded}} \tag{12}$$

This is at the base, for example, of the MARTINI force field, which is to date one of the most popular and broadly used CG models in the literature.[47] In particular, this model uses strictly the same harmonic or periodic analytical form for the stretching, bending and torsional distortion associated to bonds, angles and dihedral structural parameters, and Coulomb and Lennard-Jones like potentials for the non-bonded parts.

This approach has the advantage of being very easily implemented in all molecular dynamics codes, provides a direct estimate of the reduction of the computational requirements with respect to corresponding all-atom systems and allows the definition of highly-transferable single units that can be parameterised independently and used as building blocks to set up calculation of any protein system of interest. The MARTINI force field has been applied to the investigation of numerous processes involving protein-membrane interactions, including membrane proteins dimerization/oligomerization,[108–113] recognition of specific lipids by peripheral proteins[114–116] or the role of peptides in membrane fusion.[117,118]

As just mentioned, molecular-type CG force fields have a relatively broad range of applicability, as they were successfully used to reproduce the qualitative behaviour of protein localization at membranes, or the relative displacement of folded subdomains of large protein aggregates upon mechanical stress.[61] On the other hand, this description results too simplified to guarantee reliable transferability over different systems and cannot be used to study several properties of crucial interest in protein science. In particular, simple harmonic potentials for the backbone are not adequate for the description of secondary structure elements, making these kind of potential inadequate to study conformational changes in proteins, or folding events (as discussed in detail in the next paragraph). Moreover, the use of bead-types, in similarity with the atom-type concept of atomistic simulations, to calibrate non-bonded interactions for the various CG bodies heavily reduces the chemical variability, and thus reduces the capability to describe subtle but biologically relevant modifications of the interactions at specific sites.

For example, in a recent computational study on the mechanisms of membrane recognition and binding by α-Tocopherol Transfer Protein (α-TTP), simulations with the MARTINI force field were well able to capture the primary mechanism of α-TTP recruiting at the plasma membrane by phosphoinositides through contact formation with an evolutionarily well conserved basic patch at the protein surface.[119] On the contrary, CG simulations were not capable of determining the interaction pattern between the membrane and the sequence-variable N-terminal domain, which is known to be crucial for biological discrimination and targeting of different cellular compartments by different members of this protein family. It was instead possible to describe such interactions by atomistic models of the same system.[119]

Even though the choice of a bead type approach may intrinsically yield a reduced sensitivity of the model to both the chemical variability of the protein sequence and the surrounding environment, it is nonetheless reasonable to pose the question whether CG models of proteins should be meant to capture such effects at all. In fact, on one hand, any coarse grain mapping from an atomistic model intrinsically decreases the resolution to a less-than-chemical detail.

Opposite to this consideration there is the evidence that interactions at the chemical resolution ultimately drive the behaviour of all objects at the coarser one. Therefore, coarse models should be able to reproduce at least the main features of such interactions to guarantee that the properties of interest are reliably reproduced. The struggle between simplification of the representation and transferability/universality of the model is to date not resolved;[57,65,66] therefore, it is crucial for the user to understand the physical principles at the base of each model proposed in the literature, and to choose the most appropriate one according to the type of study of interest.

3.3.2 Refining the molecular potential terms. Since several years, several research groups have been investigating the possibility of improving the quality of CG potentials by moving beyond the molecular potential of eqn (12). Different modifications can be applied to both bonded and non-bonded terms.

Bonded-interactions. Figure 5 reports the atomistic representation of the backbone connecting three consecutive amino acids in peptide chain, and the corresponding CG representation that one would get in a hypothetical model where each amino acid is reduced to a single bead located at its C_α position. Assuming as first approximation rigid peptide bonds, relative displacement of the backbone can be achieved only by rotation around the (ϕ, ψ) torsional angles. The well-known Ramachandran plot, reporting the statistical abundance of the (ϕ, ψ) angles in polypeptides and proteins, clearly shows the existence of multiple regions of minimal free energy, corresponding to the different secondary structure elements characterizing folded polypeptides.

In C_α-based CG representations of backbone chains, the two-dimensional Ramachandran space is reduced to one bending parameter (angle γ in Fig. 5).[65] The most common secondary structure elements: α-helices and β-sheets correspond to two well-separated values of γ ($\sim 90°$ for the α-helix and between 120° and 140° for the β-sheet). This simple example is sufficient to explain how a CG potential that aims at exploring thermodynamically accessible protein conformations and/or folding events cannot use a simple harmonic approximation for the angular terms, but it requires more complicated functional forms allowing multiple minima.

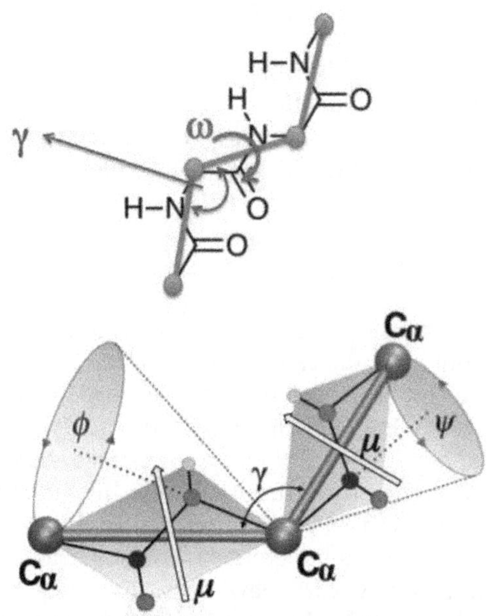

Fig. 5 Bonded structural parameters in AA and CG representations of a protein's backbone. The *top panel* evidences the angle γ and dihedral ω required to describe the relative position of consecutive C_α's along the protein backbone trace. The *bottom panel* highlights the connection between the angle γ and the AA Ramachandran angles (ϕ,ψ). Permanent electrostatic dipoles m associated to peptide bonds are evidenced with yellow arrows.

For example, a work by Tozzini *et al.* in 2006 showed that it is possible to map the (ϕ, ψ) space in an analytical way using the angle γ in combination with the dihedral angle ω formed by four consecutive beads[56] (Fig. 5).

Non-bonded interactions. Proper choice of bonded interactions can be applied to improve the specific elastic properties of the polymer chain at local sites. Similarly, more sophisticated modelling schemes can be applied to the non-bonded interaction terms of the potential function. In condensed-phase systems, non-bonded terms of the potential function constitute the most expensive part of the calculations. Therefore, it is desirable that these contributions are expressed in computationally affordable functional forms.

In atomistic models, non-bonded interactions are usually divided in electrostatic and van der Waals interactions, represented by Coulomb and Lennard-Jones potentials (eqn (3)). The computational cost of an all-atom simulation is largely due to calculation of the long-ranged electrostatic interactions, which decay with a R^{-1} power only. It is widely assumed in several CG models that in most cases non-bonded interactions may be represented by shorter-range potentials than the Coulomb one.

The electrostatic potential produced by a group of N charges in a point of space $\boldsymbol{R}_j$ is given by the Coulomb sum:

$$V^{\mathrm{C}}(\boldsymbol{R}_j) = \sum_{i=1}^{N} \frac{q_i}{|\boldsymbol{r}_i - \boldsymbol{R}_j|} \tag{13}$$

where $\boldsymbol{r}_i$ describes the position of each charge q_i. The same potential can be approximated as a function of the only vector $\boldsymbol{R} = (\boldsymbol{R}_j - \boldsymbol{R}_q)$ describing the distance of the point $\boldsymbol{R}_j$ from the centre of charge distribution $\boldsymbol{R}_q$. This approximation is the well-known expansion in multipoles:

$$V^{\mathrm{C}}(\boldsymbol{R}_j) \approx \frac{Q}{|\boldsymbol{R}|} + \frac{\boldsymbol{\mu} \cdot \boldsymbol{R}}{|\boldsymbol{R}|^3} + V(\mathbb{Q}) + \cdots \tag{14}$$

where Q is the total charge of the system, $\boldsymbol{\mu}$ is the electrostatic dipole of the charge distribution, and $\mathbb{Q}$ is its electrostatic quadrupole. The expansion extends to higher order multipoles, here not explicitly written. Disregarding the side-chains of charged amino acids, grouping of atoms into CG beads usually leads to grouping of charge distributions of negligible total charge. Therefore, the corresponding long-range potential should decay with a power law >1. In particular, interaction among not charged beads should be analogous to a dipole-dipole potential, and therefore, it should follow a R^{-6} law. If so, CG electrostatic potentials should have similar asymptotic behaviour as the van der Waals ones, thus, it should be possible to implement the two into one single effective potential. Generalised $V^{n,m}(R)$ potentials of formula:

$$V^{n,m}(R) = \left(\frac{A}{R^n} - \frac{B}{R^m}\right) \tag{15}$$

have been tested and implemented in the literature. The use of values of $n < 12$ to describe the repulsive part of the potential is usually justified by the fact that CG beads, representing a molecular moiety, are expected to be softer than the electronic cloud of single atoms, while attractive terms usually maintain $m = 6$, consistently with dipole–dipole attraction law.

The use of a potential of the $V^{n,m}(R)$ form is advantageous because it is analytical and with analytical derivatives, therefore, it guarantees cheap and accurate calculations of the forces, required for integration of the equations of motion *via* Verlet algorithm, or any other similar one. Nonetheless, such potentials may lead to rather unsatisfactory results.

In fact, proteins are expressed and exploit their function in the condensed-phase. Therefore, long-range interactions between non-bonded groups are never direct, but they are constantly mediated by the solvent. In atomistic simulations, the solvent is present in the system, either by explicit representation of the solvent molecules, or by introducing some form of dielectric continuum response. In any case, a portion of the potential function is directly taking into account the presence of the solvent. In CG representations, the situation is not so straightforward. For example, a simple dielectric response cannot be easily implemented for all those models that incorporate at least to a certain extent electrostatics into effective long-range potentials. The use of explicit CG beads for the solvent, as it is often done, not necessarily leads to reliable solvation models. This is particularly true for solute hydrophilic CG beads, where directional interactions, like H-bonds, which drive the local solvation structures, are missing. Tackling this problem implies the formulation of non-bonded effective potentials using analytical expressions that take into account the screening effect of the solvent. For example, the potential can be calibrated over radial distribution functions, thus incorporating the presence of regions of meta-stability due to the presence of solvation shells (see for example: ref. 51).

3.3.3 Radial *vs.* non-radial forms of the long-range interactions. So far, we have discussed possible modelling of the non-bonded interactions between CG beads that are strictly dependent on the distance between the two particles. This is straightforward for point or spherically symmetric objects, like atoms. This assumption is in fact not so obvious for CG beads, which are effective representation of molecular fragments.

Let's consider the multipolar expansion in eqn (14). Apart from the monopolar term, which strictly depends on the distance R only, all other terms depend on the relative orientation of the charge distribution generating the potential. Therefore, in the most general case, the long-range potential associated to a CG bead should not be dependent on the distance only.

$$V^{\mathrm{CG}}_{\text{long-range}}(\boldsymbol{R}) \neq V^{\mathrm{CG}}_{\text{long-range}}(|\boldsymbol{R}|) \tag{16}$$

One can imagine three cases when the formulation of $V^{CG}_{long-range}(\boldsymbol{R})$ as a radial potential is valid if:

(i) **The interacting particles have a neat electrostatic charge**; in that case the dominating term is the Coulomb potential;
(ii) **Both particles have either a highly symmetric charge distribution, or are weakly charged.** In these cases, low order less symmetric multipoles are zero or negligible, thus the dominating terms are practically spherical, or anyway weaker or shorter-ranged than van der Waals interactions. This occurs for example in hydrophobic groups;
(iii) **The internal dynamics of the particles is fast enough**, so that the reorientation of the multipoles along the electrostatic field is much faster than translation of the particle, for example, in free solvent particles.

It is evident that these conditions do not apply to neutral polymeric chains containing strongly polar groups. In fact, the same polymer structure imposes topological restraints that prevent free and fast relaxation of the polar moieties. Polypeptide chains fall in this last category, as each peptide bond connecting two consecutive amino acids brings a permanent dipole of about 3.5 D oriented roughly perpendicular to the main-chain direction. The orientation of the backbone dipoles is strictly associated to the Ramachandran angles, and therefore it cannot be assumed to change unless local conformational changes occur.

3.3.4 Example of a Hamiltonian beyond the molecular form: the UNRES potential. The UNRES model by Scheraga and co-workers is one of the most successful CG models for folding prediction available to date.[45,46] The possibility of exploring efficiently the conformational space of a protein sequence is achieved by keeping the number of interaction sites as small as possible: typically, one bead for the backbone, located at the peptide bond centres and one spheroid representing the side-chain. Auxiliary non-interacting sites are located at the C_α's, to facilitate the geometric definition of the chain structure (Fig. 6). The potential function of the UNRES models reads as:

$$\begin{aligned} V^{\mathrm{UNRES}} = {} & k_{\mathrm{SCSC}} \sum_{i<j} V(\mathrm{SC}_i, \mathrm{SC}_j) + k_{\mathrm{SC}_p} \sum_{i \neq j} V(\mathrm{SC}_i, p_j) + k^{el}_{pp} \sum_{i<j-1} V(p_i, p_j) \\ & + k^{vdW}_{pp} \sum_{i<j} V(p_i, p_j) + k_{\varphi} f_2(T) \sum_i V(\varphi_i) + k_{\varphi\varphi} f_3(T) \sum_i V(\varphi_i, \varphi_{i+1}) \\ & + k_{\gamma} \sum_i V(\gamma_i) + k_{\alpha\beta} \sum_i V(\alpha_i, \beta_i) + k_{\mathrm{bond}} \sum_i V_b(d_i) + \mathcal{O}(\mathrm{multibody})\ldots \end{aligned} \tag{17}$$

This potential introduces several additive contributions to the energy, taking into account the usual stretching, bending and torsional terms $V_b(d_i)$, $V(\gamma_i)$, $V(\phi_i)$, but includes also explicit correlation terms for consecutive torsional angles (ϕ_i, ϕ_{i+1}) and the angles (α, β) defining the

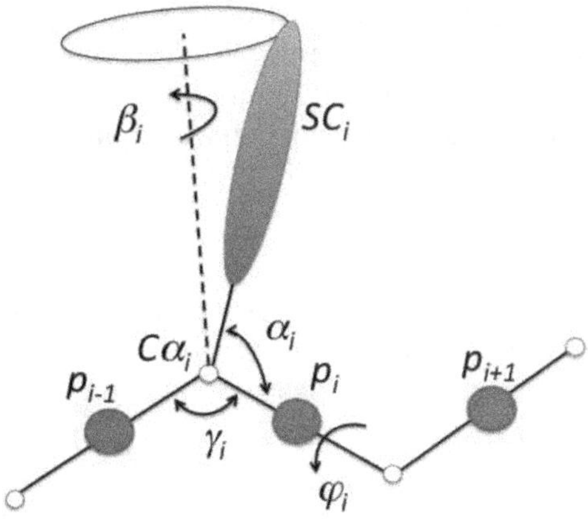

Fig. 6 Protein mapping by the UNRES model.[45,46] Each amino acid *i* is reduced to a non-interacting $C_{\alpha i}$ bead, a side-chain spheroid (SC_i) and connected to consecutive amino acid $i+1$ through an interacting bead representative of the backbone peptide bond p_i. Conformational arrangement is controlled by structural parameters (α, β, γ, ϕ).

position and orientation of each side chain group. Moreover, explicit temperature factors are included as screening constants for several terms of the Hamiltonian according to progressive terms of the Kubo cumulant-cluster expansion.[120,121]

The UNRES potential adds explicit long-range contributions to take into account orientation and alignment of the backbone groups mimicking dipole–dipole interactions, as well as multi-body contact terms to facilitate stabilisation and folding of secondary structure elements. Specifically, contact terms are used to induce the alignment of backbone groups according to experimentally known folded structures, in particular α-helices and β-sheets.

CG terms are calibrated on Potential of Mean Forces. These quantities can be extracted either from statistical distributions from the PDB data bank, or from atomistic MD simulations of model systems.[45,46] Currently, the UNRES force field provides two sets of parameters.[122] The first is calibrated over both structural and folding thermodynamics of G-related albumin-binding module (PDB code: 1GAB),[123] a triple α-helix bundle. The second set of parameters is calibrated over folding data for tryptophan cage and tryptophan zipper 2 (PDB codes 1L2Y, 1LE1),[124,125] small protein constructs displaying both α-helical and extended structural elements.

The UNRES model represents one of the most accurate CG force fields for protein folding, as confirmed by several assessments at CASP (Critical Assessment of Techniques for Protein Structure Prediction).[126] Recently the UNRES model proved to have good reliability also in predicting protein oligomerisation.[127–129]

3.4 Calibrating a coarse grained model

Definition of an appropriate functional form for a CG potential constitutes only one step of the problem. In fact, however it is expressed, the CG

Hamiltonian will, in general, depend on a set of free parameters that require appropriate calibration. Multiple strategies are present in the literature, that can be grouped into two large families: *top-down* approaches that match to global thermodynamic quantities, and *bottom-up* approaches that build parameters from the mechanical properties of corresponding atomistic models.

3.4.1 The Boltzmann inversion method.

In top-down methods, the statistical distribution $\mathscr{P}(\xi)$ of a specific order parameter ξ and the associated Free energy constitutes the typical starting point:

$$F(\xi) = -kT \ln \mathscr{P}(\xi) + \text{const.} \tag{18}$$

This expression is referred to as Boltzmann inversion, and it is the starting point to build several CG potentials. The main issue related to this approach is that the statistical distributions of different CG parameters ξ are not uncorrelated. Therefore, it is not possible to use additive potential of mean forces derived *tout-court* from the Boltzmann inversion for the different variables of interest to build a CG potential without introducing significant bias. Separation of *genuine* contributions from spurious one can be rigorously obtained by the *iterative Boltzmann inversion* procedure as introduced by Reith *et al.*[130] In this protocol, one starts from an initial potential of mean function $U_1(\xi)$, which will generate a probability distribution for the parameter ξ equal to $\mathscr{P}_1(\xi)$. The potential is then iteratively corrected according to the formula:

$$U_{i+1}(\xi) = U_i(\xi) - kT \ln\left(\frac{\mathscr{P}(\xi)}{\mathscr{P}_i(\xi)}\right) \tag{19}$$

where $\mathscr{P}(\xi)$ is the target distribution of interest. This procedure guarantees that, at convergence, the statistical distribution of ξ is matching the corrected one. Unfortunately, the number of free parameters in a CG Hamiltonian can be quite large, for example, in her review of 2010, Tozzini estimates roughly $\sim$400 terms only to calibrate bonded interactions.[65] Iterative Boltzmann inversion is in fact efficiently used in organic polymer simulations, even though some examples of application to biopolymers are now appearing (for example: ref. 131).

3.4.2 The force-matching method.

The force-matching method represents a rigorous way of building up CG parameters bottom-up from atomistic simulations. The procedure introduced by Ercolessi and Adams in 1994[132] consists in optimising the free parameters of a mapping Hamiltonian so that the forces acting on all the particles of a system reproduce those acting on the same particle described by the mapped Hamiltonian. For example, a molecular Hamiltonian may be built by matching the forces on atoms to the nuclear forces obtained from a quantum-mechanical calculation.[133] The procedure is very general, and it has been widely extended through the years by Voth and co-workers and others as a rigorous protocol to derive CG potentials from AA simulations.[43,44,48,49,55,134,135]

Briefly, the force matching procedure implies definition of a penalty function:

$$\chi^2(\alpha) = \mathrm{N} \sum_{\Gamma} \sum_{i=1}^{\#\ \text{CG beads}} \left| F_i^{\mathrm{CG}}(\Gamma;\ \alpha) - F_i^{\mathrm{AA}}(\Gamma) \right| \tag{20}$$

where Γ is a configuration of the system, α are the set of parameters defining the CG Hamiltonian, $F_i^{\mathrm{CG}}(\Gamma;\ \alpha)$ is the CG force acting on the ith CG bead, $F_i^{\mathrm{AA}}(\Gamma)$ is the corresponding force according to the AA potential, and N is a normalisation constant. Minimisation of the penalty function $\chi^2(\alpha)$ will determine the best set of parameters α.

If the potential is linearly dependent on the parameters α, the problem transforms in an over-conditioned set of linear equations, which can be minimised by several fitting protocols. Typical of such problems, the major issue of the force-matching procedure relies on the fact that minimisation may lead to several equivalent numerical solutions; therefore, solid optimisation algorithms are required, along with the use of restraints to confine the numerical solution on physically sound values. More importantly, the matching procedure requires sampling on a subset of structures typically obtained from AA simulations. As a consequence, parameter sets may be biased toward those structures that are observed in the relatively short time of an AA simulation, and therefore they may lead to inaccuracies for very long simulation times if the system drifts over conformational basins that were not explored during calibration of the potential.

To date, the force-matching procedure has been validated on several systems and at different scales. In fact, Voth and co-workers showed that this method can be reliably applied to diverse biochemical systems, spanning from lipid bilayers[44,136,137] and cholesterol/lipid interactions,[138] to monosaccharides,[139] peptides,[140] to large protein assemblies.[141–143]

3.5 Integrative multi-scale modelling

3.5.1 Hybrid AA/CG Hamiltonian. However accurate CG models can be, several phenomena associated in general to molecular recognition (like, receptor-ligand binding, adhesion at functionalised surfaces, substrate recognition, ligand-induced conformational changes, allosteric response *etc.*) strictly require an accurate description with atomistic detail. It is therefore crucial to develop multi-scale methods that allow exchange of information at multiple resolutions, thus being capable of exploiting the efficiency of CG descriptions with the accuracy of atomistic ones. Several possible approaches have been proposed. An excellent review is for example given as ref. 57.

As an example, here we present the hybrid all-atom/coarse grained (AA/CG) model introduced by Neri *et al.* in 2005.[144] Similarly to QM/MM, the AA/CG model explores the possibility of building a hybrid multi-scale

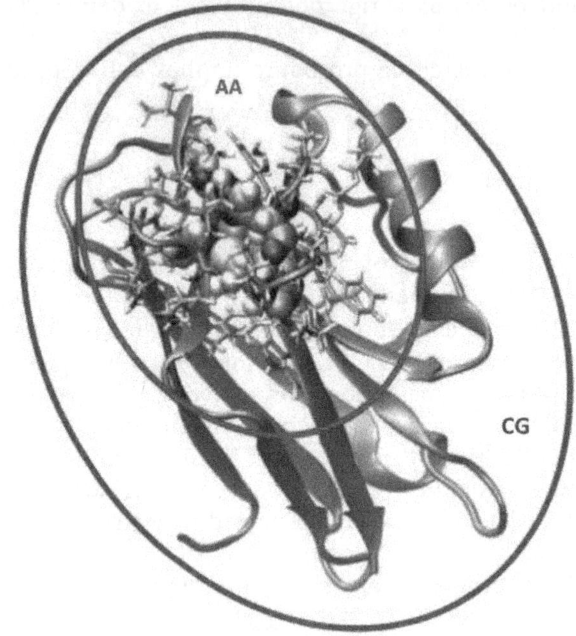

Fig. 7 The AA/MG idea from ref. 144: a portion of interest of the protein is described at the all-atom level and coupled to the rest of the system described by a Coarse Grained potential.

model where any portion of interest is treated at the more accurate atomistic resolution, while the rest is described at the CG level (Fig. 7). The original formulation as presented in the 2005 paper made use of a combination of the Gromos united-atom force field[21] and an elastic network model with anharmonic restraints:

$$V^{AA/CG} = V^{AA} + V^{CG} + V^{I} + V^{AA/I} + V^{CG/I} + V^{SD} \tag{21}$$

where V^{AA} is the atomistic potential function describing the all-atom region, V^{CG} is the anharmonic elastic network model applied to the CG part:

$$V^{CG} = \frac{1}{4}\sum_{i} k_b\left(\left|\boldsymbol{R}_{i,i+1}\right|^2 - d_0^2\right)^2 + \sum_{j>i+1} V_0\left[1 - e^{-B\left(\left|\boldsymbol{R}_{ij}\right| - b_{ij}\right)}\right]^2 \tag{22}$$

with $\boldsymbol{R}_{i,i+1}$ and d_0 being the distance of two consecutive CG particles and the average distance of two C_α's in a peptide bond, respectively; while $\boldsymbol{R}_{ij}$ and b_{ij} represent the distances of two non consecutive CG particles, and the corresponding equilibrium distance. Coupling between the atomistic and CG parts is mediated by a 6 Å thick interface shell around the atomistic portion of the system. This region maintains its all-atom details, and interacts through it with the proper all-atom portion; on the other hand, the centre of mass of each residue present in this region is connected to the elastic network and responds mechanically to it.

Different studies proved that the model is capable of describing thermal fluctuations of the protein scaffold with the same accuracy of pure elastic network models. Moreover, the atomistic region responds to the mechanical strain induced by the same fluctuations consistently with what is observed in all-atom simulations.[145–147]

The absence of direct coupling between the all-atom and the CG parts constitutes the major drawback of the model. In fact, several proteins are characterized by a strong long-range electrostatic potential, which influences local properties like the pK_a of titrable groups, the relative orientation of polar/charged moieties, the binding affinity of ligands, or the redox potential. Improvement of the model necessarily passes through the establishment of a reliable CG force field that includes explicit and accurate treatment of long-range electrostatics.

Cascella and co-workers demonstrated that simple structural information, namely, the positions of the C_α is sufficient to reconstruct the electrostatic field of a protein.[148] The model makes use of the topological correspondence between the orientation of the dipoles and the CG angle γ between three consecutive C_α's, and by reconstructing the position of the electrostatic multipoles associated to the side-chains of an amino acid based on restraints from chirality and flexibility of the side-chain similarly to the protocol by Park and Levitt.[106] Although very simple, the reconstruction of the electrostatic features of a protein is systematically more accurate than any point charge models as, for fields of intensity superior to $\approx 5.0 \cdot 10^{-5}$ a.u., the relative error with respect to the field generated by all-atom charge distributions is negligible. Moreover, the model is capable of reproducing the strong changes of the macromolecular dipole of a peptide during unfolding from helical to elongated conformations (Fig. 8).[148]

Dynamic models derived from the same assumption verified that, like in the UNRES model, the presence of the backbone dipoles is sufficient to naturally stabilise secondary structure elements and to predict assembly of secondary structure elements into super-secondary assemblies.[39] For example, the model describes accurately the presence of a macromolecular dipole along the axis of an α-helix, which in turn is sufficient to predict stabilisation of helix-coil-helix super-structures by advantageous alignment of the two respective dipoles.[39]

More recent developments of this approach have begun exploring the possibility of extending the model toward a universal force field with good transferability to all protein sequences. Spiga *et al.* showed that combination of the electrostatic model for the protein backbone with molecular-type CG potentials yielded systematically reliable predictions for different proteins with different folded structures.[55] Nonetheless, ad hoc calibration of the several parameters composing the potential function for each specific protein over atomistic simulations of the same system by force-matching procedure yielded systematically better results than a generic set of parameters calibrated over multiple targets.[55] This indicates that inclusion of accurate electrostatics into molecular-like potential functions, although improving their quality, is anyway not enough to achieve full transferability. For that, more sophisticated forms of the potential energy function might be explored.

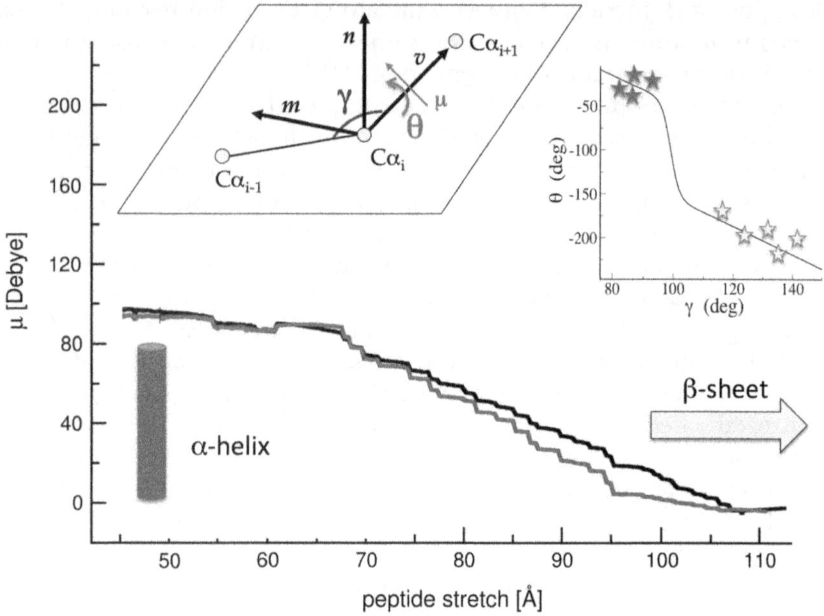

Fig. 8 Topological reconstruction of the backbone dipoles from ref. 148. *Top inset*: the relative orientation (θ) of a backbone dipole μ with respect to the internal reference system by three consecutive C_α's is determined by the angle γ, according to the relationship represented in the *right inset*. The stars indicate statistical distribution of (γ,θ) angles from the PDB databank (green = helical coiling, yellow = extended structures). *Main graph:* macromolecular electrostatic dipole along the longitudinal axis of a helical peptide undergoing unfolding by application of a longitudinal stretching force. The black line represents the estimate of the macromolecular dipole from an AA model, the red line is the prediction by the CG reconstruction.

3.5.2 Integrative experimental–computational approaches. The models presented so far aim at developing protocols that allow the investigation of bio-molecular phenomena from a purely computational perspective. In this respect, potential functions are built from sound physical principles, and experimental data are eventually used for initial calibration of parameters only. In more recent years, different groups have been developing protocols aimed at integrating experimental data with simulations *on the fly*. These are very pragmatic ways of training simulations to reproduce/interpret/rationalise in the most accurate way the data coming from the experiment. Although they may not be useful to build up transferable models, these may nonetheless produce some outstanding results. Here we present few examples of such approaches.

Integrative experimental-multiscale computational tools for refinement of cryo-EM data. Cryo-electron microscopy has revolutionised the field of protein imaging in the last years.[150] This technique presents several advantages with respect to traditional X-ray diffraction spectroscopy. First, it captures samples in an aqueous environment, which is intrinsically more "natural" than a highly concentrated condition like in a packed crystal. Second, it does not require the presence of ionic buffers for co-crystallisation, which may also induce significant local deformation of the protein structure. Third, glasses are obtained by rapid quenching of the solution; therefore, multiple room-temperature accessible conformations of the protein may be captured in one single experiment. Last but not least, the technique allows determination of large macromolecular complexes naturally present in solution but that may not be able to crystallise.

The major drawback of this technique lies in the typical resolutions that can be achieved. Especially for macromolecular complexes, it more commonly remains within 5 and 10 Å, which does not facilitate enough understanding of the specific intermolecular interactions regulating complex formation and functioning, even though outstanding structures down to 3.5 Å resolution are sometimes possible.[151]

Tama *et al.* developed in the past years a protocol called *molecular dynamics-flexible fitting*.[152,153] In this procedure, the normal modes of a protein of known structure are computed using simple elastic network model approaches (as described in Section 3.2). Then, the protein scaffold is deformed along the computed normal modes in order to fit the electron density map coming from an EM experiment. Once the structure of the protein that best fit the experimental data is found, the atomistic-detailed structure is mapped back starting from the original high-resolution data. In this way, the different conformations of a protein present in a Cryo-EM data set can be resolved at atomistic resolution.

The same protocol can be adapted to study the conformations of large macromolecular complexes for which the high-resolution structures of the single components are known. The most prominent example is the refinement of the HIV-1 capsid by Zhao *et al.*,[38] which also represents, at present, one of the largest atomistic calculations ever performed.

A similar method has been used in the past few years by Dal Peraro and co-workers to resolve several large *trans*-membrane multi-protein complexes.[149,154–156] Compared to original *molecular dynamics-flexible fitting*,[152,153] Dal Peraro and co-workers employed a *Swarm Intelligence Dynamic Modelling* search to explore efficiently the large conformational space spanned in protein complexes characterised by dramatically large conformational changes.[157]

The most prominent example is given by the resolution of the pore-forming toxin aerolysin (Fig. 9).[149] Such protein is the prototypic representative of the β-pore-forming toxin class, for which the structure was not known. Determination of low-resolution images by cryo-EM of the heptameric complex posed an even bigger puzzle, as different images were showing significantly different assemblies. Dal Peraro and co-workers proceeded with systematic investigation at both the experimental and computational levels. First, they determined at high-resolution the monomeric component of the toxin; then, atomistic simulations were employed to reveal large-scale motions accessible to these single proteins. Finally, the structure of the complex in its resting state was determined.

Analysis of the large-scale fluctuations revealed that a simple concerted twisting motion of the seven monomers allows formation of a trans-membrane β-pore, by extrusion of long β-domains from the core of the

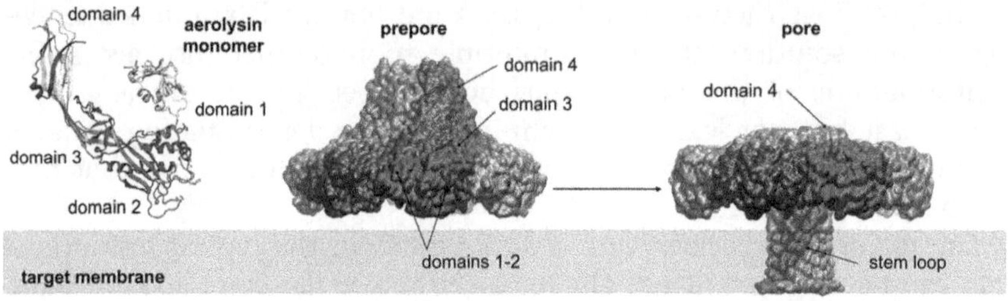

Fig. 9 Determination of macromolecular assembly of the pore-forming toxin aerolysin.[149] Thermally accessible conformational changes for monomeric units are first determined by atomistic simulations on high-resolution structures (left), and then used to fit the low-resolution cryo-EM images of the complex in different functional states (centre, right).

assembly. The existence of such conformation was confirmed by the almost perfect matching with the cryo-EM images, revealing the existence of structurally significantly different states.[149]

Integrative experimental-CG protocol for determination of macro-assemblies. Cellular protein complexes can span over dimensions that are beyond reach even for mid-/low-resolution techniques like cryo-EM, or they may have an intrinsic dynamical nature, thus preventing stable protein co-localisation for sufficient long time to allow direct probing. In this case, there is no possibility of obtaining structural data of the whole structure from single experimental methods. Still, smart combination of experimental insights and simulations can lead to surprisingly outstanding results.

The example that we report here is the determination of the three-dimensional structure of the nuclear-pore complex (NPC) of *Saccharomyces cervisiae* by Alber *et al.* achieved in 2007.[158,159]

This macromolecular assembly localises at the nuclear membrane, and it is crucial for regulation of molecular transport inside/outside the cellular nucleus. The construct is very complex; the NPC is constituted by several units of about 30 different proteins, up to 456 individual molecules, for a total mass of roughly 50 MDa. The different proteins localize dynamically at different regions of the nuclear membrane (nuclear, transmembrane, or cytpoplasmic sides), and arrange in eight rather symmetric units around the pore axis, which is filled by filamentous protein domains providing the docking sites for the transport factors. Determination at the (*quasi*)-atomistic resolution of the whole assembly by traditional structural biology methods is not feasible, both because of its huge size, and of the dynamical character of the same assembly.

At the time of the work by Alber *et al.* only low resolution images by cryo-EM were available, and only roughly 5% of the individual protein sequences were structurally characterised at the atomistic level.[158]

The integrative experimental/computational approach devised by Alber *et al.*[159] makes use of massive data coming from several proteomics techniques capable of capturing any possible structural or protein–protein interaction data. These may include sedimentation analysis to determine coarse shapes of each protein component, immuno-EM to infer localization of proteins in the assembly, affinity purification of tagged proteins to determine pair or oligomeric interactions among protein partners in the construct. *Per se*, data coming from each single technique may yield very poor information about the global structure of the complex; nonetheless, taken altogether, they should infer sufficient information to reconstruct the global structure of a macromolecular assembly.

In the approach by Alber *et al.* (Fig. 10), the first step is to translate information from experiments mentioned above (plus others) into spatial

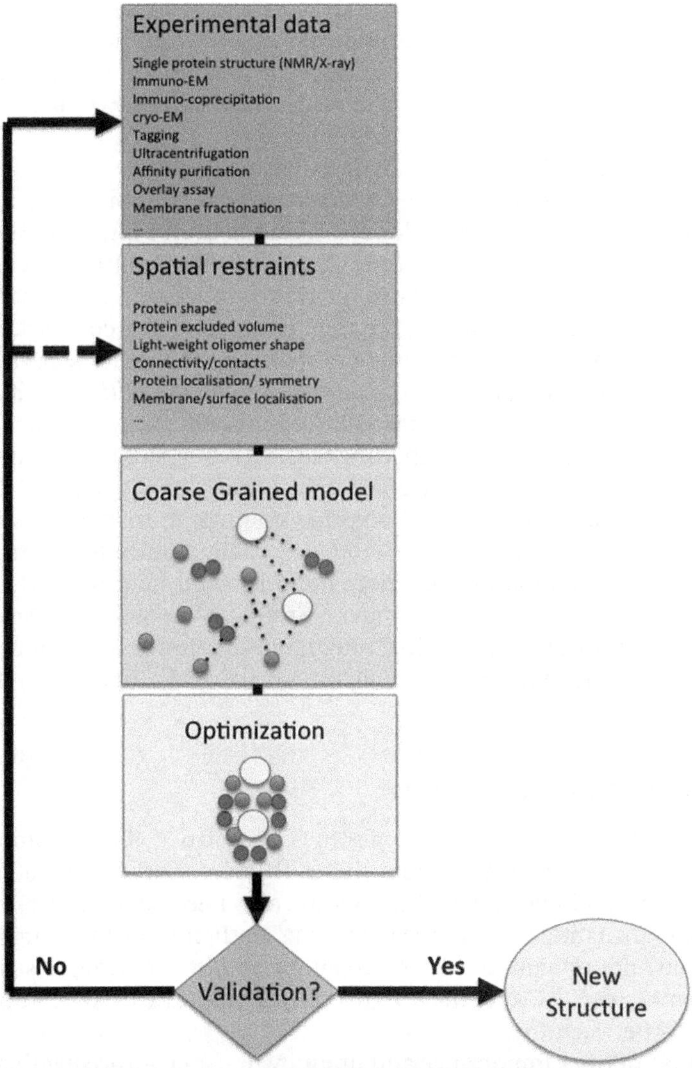

Fig. 10 Integrative modelling of large macromolecular complexes.[159] An initial large set of experimental data is translated into spatial restraints. A coarse-grained model representing the stoichiometric composition of the protein complex is used to minimise a scoring function representative of all the restraints. Optimised structures are validated against scoring result and cluster analysis. The procedure is iterated until one single solution is found.

restraints. Some restraints are applied to single proteins, to define their shape, volume occupation *etc.*; others to entire protein groups, to facilitate or destabilise protein proximity in the assembly.

Then, a computational protocol is defined where proteins are represented by a minimal set of spheroidal beads topologically arranged to grossly reproduce the respective shape, and initially randomly distributed in a confined volume. All spatial restraints are combined in one scoring function, which tests how much a given structure is compatible with the collected restraints coming from the experiment. Additional external factors, like global shape or symmetry of the complex are added as well to the scoring function. The structure is then relaxed with respect to the scoring function until a minimum is reached. The procedure can be repeated several times until a statistically meaningful set of structures is found. Finally, all best structures that satisfy similarly well the initial restraints are processed by cluster analysis.

The procedure allows three possible outputs: (i) *there is only one clustered structure that respects the initial restraints.* In this case, a likely structure for the macromolecular complex has been found. (ii) *There are multiple structures respecting initial restraints.* Then, either there exist multiple native aggregation states, or the number of restraints used is insufficient to uniquely determine the structure of the complex. In any case, additional experiments adding new restraints should be performed. (iii) *The algorithm cannot find structures satisfying the initial restraints.* In this case, either the experimental data set is biased, or there is some issue in the interpretation of the experiment in terms of spatial restraints.

This approach is surely extremely fascinating, but it requires a large number of experimental data, possibly coming from diverse sources, so to balance bias, correlation, and eventually spurious information. To date, accumulation of such large quantity (and high quality) of data may still require large investment of time and resources by scientific consortia. Advances in high-throughput proteomics methods may nonetheless facilitate the establishment of this approach in the coming years.

4 Lipids and membranes

In addition to proteins, biomembranes are also a major component of cells. They not only constitute the cell boundary that separates cells from the extracellular milieu, but they are also necessary for intracellular sub-compartmentalization, encapsulating both intracellular organelles (endoplasmic reticulum, Golgi apparatus, mitochondria, lysosomes, endosomes, *etc.*...) and the cargo vesicles that are used to transport materials between them.

Because of their importance and ubiquity, it is not surprising that large efforts have been made to accurately describe membranes using theoretical and computational approaches, and especially in order to keep up with the exploding wealth of experimental data available on the subject in the last few decades. Yet, chemical and physical properties of membranes are remarkably different from those of proteins, and modelling attempts to accurately describe and predict membrane behaviour have followed a significantly different path than those focusing on proteins.

The physicochemical differences between proteins and membranes originate from their basic constituents. While proteins are composed of amino acids, membranes are made of lipids: amphipathic molecules characterized by having a polar head at one extremity and a highly hydrophobic moiety at the other end, typically in the form of hydrocarbon chains, with the two ends usually connected by a small linker (most frequently glycerol or sphingosin).

Even though the chemistry of lipids can be as rich as one can imagine, thanks to the numerous combinatorial ways in which polar heads and acyl chains can combine, the most glaring property of lipids is their capability of self-assembly into membranes, usually in a bilayer conformation where two lipid monolayers come into contact through their hydrophobic moieties, thus preventing the passage of both water and polar solutes across the bilayer.

This behaviour is at the root of the cellular role of membranes as a protecting barrier, and different theoretical approaches, ranging from mesoscopic to microscopic models, have been developed throughout the years to investigate the unique properties of this fascinating biological structure. The particular choice of a specific methodology (continuum, atomistic, coarse grained) to investigate membrane properties using numerical methods essentially depends on the scale/size and accuracy needed to address a specific question, and in the next sections we will discuss the main characteristics of the various approaches.

4.1 Continuum models

Lipid bilayers are only few nanometres thick, but they can stably span macroscopic lateral scales (Fig. 11). Considering that each lipid occupies a surface area of approximately 0.6–0.7 nm^2 and consists of approximately 40–50 heavy atoms, it is evident that an atomistic description of an entire membrane would be rather prohibitive from a computational point of view.

However, for lateral scales larger than the bilayer thickness, membranes can be simply described as a two-dimensional surface in three-dimensional space using the classical formalism of differential geometry. If the internal degrees of freedom of the lipids constituting the membrane are sufficiently disentangled from the large-scale properties of this two-dimensional object, then the energy of our system will depend only on large-scale observables and it will be possible to describe its behaviour with a relatively simple functional.

This assumption relies on scale-separation, and it is at the root of the phenomenological Hamiltonian proposed by Canham in 1970,[160] Helfrich in 1973,[161] and Evans in 1975,[162] which identifies stretching and bending as the two main energy components of membranes:

$$H = H_{\text{stretch}} + H_{\text{bend}} = \sigma \int_{\text{memb}} \text{d}A + \int_{\text{memb}} \text{d}A \left\{ \frac{1}{2}\kappa (K - K_0)^2 + \bar{\kappa} K_{\text{G}} \right\} \quad (23)$$

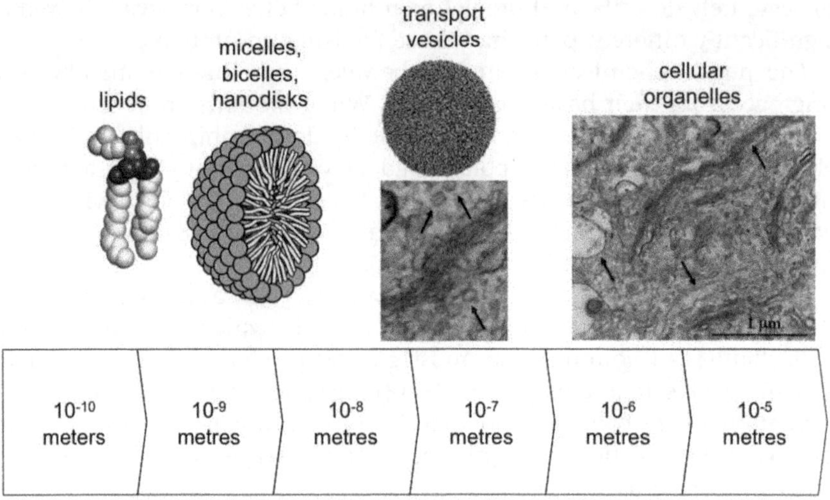

Fig. 11 Size scales of cellular lipid assemblies. Lipids (average length ≈ 2 nm and cross section ≈ 0.7 nm^2) can form high-order structures of different sizes, such as micelles (≈ 5 nm diameter), transport vesicles (50–200 nm diameter), or long lipid bilayers that define the boundaries of entire organelles and of the whole cell (up to several µm). Molecular schemes: yellow, hydrophobic moieties; grey, polar moieties. Electron microscopy images of HeLa cells courtesy of Bruno Mesmin and Sandra Lacas-Gervais.

Here, σ is surface tension, *i.e.*, the empirical parameter describing the energy penalty for area variations, κ is the bending modulus, $\bar{\kappa}$ is the Gaussian modulus and $K = (c_1 + c_2)$, $K_G = c_1c_2$ and K_0 are the extrinsic, Gaussian and spontaneous curvature of the lipid bilayer, respectively, with c_1 and c_2 as the inverse of the two principal curvature radii R_1 and R_2.

This equation is at the centre of all continuum theories of lipid membranes and has been described and reviewed in length in several occasions and we refer the reader to several remarkable texts for further details.[163–165]

The historical importance of the Canham-Helfrich formulation cannot be overstated. It not only led to the correct prediction of the topology of the membrane shape of red blood cells,[166] but it also provided a simple theoretical framework for experimentalists to interpret a wide range of observations, leading to a golden era of membrane science.

By measuring the empirical parameters in eqn (23) using various experimental techniques,[167–172] a remarkable property of membranes emerges naturally: while stretching membranes is relatively expensive in terms of energetic cost (especially since it linearly depends on the size of the membrane), membrane deformations are only one order of magnitude bigger than thermal energy, with bending moduli of cellular membranes of approximately 20 k_bT.[169,171] Hence, membranes are stable against thermal fluctuations (*i.e.* they do not break or fall apart) but soft enough to be deformed and remodelled by proteins. This property is crucial for several cellular processes, and especially signalling, allowing the formation of transport vesicles that transfer materials both between intracellular compartments and between different cells.

Despite being formulated in the 70's, the modelling of membranes using continuum approaches remains nowadays an active area of research and developments have mostly focused on the ability to incorporate microscopic details into what is essentially a phenomenological description. For example, the derivation of the expression for membrane fluctuations starting from the Helfrich Hamiltonian has allowed the derivation of the bending modulus κ using flickering spectroscopy[173–178] while other derivations allow its determination *via* scattering techniques[179–181] or physical pulling of membrane tethers.[182–184] In addition, internal degrees of freedom of lipids, including lipid tilting[185–191] or lipid shape[192–195] have been included to derive higher-order energetic terms. Also, protein-membrane interactions, especially for what concerns the role of protein insertion in membrane fission or membrane curvature sensing, have been also investigated in the last few years.[187,196,197]

Another interesting area that is receiving some attention recently is the use of MD simulations to derive the model parameters of the continuum treatment.[193,198–206] These efforts look very promising, not only to further expand our understanding of the properties of lipid membranes, but also to expose current limitations of available MD force fields to describe large-scale membrane properties (these limitations will be described in the next Sections).

In summary, continuum models of membranes have been instrumental to understand the physical properties of lipid assemblies and they have dramatically helped the development of this research field. They nowadays represent a gold standard to interpret both *in vitro* and *in vivo* experiments and to test the accuracy of bottom-up microscopic approaches on large-scale membrane properties. Yet, despite massively elaborated mathematical attempts to incorporate microscopic details into the equations, the inevitable assumptions that are intrinsic of phenomenological approaches prevent a faithful and accurate description of the chemical properties of lipid assemblies.

4.2 Atomistic models

To overcome the intrinsic limitations of continuum models to investigate the chemical properties of lipid and membranes, molecular simulations are an obvious and powerful alternative methodology. The first MD simulations of lipid assemblies date back to the early 80's,[207–209] and all of them shared some level of "coarsening" with respect to fully atomistic models, with simplified lipid head groups and hydrocarbon chains and no explicit solvent. Even though very limited in size (<100 lipids) and time (few picoseconds) scales, these seminal works allowed the investigation of the microscopic properties of individual lipids, and were already quite successful in reproducing key properties of lipid tails.

The first simulations of fully solvated phospholipid bilayers appeared in the early 90's.[210–217] These works showed that atomistic MD simulations of membranes were feasible (up to hundreds of picoseconds) and represented a powerful tool to study microscopic properties of membranes, including the slow-down of water diffusion in the proximity of lipids, membrane properties in different thermodynamic phases (liquid-crystalline, gel), or the ordering of lipid chains in agreement with deuterium NMR experiments.[218,219]

Since then, atomistic MD simulations have become a standard tool to investigate molecular properties of lipids with Ångström-level accuracy.[67,220–222] Current atomistic simulations span scales of the order of tens of nanometres and up to microseconds in length, and they have shown the capability of accurately reproducing several experimental parameters, including area per lipid, deuterium order parameter, lipid packing or lipid diffusion. Perhaps more importantly, they have become a standard approach to explicitly investigate the interactions between both integral membrane proteins and peripheral proteins with membranes and lipids.[223–226]

Yet, despite their widespread use, atomistic simulations of membranes are far from trivial to set-up and run, with multiple hidden potential pitfalls. For the uninitiated and the "expert" alike, the most discouraging aspect of atomistic simulations of lipid assemblies is the prohibitively large number of available force fields. A not comprehensive list includes Berger,[227] CHARMM36,[228–230] CHARMM36 United-Atom,[231,232] GAFF,[88] Lipid14,[87] Slipid,[233] MacRog,[234] Poger,[235] Kukol,[236] Chiu *et al.*,[237] Rabinovich *et al.*,[238] Högberg *et al.*,[89] Ulmschneider *et al.*,[239] Tjörnhammar *et al.*[240]

These force fields vary in the strategy used for their parameterization and, as a consequence, they give slightly different results, having been optimized to reproduce some specific membrane properties and inevitably providing less-than-ideal results for other non-target properties. In addition, their use with non-native software platforms may be problematic,[241] mostly because of the different ways in which energy interactions (and especially long-range terms) are numerically treated in the different software applications for molecular dynamics.

While a consensus on the performance of these force fields is far from being established, there is some room for optimism: first, sustained efforts by a large numbers of researchers are contributing to the their constant improvement, thus providing hope for a possible convergence in their accuracy; second, community-wide efforts to objectively assess their performances are starting to emerge (see ref. 241 or http://nmrlipids.blogspot.fr for more details) and it is expected that these studies will provide indications for further developments.

In general, because of the larger number of available experimental data and of a somewhat simpler chemistry, microscopic properties of the hydrophobic core of lipids are better reproduced than those of their polar counterpart.[242,243] This observation hints at a much larger problem, *i.e.* the accuracy of lipid force fields in combination with other, independently developed, parameterizations, such as those for water molecules, ions or amino acids.

Since each parameterization has intrinsic errors *per se* (consider, as an example, the large number of available water models and their associated systematic errors),[244,245] there is no guarantee that combining different force fields will lead to a serendipitous cancellation of errors. As such, membrane properties of the lipid-water interface, *i.e.* of the lipid polar heads, will depend significantly on the particular water model used. To partially overcome this problem, each lipid force field is generally developed to work in combination with a specific water model and to counterbalance its pre-existing inaccuracies.[228,246]

This approach is suitable for lipid-water interactions but generates ripple effects when dealing with other molecules such as ions. For example, Na^+ ions appear to have a higher binding affinity for membranes in simulations in comparison with experiments for almost all force fields (http://nmrlipids.blogspot.fr). This may lead to substantial errors in the phase transition temperature of bilayers since ions may promote lipid ordering *via* their (in this case, excessive) binding.[247–249]

Even though frequently and willingly neglected, this issue becomes most prominent in the case of MD simulations of protein-membrane interactions. This kind of simulations, that were first reported in the 90's,[250] are nowadays very common.[226,251] Yet, both the compatibility and accuracy of MD simulations using different lipid and protein force fields has been rarely thoroughly investigated,[252] mostly due to the huge sampling required to converge free energies in such complex systems.

Beyond the obvious choice of mixing force fields with similar parameterization strategies (*i.e.* CHARMM lipids with CHARMM proteins or AMBER lipids with AMBER proteins) some mixed combinations have been suggested (such as OPLS[253] or AMBER proteins[254] with Berger lipids). Recent works wishing to estimate the translocation free energy of individual peptides into model lipid bilayers appear the most promising avenue to gain insight into this issue, but published results appear somewhat contradictory so far.[255–259]

In summary, MD simulations of lipid assemblies are nowadays an established tool that can provide remarkable insights into the molecular structure and dynamics of model membranes. Even though some limitations persist, their ability to reproduce experimental data on membrane properties is truly amazing. Most importantly, the remaining drawbacks are under intense scrutiny, suggesting that future developments will lead to further improvements in their accuracy. In this context, the most challenging area appears to be that of protein–membrane interactions, a field that will probably remain the most significant application area for atomistic simulations of lipid assemblies.

4.3 Coarse-grained models

Remarkable progress in both software and hardware has allowed expanding the time-scales for atomistic MD simulations of membranes by four orders of magnitude in two decades, from the hundreds of picoseconds reported in the first studies of fully solvated phospholipid bilayers,[210–217] up to microseconds in current simulations.[93,94,260,261]

On the other hand, the lateral dimensions of MD simulations of lipid bilayers have only marginally increased, remaining confined to box sizes of approximately 10×10 nanometres. This limitation not only prevents investigating large-scale membrane remodelling phenomena that are crucial in cellular processes but it also does not permit a direct comparison between atomistic bottom-up numerical simulations and those continuum theories that had been historically so successful in investigating membrane properties at larger scales.

To overcome this shortcoming of atomistic simulations, and in order to retain sufficient chemical accuracy in the description of mesoscopic membrane properties, coarse-grain MD simulations have been extensively developed in the last two decades, and they have proven to be remarkably successful towards the investigation of multiple crucial membrane-mediated cellular phenomena.[40,50,53,61,262]

The first step in CG MD simulations of lipid assemblies is the choice of a suitable mapping between a fully atomistic representation and a coarse grain one. This mapping is not univocal and several choices have been proposed, with the minimal requirement of being able to distinguish between the lipid polar head and the acyl chains with different CG beads. Regardless of the mapping, however, all CG models share some intrinsic limitations that must be taken into account before indulging into a more detailed description of the different CG strategies.

First of all, the mapping between an atomistic representation and a coarser one involves the loss of degrees of freedom. This implies that certain physical interactions between system components are no longer present. In the case of a water molecule, for example, even the simple (and computationally still quite expensive) representation with a single bead implies that we lose the capability of describing the dipolar moment that is characteristic of water molecules. A two-beads representation would be able to retain dipolar interactions but would still inevitably lose the energetic terms deriving from hydrogen bonding. Thus, even though a careful parameterization of the interactions between CG beads may compensate for those missing interactions, special care must be taken when interpreting the results from CG simulations, and especially when establishing their range of validity and the systematic errors associated to the results.

In addition, the neglect of atomic degrees of freedom in the CG simulations results in smoother energy landscapes, giving rise to a much faster system dynamics. While this aspect is beneficial with respect to the exploration of the phase space of the system, it presents the inevitable drawback of preventing a straightforward interpretation of time scales in CG simulations. Even though strategies to recover system dynamics from the CG simulations have been proposed, usually involving the comparison with atomistic simulations, it is important to remember that the lower-dimensional energy landscape of CG simulations is not meant to reproduce realistically dynamical processes.

An analogous situation arises for what concerns the ability of CG models to correctly reproduce system entropy. According to the Boltzmann formalism, the entropy of a system depends on the number of available microscopic phase space microstates. Since several degrees of freedom are discarded in the CG representation, the entropy of a system may thus be (sometimes dramatically) wrong. Even though the overall free energy can be set as to reproduce that of atomistic systems or of experimental measurements (essentially thanks to enthalpy compensation), special care must be taken when interpreting entropy-dependent properties such as, for example, temperature dependent phenomena.

Finally, a crucial open question regarding CG methods is their portability. Since the parameterization of CG interactions involves a specific training set, regardless of the strategy used, it is up to debate whether these interactions would retain their validity in significantly different conditions, such as temperature, ionic strength or, especially for lipid bilayers, phase transitions. Thus, changing the simulation conditions may require, if not an entire reparameterisation, at least the careful validation of the underlying assumptions, usually *via* comparison with atomistic MD data or, better, relevant experimental observations.

As CG simulations of lipids and membranes become more and more widespread and accessible ("routine"), it is easier to lose track of how a specific model has been parameterized and of its main capabilities and limitations. We will thus briefly review the history of CG modelling approaches to study lipid assemblies and we will especially focus on the most popular CG force fields that are somewhat able to retain chemical accuracy, providing a direct link between atomistic properties of lipid bilayers and mesoscale properties of membranes that can be investigated using continuum approaches.

The first CG MD simulations of membranes described lipids as a set of Lennard-Jones particles, distinguishing between polar (for water and polar heads) and apolar (for acyl chains) particles.[263,264] These early simulations were able to observe the spontaneous self-assembly of lipids into micellar structures and their arrangement at the water/oil interface, thus characterizing structurally their behaviour as surfactants. Despite the capability to observe spontaneous self-assembly and to qualitatively reproduce basic membrane properties such as bending rigidity or protrusion modes,[41] these models were not able to investigate length scales much larger than those available for atomistic MD simulations.

A promising alternative to reach larger time and length scales was dissipative particle dynamics (DPD), a methodology akin to MD simulations where only three types of forces are present: a conserved soft repulsion force, pairwise dissipation forces and pairwise random forces. Thanks to the use of soft repulsion forces, DPD simulations can tolerate a much larger time step while still preserving momentum and providing the correct hydrodynamic behaviour, thus allowing for longer simulations times and bigger system sizes. However, the mapping used in these simulations remains really coarse.[42,262,265,266]

Alternatively, solvent-free models able to describe large-scale membrane remodelling processes using MD simulations have been developed by several groups.[54,137,199,267–271]

While these models have been successfully applied to curvature mediated-interactions (also in the presence of proteins or nanoparticles), thus to scales of the order of several tens (and sometimes up to hundreds) nanometres[69,193,272–275] they still lack the chemical accuracy to establish a direct link with the different properties of the various lipid species that are typically investigated using atomistic simulations.

A recently proposed promising alternative is the solvent-free Highly Coarse-Grained (HCG) lipid model proposed by Voth and coworkers.[276,277] This method can distinguish between different lipid species, such as DLPC, DOPC and DOPS, and is able to reproduce both elastic and structural properties of the various lipid species in comparison with reference all-atom simulations. This method is based on the so-called multi-scale coarse grain (MS-CG) methodology previously proposed by the same authors,[44,48,49] where the interactions between CG particles are derived from all-atom simulations using a force matching approach based on a variational protocol.

This methodology has been recently applied to investigate membrane-remodelling processes by peripheral proteins[278] and appears to be a promising compromise to investigate large-scale properties while still being able to distinguish between different lipid species. However, given the low resolution of the method, with individual lipids described using only three or four beads, several chemical-dependent properties of lipid assemblies remain beyond the capability of this approach.

More generally, all solvent-free CG methods have intrinsic limitations when it comes to describing explicitly processes that ultimately depend on the partitioning of molecules between water and oil, a key characteristic of lipids and surfactants. Thus, even though solvent-free CG methods have shown the capability to reproduce structural properties of membranes, more accurate thermodynamic properties require the explicit treatment of water–lipids interactions.

The two most successful CG force fields that have been trying to mimic most of the chemical properties of lipids are the MARTINI,[279] developed by Marrink and coworkers and the SDK model[280] developed by Klein and colleagues. Both methods use a finer mapping, with phospholipids described with approximately ten to twenty CG beads, depending on the acyl chain length, and an explicit treatment of the surrounding solvent (Fig. 12).

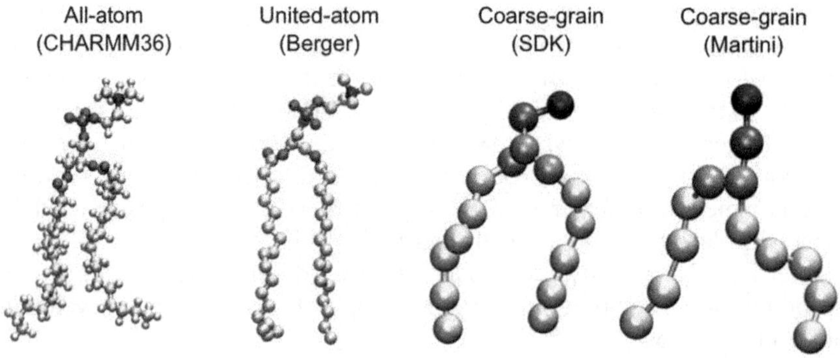

Fig. 12 Particle-representation of 1-palmitoyl-2-oleoyl-sn-glycero-3-phosphocholine (POPC) with different force fields described in the text. From left to right: all-atom representation (CHARMM36), united-atom representation (Berger), CG representation using the SDK force field, CG representation using the MARTINI force field. AA: Hydrogen atoms, white; carbon atoms, yellow; oxygen atoms, red, nitrogen atoms, blue. CG: acyl chain beads, yellow; glycerol beads, pink; phosphate beads, brown; choline beads, blue.

The MARTINI force field, developed as a substantial improvement from previous CG attempts[281] mostly uses a mapping of four heavy atoms into one CG bead, including for water molecules. A finer mapping is used for chemical moieties of subtle structural properties, such as ring compounds that are important in some lipid species, including cholesterol and phosphoinositides.

In the model, four main types of CG beads are introduced: polar, nonpolar, apolar and charged. Each of these particle types is in addition characterized by some subtype specifications, mainly its hydrogen-bond capability (acceptor, donor, both or none) and its overall polarity (on a 1–5 scale). With few exceptions, all particles have the same size and the interactions between them depend solely on the pair-pair strength interaction (in a classical 6–12 Lennard-Jones fashion) and on their charges (in the electrostatic term).

This approach allows distinguishing between several chemical compounds, while still keeping the number of different CG beads manageable. As an example, a phospholipid molecule (phosphocholine for example, see Fig. 12) is described by two charged particles representing the choline and phosphate groups, respectively, two polar particles describing the glycerol moieties, and several apolar particles modelling the two acyl chains, with the total number depending on the length of the acyl chains and consistent with the 4 to 1 mapping.

Thanks to the relatively high number of different particles, subtler chemical properties that are crucial in lipid chemistry can be investigated, such as the influence of lipid mono- or poly-unsaturation, or the properties of different lipid polar heads (even fine ones such as the difference between phosphocholine and phosphoethanolamine).

Once the atomistic to CG mapping and the potential energy terms (harmonic potential for bonds and angles, 6–12 Lennard-Jones and electrostatic with a relative dielectric constant of 15) are set, the crucial aspect of this CG procedure is the determination of the numerous intra- and inter-molecular parameters.

In the MARTINI strategy, bonded terms are parameterized such as to reproduce values obtained from atomistic simulations, while non-bonded parameters are set to reproduce, for all the individual particle types specified in the mapping, experimentally-derived thermodynamic properties (sometimes replaced by atomistic free energy calculations if the experimental data are not available), and in particular free energies of hydration, free energies of vaporization and the partitioning free energies between water and a number of organic phases.

The main advantage of this empirical approach is its immediate transferability: since any new molecule is built from the same set of building blocks, it is reasonable to assume that the force field will be able to describe, at least semi-quantitatively, its thermodynamic properties. On the other hand, the main drawback is that if a specific property (and especially structural ones) needs to be accurately described, there is no *a priori* guarantee that the MARTINI force field will be able to do so with the desired level of accuracy.

In particular, this parameterization approach has shortcomings in reproducing correctly the phase diagram of water, including, for example, water/vapor interfacial tension,[279] and, as a consequence, it has flaws for what pertains the investigation of surfactant properties of lipid monolayer at the air/water interface.[61,280] Improvements of the MARTINI water model have been proposed *via* the use of a polarizable force field,[282] but polarizable CG force fields remain poorly investigated so far in the field of CG simulations.

Despite its limitations, the MARTINI model, especially thanks to its connections with the GROMACS molecular dynamics software package[283,284] and to its user-friendly implementation (see http://md.chem.rug.nl/cgmartini), is by far the most used CG model for lipids assemblies to date and it has been proficiently used to investigate a large number of membrane and lipid properties. The successes of the MARTINI force field have been recently described in an excellent review;[61] for our purposes, it is important to underline that, in large part thanks to the remarkable efficiency of modern MD software, this model has been recently shown the ability to investigate large-scale membrane remodelling processes with unprecedented chemical accuracy, thus providing a more accurate description of phenomena that were either traditionally investigated using continuum approaches, with low-resolution CG models or simply not addressed using theoretical or numerical approaches.

Amongst them, the investigation of realistic, multi-component membranes,[68,285] entire influenza virions,[286] curvature-dependent properties,[200,201,287–290] monolayer collapse[291–293] and membrane remodelling processes such as membrane fusion[294–298] or fission.[299]

Despite those early successes, it is important to remember that the four to one mapping poses intrinsic limitations, and especially for what pertains those phenomena in which large scale properties emerge as the macroscopic read-out of much finer and detailed molecular properties, such as, for example, those involving hydrogen bonding. Thus, care must always be taken when assessing the predictive power of CG simulations and further validation against experimental properties is recommended whenever possible.

Another chemically accurate CG force field for MD simulations of lipid assemblies is the Shinoda-DeVane-Klein (SDK) force field.[280] This model is the improved version of the earlier CG model developed by Shelley *et al.*[51] and utilizes a three heavy atoms to one CG bead mapping, thus providing a slightly higher resolution than the MARTINI (Fig. 12).

While the initial model by Shelley *et al.* was solely parameterised on atomistic simulations, the SDK model uses atomistic simulations exclusively for the parameterisation of bonded interactions (analogously to the MARTINI strategy). Non-bonded interactions, on the other hand, are parameterised not only with transfer free energies (like in the MARTINI), but also on density and interfacial properties such as surface tension.

Unlike MARTINI, however, the SDK model utilizes a 12–4 Lennard-Jones potential for all pair interactions involving water beads[52] and a 9–6 Lennard-Jones potential for all other pairs. With this choice, together with the explicit treatment of long-range Coulomb interactions, water behaviour, and especially its phase properties, density and compressibility, is reproduced fairly well within the intrinsic limitations of the three to one mapping.

In addition, instead of defining a number of different CG beads and building up each individual molecule choosing from these basic building blocks, the SDK approach consists in subdividing each molecule into its main chemical constituents (while keeping the usual three to one mapping) and then fitting to the interactions of the new resulting parameters based on a mix of thermodynamic (mainly density, surface tension and solvation free energies) and atomistic simulations data.

As it is expected from its finer mapping and more sophisticated parameterization procedure, this model is generally more accurate than MARTINI, especially for what pertains properties at the lipid-water interface. On the other hand, the unusual functional form of Lennard-Jones interactions (that is not supported by most current MD software) and the need for a careful parameterization every time a new molecule is introduced, make this force field less appealing to the non-initiated.

Nevertheless, the authors have reported so far a large range of applications on lipid and membrane systems of sizes comparable to those studied using MARTINI, including spontaneous vesicle self-assembly,[53] monolayer collapse,[280] vesicle to micelle transformation[300] and membrane fusion.[301,302] The SDK model thus represents a powerful and promising approach to investigate finely tuned chemical properties of large lipid assemblies when cheaper but less accurate CG models turn out to be unsuccessful.

4.4 A prototypical example: membrane curvature

The investigation of cellular processes in which membrane curvature plays a prominent role is not only a relevant topic in membrane biology, but also a perfect scenario to assess the performances of continuum models and of bottom-up molecular simulations.

In the last few years, a large number of *in vitro* and *in vivo* studies have suggested that membrane curvature plays a major role in regulating multiple cellular phenomena,[303–305] including vesicular trafficking, nuclear pore assembly or autophagocytosis.

At the same time, recent MD simulations have shown the capability of dealing with system sizes and curvatures that are comparable with those found in intracellular transport vesicles and that have been traditionally addressed solely using continuum methods.[306]

Because of system sizes, these investigations are clearly not feasible with atomistic MD simulations: for example a lipid vesicle of a radius of approximately 25 nm would contain around 18 000 lipids, corresponding to approximately 2 500 000 atoms using a fully atomistic force field, even before taking into account the unavoidable presence of water molecules. However, thanks to progresses both in hardware and in software technologies, such systems are nowadays computationally tractable using CG models such as the MARTINI or the SDK.

The main advantage of particle-based models over continuum treatments is their capability of being able to describe simultaneously membrane curvature and lipid chemistry, as well as their interplay: how do specific lipids promote/prevent curvature stresses? And how do lipids respond to curvature stresses, for example by partitioning between the inner and the outer monolayer of highly curved vesicles?

On the other hand, since the investigation of the interplay between curvature and lipid chemistry is not straightforward using AA simulations, validation of the CG methodology in this context requires a special protocol.

To do so, Vanni and colleagues took advantage of the unique properties of a class of amphipathic peptides, the Amphipathic Lipid Packing Sensors (ALPS) motifs, which extensive biochemical analyses had identified as accurate sensors of both lipid chemistry and membrane curvature.[305]

In particular, these peptides showed a marked sensitivity to both the number of monounsaturated acyl chains, with binding increasing moving from saturated acyl chains to monounsaturated ones, and to the "size" of the lipid polar head, with higher binding in the presence of lipids containing smaller polar heads, such as diacylglycerol or phosphatidylethanolamine. Using atomistic simulations, the authors managed to show that these variations in lipid chemistry promote the formation of lipid-packing defects in membranes[307] and that ALPS motifs specifically bind to pre-existing packing-defects *via* the insertion of hydrophobic residues.[308]

This observation paved the way for the careful validation of CG simulations (in this particular case using the MARTINI force field) by evaluating their ability to reproduce lipid-packing defects measured in atomistic simulations for several flat bilayers of different compositions.[290]

On the one hand, the high correlation between the data obtained from CG and atomistic simulations indicated that the used CG model was, in this case, able to reproduce with high accuracy this specific property of lipid assemblies (Fig. 13). On the other hand, the very good agreement between the simulations and the observed binding of the ALPS motif to lipid bilayers of identical compositions *in vitro*, suggested a possible methodology to validate the CG force field also for curved bilayers, where comparison against all-atom simulations would have not been possible.

Analysis of lipid-packing defects in curved bilayers as a function of lipid composition revealed the crucial role of lipid chemistry in the adaptation of lipid bilayers to membrane curvature and, in turn, of ALPS binding to curved membranes.

In particular, CG simulations suggested, in agreement with experiments, that membrane curvature and lipid composition synergize to promote the formation of lipid packing defects, *i.e.* those lipid compositions that promoted packing defects in flat bilayers also did so in highly curved bilayers. This observation further confirmed only weak coupling between lipid shape and leaflet curvature, in agreement with fluorescence measurements[309] and in contrast with several predictions from continuum models.

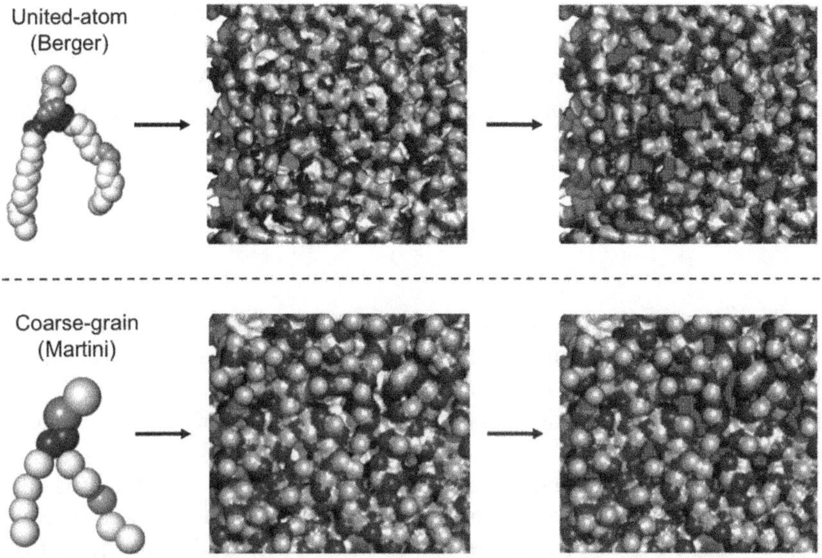

Fig. 13 Lipid-packing defects in atomistic and CG simulations. Left: Molecular models of lipids using a united-atom representation (top) or a CG representation (bottom). Polar atoms/beads, grey; hydrophobic atoms/beads, yellow/orange. Middle: Top view of lipid bilayers using a surface representation in AA ad CG simulations. Right: Top view of the lipid bilayers in the middle panel plus detected lipid-packing defects (blue, see Vanni *et al.*[290] for additional details on the methodology to identify them) for the two bilayers. Moving from a united-atom (approximately 50 atoms per lipid) to a CG representation (approximately 15 beads per lipid) does not prevent a correct quantification of this microscopic membrane property.

4.5 Multi-scale hybrid particle/field models

Even though the molecular representation of lipids with molecular-type approaches (atomistic or CG) has intrinsic advantages, as shown in the previous sections, mesoscale models have the possibility of exploring significantly larger time-space dimensionalities. We present here a very brief resume of hybrid approaches combining CG and density-field approaches, which represent the counterpart of QM/MM and AA/CG models introduced in the previous sections.

In these approaches, soft matter systems are modelled using self-consistent field (SCF) theory. In detail, systems are still formed by explicit molecular objects. The bodies, however, are not interacting directly with each other *via* usual two-body terms. Instead, molecular segments are decoupled from each other, and posed under the effect of a static external field.[310] A molecule in SCF theory is considered to be interacting with neighbouring partners only through an averaged density field. Marcelja was the first to propose a simple field model for lipid molecules in 1975.[311] There, head groups of the lipids are treated as a boundary to which the tails of the lipid molecules are anchored. The sampling of degrees of freedom corresponding to intra-molecular rearrangement is made using Rotational Isomeric State. The molecular segments are coupled to an anisotropic aligning potential, which determines the organization of the lipid layer.[311]

In a rigorous formulation, the Hamiltonian of a system composed of N molecules is split into two parts as $H_0(\Gamma) + W(\Gamma)$. Where H_0 is the reference Hamiltonian for an ideal system composed by N non-interacting molecules, but still responding to all intra-molecular forces, and W is an external potential causing perturbations from the ideal behaviour. The Canonical partition function is thus written as:

$$Z = \frac{1}{N!}\int e^{-\beta[H_0(\Gamma)+W(\Gamma)]}\mathrm{d}\Gamma \qquad (24)$$

The statistical distribution of particles $\phi(\boldsymbol{r};\Gamma)$ at position $\boldsymbol{r}$ for a given microscopic configuration Γ will have a probability that depends on W. W can thus be defined as a functional of ϕ ($W = W[\phi]$), such that a statistical distribution of "ideal" particles equal to that of the real system minimises the energy.

Equilibration and dynamics of soft-matter systems can be thus obtained by free particles that move in the external field W that self consistently responds to reproduce equilibrated particle distributions (for more technical details, one can refer to specialised reviews like[312–314]).

Combination of particle and field representations has been investigated by Daoulas *et al.*[315,316] and applied to Monte Carlo simulations of CG models of synthetic polymers.[317,318] More recently, also molecular dynamics simulations have been combined with self-consistent field description.[319,320]

The use of such models can be coupled straightforwardly with CG representation of lipids. In fact, CG reduces the number of different species present in the system, and thus facilitates the calibration of field responses for the different particles.[321,322] Both lamellar and non lamellar (including reverse micelle and micellar phases) phases are well reproduced with these models without any geometrical assumptions.

The computational efficiency of this approach allows at the same time models with a resolution close to atomistic and simulations on very large length and timescales.[323] For example, these models have been applied to study the interaction and the dynamical exchange of block copolymer chains between a spherical micelle (functioning as drug nanocarrier) and a lipid bilayer (as model of cell surface) also in the presence of drug molecules (iboprufen) in the micelle core. Simulations of 12 nm large micelles with membrane bilayers over several microseconds of simulations could be achieved on home-cluster facilities.[324]

Due to their computational efficiency, particle-field approaches are gaining popularity in the field.[312–314] Hybrid models can be more generally used with respect to pure field models and much less assumptions need to be adopted. Several CG models using this hybrid representations for biocompatible polymers,[325] biomembranes,[314,321,322,325,326] vesicles,[327] proteins,[310,328] *block* copolymers,[329–332] and colloidal particles[333,334] have been reported in the recent literature, paving a new way toward closing the gap between atomistic and mesoscopic dimensionalities.

5 Final remarks

Molecular simulations are one of the most promising tools for basic understanding of biological phenomena at the sub-cellular dimensionality. In these years the combined efforts of several groups are rapidly pushing forward the frontier of the field. All-atom simulations remain the most reliable tool, allowing the description of the complex interplay between direct two-body and indirect many body interactions at the chemical resolution level. This methodology is currently a mature research field, able to faithfully reproduce a wide range of protein and membrane properties that can be investigated experimentally and providing a microscopic structural view of membranes that would not be possible otherwise.

However, even though efforts to improve the accuracy of atomistic simulations are ongoing, and especially for what concerns the interaction of lipids with polar molecules and proteins, the main limitation of atomistic simulations is their inability, at least so far, to investigate mesoscopic scales and thus establish a direct connection between molecular properties and large-scale phenomena.

Several approaches are being proposed to go beyond the yet continuously expanding time- and size-bottlenecks that limits the applicability of atomistic models, mostly on a case-by-case basis. For example, continuum models have been crucial to the development of the field of theoretical investigations of membranes, not only establishing a conceptual framework to interpret a large number of experiments, but also providing the basic vocabulary through which membranes are still described nowadays.

To provide a direct, bottom-up, representation of large-scale phenomena starting from the individual properties of their chemical components, CG models are an area currently undergoing intense and fascinating development. Since the mapping from an atomistic representation to a coarser one can be implemented in several ways, several CG models have been proposed so far, with each one of them possessing its own advantages and limitations.

The most promising approaches in this context appear to be those CG models that are able to retain the key aspects of protein and lipid chemistry while still being sufficiently computationally-efficient to investigate large-scale processes that have been traditionally investigated with more drastic coarse-graining approaches or with continuum methodologies.

These methods are allowing the description of large molecular assemblies with an unprecedented resolution, not only as a complement to experimental studies, but also suggesting new investigations of cellular processes at all levels. While limitations in describing water-mediated processes and membrane-protein interactions still remain, CG simulations of protein and lipid assemblies are quickly becoming a mature research field.

Finally, integration of multiple scales at different resolution may provide a best-compromise strategy to include the required information at the necessary detail without increasing costs. Implementations both at the hybrid AA/CG and at the particle/field levels may yield very interesting results in the future.

Whether the next progresses in addressing current limitations of CG methods will originate from new CG approaches or from improved CG parameters and multi-resolution potentials to investigate proteins and membranes will be a fascinating thing to watch.

Acknowledgements

MC acknowledges the support of the Norwegian Research Council through the CoE Centre for Theoretical and Computational Chemistry (CTCC) Grant Nos. 179568/V30 and 171185/V30. SV acknowledges Sandra Lacas-Gervais and the Centre Commun de Microscopie Appliquée, Université Nice Sophia Antipolis for providing the electron microscopy images in Fig. 11. We thank Matteo Dal Peraro for providing Fig. 9, and Giuseppe Milano for support on particle/field simulations of membranes. We finally thank Patrick Fuchs and Giacomo Fiorin for critically reading the manuscript.

References

1 D. Russel, K. Lasker, J. Phillips, D. Schneidman-Duhovny, J. A. Velazquez-Muriel and A. Sali, *Curr. Opin. Cell Biol.*, 2009, **21**, 97.
2 R. Car and M. Parrinello, *Phys. Rev. Lett.*, 1985, **55**, 2471.
3 J. VandeVondele, U. Borstnik and J. Hutter, *J. Chem. Theory Comput.*, 2012, **8**, 3565.
4 I. S. Ufimtsev, N. Luehr and T. J. Martinez, *J. Phys. Chem. Lett.*, 2011, **2**, 1789.
5 J. F. Capitani, R. F. Nalewajski and R. G. Parr, *J. Chem. Phys.*, 1982, **76**, 568.
6 A. Dreuw and M. Head-Gordon, *Chem. Rev.*, 2005, **105**, 4009.
7 M. Elstner, D. Porezag, G. Jungnickel, J. Elsner, M. Haugk, T. Frauenheim, S. Suhai and G. Seifert, *Phys Rev B: Condens. Matter Mater. Phys.*, 1998, **58**, 7260.
8 T. Helgaker, P. Jorgensen and J. Olsen, *Molecular Electronic-Structure Theory*, Wiley, New York, 2013.
9 J. K. Hwang and A. Warshel, *J. Am. Chem. Soc.*, 1996, **118**, 11745.
10 N. Makri, *J. Chem. Phys.*, 1991, **94**, 4949.
11 J. Z. Pu, J. L. Gao and D. G. Truhlar, *J. Phys. Chem. A*, 2004, **108**, 632.
12 W. Thiel, *Wiley Interdiscip. Rev.: Comput. Mol. Sci.*, 2014, **4**, 145.
13 S. P. Webb, T. Iordanov and S. Hammes-Schiffer, *J. Chem. Phys.*, 2002, **117**, 4106.
14 J. C. Phillips, R. Braun, W. Wang, J. Gumbart, E. Tajkhorshid, E. Villa, C. Chipot, R. D. Skeel, L. Kale and K. Schulten, *J. Comput. Chem.*, 2005, **26**, 1781.
15 L. C. T. Pierce, R. Salomon-Ferrer, C. A. F. de Oliveira, J. A. McCammon and R. C. Walker, *J. Chem. Theory Comput.*, 2012, **8**, 2997.
16 T. Schlick, R. Collepardo-Guevara, L. A. Halvorsen, S. Jung and X. Xiao, *Q. Rev. Biophys.*, 2011, **44**, 191.
17 J. E. Stone, J. C. Phillips, P. L. Freddolino, D. J. Hardy, L. G. Trabuco and K. Schulten, *J. Comput. Chem.*, 2007, **28**, 2618.

18 B. R. Brooks, C. L. Brooks, A. D. Mackerell, L. Nilsson, R. J. Petrella, B. Roux, Y. Won, G. Archontis, C. Bartels, S. Boresch, A. Caflisch, L. Caves, Q. Cui, A. R. Dinner, M. Feig, S. Fischer, J. Gao, M. Hodoscek, W. Im, K. Kuczera, T. Lazaridis, J. Ma, V. Ovchinnikov, E. Paci, R. W. Pastor, C. B. Post, J. Z. Pu, M. Schaefer, B. Tidor, R. M. Venable, H. L. Woodcock, X. Wu, W. Yang, D. M. York and M. Karplus, *J. Comput. Chem.*, 2009, **30**, 1545.
19 B. R. Brooks, R. E. Bruccoleri, B. D. Olafson, D. J. States, S. Swaminathan and M. Karplus, *J. Comput. Chem.*, 1983, **4**, 187.
20 W. L. Jorgensen, D. S. Maxwell and J. Tirado-Rives, *J. Am. Chem. Soc.*, 1996, **118**, 11225.
21 W. R. P. Scott, P. H. Hunenberger, I. G. Tironi, A. E. Mark, S. R. Billeter, J. Fennen, A. E. Torda, T. Huber, P. Kruger and W. F. van Gunsteren, *J. Phys. Chem. A*, 1999, **103**, 3596.
22 J. M. Wang, R. M. Wolf, J. W. Caldwell, P. A. Kollman and D. A. Case, *J. Comput. Chem.*, 2004, **25**, 1157.
23 M. Rueda, C. Ferrer-Costa, T. Meyer, A. Perez, J. Camps, A. Hospital, J. L. Gelpi and M. Orozco, *Proc. Natl. Acad. Sci. U. S. A.*, 2007, **104**, 796.
24 E. Brunk and U. Rothlisberger, *Chem. Rev.*, 2015, **115**, 6217.
25 A. Warshel and M. Levitt, *J. Mol. Biol.*, 1976, **103**, 227.
26 F. Claeyssens, J. N. Harvey, F. R. Manby, R. A. Mata, A. J. Mulholland, K. E. Ranaghan, M. Schutz, S. Thiel, W. Thiel and H. J. Werner, *Angew. Chem., Int. Ed.*, 2006, **45**, 6856.
27 Q. Cui, M. Elstner, E. Kaxiras, T. Frauenheim and M. Karplus, *J. Phys. Chem. B*, 2001, **105**, 569.
28 R. A. Friesner and V. Guallar, *Annu. Rev. Phys. Chem.*, 2005, **56**, 389.
29 J. L. Gao and D. G. Truhlar, *Annu. Rev. Phys. Chem.*, 2002, **53**, 467.
30 S. C. L. Kamerlin and A. Warshel, *Faraday Discuss.*, 2010, **145**, 71.
31 J. Kongsted, A. Osted, K. V. Mikkelsen and O. Christiansen, *J. Chem. Phys.*, 2003, **118**, 1620.
32 A. Laio, J. VandeVondele and U. Rothlisberger, *J. Chem. Phys.*, 2002, **116**, 6941.
33 R. A. Mata, H. J. Werner, S. Thiel and W. Thiel, *J. Chem. Phys.*, 2008, **128**, 025104.
34 A. J. Mulholland, *Chem. Cent. J.*, 2007, **1**, 19.
35 M. Svensson, S. Humbel, R. D. J. Froese, T. Matsubara, S. Sieber and K. Morokuma, *J. Phys. Chem.*, 1996, **100**, 19357.
36 H. L. Woodcock, M. Hodoscek, A. T. B. Gilbert, P. M. W. Gill, H. F. Schaefer and B. R. Brooks, *J. Comput. Chem.*, 2007, **28**, 1485.
37 D. E. Shaw, P. Maragakis, K. Lindorff-Larsen, S. Piana, R. O. Dror, M. P. Eastwood, J. A. Bank, J. M. Jumper, J. K. Salmon, Y. B. Shan and W. Wriggers, *Science*, 2010, **330**, 341.
38 G. P. Zhao, J. R. Perilla, E. L. Yufenyuy, X. Meng, B. Chen, J. Y. Ning, J. Ahn, A. M. Gronenborn, K. Schulten, C. Aiken and P. J. Zhang, *Nature*, 2013, **497**, 643.
39 D. Alemani, F. Collu, M. Cascella and M. Dal Peraro, *J. Chem. Theory Comput.*, 2010, **6**, 315.
40 S. V. Bennun, M. I. Hoopes, C. Xing and R. Faller, *Chem. Phys. Lipids*, 2009, **159**, 59.
41 R. Goetz, G. Gompper and R. Lipowsky, *Phys. Rev. Lett.*, 1999, **82**, 221.
42 R. D. Groot and P. B. Warren, *J. Chem. Phys.*, 1997, **107**, 4423.
43 S. Izvekov, M. Parrinello, C. J. Burnham and G. A. Voth, *J. Chem. Phys.*, 2004, **120**, 10896.
44 S. Izvekov and G. A. Voth, *J. Phys. Chem. B*, 2005, **109**, 2469.

45 A. Liwo, M. Khalili and H. A. Scheraga, *Proc. Natl. Acad. Sci. U. S. A.*, 2005, **102**, 2362.
46 A. Liwo, J. Lee, D. R. Ripoll, J. Pillardy and H. A. Scheraga, *Proc. Natl. Acad. Sci. U. S. A.*, 1999, **96**, 5482.
47 L. Monticelli, S. K. Kandasamy, X. Periole, R. G. Larson, D. P. Tieleman and S.-J. Marrink, *J. Chem. Theory Comput.*, 2008, **4**, 819.
48 W. G. Noid, J. W. Chu, G. S. Ayton, V. Krishna, S. Izvekov, G. A. Voth, A. Das and H. C. Andersen, *J. Chem. Phys.*, 2008, **128**, 244114.
49 W. G. Noid, P. Liu, Y. Wang, J. W. Chu, G. S. Ayton, S. Izvekov, H. C. Andersen and G. A. Voth, *J. Chem. Phys.*, 2008, **128**, 244115.
50 L. Saiz and M. L. Klein, *Acc. Chem. Res.*, 2002, **35**, 482.
51 J. C. Shelley, M. Y. Shelley, R. C. Reeder, S. Bandyopadhyay and M. L. Klein, *J. Phys. Chem. B*, 2001, **105**, 4464.
52 W. Shinoda, R. DeVane and M. L. Klein, *Mol. Simul.*, 2007, **33**, 27.
53 W. Shinoda, R. DeVane and M. L. Klein, *Curr. Opin. Struct. Biol.*, 2012, **22**, 175.
54 A. J. Sodt and T. Head-Gordon, *J. Chem. Phys.*, 2010, **132**, 205103.
55 E. Spiga, D. Alemani, M. T. Degiacomi, M. Cascella and M. Dal Peraro, *J. Chem. Theory Comput.*, 2013, **9**, 3515.
56 V. Tozzini, W. Rocchia and J. A. McCammon, *J. Chem. Theory Comput.*, 2006, **2**, 667.
57 G. S. Ayton, W. G. Noid and G. A. Voth, *Curr. Opin. Struct. Biol.*, 2007, **17**, 192.
58 M. Cascella and M. Dal Peraro, *Chimia*, 2009, **63**, 14.
59 S. C. L. Kamerlin, S. Vicatos, A. Dryga and A. Warshel, *Annu. Rev. Phys. Chem.*, 2011, **62**, 41.
60 M. L. Klein and W. Shinoda, *Science*, 2008, **321**, 798.
61 S. J. Marrink and D. P. Tieleman, *Chem. Soc. Rev.*, 2013, **42**, 6801.
62 S. Riniker, J. R. Allison and W. F. van Gunsteren, *Phys. Chem. Chem. Phys.*, 2012, **14**, 12423.
63 M. G. Saunders and G. A. Voth, *Curr. Opin. Struct. Biol.*, 2012, **22**, 144.
64 V. Tozzini, *Curr. Opin. Struct. Biol.*, 2005, **15**, 144.
65 V. Tozzini, *Q. Rev. Biophys.*, 2010, **43**, 333.
66 V. Tozzini, *Acc. Chem. Res.*, 2010, **43**, 220.
67 I. Vattulainen and T. Rog, *Cold Spring Harbor Perspect. Biol.*, 2011, **3**, a004655.
68 H. I. Ingolfsson, M. N. Melo, F. J. van Eerden, C. Arnarez, C. A. Lopez, T. A. Wassenaar, X. Periole, A. H. de Vries, D. P. Tieleman and S. J. Marrink, *J. Am. Chem. Soc.*, 2014, **136**, 14554.
69 G. S. Ayton and G. A. Voth, *J. Phys. Chem. B*, 2009, **113**, 4413.
70 R. D. Hills, L. Y. Lu and G. A. Voth, *PLoS Comput. Biol.*, 2010, **6**, e1000827.
71 L. P. Kadanoff, *Statistical Physics: Statics, Dynamics, and Renormalization*, World Scientific Publishing Co., Singapore, 2000.
72 B. Widom, *J. Chem. Phys.*, 1965, **43**, 3898.
73 B. Widom, *J. Chem. Phys.*, 1965, **43**, 3892.
74 T. X. Hoang, A. Trovato, F. Seno, J. R. Banavar and A. Maritan, *Proc. Natl. Acad. Sci. U. S. A.*, 2004, **101**, 7960.
75 J. R. Banavar, T. X. Hoang, A. Maritan, F. Seno and A. Trovato, *Phys. Rev. E.: Stat., Nonlinear, Soft Matter Phys.*, 2004, **70**, 041905.
76 K. A. Dill, *Biochemistry*, 1985, **24**, 1501.
77 F. Muller-Plathe, *ChemPhysChem*, 2002, **3**, 754.
78 G. D. Rose, P. J. Fleming, J. R. Banavar and A. Maritan, *Proc. Natl. Acad. Sci. U. S. A.*, 2006, **103**, 16623.

79 N. D. Socci, W. S. Bialek and J. N. Onuchic, *Phys. Rev. E.: Stat. Phys., Plasmas, Fluids, Relat. Interdiscip. Top.*, 1994, **49**, 3440.
80 M. Levitt and A. Warshel, *Nature*, 1975, **253**, 694.
81 E. Adman, K. D. Watenpaugh and L. H. Jensen, *Proc. Natl. Acad. Sci. U. S. A.*, 1975, **72**, 4854.
82 O. Mayans, P. F. M. van der Ven, M. Wilm, A. Mues, P. Young, D. O. Furst, M. Wilmanns and M. Gautel, *Nature*, 1998, **395**, 863.
83 T. Okada, O. P. Ernst, K. Palczewski and K. P. Hofmann, *Trends Biochem. Sci.*, 2001, **26**, 318.
84 R. W. Schoenlein, L. A. Peteanu, R. A. Mathies and C. V. Shank, *Science*, 1991, **254**, 412.
85 C. S. Bolze, R. E. Helbling, R. L. Owen, A. R. Pearson, G. Pompidor, F. Dworkowski, M. R. Fuchs, J. Furrer, M. Golczak, K. Palczewski, M. Cascella and A. Stocker, *J. Am. Chem. Soc.*, 2014, **136**, 137.
86 W. L. Jorgensen, *Science*, 2004, **303**, 1813.
87 C. J. Dickson, B. D. Madej, A. A. Skjevik, R. M. Betz, K. Teigen, I. R. Gould and R. C. Walker, *J. Chem. Theory Comput.*, 2014, **10**, 865.
88 C. J. Dickson, L. Rosso, R. M. Betz, R. C. Walker and I. R. Gould, *Soft Matter*, 2012, **8**, 9617.
89 C. J. Högberg, A. M. Nikitin and A. P. Lyubartsev, *J. Comput. Chem.*, 2008, **29**, 2359.
90 G. R. Bowman, V. A. Voelz and V. S. Pande, *Curr. Opin. Struct. Biol.*, 2011, **21**, 4.
91 K. Lindorff-Larsen, S. Piana, R. O. Dror and D. E. Shaw, *Science*, 2011, **334**, 517.
92 V. A. Voelz, G. R. Bowman, K. Beauchamp and V. S. Pande, *J. Am. Chem. Soc.*, 2010, **132**, 1526.
93 R. O. Dror, H. F. Green, C. Valant, D. W. Borhani, J. R. Valcourt, A. C. Pan, D. H. Arlow, M. Canals, J. R. Lane, R. Rahmani, J. B. Baell, P. M. Sexton, A. Christopoulos and D. E. Shaw, *Nature*, 2013, **503**, 295.
94 R. Nygaard, Y. Z. Zou, R. O. Dror, T. J. Mildorf, D. H. Arlow, A. Manglik, A. C. Pan, C. W. Liu, J. J. Fung, M. P. Bokoch, F. S. Thian, T. S. Kobilka, D. E. Shaw, L. Mueller, R. S. Prosser and B. K. Kobilka, *Cell*, 2013, **152**, 532.
95 I. Bahar, A. R. Atilgan and B. Erman, *Folding Des.*, 1997, **2**, 173.
96 I. Bahar, M. Kaplan and R. L. Jernigan, *Proteins*, 1997, **29**, 292.
97 I. Bahar and R. L. Jernigan, *J. Mol. Biol.*, 1997, **266**, 195.
98 B. Brooks and M. Karplus, *Proc. Natl. Acad. Sci. U. S. A.*, 1985, **82**, 4995.
99 K. Hinsen, *Proteins*, 1998, **33**, 417.
100 J. A. Mccammon, B. R. Gelin, M. Karplus and P. G. Wolynes, *Nature*, 1976, **262**, 325.
101 C. Micheletti, P. Carloni and A. Maritan, *Proteins*, 2004, **55**, 635.
102 M. Neri, M. Cascella and C. Micheletti, *J. Phys.: Condens. Matter*, 2005, **17**, S1581.
103 T. Noguti and N. Go, *Nature*, 1982, **296**, 776.
104 S. Swaminathan, T. Ichiye, W. Vangunsteren and M. Karplus, *Biochemistry*, 1982, **21**, 5230.
105 M. M. Tirion, *Phys. Rev. Lett.*, 1996, **77**, 1905.
106 B. Park and M. Levitt, *J. Mol. Biol.*, 1996, **258**, 367.
107 C. Micheletti, *Phys. Life Rev.*, 2013, **10**, 1.
108 D. H. De Jong, L. V. Schafer, A. H. De Vries, S. J. Marrink, H. J. Berendsen and H. Grubmuller, *J. Comput. Chem.*, 2011, **32**, 1919.
109 J. M. Johnston, H. Wang, D. Provasi and M. Filizola, *PLoS Comput. Biol.*, 2012, **8**, e1002649.

110 D. Provasi, M. B. Boz, J. M. Johnston and M. Filizola, *PLoS Comput. Biol.*, 2015, **11**, e1004148.
111 E. Psachoulia, P. W. Fowler, P. J. Bond and M. S. Sansom, *Biochemistry*, 2008, **47**, 10503.
112 E. Psachoulia, D. P. Marshall and M. S. Sansom, *Acc. Chem. Res.*, 2010, **43**, 388.
113 T. A. Wassenaar, K. Pluhackova, A. Moussatova, D. Sengupta, S. J. Marrink, D. P. Tieleman and R. A. Böckmann, *J. Chem. Theory Comput.*, 2015, **11**, 2278.
114 A. C. Kalli, G. Morgan and M. S. Sansom, *Biophys. J.*, 2013, **105**, 137.
115 A. C. Kalli and M. S. Sansom, *Biochem. Soc. Trans.*, 2014, **42**, 1418.
116 L. Picas, J. Viaud, K. Schauer, S. Vanni, K. Hnia, V. Fraisier, A. Roux, P. Bassereau, F. Gaits-Iacovoni, B. Payrastre, J. Laporte, J. B. Manneville and B. Goud, *Nat. Commun.*, 2014, **5**, 5647.
117 S. Baoukina and D. P. Tieleman, *Biophys. J.*, 2010, **99**, 2134.
118 M. Fuhrmans and S. J. Marrink, *J. Am. Chem. Soc.*, 2012, **134**, 1543.
119 C. Lamprakis, A. Stocker and M. Cascella, *Front. Mol. Biosci.*, 2015, **2**, 36.
120 R. Kubo, *J. Phys. Soc. Jpn.*, 1962, **17**, 1100.
121 A. Liwo, M. Khalili, C. Czaplewski, S. Kalinowski, S. Oldziej, K. Wachucik and H. A. Scheraga, *J. Phys. Chem. B*, 2007, **111**, 260.
122 Y. He, Y. Xiao, A. Liwo and H. A. Scheraga, *J. Comput. Chem.*, 2009, **30**, 2127.
123 M. U. Johansson, M. deChateau, M. Wikstrom, S. Forsen, T. Drakenberg and L. Bjorck, *J. Mol. Biol.*, 1997, **266**, 859.
124 A. G. Cochran, N. J. Skelton and M. A. Starovasnik, *Proc. Natl. Acad. Sci. U. S. A.*, 2001, **98**, 5578.
125 J. W. Neidigh, R. M. Fesinmeyer and N. H. Andersen, *Nat. Struct. Biol.*, 2002, **9**, 425.
126 J. Moult, K. Fidelis, A. Kryshtafovych, T. Schwede and A. Tramontano, *Proteins*, 2014, **82**, 1.
127 A. Rojas, A. Liwo, D. Browne and H. A. Scheraga, *J. Mol. Biol.*, 2010, **404**, 537.
128 A. V. Rojas, A. Liwo and H. A. Scheraga, *J. Phys. Chem. B*, 2011, **115**, 12978.
129 A. K. Sieradzan, A. Liwo and U. H. E. Hansmann, *J. Chem. Theory Comput.*, 2012, **8**, 3416.
130 D. Reith, M. Putz and F. Muller-Plathe, *J. Comput. Chem.*, 2003, **24**, 1624.
131 P. Majek and R. Elber, *Proteins*, 2009, **76**, 822.
132 F. Ercolessi and J. B. Adams, *Europhys. Lett.*, 1994, **26**, 583.
133 P. Maurer, A. Laio, H. W. Hugosson, M. C. Colombo and U. Rothlisberger, *J. Chem. Theory Comput.*, 2007, **3**, 628.
134 A. Das and H. C. Andersen, *J. Chem. Phys.*, 2009, **131**, 034102.
135 Y. T. Wang, W. G. Noid, P. Liu and G. A. Voth, *Phys. Chem. Chem. Phys.*, 2009, **11**, 2002.
136 S. Izvekov and G. A. Voth, *J. Chem. Phys.*, 2005, **123**, 134105.
137 S. Izvekov and G. A. Voth, *J. Phys. Chem. B*, 2009, **113**, 4443.
138 S. Izvekov and G. A. Voth, *J. Chem. Theory Comput.*, 2006, **2**, 637.
139 P. Liu, S. Izvekov and G. A. Voth, *J. Phys. Chem. B*, 2007, **111**, 11566.
140 J. Zhou, I. F. Thorpe, S. Izvekov and G. A. Voth, *Biophys. J.*, 2007, **92**, 4289.
141 J. M. A. Grime and G. A. Voth, *Biophys. J.*, 2012, **103**, 1774.
142 E. Lyman, J. Pfaendtner and G. A. Voth, *Biophys. J.*, 2008, **95**, 4183.
143 J. W. Chu and G. A. Voth, *Biophys. J.*, 2006, **90**, 1572.
144 M. Neri, C. Anselmi, M. Cascella, A. Maritan and P. Carloni, *Phys. Rev. Lett.*, 2005, **95**, 218102.
145 V. Carnevale, S. Raugei, M. Neri, S. Pantano, C. Micheletti and P. Carloni, *J. Mol. Struct.: THEOCHEM*, 2009, **898**, 97.

146 M. Neri, C. Anselmi, V. Carnevale, A. V. Vargiu and P. Carloni, *J. Phys.: Condens. Matter*, 2006, **18**, S347.
147 M. Neri, M. Baaden, V. Carnevale, C. Anselmi, A. Maritan and P. Carloni, *Biophys. J.*, 2008, **94**, 71.
148 M. Cascella, M. A. Neri, P. Carloni and M. Dal Peraro, *J. Chem. Theory Comput.*, 2008, **4**, 1378.
149 M. T. Degiacomi, I. Lacovache, L. Pernot, M. Chami, M. Kudryashev, H. Stahlberg, F. G. van der Goot and M. Dal Peraro, *Nat. Chem. Biol.*, 2013, **9**, 623.
150 M. Adrian, J. Dubochet, J. Lepault and A. W. Mcdowall, *Nature*, 1984, **308**, 32.
151 B. S. Schuwirth, M. A. Borovinskaya, C. W. Hau, W. Zhang, A. Vila-Sanjurjo, J. M. Holton and J. H. D. Cate, *Science*, 2005, **310**, 827.
152 F. Tama, M. Valle, J. Frank and C. L. Brooks, *Proc. Natl. Acad. Sci. U. S. A.*, 2003, **100**, 9319.
153 F. Tama, O. Miyashita and C. L. Brooks, *J. Mol. Biol.*, 2004, **337**, 985.
154 S. Wagner, I. Sorg, M. Degiacomi, L. Journet, M. Dal Peraro and G. R. Cornelis, *Mol. Microbiol.*, 2009, **71**, 692.
155 T. Lemmin, C. S. Soto, G. Clinthorne, W. F. DeGrado and M. Dal Peraro, *PLoS Comput. Biol.*, 2013, **9**, e1002878.
156 M. Kudryashev, M. Stenta, S. Schmelz, M. Amstutz, U. Wiesand, D. Castano-Diez, M. T. Degiacomi, S. Munnich, C. K. E. Bleck, J. Kowal, A. Diepold, D. W. Heinz, M. Dal Peraro, G. R. Cornelis and H. Stahlberg, *eLife*, 2013, **2**, e00792.
157 M. T. Degiacomi and M. Dal Peraro, *Structure*, 2013, **21**, 1097.
158 F. Alber, S. Dokudovskaya, L. M. Veenhoff, W. H. Zhang, J. Kipper, D. Devos, A. Suprapto, O. Karni-Schmidt, R. Williams, B. T. Chait, A. Sali and M. P. Rout, *Nature*, 2007, **450**, 695.
159 F. Alber, S. Dokudovskaya, L. M. Veenhoff, W. Z. Zhang, J. Kipper, D. Devos, A. Suprapto, O. Karni-Schmidt, R. Williams, B. T. Chait, M. P. Rout and A. Sali, *Nature*, 2007, **450**, 683.
160 P. B. Canham, *J. Theor. Biol.*, 1970, **26**, 61.
161 W. Helfrich, *Z. Naturforsch., C: Biochem., Biophys., Biol., Virol.*, 1973, **28**.
162 E. A. Evans and P. L. La Celle, *Blood*, 1975, **45**, 29.
163 M. Deserno, *Chem. Phys. Lipids*, 2015, **185**, 11.
164 R. Lipowsky and E. Sackmann, Structure and Dynamics of Membranes, http://www.sciencedirect.com/science/book/9780444819758.
165 S. A. Safran, *Statistical Thermodynamics of Surfaces, Interfaces, and Membranes*, Addison-Wesley Pub., Reading, Mass., 1994.
166 H. J. Deuling and W. Helfrich, *Biophys. J.*, 1976, **16**, 861.
167 P. Bassereau, B. Sorre and A. Levy, *Adv. Colloid Interface Sci.*, 2014, **208**, 47.
168 R. Dimova, *Adv. Colloid Interface Sci.*, 2014, **208**, 225.
169 E. Evans, V. Heinrich, F. Ludwig and W. Rawicz, *Biophys. J.*, 2003, **85**, 2342.
170 E. Evans and W. Rawicz, *Phys. Rev. Lett.*, 1990, **64**, 2094.
171 J. F. Nagle, *Faraday Discuss.*, 2013, **161**, 11.
172 W. Rawicz, K. C. Olbrich, T. McIntosh, D. Needham and E. Evans, *Biophys. J.*, 2000, **79**, 328.
173 F. Brochard, P. G. Degennes and P. Pfeuty, *J. Physiol.*, 1976, **37**, 1099.
174 F. Brochard and J. F. Lennon, *J. Physiol.*, 1975, **36**, 1035.
175 J. F. Faucon, M. D. Mitov, P. Meleard, I. Bivas and P. Bothorel, *J. Physiol.*, 1989, **50**, 2389.
176 J. Henriksen, A. C. Rowat and J. H. Ipsen, *Eur. Biophys. J.*, 2004, **33**, 732.

177 J. R. Henriksen and J. H. Ipsen, *Eur. Phys. J. E: Soft Matter Biol. Phys.*, 2004, **14**, 149.
178 M. B. Schneider, J. T. Jenkins and W. W. Webb, *Biophys. J.*, 1984, **45**, 891.
179 Y. Liu and J. F. Nagle, *Phys. Rev. E.*, 2004, **69**, 040901.
180 J. Pan, T. T. Mills, S. Tristram-Nagle and J. F. Nagle, *Phys. Rev. Lett.*, 2008, **100**, 198103.
181 J. Pan, S. Tristram-Nagle, N. Kucerka and J. F. Nagle, *Biophys. J.*, 2008, **94**, 117.
182 T. Baumgart, B. R. Capraro, C. Zhu and S. L. Das, *Annu. Rev. Phys. Chem.*, 2011, **62**, 483.
183 D. Cuvelier, I. Derenyi, P. Bassereau and P. Nassoy, *Biophys. J.*, 2005, **88**, 2714.
184 M. Heinrich, A. Tian, C. Esposito and T. Baumgart, *Proc. Natl. Acad. Sci. U. S. A.*, 2010, **107**, 7208.
185 M. Hamm and M. M. Kozlov, *Eur. Phys. J. B*, 1998, **6**, 519.
186 M. Hamm and M. M. Kozlov, *Eur. Phys. J. E*, 2000, **3**, 323.
187 E. R. May, D. I. Kopelevich and A. Narang, *Biophys. J.*, 2008, **94**, 878.
188 E. R. May, A. Narang and D. I. Kopelevich, *Phys. Rev. E.*, 2007, **76**, 021913.
189 M. C. Watson and F. L. Brown, *Biophys. J.*, 2010, **98**, L9.
190 M. C. Watson, A. Morriss-Andrews, P. M. Welch and F. L. Brown, *J. Chem. Phys.*, 2013, **139**, 084706.
191 M. C. Watson, E. S. Penev, P. M. Welch and F. L. Brown, *J. Chem. Phys.*, 2011, **135**, 244701.
192 F. Campelo and M. M. Kozlov, *PLoS Comput. Biol.*, 2014, **10**, e1003556.
193 I. R. Cooke and M. Deserno, *Biophys. J.*, 2006, **91**, 487–495.
194 J. N. Israelachvili, D. J. Mitchell and B. W. Ninham, *J. Chem. Soc., Faraday Trans.*, 1976, **72**, 1525.
195 S. May, Y. Bohbot and A. BenShaul, *J. Phys. Chem. B*, 1997, **101**, 8648.
196 F. Campelo, H. T. McMahon and M. M. Kozlov, *Biophys. J.*, 2008, **95**, 2325.
197 M. M. Kozlov, F. Campelo, N. Liska, L. V. Chernomordik, S. J. Marrink and H. T. McMahon, *Curr. Opin. Cell Biol.*, 2014, **29**, 53.
198 F. Campelo, C. Arnarez, S. J. Marrink and M. M. Kozlov, *Adv. Colloid Interface Sci.*, 2014, **208**, 25.
199 I. R. Cooke and M. Deserno, *J. Chem. Phys.*, 2005, **123**, 224710.
200 M. Hu, D. H. de Jong, S. J. Marrink and M. Deserno, *Faraday Discuss.*, 2013, **161**, 365.
201 M. Hu, P. Diggins and M. Deserno, *J. Chem. Phys.*, 2013, **138**, 214110.
202 G. Khelashvili and D. Harries, *J. Phys. Chem. B*, 2013, **117**, 2411.
203 G. Khelashvili and D. Harries, *Chem. Phys. Lipids*, 2013, **169**, 113.
204 G. Khelashvili, N. Johner, G. Zhao, D. Harries and H. L. Scott, *Chem. Phys. Lipids*, 2014, **178**, 18.
205 G. Khelashvili, B. Kollmitzer, P. Heftberger, G. Pabst and D. Harries, *J. Chem. Theory Comput.*, 2013, **9**, 3866.
206 G. Khelashvili, M. Rappolt, S. W. Chiu, G. Pabst and D. Harries, *Soft Matter*, 2011, **7**, 10299.
207 A. J. Kox, J. P. J. Michels and F. W. Wiegel, *Nature*, 1980, **287**, 317.
208 P. Vanderploeg and H. J. C. Berendsen, *J. Chem. Phys.*, 1982, **76**, 3271.
209 P. Vanderploeg and H. J. C. Berendsen, *Mol. Phys.*, 1983, **49**, 233.
210 D. Bassolino-Klimas, H. E. Alper and T. R. Stouch, *Biochemistry*, 1993, **32**, 12624.
211 K. V. Damodaran, K. M. Merz, Jr. and B. P. Gaber, *Biochemistry*, 1992, **31**, 7656.

212 E. Egberts, S. J. Marrink and H. J. Berendsen, *Eur. Biophys. J.*, 1994, **22**, 423.
213 H. Heller, M. Schaefer and K. Schulten, *J. Phys. Chem.*, 1993, **97**, 8343.
214 S. J. Marrink, M. Berkowitz and H. J. C. Berendsen, *Langmuir*, 1993, **9**, 3122.
215 R. W. Pastor, *Curr. Opin. Struct. Biol.*, 1994, **4**, 486.
216 K. Raghavan, M. R. Reddy and M. L. Berkowitz, *Langmuir*, 1992, **8**, 233.
217 R. M. Venable, Y. Zhang, B. J. Hardy and R. W. Pastor, *Science*, 1993, **262**, 223.
218 J. Seelig, *Q. Rev. Biophys.*, 1977, **10**, 353.
219 J. Seelig and A. Seelig, *Q. Rev. Biophys.*, 1980, **13**, 19.
220 H. J. Berendsen and D. P. Tieleman, *Encycl. Comput. Chem.*, 1998, **3**, 1639.
221 S. E. Feller, *Curr. Opin. Colloid Interface Sci.*, 2000, **5**, 217.
222 D. P. Tieleman, S.-J. Marrink and H. J. Berendsen, *Biochim. Biophys. Acta., Rev. Biomembr.*, 1997, **1331**, 235.
223 W. L. Ash, M. R. Zlomislic, E. O. Oloo and D. P. Tieleman, *Biochim. Biophys. Acta*, 2004, **1666**, 158.
224 A. Grossfield, S. E. Feller and M. C. Pitman, *Proteins*, 2007, **67**, 31.
225 C. Kandt, W. L. Ash and D. P. Tieleman, *Methods*, 2007, **41**, 475.
226 E. Lindahl and M. S. Sansom, *Curr. Opin. Struct. Biol.*, 2008, **18**, 425.
227 O. Berger, O. Edholm and F. Jahnig, *Biophys. J.*, 1997, **72**, 2002.
228 J. B. Klauda, R. M. Venable, J. A. Freites, J. W. O'Connor, D. J. Tobias, C. Mondragon-Ramirez, I. Vorobyov, A. D. MacKerell, Jr. and R. W. Pastor, *J. Phys. Chem. B*, 2010, **114**, 7830.
229 J. B. Klauda, R. M. Venable, A. D. MacKerell and R. W. Pastor, *Curr. Top. Membr.*, 2008, **60**, 1.
230 J. B. Lim, B. Rogaski and J. B. Klauda, *J. Phys. Chem. B*, 2012, **116**, 203.
231 J. Henin, W. Shinoda and M. L. Klein, *J. Phys. Chem. B*, 2008, **112**, 7008.
232 S. Lee, A. Tran, M. Allsopp, J. B. Lim, J. Henin and J. B. Klauda, *J. Phys. Chem. B*, 2014, **118**, 547.
233 J. P. Jambeck and A. P. Lyubartsev, *J. Phys. Chem. B*, 2012, **116**, 3164.
234 A. Maciejewski, M. Pasenkiewicz-Gierula, O. Cramariuc, I. Vattulainen and T. Rog, *J. Phys. Chem. B*, 2014, **118**, 4571.
235 D. Poger, W. F. Van Gunsteren and A. E. Mark, *J. Comput. Chem.*, 2010, **31**, 1117.
236 A. Kukol, *J. Chem. Theory Comput.*, 2009, **5**, 615.
237 S. W. Chiu, S. A. Pandit, H. L. Scott and E. Jakobsson, *J. Phys. Chem. B*, 2009, **113**, 2748.
238 A. L. Rabinovich and A. P. Lyubartsev, *J. Phys.: Conf. Ser.*, 2014, 012022IOP Publishing.
239 J. P. Ulmschneider and M. B. Ulmschneider, *J. Chem. Theory Comput.*, 2009, **5**, 1803.
240 R. Tjörnhammar and O. Edholm, *J. Chem. Theory Comput.*, 2014, **10**, 5706.
241 T. J. Piggot, A. Pineiro and S. Khalid, *J. Chem. Theory Comput.*, 2012, **8**, 4593.
242 D. Marsh, *Handbook of Lipid Bilayers*, Taylor & Francis, 2nd edn, 2013.
243 O. O. H. Samuli, arXiv e-prints, 2013, **1309**, 2131.
244 A. Brodsky, *Chem. Phys. Lett.*, 1996, **261**, 563–568.
245 D. van der Spoel, P. J. van Maaren and H. J. Berendsen, *J. Chem. Phys.*, 1998, **108**, 10220.
246 R. W. Pastor and A. D. Mackerell, Jr., *J. Phys. Chem. Lett.*, 2011, **2**, 1526.
247 S. K. Kandasamy and R. G. Larson, *Biochim. Biophys. Acta*, 2006, **1758**, 1274.
248 P. Mukhopadhyay, L. Monticelli and D. P. Tieleman, *Biophys. J.*, 2004, **86**, 1601.
249 S. A. Pandit, D. Bostick and M. L. Berkowitz, *Biophys. J.*, 2003, **84**, 3743.

250 O. Edholm, O. Berger and F. Jahnig, *J. Mol. Biol.*, 1995, **250**, 94.
251 J. L. Klepeis, K. Lindorff-Larsen, R. O. Dror and D. E. Shaw, *Curr. Opin. Struct. Biol.*, 2009, **19**, 120.
252 M. Paloncyova, G. Fabre, R. H. DeVane, P. Trouillas, K. Berka and M. Otyepka, *J. Chem. Theory Comput.*, 2014, **10**, 4143.
253 D. P. Tieleman, J. L. MacCallum, W. L. Ash, C. Kandt, Z. Xu and L. Monticelli, *J. Phys.: Condens. Matter*, 2006, **18**, S1221.
254 A. Cordomí, G. Caltabiano and L. Pardo, *J. Chem. Theory Comput.*, 2012, **8**, 948.
255 A. E. Cardenas, R. Shrestha, L. J. Webb and R. Elber, *J. Phys. Chem. B*, 2015, **119**, 6412.
256 J. Gumbart and B. Roux, *Biophys. J.*, 2012, **102**, 795.
257 A. C. Johansson and E. Lindahl, *Proteins*, 2008, **70**, 1332.
258 J. L. MacCallum, W. D. Bennett and D. P. Tieleman, *J. Gen. Physiol.*, 2007, **129**, 371.
259 J. L. MacCallum, W. F. Bennett and D. P. Tieleman, *Biophys. J.*, 2008, **94**, 3393.
260 J. S. Burg, J. R. Ingram, A. J. Venkatakrishnan, K. M. Jude, A. Dukkipati, E. N. Feinberg, A. Angelini, D. Waghray, R. O. Dror, H. L. Ploegh and K. C. Garcia, *Science*, 2015, **347**, 1113.
261 C. Hong, D. P. Tieleman and Y. Wang, *Langmuir*, 2014, **30**, 11993.
262 J. C. Shillcock and R. Lipowsky, *J. Phys.: Condens. Matter*, 2006, **18**, S1191.
263 R. Goetz and R. Lipowsky, *J. Chem. Phys.*, 1998, **108**, 7397.
264 B. Smit, P. A. J. Hilbers, K. Esselink, L. A. M. Rupert, N. M. Vanos and A. G. Schlijper, *Nature*, 1990, **348**, 624.
265 M. Venturoli and B. Smit, *PhysChemComm*, 1999, **2**, 45.
266 M. Venturoli, B. Smit and M. M. Sperotto, *Biophys. J.*, 2005, **88**, 1778.
267 G. Brannigan and F. L. Brown, *J. Chem. Phys.*, 2004, **120**, 1059.
268 I. R. Cooke, K. Kremer and M. Deserno, *Phys. Rev. E.: Stat., Nonlinear, Soft Matter Phys.*, 2005, **72**, 011506.
269 O. Farago, *J. Chem. Phys.*, 2003, **119**, 596.
270 O. Lenz and F. Schmid, *J. Mol. Liquids*, 2005, **117**, 147.
271 H. Noguchi, *J. Chem. Phys.*, 2011, **134**, 055101.
272 G. Brannigan and F. L. Brown, *Biophys. J.*, 2007, **92**, 864.
273 V. A. Harmandaris and M. Deserno, *J. Chem. Phys.*, 2006, **125**, 204905.
274 S. Meinhardt, R. L. Vink and F. Schmid, *Proc. Natl. Acad. Sci. U. S. A.*, 2013, **110**, 4476.
275 B. J. Reynwar, G. Illya, V. A. Harmandaris, M. M. Muller, K. Kremer and M. Deserno, *Nature*, 2007, **447**, 461.
276 A. Srivastava and G. A. Voth, *J. Chem. Theory Comput.*, 2013, **9**, 750.
277 A. Srivastava and G. A. Voth, *J. Chem. Theory Comput.*, 2014, **10**, 4730.
278 M. Simunovic, A. Srivastava and G. A. Voth, *Proc. Natl. Acad. Sci. U. S. A.*, 2013, **110**, 20396.
279 S. J. Marrink, H. J. Risselada, S. Yefimov, D. P. Tieleman and A. H. de Vries, *J. Phys. Chem. B*, 2007, **111**, 7812.
280 W. Shinoda, R. DeVane and M. L. Klein, *J. Phys. Chem. B*, 2010, **114**, 6836.
281 A. H. de Vries, A. E. Mark and S. J. Marrink, *J. Am. Chem. Soc.*, 2004, **126**, 4488.
282 S. O. Yesylevskyy, L. V. Schafer, D. Sengupta and S. J. Marrink, *PLoS Comput. Biol.*, 2010, **6**, e1000810.
283 S. Pronk, S. Pall, R. Schulz, P. Larsson, P. Bjelkmar, R. Apostolov, M. R. Shirts, J. C. Smith, P. M. Kasson, D. van der Spoel, B. Hess and E. Lindahl, *Bioinformatics*, 2013, **29**, 845.

284 D. Van Der Spoel, E. Lindahl, B. Hess, G. Groenhof, A. E. Mark and H. J. Berendsen, *J. Comput. Chem.*, 2005, **26**, 170.
285 F. J. van Eerden, D. H. de Jong, A. H. de Vries, T. A. Wassenaar and S. J. Marrink, *Biochim. Biophys. Acta*, 2015, **1848**, 1319.
286 T. Reddy, D. Shorthouse, D. L. Parton, E. Jefferys, P. W. Fowler, M. Chavent, M. Baaden and M. S. Sansom, *Structure*, 2015, **23**, 584.
287 H. Koldso, D. Shorthouse, J. Helie and M. S. Sansom, *PLoS Comput. Biol.*, 2014, **10**, e1003911.
288 H. J. Risselada and S. J. Marrink, *Phys. Chem. Chem. Phys.*, 2009, **11**, 2056.
289 H. J. Risselada, S. J. Marrink and M. Muller, *Phys. Rev. Lett.*, 2011, **106**, 148102.
290 S. Vanni, H. Hirose, H. Barelli, B. Antonny and R. Gautier, *Nat. Commun.*, 2014, **5**, 4916.
291 S. Baoukina, L. Monticelli, M. Amrein and D. P. Tieleman, *Biophys. J.*, 2007, **93**, 3775–3782.
292 S. Baoukina, L. Monticelli, H. J. Risselada, S. J. Marrink and D. P. Tieleman, *Proc. Natl. Acad. Sci. U. S. A.*, 2008, **105**, 10803.
293 S. Baoukina, D. Rozmanov, E. Mendez-Villuendas and D. P. Tieleman, *Biophys. J.*, 2014, **107**, 1136.
294 A. J. Markvoort and S. J. Marrink, *Curr. Top. Membr.*, 2011, **68**, 259.
295 M. Pannuzzo, D. H. De Jong, A. Raudino and S. J. Marrink, *J. Chem. Phys.*, 2014, **140**, 124905.
296 H. J. Risselada, G. Bubnis and H. Grubmuller, *Proc. Natl. Acad. Sci. U. S. A.*, 2014, **111**, 11043.
297 H. J. Risselada, G. Marelli, M. Fuhrmans, Y. G. Smirnova, H. Grubmuller, S. J. Marrink and M. Muller, *PLoS One*, 2012, **7**, e38302.
298 H. J. Risselada, Y. Smirnova and H. Grubmuller, *Biophys. J.*, 2014, **107**, 2287.
299 M. Pinot, S. Vanni, S. Pagnotta, S. Lacas-Gervais, L. A. Payet, T. Ferreira, R. Gautier, B. Goud, B. Antonny and H. Barelli, *Science*, 2014, **345**, 69.
300 W. Shinoda, T. Nakamura and S. O. Nielsen, *Soft Matter*, 2011, **7**, 9012.
301 S. Kawamoto and W. Shinoda, *Soft Matter*, 2014, **10**, 3048.
302 W. Shinoda and M. L. Klein, *Pure Appl. Chem.*, 2014, **86**, 215.
303 H. T. McMahon and J. L. Gallop, *Nature*, 2005, **438**, 590.
304 J. Zimmerberg and M. M. Kozlov, *Nat. Rev. Mol. Cell Biol.*, 2006, **7**, 9.
305 J. Bigay and B. Antonny, *Dev. Cell*, 2012, **23**, 886.
306 H. J. Risselada, A. E. Mark and S. J. Marrink, *J. Phys. Chem. B*, 2008, **112**, 7438.
307 L. Vamparys, R. Gautier, S. Vanni, W. F. D. Bennett, D. P. Tieleman, B. Antonny, C. Etchebest and P. F. J. Fuchs, *Biophys. J.*, 2013, **104**, 585.
308 S. Vanni, L. Vamparys, R. Gautier, G. Drin, C. Etchebest, P. F. J. Fuchs and B. Antonny, *Biophys. J.*, 2013, **104**, 575.
309 M. M. Kamal, D. Mills, M. Grzybek and J. Howard, *Proc. Natl. Acad. Sci. U. S. A.*, 2009, **106**, 22245.
310 T. Kawakatsu, *Statistical Physics of Polymers*, Springer, Berlin, 2004.
311 S. Marcelja, *Nature*, 1973, **241**, 451.
312 P. G. Khalatur, in *Polymer Science: A Comprehensive Reference*, ed. K. Matyjaszewski and M. Möller, Elsevier, Amsterdam, 2012, pp. 417–460.
313 K. M. Langner and G. J. A. Sevink, *Soft Matter*, 2012, **8**, 5102.
314 G. Milano and T. Kawakatsu, *Phys. Biol.*, 2013, **10**, 45007.
315 K. C. Daoulas and M. Muller, *J. Chem. Phys.*, 2006, **125**, 184904.
316 K. C. Daoulas, M. Muller, J. J. de Pablo, P. F. Nealey and G. D. Smith, *Soft Matter*, 2006, **2**, 573.

317 M. P. Stoykovich, M. Muller, S. O. Kim, H. H. Solak, E. W. Edwards, J. J. de Pablo and P. F. Nealey, *Science*, 2005, **308**, 1442.
318 A. Ramirez-Hernandez, H. S. Suh, P. F. Nealey and J. J. de Pablo, *Macromolecules*, 2014, **47**, 3520.
319 G. Milano and T. Kawakatsu, *J. Chem. Phys.*, 2009, **130**, 214106.
320 G. Milano and T. Kawakatsu, *J. Chem. Phys.*, 2010, **133**, 214102.
321 A. De Nicola, Y. Zhao, T. Kawakatsu, D. Roccatano and G. Milano, *J. Chem. Theory Comput.*, 2011, **7**, 2947.
322 A. De Nicola, T. Kawakatsu and G. Milano, *J. Chem. Theory Comput.*, 2014, **10**, 5651.
323 Y. Zhao, A. De Nicola, T. Kawakatsu and G. Milano, *J. Comput. Chem.*, 2012, **33**, 868.
324 A. De Nicola, S. Hezaveh, Y. Zhao, T. Kawakatsu, D. Roccatano and G. Milano, *Phys. Chem. Chem. Phys.*, 2014, **16**, 5093.
325 A. De Nicola, T. Kawakatsu and G. Milano, *Macromol. Chem. Phys.*, 2013, **214**, 1940.
326 A. De Nicola, Y. Zhao, T. Kawakatsu, D. Roccatano and G. Milano, *Theor. Chem. Acc.*, 2012, **131**, 1167.
327 G. J. A. Sevink, M. Charlaganov and J. G. E. M. Fraaije, *Soft Matter*, 2013, **9**, 2816–2831.
328 E. Dickinson, V. J. Pinfield, D. S. Horne and F. A. M. Leermakers, *J. Chem. Soc., Faraday Trans.*, 1997, **93**, 1785.
329 M. W. Matsen and M. Schick, *Phys. Rev. Lett.*, 1994, **72**, 2660.
330 F. Drolet and G. H. Fredrickson, *Phys. Rev. Lett.*, 1999, **83**, 4317.
331 G. H. Fredrickson, V. Ganesan and F. Drolet, *Macromolecules*, 2002, **35**, 16.
332 Y. Lauw, F. A. M. Leermakers and M. A. C. Stuart, *J. Phys. Chem. B*, 2006, **110**, 465.
333 J. R. Roan and T. Kawakatsu, *J. Chem. Phys.*, 2002, **116**, 7283.
334 J. R. Roan and T. Kawakatsu, *J. Chem. Phys.*, 2002, **116**, 7295.

Chemical bonding in solids: recovering chemical concepts in the realm of infinite periodic structures

Alexey I. Baranov,*[a,b] Robert Ponec[c] and Miroslav Kohout[a]
DOI: 10.1039/9781782622703-00053

1 Introduction

Materials science as the realm of the design and the synthesis of new advanced materials with often unique physical, chemical or biological properties is one of the most rapidly expanding fields of contemporary research. Because of its inherently interdisciplinary nature, the progress of materials science considerably relies on dialogue between the solid state physicists, who possess the mathematical tools for the description of the solid state and chemists whose task is the discovery and actual synthesis of these materials. For many years the description of the electronic structure and properties of solids was dominated by the physicist's approach in terms of band theory, Brillouin zones, *k*-states *etc.*[1] Such an abstract mathematical description is however somewhat unfamiliar to chemists, whose understanding of molecular structures is much more intuitive and relies on bonding paradigms such as chemical bonds, bond orders, valences *etc.* In order to overcome this barrier a considerable effort has been devoted in recent years to the search for the possible link between the abstract mathematical language of the band theory and the more intuitive chemical approach. An example in this respect can be the books by Harrison,[2] Cox[3] and Hoffmann[4] in which the principles of solid state bonding are explained with the chemists audience in mind. The need for more efficient dialog between solid state physicists who are familiar with the mathematical formalism of the band theory and chemists whose experience and intuition is of invaluable help in the actual synthesis of new materials resulted in the emergence of recent studies trying to deduce the relation between the structure and the properties of the solids from the understanding of the bond arrangement.[5–8]

Into the framework of these approaches can be included also our recent studies[9–12] in which the tools originally designed for the description of the bonding of isolated molecules were generalized so as to be applicable to the systems consisting of infinite periodic arrays of atoms, molecules and ions. In particular the extension concerning the

[a]*Max Planck Institute for Chemical Physics of Solids, Nöthnitzer Strasse 40, 01187 Dresden, Germany*
[b]*TU Dresden, Department of Chemistry and Food Chemistry, Inorganic Chemistry II, Bergstrasse 66, 01062 Dresden, Germany. E-mail: Alexey.Baranov@chemie.tu-dresden.de*
[c]*Institute of Chemical Process Fundamentals, Czech Academy of Sciences, Rozvojová 135, 165 02 Prague 6, Czech Republic*

generalization of the classical concept of bond order and/or bond index and the methodology known as the analysis of domain averaged Fermi holes. Although the main goal of the present study is to summarize in concise form the mathematical background of both the above methodologies and to report the results of their application to both the model systems and realistic crystals, we consider it useful to first remember the original chemical definition of these concepts so as to make apparent the link with subsequent solid-state generalizations.

There have been many attempts to build a link between modern electronic structure calculations and various intuitive models of chemical bonding. Among the concepts that proved to be especially useful in this respect is the idea of bond order and/or the bond index as a theoretical counterpart of the classical concept of bond multiplicity that reflects the number of shared electron pairs involved according to Lewis model in the bonding between the given pair of atoms. There is a plethora of various definitions of bond order[13–17] but of special relevance for our purposes is the so-called Wiberg–Mayer bond order defined by eqn (1)

$$W(A,B)=\sum_{\mu}^{A}\sum_{\nu}^{B}(PS)_{\mu\nu}(PS)_{\nu\mu} \tag{1}$$

where P and S denote the charge density–bond order matrix and overlap matrix respectively. The summation runs over the basis functions localized on atoms A and B. This original intuitive definition was put on more safe theoretical footing by revealing the close link between the original index and the quantity $F(A, B)$ derived at SCF and formally[†] also Kohn–Sham level of the theory from the restricted pairwise integration of the exchange–correlation part of the pair density (eqn (2))

$$\begin{aligned} F(A,B) &= \int_{\Omega_A} d\mathbf{r}_1 \rho(\mathbf{r}_1) \int_{\Omega_B} d\mathbf{r}_2 \rho(\mathbf{r}_2) - 2\int_{\Omega_A} d\mathbf{r}_1 \int_{\Omega_B} d\mathbf{r}_2 \rho_2(\mathbf{r}_1,\mathbf{r}_2) \\ &= \int_{\Omega_A} d\mathbf{r}_1 \int_{\Omega_B} d\mathbf{r}_2 \rho_{xc}(\mathbf{r}_1,\mathbf{r}_2) \end{aligned} \tag{2}$$

$$\delta(A,B)=F(A,B)+F(B,A)$$

Although in the original definition the explicit integration was replaced by the Mulliken-like approximation of the corresponding integrals, the approach was more recently generalized into the framework of quantum theory of atoms in molecules (QTAIM) where the integrations are performed over the real 3D domains resulting from the virial partitioning of electron density.[18] In such a case the eqn (2) can be rewritten in case of a closed shell system as

$$\delta(A,B)=4\sum_{i}^{\text{occ}}\sum_{j}^{\text{occ}}S_{ij}^{A}S_{ji}^{B} \tag{3}$$

[†]Neither the pair density nor its hole part are explicitly available within DFT. The practical approach is to construct them from Kohn–Sham orbitals and their occupations implying formula for *e.g.* a one-determinantal ansatz.

where S_{ij}^X denotes the overlap integral of canonical orbitals ψ_i and ψ_j over the QTAIM domain of the atom X. It has been shown,[19] that $\delta(A,B)$ is proportional to the monopole term in the multipole expansion of the stabilizing exchange energy between corresponding atoms within the framework of interacting quantum atoms (IQA).[20] From the normalization condition of exchange–correlation part[21] to the total number of electrons in the system $\int d\mathbf{r}_1 \int d\mathbf{r}_2 \rho_{xc}(\mathbf{r}_1,\mathbf{r}_2) = N$ it follows that

$$\sum_A \sum_B F(A,B) = \sum_A F(A,A) + \frac{1}{2} \sum_{A \neq B} \delta(A,B) = N \tag{4}$$

The quantity $F(A, A) = \lambda(A)$ is a localization index,[22] equal to the number of self-pairs formed in the basin Ω_A. The difference between the average population of basin $N(A) = \int_{\Omega_A} d\mathbf{r}\rho(\mathbf{r})$ and the localization index is the statistical variance σ^2 of the basin population (*i.e.* its fluctuation).[23] When $\lambda(A) = N(A)$, all electrons are confined in that basin and the statistical variance is equal to zero. Thus the value of localization index reflects the number of electrons confined in the basin.

The original formula (eqn (3)) was later extended by Ángyán *et al.*[17] by releasing the condition of integer occupancy of canonical orbitals and replacing the doubly occupied orbitals by fractionally occupied orbitals (*e.g.* natural orbitals resulting from the diagonalization of the correlated first order density matrix). In such a case the formula in eqn (3) transforms to

$$\delta^{\mathscr{A}}(A,B) = 2 \sum_i \sum_j \theta_i \theta_j S_{ij}^A S_{ji}^B \tag{5}$$

where θ_i is the occupation of the i-th orbital. Similar generalization of bond index was proposed by Fulton,[24] whose definition is given by

$$\delta^{\mathscr{F}}(A,B) = 2 \sum_i \sum_j \theta_i^{1/2} \theta_j^{1/2} S_{ij}^A S_{ji}^B \tag{6}$$

The above two definitions are of special relevance for solid state where one often encounters the fractional occupations so that the formulae in eqn (5) and (6) represent a natural choice for the calculation of bond indices in solid state using DFT level of theory.[‡] In case of integer orbital occupations both formulations reduce to the original formula, given by eqn (3).

An alternative approach was suggested by Cioslowski and Mixon,[25] who introduced the so called covalent bond order $C(A,B)$, whose values are calculated from the localized natural orbitals $\tilde{\psi}_i$ and associated populations $\tilde{\theta}_i$ resulting from so-called isopycnic transformation[26] of the original natural orbitals.

$$C(A,B) = \sum_i \tilde{\theta}_i^2 \tilde{S}_{ii}^A \tilde{S}_{ii}^B \tag{7}$$

[‡]In contrast to Fulton formulation, the Ángyán delocalization indices fulfill the eqn (4) only in case of integer orbital occupations. At the same time, the exchange–correlation part of pair density used in eqn (6) may yield a negative contribution to the pair populations.

The usefulness of bond indices for structural elucidation arises from the nontrivial empirical finding that their values sensitively mimic the presence and/or absence of the bonding interactions in molecules. Thus, for instance, for molecules with well localized two-center two-electron (2c–2e) bonds of classical Lewis model, the Wiberg–Mayer bond orders attain non-negligible values only between directly bonded atoms, where their values closely coincide with classical bond multiplicities, while between classically non-bonded atoms the values are negligible. However frequent and useful is such a localized picture of the bonding in a wealth of ordinary molecules, its relevance in solid state is much more restricted and can be expected to apply rather occasionally, for example, in covalent crystals such as diamond, silicon *etc.* Such an expectation was indeed confirmed in the recent study[10] where the extension of the bond indices to solids has shown that degree of electron sharing often extends over more distant regions and can be best classified as delocalized multicenter bonding. The existence of such bonding was known also from the chemistry of electron deficient molecules and, for instance, the concept of 3c–2e bond proved remarkably useful in the elucidation of the structure of boranes.[27] In this connection it is interesting to remark, that the detection of multicenter bonding in ordinary molecules was made possible using the appropriate extension of the bond indices.[28–32] Using this extension the formula for 3-center bond index is given, apart from an appropriate normalization factor, by (eqn (8))

$$\Delta(A,B,C) \approx \sum_{i}^{\text{occ}}\sum_{j}^{\text{occ}}\sum_{k}^{\text{occ}} S_{ij}^{A} S_{jk}^{B} S_{ki}^{C} \tag{8}$$

whereby the generalization to multicenter bonding extended over more centers is straightforward.

In addition to the numerical information provided by the values of various bond indices, the family of tools for the description of molecular structure was complemented some time ago by the new approach based on the analysis of the so-called domain averaged Fermi holes (DAFH).[33–35] The most straightforward definition of these holes is *via* the restricted integration of the exchange part of the pair density (*cf.* eqn (2))

$$g^{A}(\mathbf{r}_1) = \int_{\Omega_A} d\mathbf{r}_2 \rho_{xc}(\mathbf{r}_1, \mathbf{r}_2) \tag{9}$$

In the case of Hartree–Fock and formally also Kohn–Sham approximation and integer orbital occupations the above formula reduces to the following equation, which also shows the close link of the hole analysis to the 2-center Wiberg–Mayer bond indices:

$$g^{A}(\mathbf{r}_1) = 2\sum_{i}^{\text{occ}}\sum_{j}^{\text{occ}} S_{ij}^{A} \psi_i(\mathbf{r}_1)\psi_j^{*}(\mathbf{r}_1) \tag{10}$$

The usefulness of this analysis arises from the fact that DAFH functions and associated populations (resulted after the diagonalization of the matrix from eqn (10)) provide the information about how and to what extent the electrons in a given domain are involved in interactions with other domains in the molecule. The practical experiences in earlier studies have shown that the most interesting and chemically most relevant information emerges from the analysis of domains resulting from a physically sound partitioning of electron density into regions, associated with individual atoms such as using Mulliken,[14] Hirshfeld[36] or QTAIM.[18] This can be done also for more complex domains formed by the union of multiple atomic domains. In this case the analysis yields, besides the information about the interactions with other domains, also the information about the electron pairs (chemical bonds, lone pairs *etc.*) retained in the domain.

The first step of the analysis consists in the diagonalization of the matrix that represents the hole (eqn (10)) in appropriate basis (usually canonical orbitals). This primary diagonalization yields the set of eigenvalues and eigenfunctions that are, in the second step subjected to the isopycnic transformation.[26] Its aim is to transform the primary eigenvectors, that are usually delocalized over the nearest neighborhood of the reference basin, into the set of more localized functions whose resemblance with localized orbitals often allows the association with classical concepts of bonds, lone pairs *etc.*, in terms of which chemists are used to classify the molecular structures.

After all these transformations DAFH (eqn (9)) remains unchanged and becomes expressed as the sum of the squared amplitudes of DAFH orbitals $f_i^A(\mathbf{r})$ multiplied by their occupation number n_i^A. This enables a decomposition of the delocalization index, obtained by the integration of the Fermi hole, into the sum of contributions from single DAFH orbitals[12] (or DAFH eigenvectors[37]).

$$\delta(A,B) = \int_{\Omega_B} \mathrm{d}\mathbf{r} g^A(\mathbf{r}) = \sum_i n_i^A \int_{\Omega_B} \mathrm{d}\mathbf{r} \left| f_i^A(\mathbf{r}) \right|^2 = \sum_i \delta_i(A,B) \tag{11}$$

The main advantage of this new analysis is the visually appealing description of various bonding situations valuable especially for the elucidation of the structure of molecules with nontrivial bonding pattern such as metal–metal bonding, multicenter bonding, hypervalence *etc.*[38–43] It was of interest to extend the methodology of DAFH analysis also to description of bonding situations in solids and the first results summarized in the study[12] will be complemented here by more detailed scrutiny of both the mathematical details of the generalization and additional computational results.

2 On interpretation of the hole part of the pair density

The domain-averaged hole density, which is the starting entity for the evaluation of Fermi orbitals as well as the first step in the calculation of delocalization indices (DI), is computed from the exchange–correlation part ρ_{xc} of the electron-pair density. At the one-determinantal level of theory the opposite-spin electrons are not explictly correlated, *i.e.*, only the exchange part involving the same-spin electrons is utilized in the evaluation of the DAFH and DI.

The exchange–correlation part of the pair density is the result of the (somewhat artificial) decomposition of the pair density into a quasi-independent part $\rho(\mathbf{r}_1)\rho(\mathbf{r}_2)$ and the departure from the quasi-independence $\rho_{xc}(\mathbf{r}_1,\mathbf{r}_2)$. In contrast to the expression for the pair density involving Slater determinants for orbital pairs, the decomposition into the quasi-independent part includes same-spin electrons occupying pairwise the same orbital, thus violating the Pauli principle. This violation does not disturb much as long as the full pair density is considered. The idea behind the DAFH and DI analysis is the assumption that the "independent" electron pairs, in case of the DI simply the product of electron populations of the evaluated domains $N(A)N(B)$, do not contribute to the bonding. Instead, it is the "deviation" from the quasi-independence, namely the number of shared electron pairs, *cf.* eqn (3), that plays the crucial role. If the integration is performed over the whole space, then the shared electrons are reduced to the number of self-pairs. The somewhat inconvenient situation of the consideration of quasi-independent pairs (as the fermions must obey the Fermi statistic) can be overcome using the definition of the fluctuation of the electronic population (the variance) $\sigma^2(A)$ in a chosen domain A enclosing the average number of electrons $N(A)$

$$\sigma^2(A) = \sum_n P_n[N(A) - n]^2 \tag{12}$$

with the event probability P_n[44] to find exclusively n electrons in the domain A (with all the remaining electrons located outside A). This leads to the expression $\sigma^2(A) = N(A) - \lambda(A)$ involving the localization index $\lambda(A)$, *cf.* also eqn (4), and establishing the connection to the number of shared electron pairs. Observe, that the variance (and similarly the covariance) deals with squared electron populations and not explicitly with electron pairs.

Another possibility to avoid the quasi-independence in the definition of the DAFH and DI offers the decomposition of the pair density into symmetric and antisymmetric parts, which can be also seen as a partitioning into the singlet and triplet components of the pair density.[45–47] For wavefunctions given by single normalized Slater determinant based on N orthonormal orbitals ψ the pair density of the singlet coupled electrons can be written as[48]

$$\rho_2^{(s)}(\mathbf{r}_1, \mathbf{r}_2) = \frac{1}{4}\sum_{i,j}^{N} p_{ij,ij}^{\alpha\beta} \left\| \psi_i(\mathbf{r}_1)\psi_j(\mathbf{r}_2) \right\| \left\| \psi_i^*(\mathbf{r}_1)\psi_j^*(\mathbf{r}_2) \right\| \tag{13}$$

with the permanents $\|\psi_i(\mathbf{r}_1)\psi_j(\mathbf{r}_2)\| = \psi_i(\mathbf{r}_1)\psi_j(\mathbf{r}_2) + \psi_j(\mathbf{r}_1)\psi_i(\mathbf{r}_2)$ and the discrete representation $p_{ij,ij}^{\alpha\beta}$ of the opposite-spin 2-matrix. For the pair density formed by wavefunctions based on single Slater determinant all the $p_{ij,ij}^{\alpha\beta}$ equal 1, unless an ensemble matrix is created. Similarly, the pair density of the triplet coupled electrons for the above case can be expressed as

$$\rho_2^{(t)}(\mathbf{r}_1,\mathbf{r}_2)=\frac{1}{2}\sum_{i<j}^{N}\left[p_{ij,ij}^{\alpha\alpha}+p_{ij,ij}^{\beta\beta}+\frac{1}{2}(p_{ij,ij}^{\alpha\beta}+p_{ji,ji}^{\alpha\beta})\right]\left|\psi_i(\mathbf{r}_1)\psi_j(\mathbf{r}_2)\right|\left|\psi_i^*(\mathbf{r}_1)\psi_j^*(\mathbf{r}_2)\right| \tag{14}$$

with the determinants $|\psi_i(\mathbf{r}_1)\psi_j(\mathbf{r}_2)|=\psi_i(\mathbf{r}_1)\psi_j(\mathbf{r}_2)-\psi_j(\mathbf{r}_1)\psi_i(\mathbf{r}_2)$ and the discrete representations $p_{ij,ij}^{\alpha\alpha}$ and $p_{ij,ij}^{\beta\beta}$ of the same-spin 2-matrices. The difference between the singlet-pair density and the specific fraction $1/x$ of the triplet-pair density is

$$\begin{aligned}
&\rho_2^{(s)}(\mathbf{r}_1,\mathbf{r}_2)-\frac{1}{x}\rho_2^{(t)}(\mathbf{r}_1,\mathbf{r}_2)\\
&=\frac{1}{2x}\sum_{i<j}\left[\frac{x-1}{2}\left(p_{ij,ij}^{\alpha\beta}+p_{ji,ji}^{\alpha\beta}\right)-\left(p_{ij,ij}^{\alpha\alpha}+p_{ij,ij}^{\beta\beta}\right)\right]\left[\left|\psi_i(\mathbf{r}_1)\right|^2\left|\psi_j(\mathbf{r}_2)\right|^2+\left|\psi_j(\mathbf{r}_1)\right|^2\left|\psi_i(\mathbf{r}_2)\right|^2\right]\\
&+\frac{1}{2x}\sum_{i<j}\left[\frac{x+1}{2}\left(p_{ij,ij}^{\alpha\beta}+p_{ji,ji}^{\alpha\beta}\right)+\left(p_{ij,ij}^{\alpha\alpha}+p_{ij,ij}^{\beta\beta}\right)\right]\left[\psi_i\psi_j^*(\mathbf{r}_1)\psi_j\psi_i^*(\mathbf{r}_2)+\psi_j\psi_i^*(\mathbf{r}_1)\psi_i\psi_j^*(\mathbf{r}_2)\right]\\
&+\sum_k p_{kk,kk}^{\alpha\beta}\left|\psi_k(\mathbf{r}_1)\right|^2\left|\psi_k(\mathbf{r}_2)\right|^2
\end{aligned} \tag{15}$$

In case of a single-determinantal closed-shell system $p_{ij,ij}^{\alpha\alpha}=p_{ij,ij}^{\beta\beta}=p_{ij,ij}^{\alpha\beta}=p_{ji,ji}^{\alpha\beta}$ for all $i\neq j$. Now, choosing $x=3$ removes the first term with the squared orbital amplitudes and what remains is actually the exchange–correlation term

$$\rho_{\mathrm{xc}}(\mathbf{r}_1,\mathbf{r}_2)=\rho_2^{(s)}(\mathbf{r}_1,\mathbf{r}_2)-\frac{1}{3}\rho_2^{(t)}(\mathbf{r}_1,\mathbf{r}_2). \tag{16}$$

Thus, in case of a closed-shell system the delocalization index between two domains equals the difference between the singlet-coupled and 1/3 of the triplet-coupled pairs formed between the domains (so-called effective pairon population[49]). However, such straightforward relationship is not valid for open-shell systems.

3 Analytical model of chemical bonding in solids

Consider a model system, build up by atoms of the same sort, arranged into a periodic lattice (Fig. 1). In the simplest case of hydrogen-like atoms carrying only one atomic orbital $\chi(\mathbf{r})$ (*i.e.* hydrogen atoms) the only single band appears. Each single-electron state $\psi_{\mathbf{k}}$ has the occupation $\theta(\mathbf{k})$ and assuming a closed-shell system with $\theta^{\uparrow}(\mathbf{k})=\theta^{\downarrow}(\mathbf{k})=\theta(\mathbf{k})$ the total number of the electrons is given by

$$N_{\mathrm{e}}=2\sum_{\mathbf{k}}\theta(\mathbf{k}) \tag{17}$$

Within the one-determinantal ansatz the exchange–correlation part of the spinless pair density is[50]

$$\rho_{\mathrm{xc}}(\mathbf{r}_1,\mathbf{r}_2)=2\sum_{\mathbf{k}}\sum_{\mathbf{k}'}\theta(\mathbf{k})\psi_{\mathbf{k}}(\mathbf{r}_1)\psi_{\mathbf{k}}^*(\mathbf{r}_2)\theta(\mathbf{k}')\psi_{\mathbf{k}'}(\mathbf{r}_2)\psi_{\mathbf{k}'}^*(\mathbf{r}_1) \tag{18}$$

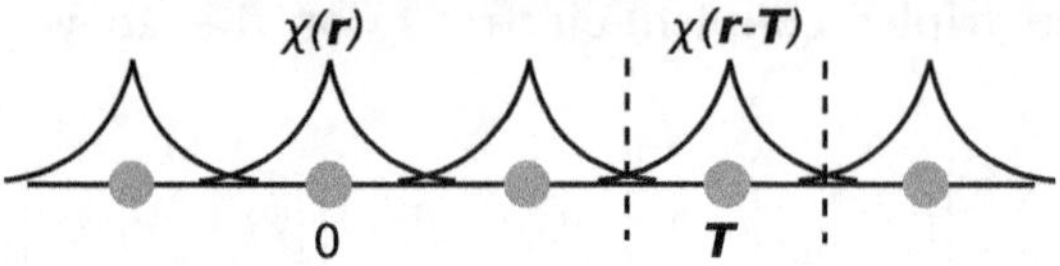

Fig. 1 One-dimensional model solid having single local basis orbital χ per site. Dashed lines show the boundaries of QTAIM basin Ω_T for the atom at the cell T [Color figure can be viewed in the electronic version.].

The integration over $\mathbf{r}_2$ in the QTAIM basin of the atom located at the cell $\mathbf{T}$, Ω_T yields the spinless domain-averaged Fermi hole

$$g^T(\mathbf{r}_1) = \int_{\Omega_T} d\mathbf{r}_2 \rho_{xc}(\mathbf{r}_1, \mathbf{r}_2) = 2\sum_{\mathbf{k}}\sum_{\mathbf{k}'} \theta(\mathbf{k})\theta(\mathbf{k}')\psi^*_{\mathbf{k}'}(\mathbf{r}_1)\psi_{\mathbf{k}}(\mathbf{r}_1)\underbrace{\int_{\Omega_T} d\mathbf{r}_2 \psi^*_{\mathbf{k}}(\mathbf{r}_2)\psi_{\mathbf{k}'}(\mathbf{r}_2)}_{S^T_{\mathbf{kk}'}}$$

$$= 2\sum_{\mathbf{k}}\sum_{\mathbf{k}'} \theta(\mathbf{k})\theta(\mathbf{k}')S^T_{\mathbf{kk}'}\psi^*_{\mathbf{k}'}(\mathbf{r}_1)\psi_{\mathbf{k}}(\mathbf{r}_1) \tag{19}$$

expressed in the basis of crystal orbitals. The crystal orbitals orthonormalized over the Born–von Kármán cluster consisting of N_k unit cells, can be expressed as

$$\psi_{\mathbf{k}}(\mathbf{r}) = \frac{1}{\sqrt{N_k}}\sum_{\mathbf{R}} \exp(-i\mathbf{kR})\chi(\mathbf{r} - \mathbf{R}) \tag{20}$$

The overlap integral between two Bloch states $\mathbf{k}$ and $\mathbf{k}'$ over the QTAIM basin of an atom in the cell $\mathbf{T}$ is

$$S^{\mathrm{T}}_{\mathbf{kk}'} = \int_{\Omega_T} d\mathbf{r}\psi^*_{\mathbf{k}}(\mathbf{r})\psi_{\mathbf{k}'}(\mathbf{r}) = \frac{1}{N_k}\sum_{\mathbf{R}}\sum_{\mathbf{R}'} \exp(i(\mathbf{kR} - \mathbf{k}'\mathbf{R}'))\underbrace{\int_{\Omega_T} d\mathbf{r}\chi^*(\mathbf{r} - \mathbf{R})\chi(\mathbf{r} - \mathbf{R}')}_{S^{\mathrm{T}}_{\mathbf{RR}'}} \tag{21}$$

Following ref. 9, let us assume that the basis orbitals are sufficiently local in the sense that the integral between them is negligible unless both local orbitals and the QTAIM basin refer to the same unit cell ($\mathbf{R} = \mathbf{R}' = \mathbf{T}$). For normalized local orbitals

$$S^{\mathrm{T}}_{\mathbf{RR}'} \approx \delta_{\mathbf{T},\mathbf{R}}\delta_{\mathbf{T},\mathbf{R}'} \tag{22}$$

so that

$$S^{\mathrm{T}}_{\mathbf{kk}'} = \frac{1}{N_k}\sum_{\mathbf{R}}\sum_{\mathbf{R}'} \exp(i(\mathbf{kR} - \mathbf{k}'\mathbf{R}'))\delta_{\mathbf{T},\mathbf{R}}\delta_{\mathbf{T},\mathbf{R}'} \approx \frac{1}{N_k}\exp(i(\mathbf{k} - \mathbf{k}')\mathbf{T}) \tag{23}$$

Substituting eqn (20) and (23) into eqn (19), an expansion of DAFH in the basis of local orbitals can be obtained

$$g^T(\mathbf{r}) = \frac{2}{N_k^2}\sum_{\mathbf{R}'\mathbf{R}}\sum_{\mathbf{kk}'} \theta(\mathbf{k})\theta(\mathbf{k}')\exp(i\mathbf{k}'(\mathbf{R}' - \mathbf{T}))\exp(-i\mathbf{k}(\mathbf{R} - \mathbf{T}))\chi^*(\mathbf{r} - \mathbf{R}')\chi(\mathbf{r} - \mathbf{R}) \tag{24}$$

Isolating the summations over $\mathbf{k}$ and $\mathbf{k}'$, one arrives at

$$g^T(\mathbf{r}) = 2\sum_{\mathbf{RR}'} \underbrace{\frac{1}{N_k}\sum_{\mathbf{k}} \theta(\mathbf{k}')\exp(i\mathbf{k}'(\mathbf{R}'-\mathbf{T}))}_{\Theta^*_{\mathbf{T}}(\mathbf{R}')} \underbrace{\frac{1}{N_k}\sum_{\mathbf{k}} \theta(\mathbf{k})\exp(-i\mathbf{k}(\mathbf{R}-\mathbf{T}))}_{\Theta_{\mathbf{T}}(\mathbf{R})} \chi^*(\mathbf{r}-\mathbf{R}')\chi(\mathbf{r}-\mathbf{R})$$

$$= \sum_{\mathbf{RR}'} \underbrace{2\Theta^*_{\mathbf{T}}(\mathbf{R}')\Theta_{\mathbf{T}}(\mathbf{R})}_{g^{\mathbf{T}}_{\mathbf{RR}'}} \chi^*(\mathbf{r}-\mathbf{R}')\chi(\mathbf{r}-\mathbf{R}) = \sum_{\mathbf{RR}'} g^{\mathbf{T}}_{\mathbf{RR}'}\chi^*(\mathbf{r}-\mathbf{R}')\chi(\mathbf{r}-\mathbf{R}) \tag{25}$$

where

$$\Theta_{\mathbf{T}}(\mathbf{R}) = \frac{1}{N_k}\sum_{\mathbf{k}} \theta(\mathbf{k})\exp(-i\mathbf{k}(\mathbf{R}-\mathbf{T})) = \Theta(\mathbf{R}-\mathbf{T}) \tag{26}$$

is the Fourier transform of the occupation number distribution $\theta(\mathbf{k})$ from the reciprocal k-space to the real space (vector $\mathbf{R}$ belongs to the crystal lattice) performed for the basin centered at the cell $\mathbf{T}$.[12] If the k-mesh is dense enough, the summation can be replaced by an integration over the Brillouin zone (BZ)

$$\Theta_{\mathbf{T}}(\mathbf{R}) = \frac{1}{V_{\mathrm{BZ}}}\int_{\mathrm{BZ}} \mathrm{d}\mathbf{k}\exp(-i\mathbf{k}(\mathbf{R}-\mathbf{T}))\theta(\mathbf{k}) = \Theta(\mathbf{R}-\mathbf{T}) \tag{27}$$

Assuming that the occupations are either unity or zero[§] one obtains from the Parseval's theorem and eqn (17) for the squared norm of $\Theta(\mathbf{R})$

$$\sum_{\mathbf{R}} |\Theta_{\mathbf{T}}(\mathbf{R})|^2 = \frac{N_e}{2N_k} \tag{28}$$

i.e. the squared length of the occupation number vector in both reciprocal and direct space is equal to the average number of electrons per unit cell.

3.1 Two-center bonding indices

The delocalization index between QTAIM basins for the atoms located in the cells $\mathbf{A}$ and $\mathbf{B}$ is

$$\begin{aligned}\delta(\mathbf{A},\mathbf{B}) &= 2\int_{\Omega_{\mathrm{B}}} \mathrm{d}\mathbf{r} g^{\mathbf{A}}(\mathbf{r}) = 4\sum_{\mathbf{RR}'} \Theta^*_{\mathbf{A}}(\mathbf{R})\Theta_{\mathbf{A}}(\mathbf{R}')\int_{\Omega_{\mathrm{B}}} \chi^*(\mathbf{r}-\mathbf{R})\chi(\mathbf{r}-\mathbf{R}') \\ &= 4\sum_{\mathbf{RR}'} \Theta^*_{\mathbf{A}}(\mathbf{R})\Theta_{\mathbf{A}}(\mathbf{R}')S^{\mathbf{B}}_{\mathbf{RR}'} = 4\Theta^*_{\mathbf{A}}(\mathbf{B})\Theta_{\mathbf{A}}(\mathbf{B})\end{aligned} \tag{29}$$

where eqn (22) has been used.

Thus for the model considered the delocalization indices can be evaluated just from the Fourier transforms $\Theta(\mathbf{R})$ (eqn (26)) of the occupation numbers $\theta(\mathbf{k})$ from the reciprocal k-space to the real space.

§In case of noninteger occupations one can employ Buijse–Baerends formula[51] instead of eqn (18).

3.2 Domain-averaged Fermi hole analysis

The first step of DAFH analysis is to find the eigenvectors of DAFH. In the basis of local orbitals the elements of DAFH matrix G^{T} are given by $g^{\mathrm{T}}_{\mathbf{R}'\mathbf{R}} = 2\Theta^{*}_{\mathrm{T}}(\mathbf{R})\Theta_{\mathrm{T}}(\mathbf{R}')$ (eqn (25)). Direct check immediately shows, that Θ_T is an eigenvector of G^{T} with the eigenvalue $N_{\mathrm{e}}/N_{\mathrm{k}}$

$$G^{\mathrm{T}}\Theta_{\mathrm{T}} = \sum_{\mathbf{R}} g^{\mathrm{T}}_{\mathbf{R}'\mathbf{R}}\Theta_{\mathrm{T}}(\mathbf{R}) = 2\Theta_{\mathrm{T}}(\mathbf{R}')\sum_{\mathbf{R}}\Theta^{*}_{\mathrm{T}}(\mathbf{R})\Theta_{\mathrm{T}}(\mathbf{R}) = \frac{N_{\mathrm{e}}}{N_{\mathrm{k}}}\Theta_{\mathrm{T}}(\mathbf{R}') \tag{30}$$

where eqn (28) has been used. This eigenvalue coincides with the trace of G^{T} matrix

$$\mathrm{Tr}(G^{\mathrm{T}}) = \sum_{\mathbf{R}} g^{\mathrm{T}}_{\mathbf{R}\mathbf{R}} = 2\sum_{\mathbf{R}}\Theta^{*}_{\mathrm{T}}(\mathbf{R})\Theta_{\mathrm{T}}(\mathbf{R}) = \frac{N_{\mathrm{e}}}{N_{\mathrm{k}}} \tag{31}$$

Therefore $\Theta_{\mathrm{T}}(\mathbf{R})$ is the only nontrivial eigenvector of DAFH and an isopycnic transformation is not necessary in this case. Thus $\Theta_{\mathrm{T}}(\mathbf{R})$ defines[12] the expansion of the DAFH orbital $f_{\mathrm{T}}(\mathbf{r})$ in terms of the local orbitals χ

$$f_{\mathrm{T}}(\mathbf{r}) = \sqrt{\frac{2N_{\mathrm{k}}}{N_{\mathrm{e}}}}\sum_{\mathbf{R}}\Theta_{\mathrm{T}}(\mathbf{R})\chi(\mathbf{r}-\mathbf{R}) \tag{32}$$

where the square root multiplier is the normalization coefficient.

3.3 Application to 1D–3D hydrogen lattices

In the case of the **1D hydrogen lattice** the occupation distribution in k-space can be written as rescaled rectangular function Π (Fig. 2)

$$\theta(k) = \Pi\left(\frac{k}{2k_F}\right) = \begin{cases} 1 & |k| \le k_F \\ 0 & |k| > k_F \end{cases} \tag{33}$$

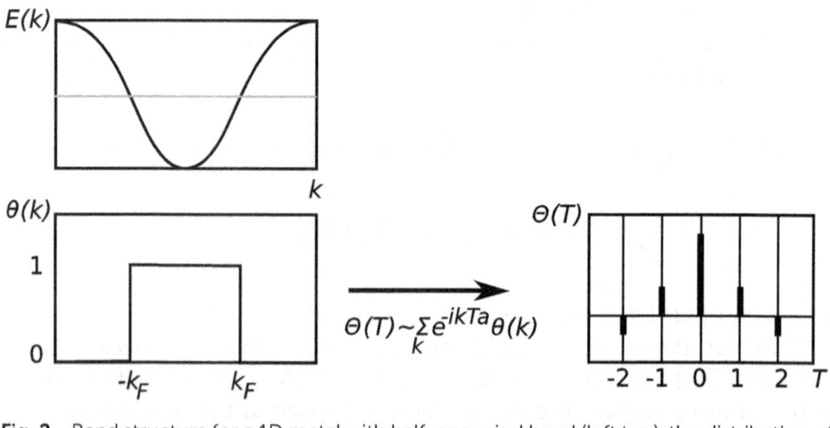

Fig. 2 Band structure for a 1D metal with half-occupied band (left top), the distribution of occupation numbers in the k-space inside the first Brillouin zone (left bottom), and its Fourier transform to the real space (right) [Color figure can be viewed in the electronic version.].

Its Fourier transform is the cardinal sine function

$$\Theta^{1\mathrm{D}}(r) = \frac{1}{V_{\mathrm{BZ}}}\int \mathrm{d}k \exp(-ikr)\theta(k) = \frac{1}{V_{\mathrm{BZ}}}\int \mathrm{d}k \exp(-ikr)\Pi\left(\frac{k}{2k_F}\right)$$
$$= \frac{2k_F}{V_{\mathrm{BZ}}}\operatorname{sinc}(k_F r) \tag{34}$$

If the distance between the nearest atoms is a, then $r = ta$ (index t labels unit cells), $V_{\mathrm{BZ}} = 2\pi/a$ and $k_F = \pi/2a$ and the delocalization index (when $t \neq 0$) amounts to

$$\delta(0,t) = 4\left[\Theta^{1\mathrm{D}}(ta)\right]^2 = \frac{4}{(\pi t)^2}\sin^2(\pi t/2) \tag{35}$$

which coincides with eqn (11) of ref. 9. In this case the delocalization indices between atoms, separated by t cells, depend on the distance proportional to $1/t^2$. The long-range behavior is due to the presence of the discontinuity of the occupation distribution in k-space, or, in other words, due to the presence of the Fermi surface.

Figure 3 shows a schematic representation of the DAFH orbital for the QTAIM basin of the zeroth unit cell in the 1D hydrogen lattice constructed according to eqn (32). Since the function $\Theta^{1\mathrm{D}}(r)$ decays slowly (proportional to $1/r$), the DAFH orbital has non-negligible contributions from unit cells around the reference atom and thus describes delocalized multicenter bonding. Again, this long-range behavior is caused by the presence of the discontinuity in the occupation distribution $\theta(k)$, whose Fourier transform is oscillating-decaying $\operatorname{sinc}(k_F r)$ function.

For a constant occupation distribution (*i.e.* fully occupied band) the corresponding Fourier transform is a delta-function like, having the only contribution from the local orbital centered at the reference basin. In

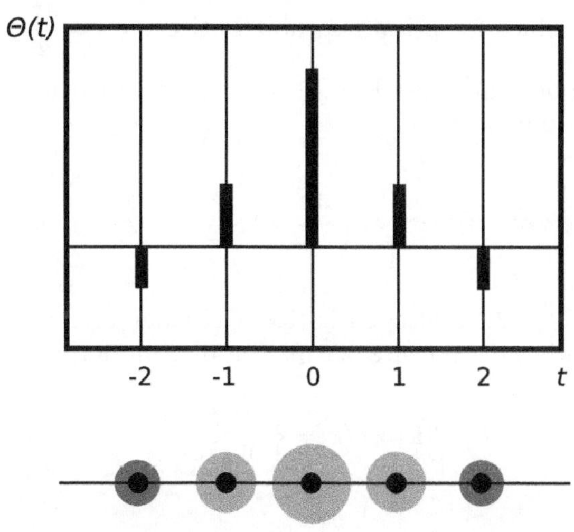

Fig. 3 Schematic representation of the Fourier transform of the occupation numbers for the 1D hydrogen lattice (top) and the corresponding DAFH orbital (bottom) [Color figure can be viewed in the electronic version.].

this case the degree of orbital localization will depend on the mutual overlap of the local orbitals, which was neglected (eqn (22)) in the current model.

For a **3D lattice of hydrogen atoms** the simplest reasonable form of Fermi surface can be chosen as a sphere of radius k_F, corresponding to the model of the homogeneous electron gas. The function $\theta(k)$ can be represented as a multipole expansion with following coefficients

$$\theta_{lm}(k) = \sqrt{4\pi}\delta_{l0}\delta_{m0}\begin{cases} 1 & k \le k_F \\ 0 & k > k_F \end{cases} \tag{36}$$

The evaluation of the Fourier transform employs Rayleigh expansion of the exponential over spherical harmonics $\exp(i\mathbf{kr}) = 4\pi\sum_{lm} i^l j_l(kr) Y_{lm}(r_\theta, r_\phi) Y^*_{lm}(k_\theta, k_\phi)$ which yields after the k-integration for $\mathbf{r} \neq 0$[¶]

$$\begin{aligned}\Theta^{3D}(\mathbf{r}) &= \frac{1}{V_{BZ}}\int \mathrm{d}\mathbf{k}\exp(-i\mathbf{kr})\theta(\mathbf{k}) \\ &= \frac{4\pi}{V_{BZ}}\sum_{lm} i^l Y_{lm}(r_\theta, r_\phi)\int k^2 \mathrm{d}k j_l(kr)\theta_{00}(k)\underbrace{\int_0^\pi \sin(k_\theta)\mathrm{d}k_\theta \int_0^{2\pi} \mathrm{d}k_\phi Y_{00}(k_\theta, k_\phi) Y^*_{lm}(k_\theta, k_\phi)}_{\delta_{l0}\delta_{m0}} \\ &= \frac{4\pi}{V_{BZ}}\int_0^{k_F} k^2 \mathrm{d}k j_0(kr) = \frac{4\pi k_F^2}{V_{BZ} r} j_1(k_F r)\end{aligned} \tag{37}$$

The spherical Bessel function of first order $j_1(r)$ decays as $1/r$ and therefore $\Theta^{3D}(r)$ decays as $1/r^2$. Thus also the DAFH orbital demonstrates inverse square decay with the distance. The delocalization indices being proportional to the squared amplitude of $\Theta(r)$ exhibit $1/r^4$ decay. Substituting the values for primitive cubic lattice $\left(V_{BZ} = \frac{8\pi^3}{a^3},\ k_F = \frac{\sqrt[3]{3\pi^2}}{a}\right)$ yields for the delocalization index with any nearest neighbor in coincidence with results of ref. 11

$$\delta(0, a) = 4(\Theta^{3D}(a))^2 = 4\left[4\pi\left(\frac{\sqrt[3]{3\pi^2}}{a}\right)^2 \frac{a^3}{8\pi^3 a} j_1(\sqrt[3]{3\pi^2})\right]^2 \approx 0.101 \tag{38}$$

Similarly, for a **2D hydrogen lattice** a circular Fermi surface can be considered with the following occupation distribution in k-space

$$\theta(\mathbf{k}) = \theta(k) = \begin{cases} 1 & k \le k_F \\ 0 & k > k_F \end{cases} \tag{39}$$

[¶]Otherwise $\Theta^{3D}(0) = \frac{4\pi k_F^3}{3V_{BZ}}$.

Its Fourier transform can be easily obtained using Jacobi–Anger expansion $e^{ix\cos\varphi} = \sum_{-\infty}^{\infty} i^n J_n(x) e^{in\varphi}$ which for $\mathbf{r} \neq 0$ yields[‖]

$$\begin{aligned}
\Theta^{2D}(\mathbf{r}) &= \frac{1}{V_{BZ}} \int \mathrm{d}\mathbf{k} \exp(-i\mathbf{kr})\theta(\mathbf{k}) \\
&= \frac{1}{V_{BZ}} \int k\mathrm{d}k \int_0^{2\pi} \mathrm{d}k_\varphi \exp(-ikr\cos(k_\varphi - r_\varphi))\theta(k) \\
&= \frac{1}{V_{BZ}} \sum_n (-i)^n \int k\mathrm{d}k\theta(k) J_n(kr) \underbrace{\int_0^{2\pi} \mathrm{d}k_\varphi \exp(-ink_\varphi)\exp(inr_\varphi)}_{2\pi\delta_{n0}} \\
&= \frac{2\pi}{V_{BZ}} \int k\mathrm{d}k\theta(k) J_0(kr) = \frac{2\pi}{V_{BZ}} \int_0^{k_F} k\mathrm{d}k J_0(kr) \\
&= \frac{2\pi k_F}{V_{BZ} r} J_1(k_F r)
\end{aligned} \tag{40}$$

The Bessel function of first order $J_1(r)$ decays as $1/\sqrt{r}$ and therefore $\Theta^{2D}(r)$ decays as $1/r^{3/2}$ yielding $1/r^3$ decay for the delocalization indices, just between inverse square and inverse fourth power decay for the 1D and 3D cases respectively.

Another possible choice of the Fermi surface for a square planar lattice, which in fact results from the nearest-neighbor tight-binding model[10] can be specified by following conditions

$$\theta(\mathbf{k}) = \begin{cases} 1 & |k_x + k_y| \leq k_F \vee |k_x - k_y| \leq k_F \\ 0 & \text{otherwise} \end{cases} \tag{41}$$

where k_x and k_y are Cartesian coordinates and k_F is the length of the largest occupied k-vector (Fig. 4).

Introducing the variables $\begin{pmatrix} q_x \\ q_y \end{pmatrix} = \begin{pmatrix} 1/\sqrt{2} & 1/\sqrt{2} \\ -1/\sqrt{2} & 1/\sqrt{2} \end{pmatrix} \begin{pmatrix} k_x \\ k_y \end{pmatrix}$ and $\begin{pmatrix} u_x \\ u_y \end{pmatrix} = \begin{pmatrix} 1/\sqrt{2} & 1/\sqrt{2} \\ -1/\sqrt{2} & 1/\sqrt{2} \end{pmatrix} \begin{pmatrix} x \\ y \end{pmatrix}$, the Fourier transform of this $\theta(\mathbf{k})$ is easily computed by factorizing it into two separate one-dimensional Fourier transforms from eqn (34)

$$\begin{aligned}
\Theta^{2D}(\mathbf{u}) &= \frac{1}{V_{BZ}} \int \mathrm{d}\mathbf{q} \exp(-i\mathbf{qu})\theta(\mathbf{q}) \\
&= \frac{1}{V_{BZ}} 2\frac{k_F}{\sqrt{2}} \mathrm{sinc}(k_F u_x/\sqrt{2}) 2\frac{k_F}{\sqrt{2}} \mathrm{sinc}(k_F u_y/\sqrt{2})
\end{aligned} \tag{42}$$

Substituting back the expressions for u_x and u_y yields

$$\Theta^{2D}(\mathbf{r}) = \frac{2k_F^2}{V_{BZ}} \mathrm{sinc}(k_F(x+y)/2)\mathrm{sinc}(k_F(-x+y)/2) \tag{43}$$

where x and y are Cartesian coordinates of vector $\mathbf{r} = (x,y)$ in position space. In this case the Θ^{2D} decay behavior is anisotropic. It is

‖ $\Theta^{2D}(0) = \frac{\pi k_F^2}{V_{BZ}}$.

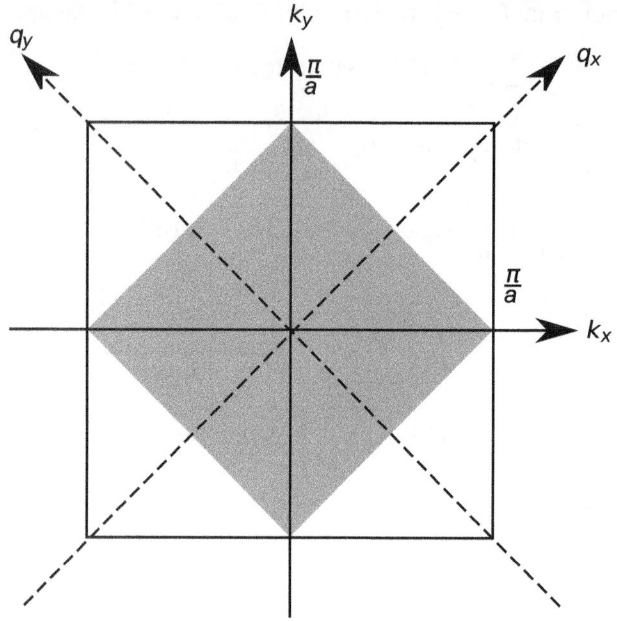

Fig. 4 1st Brillouin zone and occupied states (shaded region) for the square planar lattice. Here $k_F = \pi/a$. [Color figure can be viewed in the electronic version.].

proportional to $1/r^2$ along directions parallel to x or y axes and proportional to $1/r$ along directions $x = y$ or $x = -y$ (when max $|k| = k_F = \pi/a$, it is identically zero along these directions).

Long-range electron sharing indicated by the values of delocalization indices between QTAIM basins from different cells is therefore a typical feature of the metal. As it was shown above for the square planar lattice, this effect does not depend on the particular form of the Fermi surface and always appears, when $\theta(\mathbf{k})$ has discontinuities, which necessarily makes its Fourier transform $\Theta(\mathbf{R})$ a slowly decaying function. $\Theta(\mathbf{R})$ defines the spatial behavior of the delocalization indices (eqn (29)) and DAFH orbitals (eqn (24)), which reveal nonzero contributions even between non-neighboring cells, representing delocalized multicenter bonding with non-negligible contributions also for distant atoms.

In Sec. 4 the DAFH orbitals evaluated here are compared with the orbitals, obtained numerically from the results of DFT band structure calculations.

4 Chemical bonding analysis from DFT calculations

The bonding indices introduced in Sec. 1 can be also evaluated from band structure calculations. The atomic overlap integrals can be calculated using the shape function of the basin $\Xi^{\Omega}(\mathbf{r})$:

$$\Xi^{\Omega}(\mathbf{r}) = \begin{cases} 1 & \mathbf{r} \in \Omega \\ 0 & \mathbf{r} \notin \Omega \end{cases} \tag{44}$$

with which the integral over the basin can be transformed to the integral over the unit cell

$$S_{n\mathbf{k},n'\mathbf{k}'}(\Omega) = \int_{\Omega} d\mathbf{r}\psi_{n\mathbf{k}}(\mathbf{r})\psi^{*}_{n'\mathbf{k}'}(\mathbf{r}) = \int_{V} d\mathbf{r}\psi_{n\mathbf{k}}(\mathbf{r})\psi^{*}_{n'\mathbf{k}'}(\mathbf{r})\Xi^{\Omega}(\mathbf{r}) \tag{45}$$

The latter can be evaluated using, for instance, plane wave and multipole expansions as it was employed in (L)APW + lo + LO method.[11] In the solid state codes the Kohn–Sham orbitals are usually normalized over the volume of a single unit cell (not over the volume of the Born–von Kármán cluster) so that

$$\int d\mathbf{r}\psi_{n\mathbf{k}}(\mathbf{r})\psi^{*}_{n'\mathbf{k}'}(\mathbf{r}) = N_k\delta_{n,n'}\delta_{\mathbf{k},\mathbf{k}'} \tag{46}$$

In this case the orbitals must be rescaled by $1/\sqrt{N_k}$, which introduces the normalization factors $1/N_k$ for overlap integrals and $1/N_k^2$ for delocalization indices.

The resulting overlap integrals can be used in eqn (5)–(7) and (10) for the evaluation of delocalization indices, Cioslowski–Mixon bond orders and DAFH analysis respectively. The summation over orbitals in all these equations has to be performed over all bands and *all* the *k*-points from the whole Brillouin zone.

The delocalization indices and DAFH analysis for solids have been evaluated from the results of (L)APW + lo + LO calculations.[11,12,52–56] An overview of these results is presented below.

4.1 Hydrogen lattices

Figure 5 presents the plot of the DAFH orbital with non-negligible occupation number (0.936) from the DFT calculation for the 1D hydrogen lattice.[11] This DAFH orbital exhibits a decaying wave-like behavior showing essential multicenter character. About 50% of its norm is found in the reference QTAIM basin and 40% in the two nearest QTAIM basins. Thus about 10% of its norm is spread over the basins of the more distant cells. The rescaled $\Theta^{1D}(r)$ (eqn (34)), evaluated at the positions of nuclei is also shown on Fig. 5 for comparison. Its values define the contributions of local orbitals to the DAFH orbital in the analytical model (eqn (32)). One observes the excellent quantitative coincidence between the DFT and the analytical DAFH orbitals. Correspondingly, the values of delocalization indices from the analytical and DFT results also match well.[11]

Table 1 shows the Ángyán and Fulton delocalization indices respectively, calculated for the 2D hydrogen lattice, using various numbers of *k*-points. Cioslowski–Mixon bond indices and the indices from the analytical calculations are also shown. All the values for the shortest contacts from both analytical and numerical calculations are reasonably close. The indices evaluated between next nearest neighbor atoms are similar for the DFT results but the analytical model yields much smaller values. The reason for this discrepancy seems to stem from the approximation eqn (22) since the shape of the Fermi surface from the DFT calculation is very close to the circular one, used in the analytical model. The difference

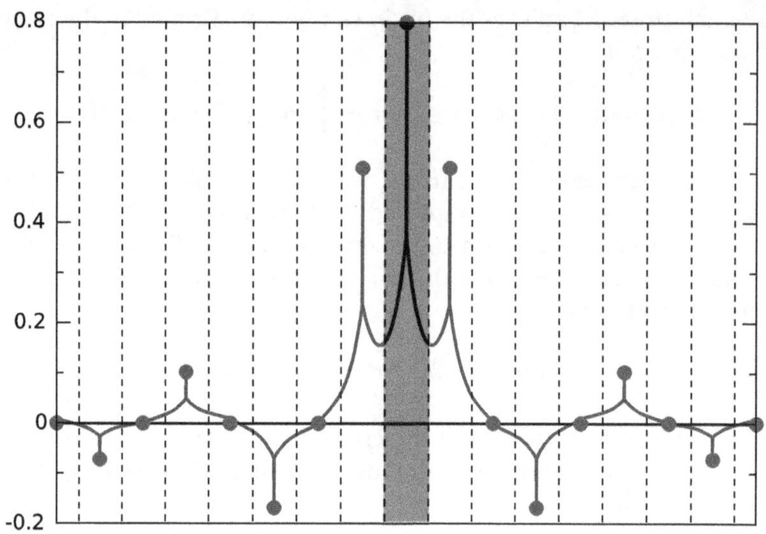

Fig. 5 DAFH orbital for the 1D hydrogen lattice. The solid line is the DAFH orbital from DFT calculation and dots show $\Theta(t)$ values from the analytical model calculation (eqn (34)), rescaled to match the absolute maximum of numerical DAFH orbital. Gray-shaded area shows the reference QTAIM basin. [Color figure can be viewed in the electronic version.] Reprinted with permission from ref. 12, Copyright 2012, AIP Publishing LLC.

Table 1 Delocalization indices from the DFT calculations and the analytical model as well as Cioslowski–Mixon bond orders for the shortest and next shortest contacts in the 2D hydrogen square lattice. The second line of the table head for the first three columns shows the number of *k*-points being used in the DFT calculations. FS stays for the Fermi surface.

	Ángyán		Fulton		Cioslowski–Mixon	Analytical	
	8^2	32^2	8^2	32^2	8^2	Circular FS	Square FS
Shortest	0.190	0.192	0.198	0.193	0.151	0.156	0.164
Next shortest	0.044	0.046	0.045	0.046	0.046	0.004	0

between Ángyán and Fulton values decreases as the number of *k*-points used in numerical calculation increases. It should be noted, that the Cioslowski–Mixon bond indices demonstrate essential variation for symmetrically equivalent contacts (*e.g.* from 0.10 to 0.20 for the nearest neighbor contacts), similar to molecular benzenoid hydrocarbons.[25] An additional reason for that could be the shortcoming of our iterative algorithm of isopycnic transformation, experiencing difficulties when localizing a large set of orbitals with fractional occupations.

Figure 6 shows the 3D representation of the DAFH orbital for 3D hydrogen lattice with the occupation number 0.777 as well as its one-dimensional plot along the space diagonal of the cubic cell with the rescaled $\Theta^{3D}(r)$ values from analytical model (eqn (37)). One observes for the 3D case a somewhat poorer matching between analytical and numerical results. A similar situation has been also observed for delocalization indices.[11] It is attributed to the enhancement of electron sharing as the system dimensionality increases, which can be seen by

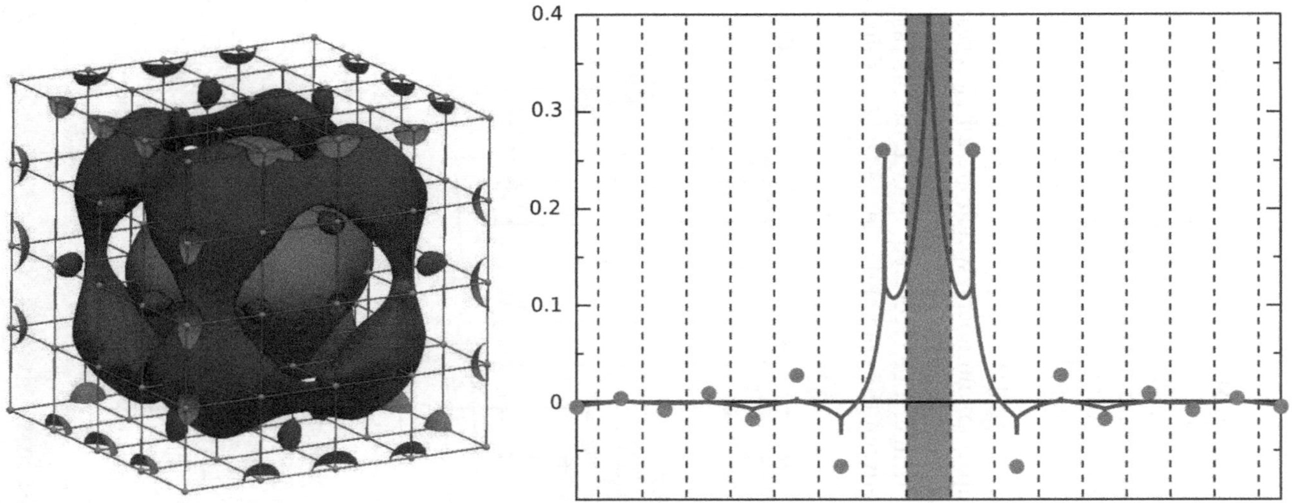

Fig. 6 DAFH orbital for the 3D hydrogen lattice as three-dimensional amplitude isosurface of level 0.0107 (left) and one-dimensional plot along cubic cell space diagonal (right, solid line), superimposed with $\Theta^{3D}(t)$ (dots), rescaled to match the absolute maximum of numerical DAFH orbital. [Color figure can be viewed in the electronic version.] Reprinted with permission from ref. 12, Copyright 2012, AIP Publishing LLC.

progressively decreased values of the localization index. However, from the analytical model calculations, employing eqn (22) the value of the localization index is equal to 0.5 for the system of any dimension from 1 to 3.

4.2 Delocalization for QTAIM basins and DAFH analysis for solids

As expected from the chemical intuition, ionic solids show high degree of electron localization. For NaCl the values of localization indices are only slightly smaller, than the basin populations (for example for Na cation $\lambda(\mathrm{Na}) = 9.92$ *vs.* $N(\mathrm{Na}) = 10.14$).[11] Correspondingly, the values of delocalization indices are low (less than 0.1 for both nearest Na–Cl and Cl–Cl contacts).

For both Na and Cl QTAIM basins there are only few DAFH orbitals with essential occupations numbers, which are all close to 2 (Fig. 7).[12] They all are well confined in their reference QTAIM basins, where more than 95% of their norm can be recovered. In fact, in this case there is no much difference between the core and valence DAFH orbitals.

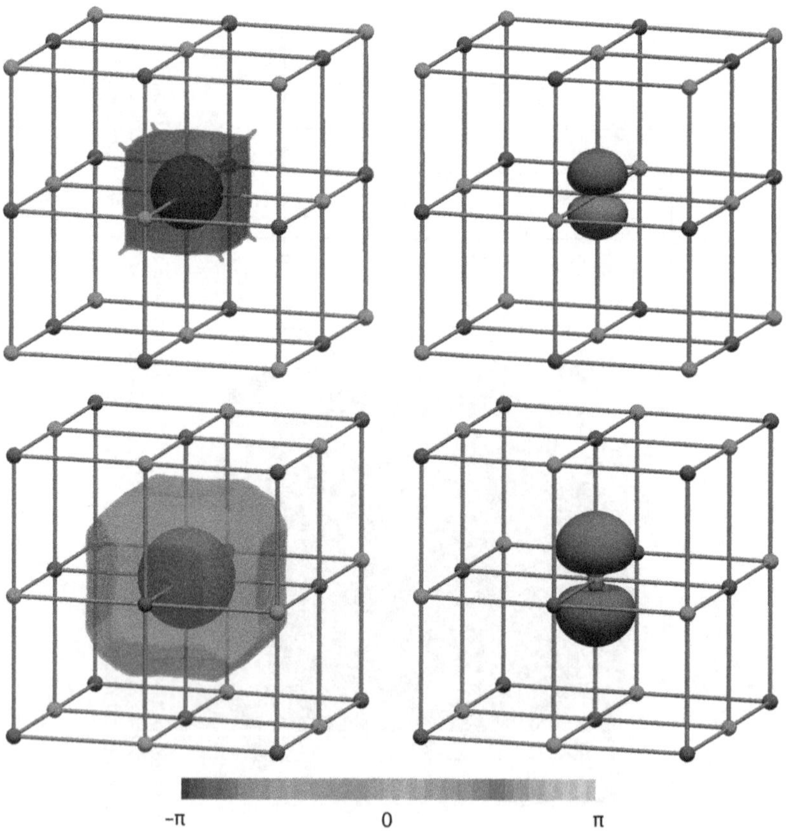

Fig. 7 DAFH orbitals in NaCl. Shown are the isosurfaces of orbital amplitude equal to 0.1 colored by orbital phase. The DAFH reference basins are shown left as transparent objects (top for Na, bottom for Cl). [Color figure can be viewed in the electronic version.] Reprinted with permission from ref. 12, Copyright 2012, AIP Publishing LLC.

In covalent solids essential electron sharing at least between nearest neighbor atoms is expected. The localization index of the C atom in diamond, equal to $\lambda(\mathrm{C})=3.81$,[11] is significantly smaller than the basin population (obviously equal to 6 e^-) assuming large fluctuation value and remarkable electron delocalization. The delocalization index between nearest neighbor QTAIM basins in diamond ($d_{\mathrm{C\text{-}C}}=1.54$ Å, $\delta=0.91$) is close to unity expected for the classical single bond. The delocalization index in diamond quickly decays with the distance between atoms: for the next nearest neighbors ($d=2.52$ Å) it is only 0.039 and for the contact with the atom of the third coordination sphere ($d=2.95$ Å) it drops to 0.008. Cioslowski–Mixon bond orders show similar values: 0.95 for the nearest neighbor contacts, 0.05 for the contact to the atoms of the second and 0.007 of the third coordination spheres.

Figure 8 shows one of the orbitals after the isopycnic transformation[26] of the canonical Kohn–Sham orbitals. Having occupation number 1 for each spin channel, it nicely represents the 2c–2e bond between the nearest carbon atoms.

For each QTAIM atom in diamond the DAFH analysis delivers 5 orbitals with essential occupations.[12] One of them has occupation very close to 2 and is almost completely confined in the QTAIM basin, thus corresponding to a core orbital. Each of the four other orbitals has occupation number 0.94 and is located between two nearest C atoms (Fig. 9, left). Around 48% of its norm is recovered in the basin of the reference QTAIM atom and 43% of the norm is in the basin of closest C atom. Therefore, more than 90% of its norm is confined in the basins of the two bonded C atoms. The complement sum of occupation numbers of two such orbitals from each participating atom yields the number very close to 2, which nicely describes the localized 2c–2e bond, similar to the isopycnic localized orbital in Fig. 8. The DAFH analysis easily recovers

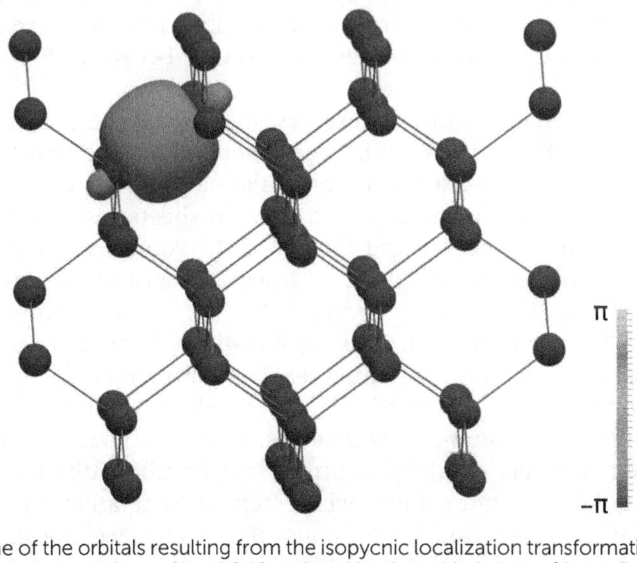

Fig. 8 One of the orbitals resulting from the isopycnic localization transformation. Shown is the amplitude isosurface of level 0.12, colored by the orbital phase. [Color figure can be viewed in the electronic version.].

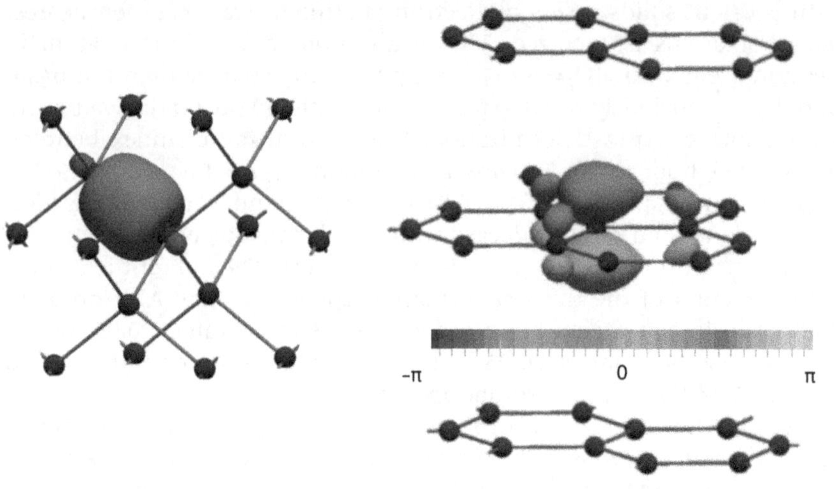

Fig. 9 DAFH orbital corresponding to the C–C bond in diamond (left) and the π-component of the C–C bond in graphite (right). Shown are the isosurfaces of orbital amplitudes of level 0.15 and 0.11 respectively, colored by orbital phase. [Color figure can be viewed in the electronic version.].

here the similarity between the bonding in diamond and in methane molecule,[57] evident to every chemist.

In graphite the nearest-neighbors (*i.e.*, in-plane) Ángyán delocalization index value is $\delta = 1.20$, which is close to the formal bond order $1\frac{1}{3}$.[11] For the shortest C···C contact between graphite sheets ($d = 3.36$ Å) the delocalization index amounts to 0.02 thus indicating only weak interaction. For the next nearest contact within the sheet ($d = 2.46$ Å) the delocalization index is 0.058 and for the contact with an atom of the third coordination sphere in the sheet ($d = 2.85$ Å) the delocalization index equals to 0.038. Comparison with the values reported above for the diamond indicates that the electron sharing in diamond is more localized than in graphite, where it decays slower as the distance between QTAIM basins becomes larger.

The Fulton delocalization indices have nearly the same values as the Ángyán ones with the deviation between them noticeable only for coarse *k*-meshes. Cioslowski–Mixon bond orders are around 1.2 for the nearest neighbor contacts within and ≈ 0.02 between sheets, ≈ 0.1 for the contact to an atom of second and 0.03 of third coordination sphere respectively within the same sheet, *i.e.* they are reasonably close to the Ángyán indices. Similar to 2D hydrogen lattices and structurally related benzenoid hydrocarbons, the Cioslowski–Mixon bond indices demonstrate essential variation for symmetrically equivalent contacts.

Graphite has two symmetrically inequivalent C atoms and, correspondingly, two different QTAIM basins. However, the DAFH orbitals of those basins are very similar (*i.e.* occupation numbers differ at most at the second digit).[12] Four orbitals with essential occupation numbers are very similar to those found for diamond. Besides one well localized core, there are three equivalent DAFH orbitals with occupations >0.9, located between pairs of nearest atoms. Inside their QTAIM basins more than

90% of the orbital norm is recovered, thus describing localized 2c–2e bond between nearest neighbors. The remaining DAFH orbital, having clear π-character, is characterized by occupation number close to 0.9 and is delocalized over four atoms (Fig. 9, right). Only 82% of its norm can be recovered in the basins of the reference atom and its nearest neighbors, which shows the significant delocalization over further coordination spheres. Thus, the DAFH orbitals clearly reflects the σ- and π-components of C–C bond in graphite. The σ-type and π-type DAFH orbitals contribute with 0.88 and 0.22, respectively, to the total value of 1.20 of the nearest neighbor delocalization index (eqn (11)).

The delocalization indices between nearest QTAIM basins in metals are quite low (for example 0.26 in *fcc* Cu).[11] From the high coordination number combined with low number of valence electrons in metals a formal bond order less than 1 would be also anticipated. The delocalization indices between distant QTAIM basins are decaying slowly with the distance and even for atoms of higher coordination spheres are non-negligible (for instance for $d_{\text{Na–Na}} = 5.97\,\text{Å}$ $\delta = 0.008$), taking into account low values for nearest neighbors ($d = 3.66\,\text{Å}$ $\delta = 0.10$, $d = 4.22\,\text{Å}$ $\delta = 0.06$). Cioslowski–Mixon bond indices in *bcc* Na show somewhat smaller values: ≈ 0.06 for nearest neighbor contacts at 3.66 Å and ≈ 0.03 for 6 next nearest atoms at 4.22 Å. As in case of 2D hydrogen lattice and graphite, they demonstrate a certain variation for equivalent contacts.

In metals, besides well localized core DAFH orbitals, a very special type of delocalized orbitals with low occupation numbers <0.7 occurs.[12] Those orbitals are weakly localized inside the reference QTAIM basin: only less than 40% of their norm can be recovered inside the basin (Fig. 10). Moreover, only 66% of its norm can be recovered inside the basins of all the surrounding atoms up to the second coordination sphere. Their form can be roughly described as the decaying wave of s- or p-type character. Fig. 11 compares the form of s-type orbital with the rescaled Θ^{3D} distribution (eqn (37)) evaluated under the assumption of a spherical Fermi surface in *bcc* Na. The nearly quantitative coincidence of

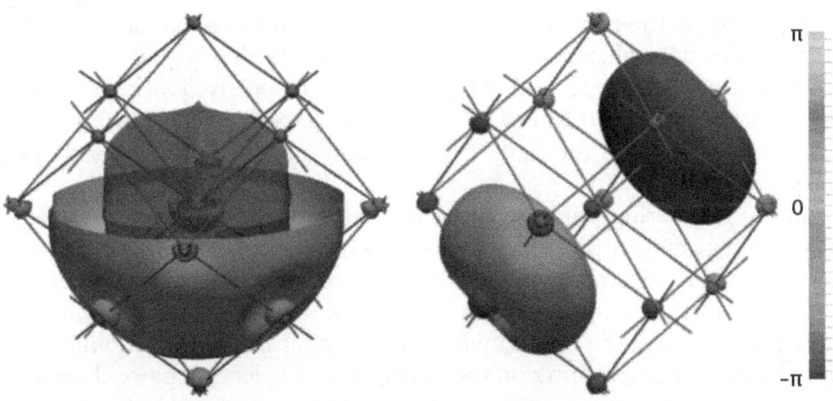

Fig. 10 The s-type and p-type DAFH orbitals in *bcc* Na shown as isosurfaces of their amplitudes of levels 0.015 and 0.025 respectively, colored by orbital phase. The reference QTAIM basin is shown by the transparent object. [Color figure can be viewed in the electronic version.].

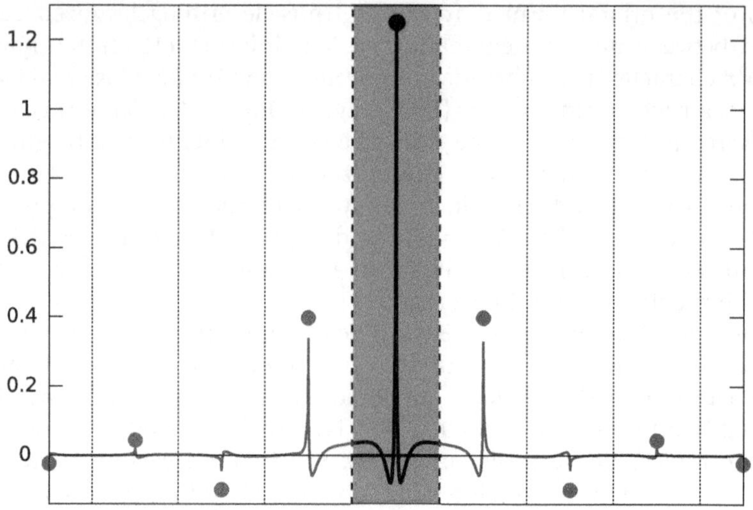

Fig. 11 The 1D plot along space diagonal of the cubic lattice of *bcc* Na of the s-type DAFH orbital (occupation number 0.60) from the DFT calculation (solid line), superimposed with Θ^{3D} (dots), rescaled to match the absolute maximum of the DFT DAFH orbital. [Color figure can be viewed in the electronic version.] Reprinted with permission from ref. 12, Copyright 2012, AIP Publishing LLC.

the orbital form illustrates thus an intimate connection between the electron delocalization in metals with the presence of the discontinuities of the occupation number distribution in *k*-space (*i.e.* the presence of the Fermi surface).

Besides core and delocalized DAFH orbitals, obviously DAFH orbitals corresponding to the d-orbitals of the penultimate shell (Fig. 12) are found in the transition metals. Their occupation numbers are slightly lower than 2 which means that the orbitals can be pretty localized: for *fcc* Cu slightly more than 90% of their norm are recovered in the reference QTAIM basin.

However, delocalized and localized d-like orbitals contribute (eqn (11)) comparably to the nearest neighbor delocalization index value. Thus, for *bcc* Na the delocalization index value $\delta = 0.1$ is due to 0.07 and 0.03 contributions of s-type and p-type orbitals, respectively. The total delocalization index of 0.26 in *fcc* Cu is due to 0.06 contribution of s-, 0.08 contribution of p- and 0.07 contribution of d-type DAFH orbitals, respectively.

In the pioneering publication on the delocalization indices for solids[9] it was suggested to partition the basin population into the "localized" and "mobile" electrons, where the electrons shared with the nearest QTAIM basins, having common interatomic surfaces were counted as "localized" and those shared with more distant basins as "mobile". Alternatively one could think on the partitioning of electron pairs shared by a basin into "close" and "distant".[11] It could be anticipated, that in ionic solids the fraction of mobile electrons or distant sharing should be low and in metals, in contrast, it should be rather high. Indeed, for NaCl it is only ~ 0.01 whereas for *bcc* Na ≈ 0.28. However, the fractions of distant

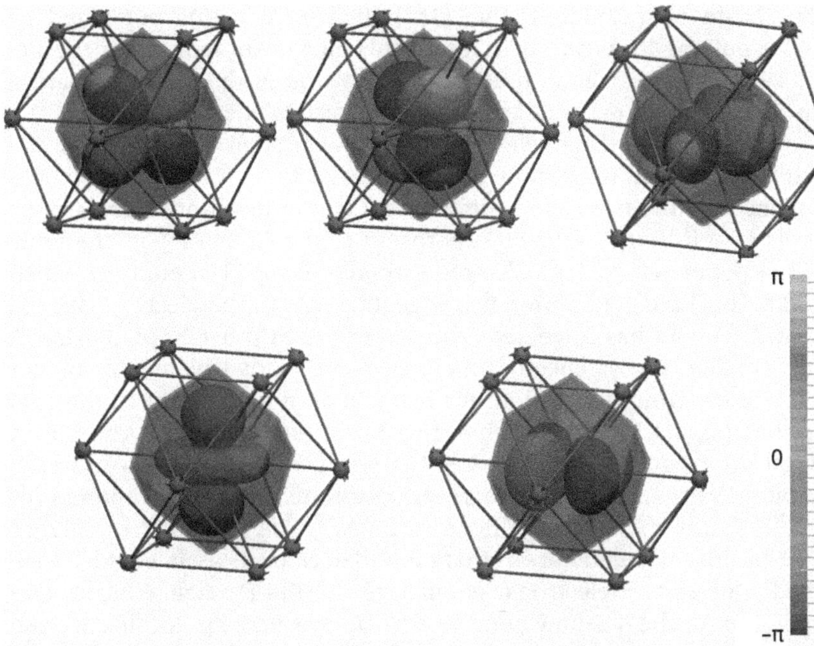

Fig. 12 DAFH orbitals in *fcc* Cu, corresponding to 3d atomic orbitals, shown as orbital amplitude isosurfaces of level 0.05 and colored by orbital phase. The reference QTAIM basin is shown as transparent object. [Color figure can be viewed in the electronic version.] Reprinted with permission from ref. 12, Copyright 2012, AIP Publishing LLC.

shared pairs in diamond and *fcc* Cu are nearly the same (≈ 0.17) and in graphite it is smaller (≈ 0.15), than in diamond.

We believe, that this indicator is insufficient to clearly differentiate between localized bonding in covalent solids from delocalized bonding in metals. What discriminates the bonding in these two classes of solids is the decay of the delocalization indices over the coordination spheres or, alternatively, the degree of localization of DAFH orbital, as it was demonstrated in the Sec. 3.3 with the example of simplest solid state systems, which is directly related to the presence or the absence of the Fermi surface within the framework of the band structure model.

4.3 Delocalization for ELI-D basins for solids

Delocalization indices have been also evaluated[11] for the topological basins of the electron localizability indicator (ELI),[58,59] whose topology defines partition of space into basins, representing various elements of chemical bonding, emerged from atomic shells: cores, penultimate shells, lone pairs and bonds.

Basins corresponding to atomic cores show high degree of electron localization and the corresponding localization index is close to the basin population. For instance 9.49 electrons are localized in the Cl core superbasin (containing the first two shells of the Cl atom) enclosing the population of 10.07 electrons. Such a high localization of core basins has been also reported for ELF core basins in molecules.[60]

The basin of the penultimate (3rd) shell in *fcc* Cu is populated by 16.40 e but the localization index is only 13.64. An analysis of the delocalization indices shows that it is mainly due to the electron sharing with its core basin ($\delta = 2.56$) as well as nearest valence basins ($\delta = 0.26$). Direct electron exchange with the penultimate shell of the nearest Cu atom is in order of magnitude smaller ($\delta = 0.03$).

Valence ELI-D basins show different degrees of localization for various classes of solids. The valence superbasin of the Cl atom** in the ionic NaCl is populated by 7.87 electrons (representing Cl^- anion) of which 6.79 are localized. The difference is mainly due to the electron sharing with the Cl core basin whereas the sharing with other ELI-D basins is much smaller. The valence basins in covalent solids have much smaller degree of electron localization: for instance in diamond out of the 1.94 electrons populating the C–C bonding basin only 0.89 are localized; in graphite the population of C–C ELI-D basin is 2.59 whereas its localization index is 1.32. Similar degree of localization has been reported for valence basins in molecular hydrocarbons.[60] The electrons in valence ELI-D basins of metals are remarkably delocalized, as it could be expected. Out of 1.06 electrons, populating the ELI-D valence basin, corresponding to the 4-center bond in *fcc* Cu, only 0.17 are localized. Even more dramatic delocalization is found in *bcc* Na, where valence basin, corresponding to the 4-center bond between Na atoms is populated by 0.15 electrons, of which only 0.01 is localized. As it can be seen from eqn (2) for the case $A = B$, the largest value of localization index $F(A,A)$ is equal to the squared basin population $N^2(A)$ (the first term), since the second term, equivalent to the number of pairs in the basin A can only reduce it and thus, for the basin population equal to 0.15 the localization index can not be higher than 0.0225. It is evident, that any basin with population less than unity has to show high degree of electron delocalization because such low average population precludes the situation that an electron is confined in that basin.

4.4 Applications to selected inorganic solids

The portfolio of successful applications of delocalization indices for solids, besides simple inorganic solids, includes borides[52,53] and intermetallic compounds[54–56] which are known to be difficult objects for classical theories of chemical bonding. For all of them delocalization indices have been evaluated from the results of (L)APW + lo + LO Kohn–Sham LDA DFT calculations.

MB_2 class of compounds with AlB_2 type structure (Fig. 13) is remarkable due to the number of representatives known to exist for different metals (M) across the periodic table reaching from Mg to W. Moreover, many other crystal structures can be related to this simple structure type. Valuable properties, like high hardness or corrosion-resistance,[61] and not

**Made up by six basins because the spherical atomic shell becomes split by the octahedral site symmetry.

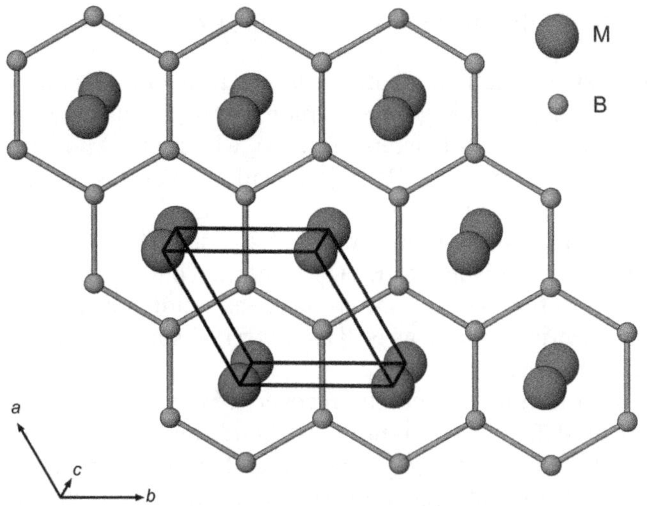

Fig. 13 Crystal structure of metal diborides with AlB_2 type structure. [Color figure can be viewed in the electronic version.] Reprinted with permission from ref. 52, Copyright 2013, Wiley-VCH Verlag GmbH & Co. KGaA.

least the superconductivity of MgB_2 stimulated the investigations of the bonding in MB_2.

From the values of delocalization indices for the QTAIM atoms it can be concluded, that in the diborides of main group metals (Mg and Al) the bonding between metal and boron is predominantly ionic. The bonding in the boron sheets is covalent with $\delta(B, B) \approx 1.0$ and resembles that in graphite, but is more delocalized in the sense, that there is essential electron sharing between more distant B atoms.

For transition metal (TM) diborides the bond order between the nearest B atoms drops along the transition metal series, which is somewhat unusual, since the interatomic distance decreases at the same time. In contrast, $\delta(M, B)$ increases along the TM series, which overweights the weakening of the B–B bond and explains the contravariance of unit cell parameter a. Increase of the covalent character of the M–B interaction has been also verified by the decrease of QTAIM effective charge magnitudes. $\delta(M, M')$ for contacts, lying both along and perpendicular to the boron sheets also increases along TM series, which is consistent with the contraction of both unit cell parameters a and c.

For the late TM diborides the $\delta(M, M') \approx 0.15$, which is comparable with the corresponding values in elementary metals. Thus, the bonding picture in these diborides can smoothly vary from covalent delocalized bonding inside the boron sheets combined with ionic one between these sheets and metals to the covalent bonding between M and B as well as between metal atoms in all three spatial directions, which explains remarkable chemical flexibility of these diboride structure type. The weakening of the B–B bonds within the sheet for the late transition metals can explain the appearance of distorted boron sheets, observed in the crystal structures of some 4d-metal diborides.[62] Enhanced interaction between metal atoms in two and three directions is likely related with the occurrence of complicated noncollinear magnetic structures of CrB_2 and MnB_2.

Hexaborides of alkaline, alkali-earth and rare-earth metals crystallize in a cubic lattice, featuring three-dimensional boron network of interconnected octahedral B_6 clusters (Fig. 14), which also known to exist as isolated anions, for example $[B_6H_6]^{2-}$. Wade's rules suggest that B_6 is a *closo* cluster in case of divalent metal having six exohedral (2c, d(B–B) = 1.63...1.78 Å) and seven endohedral (3c, d(B–B) = 1.73...1.83 Å) bonds similar to the isolated $[B_6H_6]^{2-}$ anion. The chemical bonding as well as effects of the metal variation has been studied by Börrnert *et al.*[53] using delocalization indices as well as other real-space chemical bonding analysis techniques.

Delocalization indices allow the recovery of electron counting rules in these electron-deficient compounds without any resort to orbitals. The degree of electron sharing between metal and boron atoms is low (δ(M, B) $\leq$ 0.1) thus marking the predominantly ionic bonding between the metals and the boron framework. In CaB_6 the delocalization index value $\delta(B, B)_{exo} = 0.94$ clearly marks six 2c–2e bonds (corresponding to nearly 12 e) whereas for every bond of the 12 edge B–B contacts one obtains $\delta(B, B)_{endo} = 0.6$ yielding together $12 \times 0.6 = 7.2$ effective B–B bonds holding 14 e, which sum up to $14 + 12 = 26 = 4 \times 6 + 2$ skeletal electron pairs just as stated by Wade's rule for *closo* clusters. The replacement of the metal from a divalent to mono- or trivalent one alters all delocalization indices mentioned above only slightly (the largest change does not exceed 0.05). The most visible change introduced by the metal

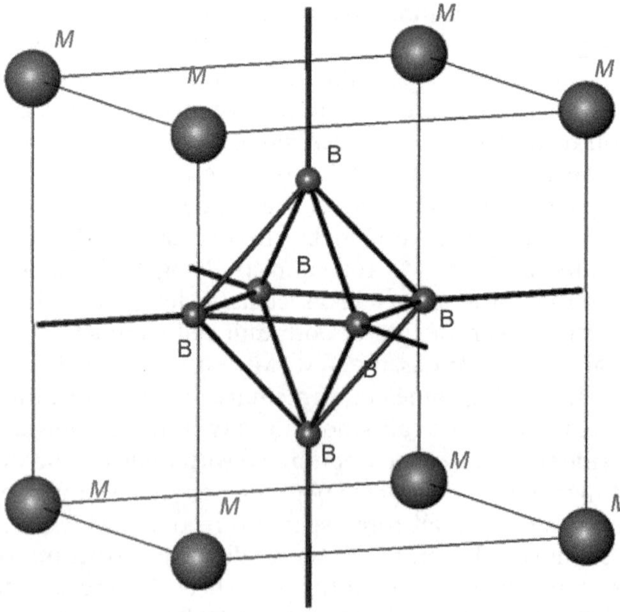

Fig. 14 Crystal structure of metal diborides with CaB_6 type structure. [Color figure can be viewed in the electronic version.] Reprinted with permission from ref. 53, Copyright 2013, Wiley-VCH Verlag GmbH & Co. KGaA.

variation is an enhancement of electron sharing of the metal QTAIM basin, as reproduced by the fluctuation of the basin population from 0.64 for CaB_6 to 1.52 for LaB_6. Interestingly, at the same time the fluctuation of the QTAIM basin of the boron atom practically does not change. This means, that the enhancement of sharing occurs due to the electron pair exchange between metal atoms, which can be confirmed by enhanced nearest neighbor $\delta(\text{M, M}) = 0.02$ values. Together with other bonding indicators this finding assumes that the formal oxidation state of rare earth elements is not 3+ but only 2+ which is sufficient to saturate the demand of boron framework and the remaining electron participates in weak delocalized M–M bonding. Additional insights into the bonding in these interesting compounds have been obtained from the analysis of three-center bonding indices,[53] evaluated in this publication for the first time from solid state DFT calculations.

The so-called **"half-Heusler" compounds having the MgAgAs structure type** are promising materials for many technological applications like optoelectronics, thermoelectrics and spintronics. Usually they contain three different elements *X*, *Y* and *Z* and their crystal structure can be described as a rock salt lattice built by elements *X* and *Y*, half of the cubic voids in which are occupied by element *Z*. Alternatively one can describe this structure as a zinc blende lattice formed by elements *X* and *Z* which is additionally filled at the edges of the cubic cell by the third one (*Y*). The elements *X*, *Y* and *Z* can be main group elements as well as transition or rare-earth metals. Depending on the composition, each type of element can prefer to occupy a particular position, which of course can influence the properties of the compound. Therefore, the rationalization of the type of atomic arrangement is an important question, which have been analyzed for some semiconductor representatives[55] with the use of delocalization indices.

It was found, that for all the semiconductor compounds of MgAgAs-type, containing exclusively main group elements,[55,56] the delocalization indices can reach significant values, for example, $\delta(\text{In, Ge}) = 0.69$ in LiInGe (when In and Ge both reside in zinc blende lattice). However, the delocalization indices between other elements are typically small indicating thus ionic interactions between metal cations and the tetrahedral framework formed by late main group elements. Indeed, QTAIM charges indicate essential charge transfer between the elements ($Q(\text{Li}) = +0.84$, $Q(\text{In}) = +0.18$, $Q(\text{Ge}) = -1.02$) and it turns out, that it is the ionic interaction and not the covalent one, which plays a decisive role for the definition of the most stable atomic arrangement. A clear tendency that the most electronegative main group element prefers to occupy the *Z* position on Fig. 15 can be explained solely from the ionic model by comparing the Madelung energies (which have been evaluated using the QTAIM charges[55]), whereas no clear correlation between the delocalization indices and the atomic arrangement has been established.

A different situation occurs for compounds containing two transition metals (early and late) and a main group element. For such compounds the covalent interactions, quantitatively characterized by the delocalization indices, play a decisive role and the total energy nicely parallels

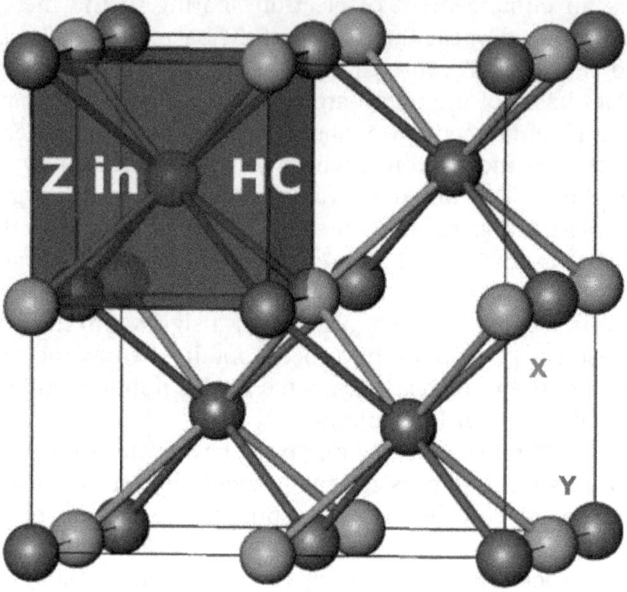

Fig. 15 The MgAgAs-type crystal structure. HC stays for the heteroatomic cubic environment. [Color figure can be viewed in the electronic version.] Reprinted with permission from ref. 56, http://pubs.acs.org/doi/pdf/10.1021/acs.inorgchem.5b00135, Copyright 2015, American Chemical Society.

the sum of the delocalization indices for the shortest contacts in the structure, occurring within the zinc blende sublattices. Then, the late transition elements prefer to occupy the *Z* position, which, like for instance the Ni in ScNiSb yields two essential nearest neighbor delocalization indices δ(Ni, Sb) = 0.67 and δ(Ni, Sc) = 0.35. The QTAIM charges are not that high (Q(V) = +1.39, Q(Fe) = −0.79 and Q(Sb) = −0.60) and the Madelung energy, evaluated from these charges, does not always follow observed structural trends.

More complicated situation occurs in compounds, containing, like in the prototypical MgAgAs, both late main group and late transition elements, each of them showing tendency to occupy the *Z* site. Then the covalent and ionic factors are opposite and in MgAgAs the latter obviously dominates since As occupies the *Z* position, disregarding the somewhat larger sum of delocalization indices for the shortest interactions with Ag at the *Z* site.

γ-**Bi_2Pt**, stable at high temperatures, has the crystal structure resembling the well known CdI_2 structure type. The major difference is the distortion of the edge-shared $PtBi_6$ polyhedra, leading to the formation of a superstructure with 3 times larger unit cell. The Pt atoms form triangles with d(Pt–Pt) = 2.994 Å and one side of the CdI_2-like layer becomes slightly puckered.

Taking the Bi atoms as a 3-electron donors, one arrives at $10 + 2\times3 = 16$ valence electrons on the Pt atom, which means, that 2 more electrons are needed in order to achieve a stable 18-electron configuration. The formation of Pt–Pt bonds could be a possibility for that.

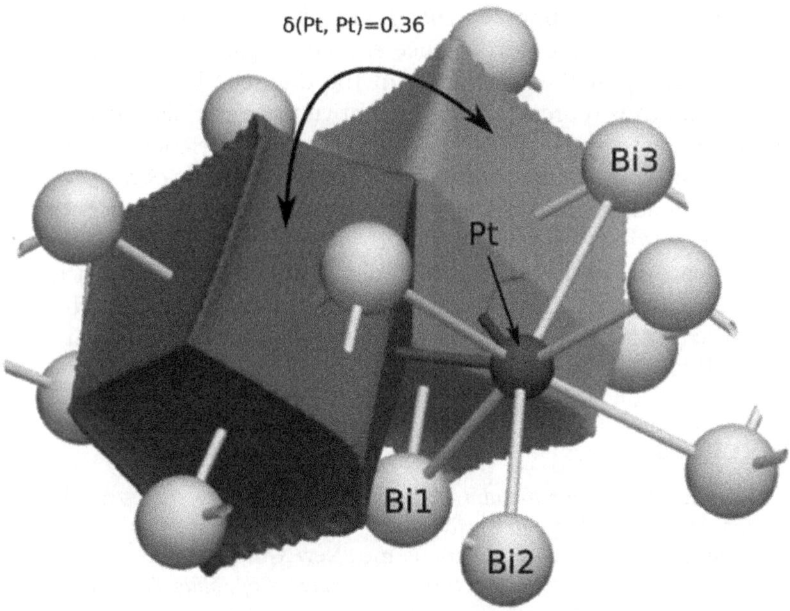

Fig. 16 QTAIM basins for Pt atoms forming a triangle with d(Pt–Pt) = 2.994 Å. Also shown is the value of the delocalization index between them. Reprinted with permission from ref. 54, Copyright 2014, Wiley-VCH Verlag GmbH & Co. KGaA.

The observed Pt–Pt distance is however considerably larger than the Pt–Pt distance in elemental Pt (d(Pt–Pt) = 2.774 Å) and the question appears, whether the Pt–Pt interaction is strong enough to introduce such a distortion of the structure.

In order to clarify this question, the delocalization indices between the QTAIM basins (Fig. 16) have been evaluated. Besides an essential electron sharing between the nearest Bi and Pt atoms (δ(Bi, Pt) ≈ 0.6), the analysis of delocalization indices yields the value δ(Pt, Pt) = 0.36, which is only somewhat lower, than δ(Pt, Pt) = 0.41 for the shortest contact in elemental Pt. Therefore indeed, the bonding between Pt atoms is quite strong and can explain the observed distortion. A comparison between the value of the QTAIM basin fluctuation and the values of delocalization indices has shown, that the bonding in this compound is not restricted to nearest neighbors only, but also involves more distant atoms, which is typical for metals. The QTAIM charges on Bi atoms were found to be positive thus confirming the donor role of the Bi atoms, mentioned above.

5 Conclusion

The chapter summarizes the results of recent studies from our laboratories aiming at the development of a chemist-oriented approach to the description of the structures and bonding of the solid state. The main motivation for these efforts originates from the fact that traditional physicist's approach to the description of the solid state in terms of band theory considerably differs from the more intuitive but remarkably successful chemical approach that relies on the traditional concept of chemical bond, bond order, valence *etc.* The reconciliation of these approaches aiming at the facilitation of the dialogue between both communities can thus represent a new incentive for materials science, whose further progress can hardly afford to ignore the experience and intuition that chemists contribute to the design and the synthesis of new advanced materials. In addition to this primary goal, we also believe that the new insights emerging from modern formulation of traditional chemical concepts can prove useful in revealing the peculiar nature of the long-range delocalized bonding interactions in metallic solids, which exceed the classical model of localized bonds applicable to the majority of covalent solids.

References

1 C. Kittel, *Introduction to Solid State Physics*, Wiley, New York, 7th edn, 1996.
2 W. A. Harrison, *Electronic Structure and the Properties of Solids. The Physics of Chemical Bond*, Dower Publications Inc., New York, 1989.
3 P. A. Cox, *The Electronic Structure and Chemistry of Solids*, Oxford University Press, Oxford, 1987.
4 R. Hoffmann, *Solids and Surfaces. A Chemist's View of Bonding in Extended Systems*, VCH, New York, Weinheim, 1988.
5 R. Hoffmann, *Angew. Chem., Int. Ed. Engl.*, 1987, **26**, 846–878.
6 B. Silvi and C. Gatti, *J. Phys. Chem. A*, 2000, **104**, 947–953.
7 V. P. Smirnov, R. A. Evarestov and D. E. Usvyat, *Int. J. Quantum Chem.*, 2002, **88**, 642–651.
8 R. A. Evarestov, V. P. Smirnov and D. E. Usvyat, *Theor. Chem. Acc.*, 2005, **114**, 19–28.
9 R. Ponec, *Theor. Chem. Acc.*, 2005, **114**, 208–212.
10 R. Ponec, *J. Comput. Chem.*, 2011, **32**, 3114–3121.
11 A. I. Baranov and M. Kohout, *J. Comput. Chem.*, 2011, **32**, 2064–2076.
12 A. I. Baranov, R. Ponec and M. Kohout, *J. Chem. Phys.*, 2012, **137**, 214109.
13 C. A. Coulson, *Proc. R. Soc. London, Ser. A*, 1939, **169**, 413–428.
14 R. S. Mulliken, *J. Chem. Phys.*, 1955, **23**, 1833–1846.
15 K. S. Wiberg, *Tetrahedron*, 1968, **24**, 1083–1096.
16 I. Mayer, *Chem. Phys. Lett.*, 1983, **97**, 270–274.
17 J. Ángyán, M. Loos and I. Mayer, *J. Phys. Chem.*, 1994, **98**, 5244–5248.
18 R. F. W. Bader, *Atoms In Molecules: A Quantum Theory*, Oxford University Press, Oxford, 1990.
19 M. Rafat and P. L. A. Popelier, in *The Quantum Theory of Atoms in Molecules: From Solid State to DNA and Drug Design*, ed. C. F. Matta and R. F. Boyd, Wiley-VCH, Weinheim, 2007, pp. 121–140.
20 M. A. Blanco, A. Martn Pendás and E. Francisco, *J. Chem. Theory Comput.*, 2005, **1**, 1096–1109.
21 R. McWeeny, *Rev. Mod. Phys.*, 1960, **32**, 335–369.
22 R. F. W. Bader and M. E. Stephens, *J. Am. Chem. Soc.*, 1975, **97**, 7391–7399.
23 R. F. W. Bader and M. E. Stephens, *Chem. Phys. Lett.*, 1974, **25**, 445–449.
24 R. L. Fulton, *J. Phys. Chem.*, 1993, **97**, 7516–7529.
25 J. Cioslowski and S. T. Mixon, *J. Am. Chem. Soc.*, 1991, **113**, 4142–4145.
26 J. Cioslowski, *Int. J. Quantum Chem.*, 1990, **38**(Suppl. 24), 15–28.
27 W. N. Lipscomb, *Science*, 1977, **196**, 1047–1055.

28 K. C. Mundim, M. Giambiagi and M. S. de Giambiagi, *J. Phys. Chem.*, 1994, **98**, 6118–6191.
29 M. Giambiagi, M. S. de Giambiagi and K. C. Mundim, *Struct. Chem.*, 1990, **1**, 423–427.
30 A. B. Sannigrahi and T. Kar, *Chem. Phys. Lett.*, 1990, **173**, 569–572.
31 A. B. Sannigrahi and T. Kar, *Chem. Phys. Lett.*, 1999, **299**, 518–526.
32 R. Ponec and I. Mayer, *J. Phys. Chem. A*, 1997, **101**, 1738–1741.
33 R. Ponec, *J. Math. Chem.*, 1997, **21**, 323–333.
34 R. Ponec and A. J. Duben, *J. Comput. Chem.*, 1999, **20**, 760–771.
35 R. Ponec, D. L. Cooper and A. Savin, *Chem.–Eur. J.*, 2008, **14**, 3338–3345.
36 F. L. Hirshfeld, *Theor. Chim. Acta*, 1977, **44**, 129–138.
37 D. Tiana, E. Francisco, M. A. Blanco, P. Macchi, A. Sironi and A. Martn Pendás, *Phys. Chem. Chem. Phys.*, 2011, **13**, 5068–5077.
38 R. Ponec and F. Feixas, *J. Phys. Chem. A*, 2009, **113**, 8394–8400.
39 R. Ponec and G. Yuzhakov, *Theor. Chem. Acc.*, 2007, **118**, 791–797.
40 R. Ponec, G. Yuzhakov and D. L. Cooper, *Theor. Chem. Acc.*, 2004, **112**, 419–430.
41 R. Ponec, G. Lendvay and J. Chaves, *J. Comput. Chem.*, 2008, **29**, 1387–1398.
42 R. Ponec and X. Girones, *J. Phys. Chem. A*, 2002, **106**, 9506–9511.
43 R. Ponec, *Comput. Theor. Chem.*, 2015, **1053**, 195–213.
44 R. Daudel, *The Fundamentals of Theoretical Chemistry*, Pergamon Press, Oxford, 1968.
45 W. Kutzelnigg, in *Explicitly Correlated Wave Functions in Chemistry and Physics: Theory and Applications*, ed. J. Rychlewski, Kluwer, The Netherlands, 2002, pp. 14–17.
46 D. L. Cooper, R. Ponec, T. Thorsteinssohn and G. Raos, *Int. J. Quantum Chem.*, 1994, **57**, 501–518.
47 V. N. Staroverov and E. R. Davidson, *Int. J. Quantum Chem.*, 2000, **77**, 651–660.
48 M. Kohout, F. R. Wagner and Y. Grin, *Theor. Chem. Acc.*, 2008, **119**, 413–420.
49 R. Ponec and M. Strnad, *Int. J. Quantum Chem.*, 1994, **50**, 43–53.
50 P.-O. Löwdin, *Phys. Rev.*, 1955, **97**, 1474–1489.
51 M. A. Buijse and E. J. Baerends, *Mol. Phys.*, 2002, **100**, 401–421.
52 F. R. Wagner, A. I. Baranov, Y. Grin and M. Kohout, *Z. Anorg. Allg. Chem.*, 2013, **639**, 2025–2035.
53 C. Börrnert, Y. Grin and F. R. Wagner, *Z. Anorg. Allg. Chem.*, 2013, **639**, 2013–2024.
54 M. Kaiser, A. I. Baranov and M. Ruck, *Z. Anorg. Allg. Chem.*, 2014, **640**, 2742–246.
55 D. Bende, Y. Grin and F. R. Wagner, *Chem.–Eur. J.*, 2014, **20**, 9702–9708.
56 D. Bende, F. R. Wagner and Y. Grin, *Inorg. Chem.*, 2015, **54**, 3970–3978.
57 R. Ponec, *J. Math. Chem.*, 1998, **23**, 85–103.
58 M. Kohout, *Int. J. Quantum Chem.*, 2004, **97**, 651–658.
59 M. Kohout, *Faraday Discuss.*, 2007, **135**, 43–54.
60 A. Savin, B. Silvi and F. Colonna, *Can. J. Chem.*, 1996, **74**, 1088–1096.
61 P. Vajeeston, P. Ravindran, C. Ravi and R. Asokamani, *Phys. Rev.*, 2001, **B63**, 045115.
62 A. Pallas and K. Larsson, *J. Phys. Chem. B*, 2006, **110**, 5367–5371.

Vibrational quantum dynamics at metallic surfaces

Jean Christophe Tremblay
DOI: 10.1039/9781782620273-00084

1 Introduction

Chemical dynamics is concerned with all transient aspects of chemistry at the molecular level. In general terms, these include, *e.g.*, the energy flow between and within molecules, the transformation of reactants to products, or the motion induced by the interaction of the species with electro-magnetic fields. For most chemically-relevant considerations, these molecules are well described as being composed of two separate entities: electrons and nuclei. Due to their large mass difference, chemical dynamics is usually explained in most textbooks in terms of the Born–Oppenheimer approximation[1]: the slow-moving, heavy nuclei evolve on an energy landscape that is due to the fast-moving, light electrons. The very concept of a potential energy surface, introduced by Eyring and Polanyi in 1931, is built around this mass mismatch, effectively separating the timescales of the electron and nuclei dynamics.[2,3] Although the idea of having electrons reacting instantaneously to the motion of the nuclei was originally met with skepticism, it is now ubiquitous in chemical dynamics and it has lead to the definition of many concepts familiar to the chemists. Static molecular structures define the shape and the orientation of nuclei, and stable configurations in the vicinity of local minima of the potential energy surface define the reactants and the products for a given reaction, as well as various metastable intermediates. Further, following the minimal energy path connecting these molecular structures, it is possible to define a so-called reaction coordinate that forms the basis of transition state theory,[4–6] in which the local maxima between stable/metastable structures are short-lived species that yield information about the rate of change between two chemical species. These ideas have been immensely successful at explaining the reactivity of molecules in the gas and in the condensed phase[7] and they will form the basis of the dynamical investigations discussed in this chapter.

Various theoretical treatments have been advocated for studying chemical dynamics and the present work will highlight some important results that can be obtained using non-relativistic quantum mechanics. In the absence of an external field, the state of a system composed of N_e

Institute for Chemistry and Biochemistry, Freie Universität Berlin, Takustraße 3, 14195 Berlin, Germany. E-mail: jc.tremblay@fu-berlin.de electrons and N_A nuclei obeys the time-independent Schrödinger equation, as defined by the total Hamiltonian

$$\begin{aligned}\hat{H}_{\text{tot}}(\mathbf{r},\mathbf{R})\Psi_k(\mathbf{r},\mathbf{R}) &= E_k\Psi_k(\mathbf{r},\mathbf{R}) \\ \hat{H}_{\text{tot}}(\mathbf{r},\mathbf{R}) &= -\sum_{A=1}^{N_A}\frac{\hbar^2}{2M_A}\vec{\nabla}_A^2 - \sum_{i=1}^{N_e}\frac{\hbar^2}{2m_e}\vec{\nabla}_i^2 + V(\mathbf{r},\mathbf{R})\end{aligned} \tag{1}$$

where **R** and **r** are vectors containing all nuclear and electronic coordinates, respectively. The first two terms on the right-hand-side are respectively the kinetic energy of the nuclei and electrons, and the potential $V(\mathbf{r},\mathbf{R})$ includes all inter-particle Coulomb interactions. The solution of the time-dependent Schrödinger equation for the total system evolving under the Hamiltonian, eqn (1), is too cumbersome to be of any practical use. To circumvent this problem, the total time-dependent wave function of the system, $\Psi(\mathbf{r},\mathbf{R},t)$, is often represented as a non-direct product of vibrational and electronic functions

$$\Psi(\mathbf{r},\mathbf{R},t) = \phi_0(\mathbf{r};\mathbf{R})\psi(\mathbf{R},t) \tag{2}$$

where $\psi(\mathbf{R},t)$ is a purely vibrational wave function which contains all the information about the time evolution of the system. The function $\phi_0(\mathbf{r};\mathbf{R})$ defines the electronic state for a given nuclear configuration **R**. Quantum dynamics is often performed using, instead of a single one, a conventional set of adiabatic electronic functions, $\phi_k(\mathbf{r};\mathbf{R})$, which are solutions of the time-independent electronic Schrödinger equation that depend parametrically on the nuclear positions

$$\begin{aligned}\hat{H}_{\text{el}}(\mathbf{r};\mathbf{R})\phi_k(\mathbf{r};\mathbf{R}) &= E_k(\mathbf{R})\phi_k(\mathbf{r};\mathbf{R}) \\ \hat{H}_{\text{el}}(\mathbf{r};\mathbf{R}) &= -\sum_{i=1}^{N_e}\frac{\hbar^2}{2m_e}\vec{\nabla}_i^2 + V(\mathbf{r};\mathbf{R})\end{aligned} \tag{3}$$

To study chemical dynamics, it is often sufficient to restrict the electronic basis to the ground state electronic wave function, $\phi_0(\mathbf{r};\mathbf{R})$, and this choice will be assumed throughout this work. The associated electronic energy, $E_0(\mathbf{R})$, is seen to depend on the particular structure of the molecule, **R**. This dependence defines the topology of the potential energy landscape on which the nuclear dynamics happens.

Within the Born–Oppenheimer approximation, the variation of the electronic wave function with respect to the nuclear coordinate is neglected altogether, leading to uncoupled equations of motion for the nuclei. The interaction between electrons and nuclei is only felt *via* the potential energy surface, $V(\mathbf{R}) = E_0(\mathbf{R})$, which is obtained by solving the time-independent Schrödinger equation for the electrons at a set of nuclear configurations, **R**. The time-dependence of the state of the system is described according to the factorization ansatz, eqn (2), by the so-called vibrational time-dependent Schrödinger equation, which takes the general form

$$\frac{\partial}{\partial t}\psi(\mathbf{R},t) = -\frac{i}{\hbar}\hat{H}_{\mathrm{nuc}}(\mathbf{R},t)\psi(\mathbf{R},t) \tag{4}$$

Here, the state of the system, $\psi(\mathbf{R},t)$, is seen to evolve under the action of the vibrational Hamiltonian, $\hat{H}_{\mathrm{nuc}}(\mathbf{R},t) = \hat{T}_{\mathrm{nuc}} + V(\mathbf{R}) + V_{\mathrm{ext}}(\mathbf{R},t)$, which is composed of a kinetic, $\hat{T}_{\mathrm{nuc}}$, and a potential term $V(\mathbf{R})$. Further, $V_{\mathrm{ext}}(\mathbf{R},t)$ is the coupling with an external time-dependent potential, such as an electric field, which is often treated in the semi-classical dipole approximation, as will be seen later. As should now be evident from the initial considerations above, proper quantum mechanical treatment of nuclear dynamics requires first and foremost an adequate characterization of a potential energy surface for the reaction investigated. The latter finds its origin in the subtle balance of the Coulomb interactions in the system: the electrostatic attraction between electrons and nuclei, and the nuclei–nuclei and electron–electron repulsive forces. Although electronic structure characterization of the potential energy surfaces is of primal importance for chemical dynamics, an in-depth description of the available methods, as well as their strengths and weaknesses, is beyond the scope of the present work. For more information, the reader is thus referred to recent reviews on the topic (see special issue ref. 8). From now on, it will be assumed that an accurate ground state potential energy surface is available for the vibrational dynamics of interest.

2 Dynamics in the condensed phase

This chapter reports on vibrational dynamics of molecules in the vicinity of metallic surfaces, which belongs to the general topic of chemical dynamics in the condensed phase. For systems of such complexity, it is impossible to take into account the infinite number of vibrational and electronic degrees of freedom explicitly, let alone write down the total Hamiltonian of the system. The solution of the electronic Schrödinger equation to generate the potential energy landscape on which reactions take place, as discussed in the previous section, thus becomes intractable. In most cases, the dynamics of interest occurs in a more or less local structure of the system, which allows for a separation of a reactive pocket and its surroundings.[9] The examples of such separations are plenty, ranging from active sites in proteins for targeted drug delivery to doped bulk materials for optical applications and defective surfaces for catalysis.[9,10] Provided the interaction between the local structure and its surroundings is weak, these can be efficiently separated and characterized using various levels of approximations.[11] Important representatives of such schemes propose a very accurate, quantum mechanical treatment of the active site, embedded in a classical environment. These are often termed QM/MM methods for interfacing quantum mechanics (QM) and classical molecular mechanics (MM), and find a wide range of applications in chemistry, biology, and biochemistry (see for example ref. 12–15). However, the interface between the quantum and classical regions can be seen to dominate the dynamical evolution of the system and care must be taken in properly defining the coupling between active pocket and the classical region.

In this chapter, we follow a purely quantum mechanical approach to the chemical dynamics in the condensed phase. To this end, we resort to ideas common to the QM/MM ansatz, namely the separation of a reactive site from its environment. This is particularly suitable for gas-surface reactions, for which the gas phase molecule typically interacts more weakly with the surface than the two fragments within themselves. Furthermore, this separation respects the physical nature of the system, as both moieties behave as two independent fragments at large gas-surface distances. The consequence of such a separation is that it renders its treatment tractable from a purely quantum mechanical perspective: the vibrational dynamics proceeds on a potential energy surface obtained from first principles, eqn (3), and its time evolution is characterized by the vibrational time-dependent Schrödinger equation, eqn (4). From an electronic perspective, the local nature of the interaction is often treated within a periodic density functional theory (DFT) framework,[16] where the regularity of the surface is taken into account by imposing periodic boundary conditions to the solution of the secular equations, but alternative approaches have been put forward (see, *e.g.*, ref. 17). Alternative choices that were used to various degrees of success are parametric and semi-empirical methods, where the potential energy surfaces are obtained from a parametrized electronic Schrödinger equation (see, *e.g.*, ref. 18–24) or derived directly from experiment (see ref. 25 and references therein). For the first principle characterization of potential energy surfaces for adsorbates in metallic environments, the Kohn–Sham formulation of periodic DFT is currently accepted as the gold standard in surface science.[26] The self-consistent solution of the effective one-electron Schrödinger equation for the Hamiltonian $h_{\text{eff}} = -\frac{\hbar^2}{2m_e}\vec{\nabla}_e^2 + v_{\text{eff}}$, in a basis of plane waves respecting the periodicity of the surface, yields the electronic density of the ground state and its associated energy. Repeating this calculation for various positions of the adsorbate at the metallic surface allows to define the ground state potential energy surface for the adsorbate vibrational dynamics of interest.

Within the Born–Oppenheimer framework, the variation of the electronic wave function and the associated electronic density with respect to nuclear motion is completely neglected. This approximation can be seen to suffer severe failures in the condensed phase, in particularly at metallic surfaces.[27–33] To fully appreciate this situation, recall that the chemical dynamics in which we are interested proceeds on a potential energy surface that is computed as the lowest energy solution of the electronic Schrödinger equation, eqn (3), or its Kohn–Sham counterpart. It is important to recognize that the electronic Schrödinger equation admits an infinite number of energetically higher-lying solutions that define other, excited potential energy surfaces.[30] This situation is

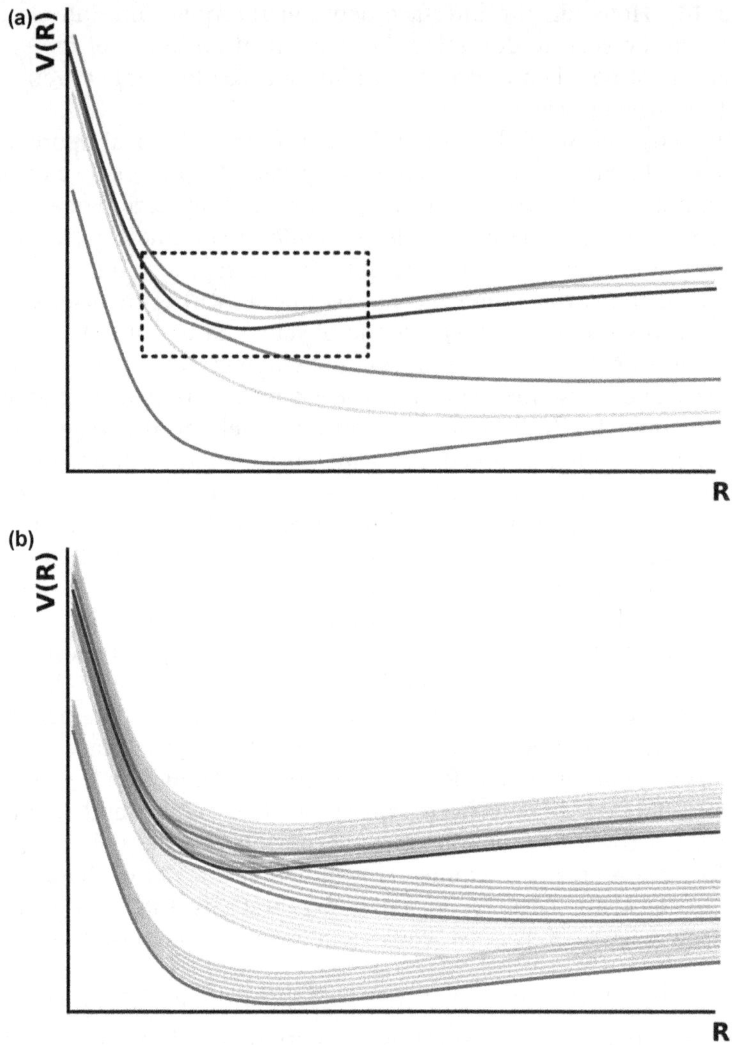

Fig. 1 Schematic representation of ground and excited states potential energy curves of a molecule along a generic coordinate *R*.

represented in Fig. 1a for a generic coordinate *R*, which could be understood as a chosen bond length or the distance between a gas phase molecule and a substrate. The many excited electronic states are ordered in energetically increasing order and connected together at varying nuclear configurations, creating multiple potential energy landscapes for the nuclear dynamics. It is important to recognize that, whenever two or more potential energy curves approach each other (see inside the dashed box), the character of the electronic wave functions associated with these adiabatic, energy-ordered PES may change abruptly. Around these so-called avoided crossings, the Born–Oppenheimer approximation fails.

Considering the nuclear dynamics on a large number of uncoupled potential energy surfaces represents a trivial extension to the adiabatic representation of the wave function, eqn (2), where the total wave function can now be written as

$$\Psi(\mathbf{r},\mathbf{R},t) = \sum_i \phi_i(\mathbf{r};\mathbf{R})\psi_i(\mathbf{R},t) \tag{5}$$

This expression is known as Born–Huang expansion[34] and can be understood as a set of nuclear wave packets, $\psi_i(\mathbf{R},t)$, evolving on different energy landscapes. All phase information between the vibronic terms coming from the dynamics are included in the vibrational terms. Substituting this expansion in the total Schrödinger equation and integrating over the electronic degrees of freedom yields the following equation of motions for the nuclear degrees of freedom

$$\begin{aligned} \imath\hbar\frac{\partial}{\partial t}\psi_i(\mathbf{R},t) = \hat{H}_{\mathrm{nuc},i}(\mathbf{R},t)\psi_i(\mathbf{R},t) + \sum_j\sum_A \vec{d}_{ij,A}(\mathbf{R})\cdot\vec{\nabla}_A\psi_j(\mathbf{R},t) \\ + \sum_j D_{ij}(\mathbf{R})\psi_j(\mathbf{R},t) \end{aligned} \tag{6}$$

Here, each nuclear wave packet is seen to evolve according to a different vibrational Hamiltonian, $\hat{H}_{\mathrm{nuc},i} = \hat{T}_{\mathrm{nuc}} + V_i(\mathbf{R})$, one for each potential energy surface $V_i(\mathbf{R})$. These PESs are coupled *via* the nuclear kinetic energy operator, leading to first and second order non-adiabatic terms of the form

$$\begin{aligned} \vec{d}_{ij,A}(\mathbf{R}) &= -\frac{\hbar^2}{M_A}\int\phi_i(\mathbf{r};\mathbf{R})\vec{\nabla}_A\phi_j(\mathbf{r};\mathbf{R})\mathrm{d}\mathbf{r} \\ D_{ij}(\mathbf{R}) &= \sum_A\frac{\hbar^2}{2M_A}\int\phi_i(\mathbf{r};\mathbf{R})\vec{\nabla}_A^2\phi_j(\mathbf{r};\mathbf{R})\mathrm{d}\mathbf{r} \end{aligned} \tag{7}$$

These non-adiabatic coupling terms generally depend on the nuclear configuration **R** and their computation can become very tedious for large systems. Often one prefers to neglect the coupling terms and investigate the vibrational dynamics of gas phase molecules on a single PES. Provided the potential energy surfaces are well separated, this adiabatic approximation is very well suited to understand the chemical dynamics of interest.

For metallic surfaces, the picture is complicated by the fact that the electronic ground and excited states are found within the conduction band of the system.[30] This means that an infinite number of potential energy surfaces lie within a small energy range around each PES. This situation is depicted schematically in Fig. 1b. The electronic wave functions associated with the states within one band can be thought of having a similar character but a different dispersion relation in momentum space. The latter is a consequence of the periodicity of the metallic substrate. From the consideration of the non-adiabatic coupling elements described above, it is clear that the states lying within a band will be non-negligibly coupled at all points in configuration space due to their infinitesimally small energy gaps. Here, two types of interaction can be considered: the strong coupling regime, which may be mediated by a charge transfer between adsorbate and surface, and the weak coupling limit. The former can be assimilated to the situation in the gas phase and the adjective "strong" emphasizes the fact that this type of coupling leads to an important change in electronic wave function character rather than its magnitude. In the vicinity of the global minimum of the PES and for small amplitude motion at the metallic surface, these strong couplings involve well-separated potential energy surfaces and they play only a marginal role in the chemical dynamics. Consequently, it is often the case that they are neglected altogether.[28]

On the other hand, care must be taken of the so-called weak couplings. These are described in the literature under various names, such as electronic friction or metallic non-adiabaticity. Whereas the character of a given band may remain unchanged in the weak coupling limit, the components within the band are strongly coupled *via* their dispersion relation. In view of the non-adiabatic coupling expressions above, see eqn (6) and (7), these can be understood as involving a transfer of momentum from the nuclei to the electrons and vice versa. Although each contribution to the momentum transfer remains rather small, the sum over j appearing in eqn (6) formally runs to infinity for a metallic environment. It thus comes as no surprise that a complete neglect of these weak couplings can lead to inaccurate vibrational or chemical dynamical simulations. A famous example is that of vibrational relaxation, which is predicted to be orders of magnitude too slow in the absence of non-adiabatic coupling.[28,30]

In the following section, the importance of metallic non-adiabaticity in chemical dynamics simulations will be illustrated through a few selected example applications. First, the vibrational relaxation of adsorbates at metallic surfaces will be treated in the perturbative regime. Second, the effect of weak non-adiabatic coupling on laser excitation simulations will be discussed. Finally, inelastic effects in scanning tunnelling microscopy of highly mobile species in metallic environments will be discussed in terms of non-adiabaticity.

3 Non-adiabatic vibrational dynamics through selected applications

3.1 Vibrational relaxation

One of the most striking consequences of non-adiabatic coupling at metallic surfaces is the rapid relaxation of vibrationally excited states. This can be understood in terms of energy exchange from the nuclear to the electronic degrees of freedom mediated by the kinetic energy operator of the nuclei, eqn (7), creating a so-called electron–hole pair.[35–42] Within the conduction band belonging to a single electronic manifold (*e.g.*, the ground state curves in Fig. 1b), there exist an infinite number of vibronic states at the same energy. Owing to the fact that these lie infinitesimally close in energy, the non-adiabatic coupling terms in eqn (6) and (7) can become large, leading to a change in character for the vibronic wave function. This situation is depicted in Fig. 2, where a

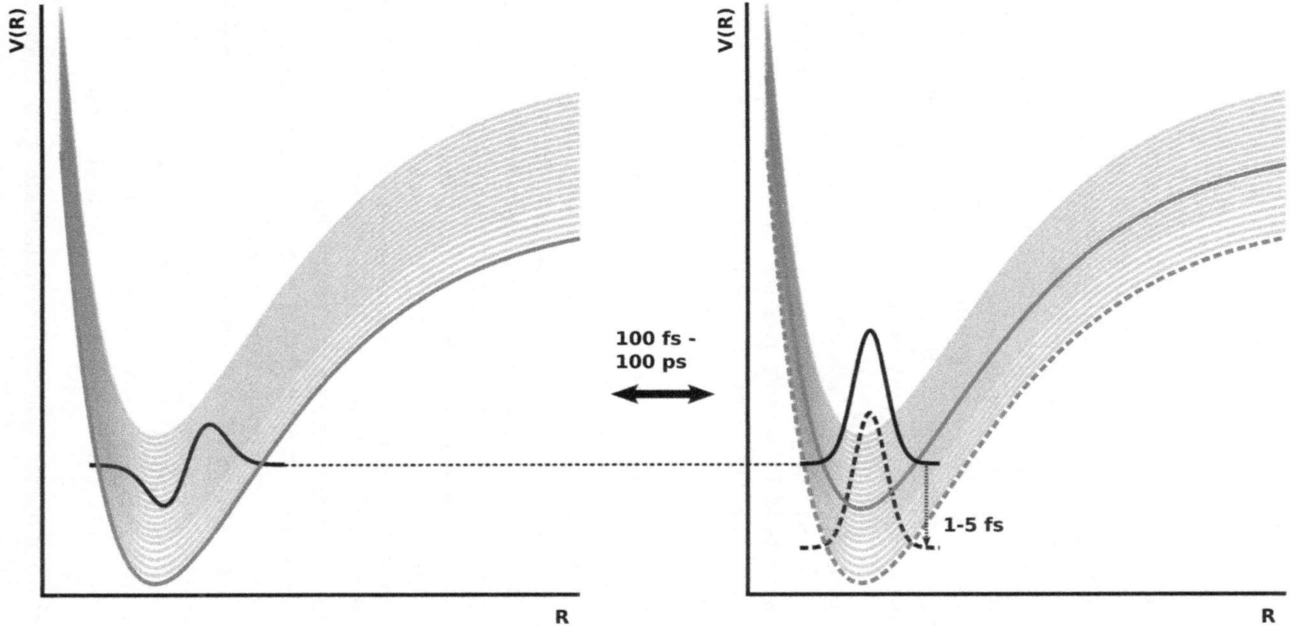

Fig. 2 Cartoon depicting vibrational relaxation at metallic surfaces. Left panel: the system is originally found in a vibrationally excited (solid black line) state of the electronic ground state (thick solid line). Right panel: Upon conversion, the system is found in a lower excited vibrational state (solid black line) of an excited state of the electronic manifold (thick solid line). Typical vibronic state conversion happens on 100 fs – 100 ps timescales. The excited electronic state relaxes irreversibly to the ground state (thick dashed line) within 1–5 fs, as the electron–hole pair diffuses in the metallic substrate.

vibrationally excited state of the electronic ground state (left panel) is converted into the vibrational ground state of an electronic excited state (right panel). Interestingly, this transition implies a drastic change in the vibrational character of the vibronic state, while its electronic character is expected to remain mostly unchanged, as it belongs to the same band as the electronic ground state. The transfer of momentum from the nuclei lands mobility to an electron that becomes separated from the positive background charges. This phenomenon is typically described as the creation of an electron–hole pair in the substrate, as the charges appear dislocated but remain bound together. The electrons of the metallic substrate are strongly delocalized, which implies that the created electron–hole pair will have an extremely short lifetime from the local perspective of the adsorbate. Within about 1–5 fs, the electron–hole pair diffuses irreversibly inside the surface and the excited electronic state is seen to relax back to the electronic ground state.[28,43] In the process, the character of the vibrational wave function remains unchanged (see dashed lines in Fig. 2). It is important to recognize that, although the energy is conserved in the total system, it is irreversibly lost from the local structure that characterizes the adsorbate-metal interaction. The environment being of infinite size, the recurrence time at which the electron–hole pair would be seen to coalesce in the vicinity of the adsorbate is infinitely long. This phenomenon is understood as vibrational energy relaxation.

3.1.1 First-principle vibrational lifetime determination. The non-adiabatic theory of chemical dynamics presented in Sections 1–2 proposes a natural framework to investigate vibrational relaxation at metallic surfaces from first principles. From an electronic structure perspective, it is obviously impossible to compute all the terms in eqn (6) and (7) required to perform chemical dynamics simulations in an infinite system. Fortunately, the phenomenon treated in this chapter does support some degree of simplification. First, the chosen examples involve only non-adiabatic coupling within the band of the electronic ground state, and second, they can be treated perturbatively. Strong coupling can become important when studying reaction dynamics involving, *e.g.*, covalent bond breaking, but remains only marginally involved when considering relaxation dynamics. This is because the vibrational states of interest remain mostly localized in a region of the potential energy surfaces where all electronic excited states are well separated from the ground state. As argued in the previous section, this condition does not hold for electronic states within the fundamental band, as they lie infinitesimally close to each other.

Taking advantage of the fact that the electronic states within a band are only weakly coupled, the non-adiabatic couplings are often treated using time-dependent perturbation theory to first-order.[37,38,40] This allows to characterize the dynamical evolution of the system implicitly, within a time-independent framework. The typical derivation for the vibrational lifetime begins with Fermi's Golden Rule[9,44,45] for the relaxation rate, here given in its canonical form as

$$\Gamma_{n' \to n} = \frac{2\pi}{\hbar} \sum_{\alpha\beta} \left| \int d\mathbf{R} \int d\mathbf{r} \phi_\alpha(\mathbf{r}; \mathbf{R}) \psi_n(\mathbf{R}) \hat{T}_{\text{nuc}} \phi_\beta(\mathbf{r}; \mathbf{R}) \psi_{n'}(\mathbf{R}) \right|^2 \times f_\beta(E_\beta)(1 - f_\alpha(E_\alpha)) \delta(E_{\alpha,n} - E_{\beta,n'}) \tag{8}$$

The energies $E_{\beta,n'}$ and $E_{\alpha,n}$ are associated with the initial and final vibronic states, respectively $\phi_\beta(\mathbf{r}; \mathbf{R})\psi_{n'}(\mathbf{R})$ and $\phi_\alpha(\mathbf{r}; \mathbf{R})\psi_n(\mathbf{R})$. The function $f_\beta(f_\alpha)$ is a Fermi distribution which ensures that the initial (final) electronic state is occupied (free) at a given temperature. Note that here, in contrast with the traditional Born–Huang expansion, eqn (5), the vibrational and electronic indices are separated, since the vibronic states within a band are treated as separable. In the spirit of first-order perturbation theory, the non-adiabatic contribution of each nuclear degree of freedom is treated as an independent relaxation channel

$$\hat{T}_{\text{nuc}} = -\sum_{q=1}^{3N_A} \frac{\hbar^2}{2M_q} \frac{\partial^2}{\partial q^2}; \quad \Gamma_{n' \to n} = \sum_q \Gamma_{n' \to n}^{(q)} \tag{9}$$

Here, the label q refers to one of the $3N_A$ nuclear degrees of freedom of the system. The associated coupling element reads

$$\begin{aligned} T_{\alpha\beta,n'n}^{(q)} &= \frac{\hbar^2}{2M_q} \int d\mathbf{R} \int d\mathbf{r} \phi_\alpha(\mathbf{r}; \mathbf{R}) \psi_n(\mathbf{R}) \frac{\partial^2}{\partial q^2} \phi_\beta(\mathbf{r}; \mathbf{R}) \psi_{n'}(\mathbf{R}) \\ &= \frac{\hbar^2}{2M_q} \int d\mathbf{R} \left\{ 2\psi_n(\mathbf{R}) \frac{\partial \psi_{n'}}{\partial q}(\mathbf{R}) \int d\mathbf{r} \phi_\alpha(\mathbf{r}; \mathbf{R}) \frac{\partial \phi_\beta}{\partial q}(\mathbf{r}; \mathbf{R}) \right. \\ &\quad + \psi_n(\mathbf{R})\psi_{n'}(\mathbf{R}) \int d\mathbf{r} \phi_\alpha(\mathbf{r}; \mathbf{R}) \frac{\partial^2 \phi_\beta}{\partial q^2}(\mathbf{r}; \mathbf{R}) \\ &\quad \left. + \psi_n(\mathbf{R}) \frac{\partial^2 \psi_{n'}}{\partial q^2}(\mathbf{R}) \int d\mathbf{r} \phi_\alpha(\mathbf{r}; \mathbf{R}) \phi_\beta(\mathbf{r}; \mathbf{R}) \right\} \end{aligned} \tag{10}$$

The last term vanishes because of the orthonormality of the electronic wave functions, while the second term can be shown to be of higher order in the non-adiabatic coupling.[37] The remaining term involves a transfer of momentum between electrons and nuclei. The expression is usually treated in the local harmonic approximation leading to a greatly simplified expression[37]

$$\begin{aligned} T_{\alpha\beta,n'n}^{(q)} &= \frac{\hbar^2}{M_q} \left\{ \int d\mathbf{R} \psi_n(\mathbf{R}) \frac{\partial \psi_{n'}}{\partial q}(\mathbf{R}) \right\} \left\{ \int d\mathbf{r} \phi_\alpha(\mathbf{r}; \mathbf{R}_0) \frac{\partial \phi_\beta}{\partial q}(\mathbf{r}; \mathbf{R}_0) \right\} \\ &= \left(\frac{n'\hbar^3 \omega_q}{2M_q} \right)^{1/2} \left\{ \int d\mathbf{r} \phi_\alpha(\mathbf{r}; \mathbf{R}_0) \frac{\partial \phi_\beta}{\partial q}(\mathbf{r}; \mathbf{R}_0) \right\} \delta_{n+1,\, n'} \end{aligned} \tag{11}$$

where ω_q is the frequency of the qth mode at the reference position $\mathbf{R}_0$. The choice of local harmonic vibrations further restricts the number of relaxation quanta to one, *i.e.*, $n' = n + 1$.

For a given dissipative channel q, eqn (11) can be resolved by expanding the operator in a suitable basis of electronic states. For adsorbates at metallic surfaces, it is generally accepted that the one-electron Kohn–Sham states obtained in the framework of periodic density functional theory represent an adequate approximation to the true N-electron states.[40] This allows to rewrite the electronic contribution as

$$\int d\mathbf{r}\phi_\alpha(\mathbf{r};\mathbf{R}_0)\frac{\partial\phi_\beta}{\partial q}(\mathbf{r};\mathbf{R}_0) \simeq \int d\mathbf{r}\phi_k(\mathbf{r};\mathbf{R}_0)\frac{\partial\phi_{k'}}{\partial q}(\mathbf{r};\mathbf{R}_0) \tag{12}$$

Using the fact that electronic and vibrational transitions must be resonant to substitute $\hbar\omega_q = (\varepsilon_k - \varepsilon_{k'})$ leads to a tractable perturbative expression for the relaxation rates

$$\Gamma^{(q)}_{n+1\to n} = \frac{2(n+1)\pi\hbar}{m_q}\sum_{kk'}(\varepsilon_k - \varepsilon_{k'})\left|\int d\mathbf{r}\phi_k(\mathbf{r};\mathbf{R}_0)\frac{\partial\phi_{k'}}{\partial q}(\mathbf{r};\mathbf{R}_0)\right|^2 \\ \times f_{k'}(\varepsilon_{k'})(1 - f_k(\varepsilon_k))\delta(\varepsilon_{k'} - \varepsilon_k - \hbar\omega_q) \tag{13}$$

Interestingly, an equivalent equation can also be obtained from the imaginary part of the dynamical self-energy as the electronic response to a weak, slow time-dependent perturbation.[35,36] The lifetime associated with a given mode q is then defined as the inverse transition rates for the fundamental transition $1\to 0$, $\tau_q = 1/\Gamma^{(q)}_{1\to 0}$, and can be related to the state energy broadening as $\Delta E = \hbar\Gamma^{(q)}_{1\to 0}$.

The numerical solution of eqn (13) remains a daunting task. First, a very fine sampling of the so-called k-space is typically required to properly represent the continuum of states belonging to the ground electronic state.[35,36,40] Second, lacking a more accurate alternative, the electronic contributions to the non-adiabatic couplings, eqn (12), are usually computed using finite differences along the modes of interest. Finally, the delta function ensuring energy conservation must be approximated in some way. For this purpose, various functional forms are preferred by different authors[35,36,40] and a simple expression can be obtained by Gaussian broadening,[33,46–48] $\delta(\varepsilon_{k'} - \varepsilon_k - \hbar\omega_q) \simeq (2\pi\sigma^2)^{-1/2}\exp\left(-\frac{(\varepsilon_{k'} - \varepsilon_k - \hbar\omega_q)^2}{2\sigma^2}\right)$. Care must be taken in choosing the Gaussian width, σ. Whereas eqn (13) is strictly valid in the limit of an infinitely narrow delta function, discrete sampling of the k-space implies a vanishing number of states around the Fermi energy as the width tends to zero. Choosing a large broadening width can cure that problem, while unphysical results can be obtained if too many off-resonant electronic contributions are included in the expression. A proper estimate of the state lifetime thus boils down to finding a compromise between those two extremes. Alternatively, it was shown that by approximating the delta function in eqn (13) with a narrow box around the Fermi energy, the relaxation rates exhibit a quadratic dependence on the inter-band energy difference[40]

$$\Gamma^{(q)}_{n+1\to n} = \frac{2(n+1)\pi\hbar}{m_q\varepsilon^2}\sum_{kk'}(\varepsilon_k - \varepsilon_{k'})^2\left|\int \mathrm{d}\mathbf{r}\phi_k(\mathbf{r};\mathbf{R}_0)\frac{\partial\phi_{k'}}{\partial q}(\mathbf{r};\mathbf{R}_0)\right|^2 f_{k'}(\varepsilon_{k'})(1-f_k(\varepsilon_k)) \tag{14}$$

where ε is a small energy window. Varying the width of this window and fitting the resulting transition rates to a quadratic expression yield a very robust procedure for the determination of lifetimes from first principles that is almost independent of discretization error. This method was employed with great success to a large number of adsorbate-metal systems, notably the CO, CN, and NO molecules on copper, nickel and platinum surfaces.[40] It was found that the internal vibration covers a wide range of lifetimes in different environments, varying from 3.3 ps for CO on a cubic Cu(100) surface to 15 ps for CN on a hexagonal Pt(111) substrate. The molecule-surface stretching mode was found to have a somewhat longer lifetime, on the order 15 ps for CO on various cubic surfaces. The perturbative expression appears to be sensitive enough to capture subtle effects from the metallic environment, showing a large discrepancy for CO on Ni(100) at 16.7 ps and Ni(111) at 5.9 ps. The frustrated translation mode, corresponding to the motion parallel to the surface, is also observed to be sensitive to the precise structure of the surface environment, whereas the libration mode, *i.e.*, the frustrated rotational mode, is more homogeneous across the tested systems. These results are in good agreement with other previously published theoretical estimates and, more importantly, are in line with experimental findings for similar systems.[49–53] Some discrepancies remain when comparing with cluster-based methods, on which we can cast severe doubts for metallic systems where the electronic states are expected to be very delocalized.[37] The bottom line is that, provided a good approximation for the delta function in eqn (13) finds the correct balance between electronic state broadening and discretization errors, the relaxation dynamics of low-lying, localized, harmonic vibrations induced by non-adiabatic coupling with metallic substrates can be studied reliably using time-dependent perturbation theory without resorting to any dynamical simulation.

3.1.2 Transient vibrational dynamics. Perturbation theory yields an intuitive picture of adsorbate relaxation: the loss of a vibrational quantum and associated nodal structure in the nuclear wave function is coupled to an irreversible transfer of momentum to the metallic electrons (see Fig. 2). To obtain time-resolved information about the dynamical processes at work, it is nonetheless necessary to go beyond this simple model. In the past decades, classical molecular dynamics has been hugely successful at shedding light on the transient vibrational evolution in a variety of adsorbate-surface systems (see, *e.g.*, ref. 54–56). The methods of choice for including non-adiabatic effects on the dynamics can be divided in two main families: friction-type theories.[39,42,57–71] and surface hopping.[54–56] The latter uses trajectories to sample the classical dynamics on multiple potential energy surfaces, while non-adiabatic couplings are treated as stochastic jumps between potential energy surfaces. In the former, the metallic environment is treated as a continuous fluid and excitation of electron–hole pairs introduces an effective friction term proportional to the molecular velocity, leading to non-conservative classical dynamics on the ground state potential energy surface. The definition of the friction tensor is not unique and the simplest model relies on the concept of stopping power in a spherically symmetric electronic jellium, or free electron gas, with a density defined as[39,42,57,61,62]

$$n(\mathbf{R}) = \frac{3}{4\pi r_s^3(\mathbf{R})} \tag{15}$$

The Wigner–Seitz radius $r_s(\mathbf{R})$ of the jellium varies with the adsorbate shape and position, as the environment offers more or less resistance to the vibrational motion in different regions of the potential energy landscape.[42] The momentum transfer to the metallic electrons specific for a given atom is then calculated from the dephasing of the jellium one-electron wave functions.[42,62] More sophisticated microscopic models typically rely on time-dependent perturbation theory,[69–71] for which a position-dependent friction tensor can be extracted in the static limit. Both friction-type theories and surface hopping are well suited for direct dynamics, where no *a priori* knowledge of the potential energy surface is required. In this case, the electronic energy of the system is calculated on-the-fly and the forces acting on the nuclei are computed in virtue of the Hellmann-Feynman theorem. These methods have been applied with different degrees of success to various problems in surface science, in particular to scattering and sticking of molecules on metals.

From a quantum mechanical perspective, the transient vibrational relaxation dynamics is often studied within a reduced density matrix formalism.[72] The dissipative dynamics of the so-called open system is described by a Liouville von Neumann equation that includes implicitly the effects of the surroundings on the dynamics

$$\dot{\hat{\rho}}(t) = -\frac{i}{\hbar}\left[\hat{H}_{\mathrm{nuc}}, \hat{\rho}(t)\right] + \mathscr{L}_{\mathrm{D}}\hat{\rho}(t) \tag{16}$$

Note that the term "open system" refers here to exchange of energy and phase with the environment, as the number of particles is conserved throughout. The reduced density matrix $\hat{\rho}(t)$ evolves coherently under the influence of the nuclear Hamiltonian, $\hat{H}_{\mathrm{nuc}}$, and the non-adiabatic effects enter the equation *via* the dissipative Liouvillian superoperator $\mathscr{L}_{\mathrm{D}}$. The latter is also termed "memory kernel", as it contains information about the entire history of the environmental evolution and its interaction with the system.[9,11,72] The definition of the memory kernel is by no means unique nor straightforward. One possible solution is to start from the microscopic Hamiltonian of the total system, eqn (1). Using the projector formalism, it is possible to separate the evolution of the system, *i.e.*, the adsorbate vibrations, from the environmental contributions, the electron–hole pairs in the metal. This leads to a standard Nakajima-Zwanzig[73,74] equation, for which the evaluation and propagation of the memory kernel is as expensive as solving the time-dependent Schrödinger equation for the total system. For this reason, perturbative expansions such as the time-convolutionless technique,[11] the auxiliary densities approach,[75] or the hierarchical equation of motions Hamiltonian[76–78] are usually preferable. These find particular application in the dynamics of vibrations coupled to phonons or other vibrations with similar masses, for which bath memory effects can play an important role.

For metallic environments, non-adiabatic effects in the form of electron–hole pair coupling is known to dominate the dissipative dynamics.[28,30,33] The bath of electron–hole pairs typically thermalizes within a few femtoseconds, which allows for a great simplification of the memory kernel. In view of the adsorbate dynamics, the bath thus remains in its equilibrium state and the memory kernel can be treated in the second-order Born–Markov approximation.[9,11,72] For Markovian dynamics, the dissipative Liouvillian is often written in its diagonal Lindblad form[79–81]

$$\mathscr{L}_{\mathrm{D}}\hat{\rho}(t) = \sum_k \frac{\Gamma_k}{2}\left(\left[\hat{C}_k, \hat{\rho}(t)\hat{C}_k^\dagger\right] + \left[\hat{C}_k\hat{\rho}(t), \hat{C}_k^\dagger\right]\right) \tag{17}$$

Each dissipation channel k is described by a so-called Lindblad operator, $\hat{C}_k$, which is used to represent the energy exchange *via* inelastic collision with the electrons of the metal. In this particular case, the operators take the form of a transition between two vibrational states of the system

$$\hat{C}_k = \hat{C}_{n\rightarrow n'} = |\, n'\rangle\langle n\,| \tag{18}$$

This form of the Lindblad operators suggests that, provided their computation remains numerically tractable, the eigenstates of the vibrational Hamiltonian would provide the most efficient basis to represent the evolution of the reduced density matrix, eqn (16). Indeed, in such a basis, the equations of motion for the matrix elements take the particularly simple and elegant form

$$\dot{\rho}_{mm'} = \begin{cases} \sum_n (\Gamma_{n\rightarrow m}\rho_{nn}(t) - \Gamma_{m\rightarrow n}\rho_{mm}) & \text{for } m = m' \\ -(i\omega_{mm'} + \tilde{\gamma}_{mm'})\rho_{mm'}(t) & \text{for } m \neq m' \end{cases} \tag{19}$$

where $\tilde{\gamma}_{mm'} = \frac{1}{2}\sum_n (\Gamma_{m\rightarrow n} + \Gamma_{m'\rightarrow n})$. The labels $\{n,n',m,m'\}$ refer to eigenstates of the full dimensional nuclear Hamiltonian, $\hat{H}_{\mathrm{nuc}}\,|\,n\rangle = E_n\,|\,n\rangle$, also termed "zeroth-order" states in the following. These can nowadays be routinely calculated up to six dimensions using simple basis set expansions,[82–85] and to even higher dimensions using more sophisticated approaches such contracted iterative methods,[86–90] multidimensional quadratures with basis pruning,[91] or multi-configuration time-dependent ansatz.[92–95] Interestingly, the same equations of motion in eqn (19) can be derived from a Redfield tensor in the strict secular approximation.[96] In both cases, the evolution of the reduced density matrix is said to provide a dynamical map of the reduced density operator, meaning that the time propagator effectively maps the density matrix of the system at a given time to the density matrix of the same system at a later time.[11] This is an extremely desirable property that is required for ensuring the probabilistic interpretation of the diagonal elements of the density matrix as population of the zeroth-order states and the off-diagonal elements as their coherences. Analytical integration of the off-diagonal matrix elements in eqn (19) reveals that the coherences decay exponentially with time. The populations obey a Pauli master equation and tend asymptotically to an equilibrium distribution determined by the balance between the microscopic rates.[9,11,72]

The relaxation rates, $\Gamma_{n\to n'}$ associated with each dissipative channel can be computed using time-dependent perturbation theory, as discussed in the previous section. Note that the brackets appearing in the definition of the Lindbald operators, eqn (18), are associated with fully-coupled multi-dimensional vibrational wave functions. These are subject to anharmonic coupling and to intermode mixing, especially for higher vibrational excited states. It was demonstrated that, upon relaxation, vibrational energy can be redistributed among the modes and this phenomenon is solely due to the topology of the potential energy surface, *i.e.*, it is independent of the choice of representation to a very good approximation.[97,98] Further, the selection rule for the harmonic case becomes invalid, opening up new multiple quanta relaxation channels.[47,99]

These two behaviors can be captured in a perturbative treatment by performing numerical integration of the non-adiabatic coupling elements, eqn (11), for the multi-dimensional vibrational wave functions along the nuclear coordinates **R**. To simulate the variation of the non-adiabatic coupling strength as a function of the adsorbate position, simplifying assumptions from electronic friction theories are used.[39,42,57,61,62] At any given point, the electron–hole pair coupling is assumed to be mediated by a spherical jellium of electronic density eqn (15), yielding a simple expression for the transition rate between arbitrary states

$$\Gamma^{(q)}_{n\to n'} = \frac{\gamma^{(q)}}{\omega_{nn'}} \left| \int \mathrm{d}\mathbf{R}\psi_n(\mathbf{R}) n^{1/3}(\mathbf{R}) \frac{\partial \psi_{n'}}{\partial q}(\mathbf{R}) \right|^2 \tag{20}$$

where the integration is performed over all nuclear coordinates for the derivative along each single mode $q \in \mathbf{R}$. Since electron–hole pair coupling is mediated by an effective free electron gas surrounding the adsorbate impurity, the jellium density is termed "embedding density". Various definitions for the embedding density were proposed without major influence on the resulting rates.[47] A conventional choice is to use the electronic density of the adsorbate-free metallic substrate to define the Wigner–Seitz radius of the embedding density $n^{1/3}(\mathbf{R})$ for each atom in the molecule individually. A closer look at the non-adiabatic coupling constant, $\gamma^{(q)}$, reveals its origin as a momentum transfer to the electrons mediated by the vibrational mode q

$$\gamma^{(q)} = \frac{2\pi\hbar}{m_q \varepsilon^2 n^{2/3}(\mathbf{R}_0)} \sum_{kk'} (\varepsilon_k - \varepsilon_{k'})^2 \left| \int \mathrm{d}\mathbf{r} \phi_k(\mathbf{r};\mathbf{R}_0) \frac{\partial \phi_{k'}}{\partial q}(\mathbf{r};\mathbf{R}_0) \right|^2 f_{k'}(\varepsilon_{k'})(1 - f_k(\varepsilon_k)) \tag{21}$$

Note that the simple and robust definition of the delta-function proposed by Krishna and Tully[40] was used in the above expression, but this choice is by no means restrictive. The anharmonicity and the intermode coupling of the vibrational states enter naturally the expression *via* the friction-like term, $n^{1/3}(\mathbf{R})$, which modulates the strength of the non-adiabatic couplings with respect to a reference position $\mathbf{R}_0$.

To appreciate the importance of multiple quanta relaxation on excited states lifetime, let us consider the dissipative dynamics of a hydrogen impurity located in the vicinity of a hexagonal palladium(111) surface.[47] The potential energy surface is highly corrugated[100,101] and, due to its peculiar topology with low diffusion barriers, it presents a multitude of local minima that can support localized vibrational states. Apart from the surface sites, labeled fcc and hcp, the hydrogen atom is seen to be stable in the first subsurface octahedral cavity, as well as in deeper octahedral bulk-like sites. These sites are labeled O_h(sub) and O_h(blk) in Fig. 3, respectively. The circles represent the inverse transition rates, here called lifetimes, between adjacent vibrational levels of the hydrogen motion perpendicular to the surface plane. The solid lines depict the expected behavior in the local harmonic approximation, *i.e.*, a scaling of $1/n$ for increasing initial quantum number n. For the localized vibration in the most stable minimum, the fcc site, the lifetime follows the harmonic scaling law up to the third excited states and the inverse transition rates increases at higher energies. This trend is also observed in other minima but for much lower excitations. The full squares depicted in Fig. 3 are lifetime estimates including the contributions from all vibrational states at an energy lower than the state considered

$$\tau_n = 1 \Bigg/ \left(\sum_q \sum_{n' < n} \Gamma^{(q)}_{n \to n'} \right) \tag{22}$$

These show a surprisingly regular behavior that matches the harmonic scaling law, albeit including multiple quanta relaxation. To recover the simple scaling for the lifetime of high-lying excited states that depart from the single-mode, local, harmonic model, it appears necessary to include all dissipation channels from a given initial state. This in turns is a proof that the harmonic scaling law is a very valuable tool to estimate vibrational state lifetimes at any energy, but the microscopic picture of the relaxation dynamics will be utterly skewed. Further, as can be seen from the top right panel of Fig. 3, the identification of a harmonic progression can become ambiguous even at low energies, as intermode coupling lead to strong mixing amongst different states, preventing their proper labelling (see circles for quantum numbers $n = 1$ and $n = 2$).

The influence of intermode coupling on transition rates can be more easily seen from an explicit simulation of relaxation dynamics. For illustrative purposes, let us consider a model system consisting of two

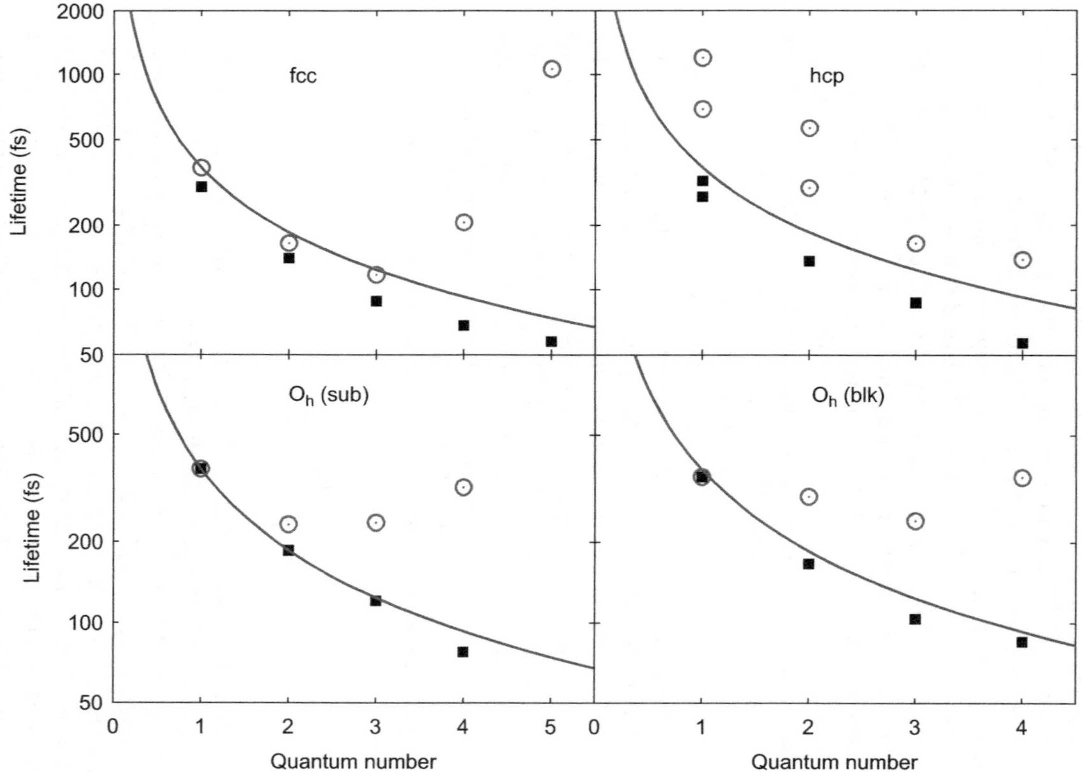

Fig. 3 Lifetimes associated with the hydrogen vibrational mode perpendicular to a Pd(111) surface in different adsorption sites. The circles are the state-to-state inverse transition rates from a given initial state. The total lifetimes are represented by full squares and the harmonic scaling law is depicted as a solid line. The quantum numbers are determined by visual inspection of one-dimensional cuts of the wave functions integrated over all remaining coordinates. For computation of the rates using eqn (20), the embedding density is chosen as that of the pure metal. Reproduced with permission from ref. 47.

carbon monoxide molecules adsorbed on a copper surface, as modelled using a recent accurate analytical ground state potential energy surface.[102] In this reduced-dimension model, only the CO stretching modes, labeled q_1 and q_2, are considered as dynamical variables, and their intermode coupling strength is varied to simulate weak to strong intermode mixing.[97,98] Figure 4 shows snapshots of the 2D probability density evolution for a system initially excited in a pure q_1 vibrational state. The situation on the left shows a system devoid of intermode coupling. The labels *full* and *fact.* refer to different choices of bases, where the former is the full dimensional basis used in the definition of eqn (19), and the latter is a factorizable form of the basis as a tensor product of one-dimensional functions, $|n\rangle = |n_1\rangle|n_2\rangle$. In this case, the equations of motion for the reduced density matrix must be modified to include intramolecular vibrational energy redistribution (IVR) in the form of a potential term that mixes the zeroth-order states. The form of this potential correction is by no means trivial but computationally tractable, and the interested reader is referred to the literature for more detail.[97,98] Using this representation, computation of the relaxation rates become much less demanding and the structure of the basis allows for greater insight in the mechanism of intermode mixing. While the equations of motion for the reduced density matrix lose their analytical integration properties, these can be solved efficiently in the interaction picture using standard numerical methods.[103]

From the initial condition on the top left panel, the probability density of the nuclei along the two stretch modes is seen to relax to the global ground state within about 10 ps in both representations. The absence of intermode coupling hinders the buildup of nodal structure in the stretching mode of the neighboring CO molecule. Consequently, the relaxation rates in eqn (20) are perfectly accounted by a factorizable, single-mode model. On the other hand, a large coupling between the two CO molecules lead to a fast beating of the population between the two stretching modes excited states (see right panels). Due to IVR, the system is found to be singly excited in both the q_1 and q_2 modes at $t = 0.48$ ps (top right panel), and the nodal structure is transferred completely to the mode q_2 at time $t = 1.92$ ps. The system undergoes coherent, Rabi-like oscillations between excited q_1 and q_2 modes while incoherently relaxing to the ground states. In the chosen example, coherent IVR is observed to noticeably slow down the relaxation process, and the outcome of the dynamical simulation leads to a ground state vibrational wave function that is distorted by the potential coupling (see bottom right panel). Most importantly, the dissipative dynamics is found to be similar using both the factorizable and fully coupled representations, even in the strong intermode coupling regime. Unfortunately, the phenomenon of IVR has been largely overlooked in early quantum mechanical studies of relaxation dynamics at metallic surface, and the examples in the literature remain scarce.[47,48,97–99,104,105] The availability of more general relaxation models such as, *e.g.*, eqn (20), and the realization that IVR can be simply treated *via* a potential term in the vibrational Hamiltonian without altering the dynamical evolution of the system will pave the way for more accurate description of dissipative dynamics of adsorbates at metal.

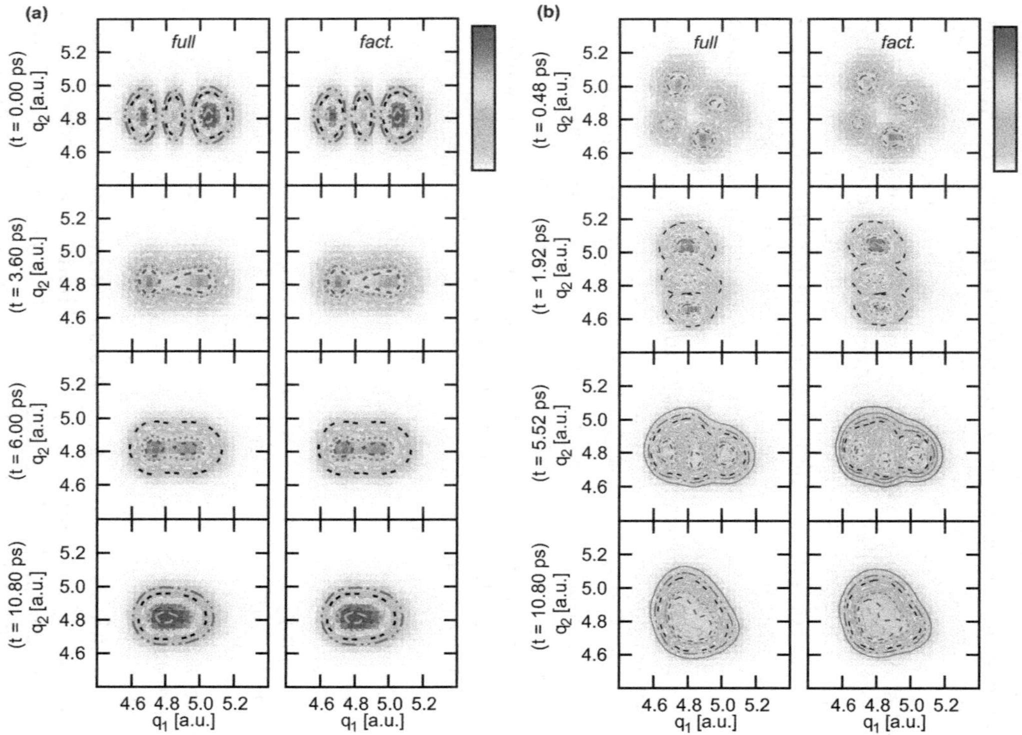

Fig. 4 Snapshots of the probability distribution evolution for two CO molecules on a copper(100) surface subject to non-adiabatic coupling. The dynamics is initiated with two quanta of vibrational excitation in one of the CO molecules, here the q_1 mode. The labels *full* and *fact.* refer to the choice of vibrational basis to represent the reduced density matrix in eqn (19). The left panels show the system in the absence of intermode coupling and the strong intermode coupling regime is depicted in the right panels. Reproduced with permission from ref. 97.

3.2 Induced vibrational dynamics

Limited by its exponential scaling with system dimension, state-of-the-art quantum mechanical treatment of adsorbate dynamics mostly remains constrained to the study of either atomic or diatomic species.[33] Even so, it is often the case that systems of reduced dimensionality or model potentials are used to explain important experimental features at a quantum mechanical level. Bearing in mind that classical treatment of adsorbate dynamics provides an accurate description of both small and large amplitude relaxation processes and of surface scattering, the necessity of using a higher level of theory may appear somewhat debatable. The advantages of a quantum mechanical treatment reside first and foremost in the proper inclusion of all inherently quantum mechanical effects, such as non-locality or tunneling, and the correct statistical representation of the result distributions that can be obtained classically only in the large sampling limit. Further, it allows to include seamlessly different excitation mechanisms, *e.g.*, the coherent interaction with an electromagnetic field or with a quasi-inert probing projectile such as helium, directly in the formalism. Despite its inherent complexity, the benefits of treating the whole system using a balanced level of theory are thus great, as will be shown in the following example applications.

3.2.1 Infrared excitations. Our first prototypical application is the infrared laser excitation of selected adsorbate modes. It was shown experimentally that pre-excitation of low-lying adsorbate vibrational states using infrared sources lead to higher desorption yields and different desorption product distributions upon subsequent UV excitation.[28] These findings could have an important impact in catalysis and for hydrogen storage and release. A more direct route for molecular abstraction from surfaces consists in photodesorption by vibrational ladder-climbing using high-power infrared laser sources, which remains a theoretical proposal up to this day.[106]

From a quantum mechanical perspective, the treatment of the laser–adsorbate interaction can be straightforwardly included in the equations of motion for the reduced density matrix. Since the molecules are only of modest size compared to the wavelength of the incoming electric field, $\vec{F}(t)$, the interaction can be treated within the semi-classical dipole approximation,[107] *i.e.* the system Hamiltonian is replaced by

$$\hat{H}_{\text{nuc}} \rightarrow \hat{H}_{\text{nuc}} - \hat{\vec{\mu}} \cdot \vec{F}(t) \tag{23}$$

Consequently, the equations of motion for the reduced density matrix in the eigenstate basis of the nuclear Hamiltonian take the form

$$\begin{aligned} \dot{\rho}_{mm'} = & \sum_{j=x,y,z} \frac{iF_j(t)}{\hbar} \sum_k \left(\mu_{mk}^{(j)} \rho_{km'}(t) - \rho_{mk} \mu_{km'}^{(j)} \right) \\ & + \begin{cases} \sum_n \left(\Gamma_{n\rightarrow m} \rho_{nn}(t) - \Gamma_{m\rightarrow n} \rho_{mm} \right) & \text{for } m = m' \\ -(i\omega_{mm'} + \tilde{\gamma}_{mm'}) \rho_{mm'}(t) & \text{for } m \neq m' \end{cases} \end{aligned} \tag{24}$$

where $\mu_{mk}^{(j)}$ is a matrix element of the *j*th component of the dipole moment. By doing so, it is assumed that the indirect interaction of the electric field *via* the environment – here, the electrons of the metal – does not affect significantly the system dynamics. This is equivalent to neglecting the fluctuations of the electron cloud which can lead to plasmon resonances in the UV regime. In the IR regime, these fluctuations can be assimilated to a simple electric field enhancement, and the reported field intensities should thus be interpreted with care. It is important to recognize that the Lindblad dynamics and the chosen representation of the adsorbate–field interaction treat the broadening of the vibrational bands as symmetric. By doing so, potentially strongly asymmetric Fano profiles are neglected. This assumption was confirmed to be valid for the systems reported below, *e.g.*, by investigation of the experimental LEED signal. As a final remark, it can be seen from eqn (24) that population transfer between zeroth-order states induced by the coupling to the electric field are mediated solely by the off-diagonal elements of the reduced density matrix. Upon detailed inspection of the populations evolution, one finds

$$\dot{\rho}_{mm} = \sum_{j=x,y,z} \frac{iF_j(t)}{\hbar} \left(\sum_{k \neq m} \left(\mu_{mk}^{(j)} \rho_{km}(t) - \rho_{mk} \mu_{km}^{(j)} \right) + \left(\mu_{mm}^{(j)} \rho_{mm}(t) - \rho_{mm} \mu_{mm}^{(j)} \right) \right) + \sum_n (\Gamma_{n \to m} \rho_{nn}(t) - \Gamma_{m \to n} \rho_{mm}) \quad (25)$$

with the last term the field coupling obviously vanishing. As was described in eqn (19) above, the off-diagonal elements of the reduced density matrix in a dissipative environment decay exponentially. This in turns imply that, at times longer than the timescales defined by the relaxation rates, the coupling with the electric field cannot induce population transfer in the system. For metallic environments where vibrational lifetimes vary from a few hundreds of femtoseconds to tens of picoseconds, it means that coherent laser excitation can only be performed in the picosecond regime. Bearing in mind that a fixed amount of energy must be injected in the system to achieve quantitative population transfer, the use of short laser fields implies that their intensity should be high. As an example, let us consider the population transfer in a dissipation-free two-level system using a short pulse of length Δt and carrier frequency ω_0 chosen resonant with the transition

$$F(t) = F_0 \sin^2(\pi t/\Delta t) \cos(\omega t) \quad (26)$$

The field intensity to achieve complete population inversion can be calculated analytically in the rotating wave approximation[107] to yield $F_0 = \frac{\pi \hbar}{\mu_0 \Delta t}$, with μ_0 being the transition dipole moment between the two states. It can thus be seen that this so-called π-pulse condition can be satisfied with weaker electric field intensities with increasing pulse duration. On the other hand, the short and intense pulses required for laser excitation in metallic environments have the disadvantage of allowing for multiple photon processes and leading to dynamical broadening of the vibrational bands. The latter is mostly due to a strong dynamical Stark effect, *i.e.* the interaction of the states permanent dipole with the oscillating field causing their energy to move up and down in the field. By doing so, multiple quasi-resonant excitation channels become accessible and the excitation loses its selectivity. Tailoring an electric field for selective vibrational excitation in a metallic environment is thus a tremendous challenge, which amounts foremost to finding the right balance between a too long pulse, where dissipation dominates, and a too short pulse, where dynamical broadening dominates.

Figure 5 shows an example of mode selective infrared excitation for D_2 on a hexagonal ruthenium(0001) surface. The model Hamiltonian is restricted to the four most important modes: the vibration perpendicular to the surface (Z-mode), the internal stretch vibration (r-mode), and the frustrated rotations (θ- and ϕ-modes). The latter are sometimes called librations, as the vibrational motion at the surface corresponds to a rotation in the gas phase. The lifetime of the first excited state in each of the modes is calculated to be about 400 fs for the r-, θ-, and ϕ-modes, while it is four times longer at 1.6 ps for the Z-mode.[105] This relation is expected from the mass-dependence of the transition rates, *e.g.*, as in eqn (14). The figure shows the time evolution of selected state populations, *i.e.*, the diagonal elements of the reduced density matrix in the eigenstate basis of the 4D vibrational Hamiltonian, for different series of π-pulses, eqn (26). The ground and first three excited states along the Z-mode are labeled $|0\rangle$, $|3\rangle$, $|12\rangle$, $|37\rangle$, respectively. The pulses are tuned at the fundamental transition frequency ($\hbar\omega = 764$ hc cm^{-1}) and the field intensity is determined to completely invert the population in a dissipation-free two-level system. The first pulse in panel a) is chosen on a timescale slightly shorter than the vibrational lifetime. Consequently, the effects of dissipation are strongly felt upon excitation and about 51% of the population returns to the ground state by the end of the first pulse. Using an even longer pulse (panel b) exacerbates this undesirable feature ($P_0 = 75\%$), while shortening the pulse to about a fourth of the reference lifetime (panel c) greatly improves excitation probability.

The first pulses of duration 1 ps, 2 ps, and 500 fs depicted in panels a–c also demonstrate the effect of dynamical broadening on short-time population dynamics. As the pulses become shorter, the higher vibrational excited states become more significantly populated and the first excited state population saturates at around 30%. The Z-mode of the D_2/Ru(0001) system is strongly harmonic and the transitions between the first few excited states have very similar energies. Hence, already a small contribution from dynamical broadening suffices to open the possibility of multiple photon excitations and ladder climbing. Interestingly, the shorter pulse yields a perfect mode selectivity, which is calculated as the ratio of excited population in the desired mode to the total excited

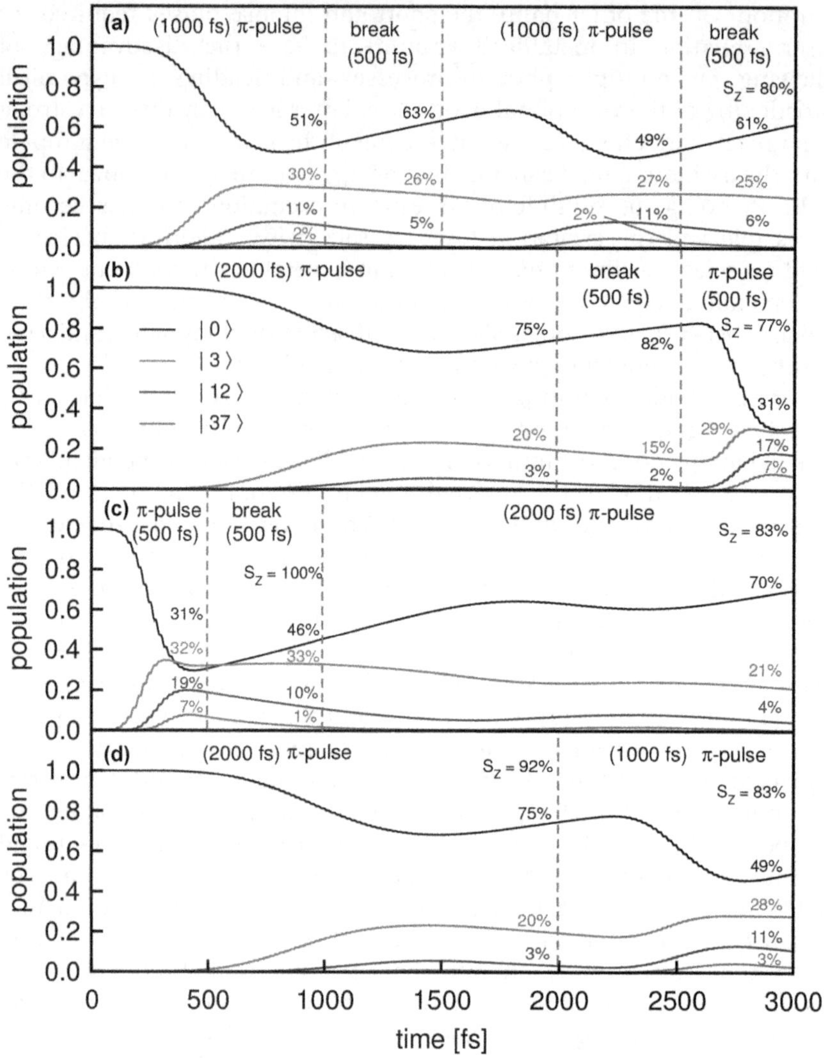

Fig. 5 Population evolution of selected states for the dissipative dynamics of D_2/Ru(0001) using different series of π-pulses. In panels a–c, the pulses are separated by free relaxation intervals. The percentual population of the states are given at the end of each interval, denoted by the vertical dashed lines. The mode selectivity S_z is reported where relevant. Reproduced with permission from ref. 105.

population, $S_Z = \frac{\sum_i P_i}{1 - P_0}$. This means that, provided the excitation time is faster than relaxation, no significant inter-mode mixing will happen. In Fig. 5, the mode selectivity is seen to be reduced from 100% after a 500 fs pulse (panel c) to 92% for a 2 ps π-pulse (panel d). In the latter case, mode selectivity is even further reduced upon re-excitation after a 500 fs break. These are positive proofs that, in this system, intermode coupling leading to intramolecular vibrational energy redistribution is solely mediated by the relaxation process. Note that, although the mode selectivity is the same in panel c–d by the end of the simulation, the sequence without relaxation is also more efficient while pumping a larger fraction of the population to the higher excited states. In the view of these various results, it appears that a short, coherent excitation performed at the end of a waiting interval allowing for the system to return to its thermodynamical equilibrium will improve mode selectivity but reduce state selectivity.

As a side note, it is important to mention other approaches to vibrational excitation at metallic surfaces using short, intense laser fields. Whereas the simple concept of parametric optimization can be used to tune pulse sequences for reaching a desired target as shown above, it can be more expedient to use the physical knowledge about the system and exploit interference patterns between various reaction paths to achieve a desired target. This approach is known under the name coherent control and has been applied successfully to a wide range of problems in the condensed phase.[108] Stochastic pulse optimizations,[109] either in parametric form as above or directly applied to the time-dependent electric field, are also a popular choice for tailoring pulses beyond the rational pulse design. Finally, optimal control algorithms, where a cost functional subject to physical penalty factors is maximized, have been applied to the problem of quantum optical control in dissipative environments with various degrees of success.[110–112]

3.2.2 Scanning tunnelling microscopy (STM). Another important application that can be treated in the weak non-adiabatic coupling limit is the inelastic contribution to the scanning tunnelling current.[113,114] In this experiment, a metal tip is placed at a given height above a surface sample and electrons are injected in/from the surface *via* the STM tip. The potential bias between the tip and the surface defines the energy of the incoming/outgoing electrons, which is then varied to yield a signal that depends on the density of states in the sample as a function of the transferred electronic energy. For this reason, the latter technique is sometimes referred to as scanning tunnelling spectroscopy. The signal is composed of two major components: the elastic current, directly probing the density of states at a given energy, and an inelastic current.[113,114] The derivative of the measured current with respect to the applied potential bias yields the spectroscopic signal. Although a rich literature covers the effect of phonons on electronic properties such as the STM current itself, the emphasis will be put here on the influence of electron–hole pair coupling on the vibrational dynamics.

In early investigations, Persson and Demuth[115] proposed that the electric field of the STM tip interacts linearly with the dipole moment induced by the tunnelling electrons on the adsorbate and its image charge in the metal. This can be understood as a dipole-mediated distortion of the potential energy surface, which can drive the nuclear dynamics out of equilibrium. Another important excitation channel originating from the coupling to the STM tip is the so-called resonance mechanism, where an electron is transferred to a short-lived resonance state localized on the adsorbate that mediates energy transfer with the sample vibrations. This latter type of non-adiabatic coupling is often the dominant contribution to the inelastic current for adsorbate at metallic surfaces.[116–120] Gao and co-workers[120] proposed a perturbative treatment of scanning tunneling microscopy relaxation at metallic surfaces in terms of non-adiabatic couplings. The model is based on the idea that the excitation of a vibrational quantum causes a linear shift in an electronic level of the adsorbates. Consequently, a simple expression for the fundamental transition rate induced by STM inelastic currents at a given potential bias U is obtained in terms of projection of the electronic states onto the resonance

$$\Gamma_{\mathrm{stm}}^{(q)} = \frac{|eU|}{\hbar\omega_q}\frac{\Delta_t}{\Delta_s}\gamma_{eh}^{(q)}; \quad \gamma_{eh}^{(q)} = \frac{2\pi\hbar}{M_q}\left(\varepsilon_a'\right)^2\rho_a^2(\varepsilon_F) \tag{27}$$

The quantities $\Delta_{s,t}$ define the resonance partial width of the adsorbate level due to the tip (t) and to the substrate (s) within an Anderson model, and $\rho_a(\varepsilon_F)$ is the local density of substrate states on the adsorbate at the Fermi energy. The factor ε_a' determines the adsorbate energy level rise due to non-adiabatic coupling. The relaxation/excitation rates at zero potential bias are simply given by the factor $\gamma_{eh}^{(q)}$ weighted by Bose–Einstein factors allowing to recover the proper detailed balance. The total upward and downward rates due to the resonance mechanism in the presence of an STM-induced electron flow is thus given by a skewed thermal distribution of the non-adiabatic coupling rates[120]

$$\begin{aligned} \Gamma_{1\to 0}^{(q)} &= \left(1 + B(\hbar\omega_0) + \frac{|eU|}{\hbar\omega_q}\frac{\Delta_t}{\Delta_s}\right)\gamma_{eh}^{(q)} \\ \Gamma_{1\leftarrow 0}^{(q)} &= \left(B(\hbar\omega_0) + \frac{|eU|}{\hbar\omega_q}\frac{\Delta_t}{\Delta_s}\right)\gamma_{eh}^{(q)} \end{aligned} \tag{28}$$

Note that these rates are only valid in the local harmonic approximation. Since the resonance partial width of the tip can be difficult to calculate accurately, one usually resorts to *a posteriori* scaling of the Δ_t term to experimental results. A somewhat more involved derivation of the non-adiabatic coupling rates in terms of source-bridge-sink terminology, where the molecular adsorbate serves as a bridge for the flowing electrons, allows to extend the above rates to anharmonic vibrations and large amplitude motion.[121] In the process, an alternative definition for the STM scaling factor can be obtained, which contains information about the density of states around the Fermi energy and introduces a current dependence in the prefactor. These generalized rates read

$$\begin{aligned} \Gamma_{n\to n'}^{(q)} &= \left(1 + B(\hbar\omega_{nn'}) + \frac{|eU|}{\hbar\omega_{nn'}}\left(\frac{I\tau_0}{e\rho_0(U)}\right)\right)\Gamma_{nn'}^{(q)}(T=0) \\ \Gamma_{n\leftarrow n'}^{(q)} &= \left(B(\hbar\omega_{nn'}) + \frac{|eU|}{\hbar\omega_{nn'}}\left(\frac{I\tau_0}{e\rho_0(U)}\right)\right)\Gamma_{nn'}^{(q)}(T=0) \end{aligned} \tag{29}$$

where I is the STM current and $\rho_0(U)$ is the number of available one-electron states per metal atom at an energy $\varepsilon_F - eU$ around the Fermi energy ε_F. The timescale τ_0 defines the residence time of an electron on the adsorbate, which can be evaluated as the local harmonic limit of eqn (20). An important advantage of the latter expression is that all involved quantities can be defined from first principles.

Although the excitation/relaxation rates are interesting *per se*, these are only indirectly related to the experimental transfer rates that are inferred from STM-induced manipulation of adsorbate at metals. To understand this fact, it is instructive to investigate the timescales involved in these experiments. For hydrogen on palladium(111), the energy relaxation at 0 K is on the order of $1/\Gamma_{10}^{(q)}(T=0)=\tau_0=400$ fs, as determined from first principles[47] (see eqn (14) and Fig. 3). The adsorbate transfer, on the other hand, happens on a much longer timescale, which can only be inferred to be faster than a few minutes from time-unresolved experiments.[122] The time-resolved experiments of Eigler, Lutz, and Rudge[123] for xenon on a nickel(110) surface yield transfer times ranging between 100 μs and 10 s at potential biases between 200 mV and 20 mV. This is in spite of a non-adiabatic coupling lifetime of $\sim$3 ns. Taking into account the largely different timescales, it appears that dephasing between vibrational states is bound to happen much faster than the transfer dynamics. Consequently, the off-diagonal elements of the reduced density matrix in the vibrational eigenstate basis, (24), will vanish before influencing the long-time dynamics. To a very good approximation, the STM-induced dynamics of adsorbates at metallic surfaces can thus be modelled as a population dynamics of zeroth-order states obeying a Pauli master equation

$$\dot{\rho}_{mm} = \sum_n (\Gamma_{n\to m}\rho_{nn}(t) - \Gamma_{m\to n}\rho_{mm}) \tag{30}$$

Note that, since the coherences of the reduced density matrix vanish on long timescales, the coupling to the STM electric field disappears from the equations of motion and only the incoherently driven evolution of the system mediated by the non-adiabatic couplings remains. This is consistent with the picture of an ensemble of localized states in quasi-thermal equilibrium that diffuses towards other local minima. Since the zeroth-order vibrational states are non-local by construction, both tunneling and above-threshold excitations are seamlessly included in this treatment.

Despite its simplicity, the model was applied with great success to, *e.g.*, the STM-driven transfer of a xenon atom on a nickel surface.[120] This pioneering experiment[123] was the very first example of an STM-controlled atomic switch, where the xenon atom was moved from the nickel surface to the tungsten STM tip. Figure 6 shows the comparison between theoretical and experimental transfer rates for different values of the ratio Δ_t/Δ_s. The computed transfer rates are inferred from an implicit dynamics between truncated harmonic oscillators, with the initial conditions chosen as the fifth excited vibrational state located on the surface.[120] This corresponds to a situation where above-threshold dynamics dominates. By construction, the rates exhibit the proper power-law dependence with increasing potential bias. It is found that, for this

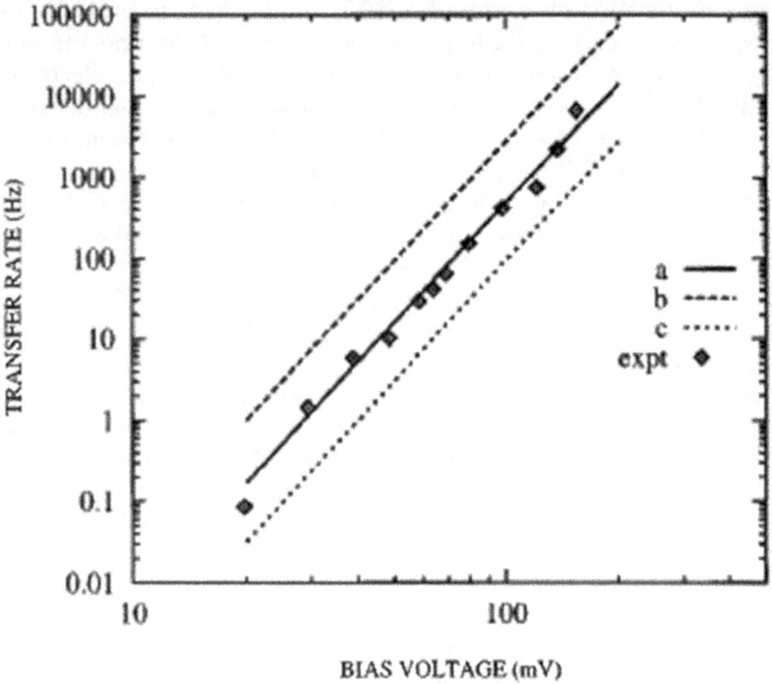

Fig. 6 Transfer rates for a xenon atom driven from a nickel surface towards a tungsten tip. The theoretical rates are computed using eqn (27) and (28) at $T=0$ K, with $h\omega_q=4$ meV, and $\gamma_{eh}^{(q)}=3.3\times10^8$ s^{-1}. The value of the ratio Δ_t/Δ_s is varied from 0.05 (dotted line) to 0.07 (solid line) and 0.1 (dashed line) to reproduce the experimental data (squares). Reproduced with permission from ref. 120.

particular case, the optimal ratio Δ_t/Δ_s is about a factor of three smaller than the theoretically predicted value. Note that electron-phonons interaction was included in the simulations. In this temperature regime, the coupling to the system phonons is found to be two orders of magnitude slower than the electron–hole pair coupling and, consequently, plays only a minor role as expected.

In other experiments for hydrogen on a palladium(111) surface under various hydrogen pre-coverages,[122,124] it was found that the presence of an STM tip can cause both depletion or clustering of the hydrogen atoms at the interface. This is revealed by bright protrusions on the surface after STM excitation. Sykes and co-workers[122] proposed that these originate from the rise of the topmost palladium layer upon population of the subsurface sites by hydrogens migrating from the bulk. Other joint experimental–theoretical results pointed rather at a clustering of the hydrogen impurities in the vicinity of the surface, stabilized by the field-induced deformation of the potential energy surface.[124] The static theoretical treatment of hydrogen migration in the bulk induced by STM in terms of non-adiabatic couplings proved an important step towards understanding these experiments. It was shown that, since the impinging electrons from the STM tip propagates like S-waves inside the substrate, only electrons belonging to s-shells of the substrate contributes to the non-adiabatic coupling rates.[125] This explained the slight asymmetry of the transfer rates observed experimentally while leaving the situation at the interface open. Real-time dynamics of the hydrogen impurity at palladium(111) for different hydrogen pre-coverages, including adiabatic relaxation of the palladium surface, could offer a more detailed picture of the resurfacing dynamics.[121,126,127] Using the generalized rates, eqn (29), the Pauli master equation, eqn (30), can be solved numerically in the basis of anharmonic vibrational eigenstates for the hydrogen impurity. Placing the impurity in a stable local minimum deep inside the surface, it is possible to observe how the hydrogen migrates under the influence of the STM perturbation. By fitting the population decay of a hydrogen impurity placed below the surface to an exponential function, it is possible to extract the transfer rates from the fully coupled dynamics beyond the harmonic approximation. The associated transfer times, *i.e.*, the inverse of the transfer rates, are reported in Fig. 7 as a function of an effective potential bias

$$U_{\text{eff}} = \left(\frac{I\tau_0}{e\rho_0(U)}\right)|U| \tag{31}$$

The latter contains all information about the STM current and applied potential bias, as well as about the substrate density of states topology around the Fermi energy.[127] These are important factors influencing the strength of the electron–hole pair coupling induced by the STM tip, which determines uniquely the skewed thermal distribution appearing in eqn (29). As for Xe/Ni(110), a power law dependence of the transfer times

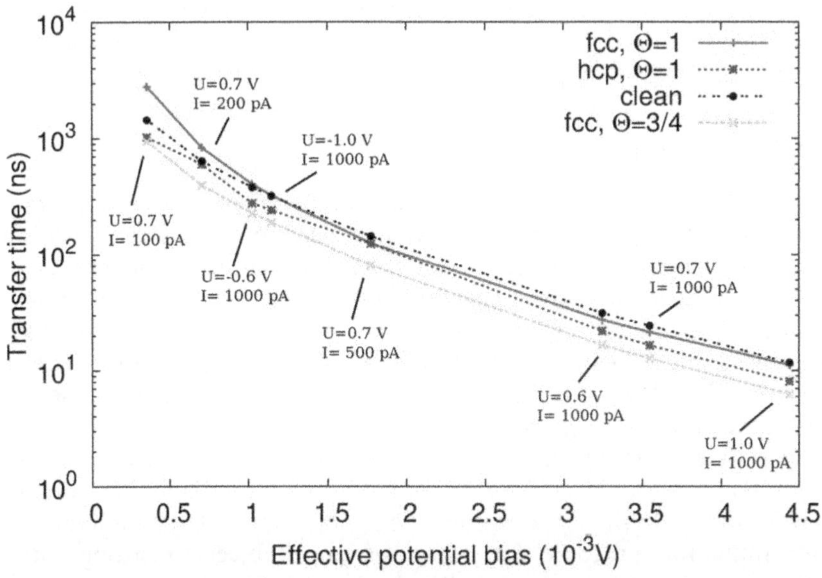

Fig. 7 Dependence of the transfer time on the effective potential bias for the migration of a hydrogen impurity in the vicinity of a palladium(111) surface. The transfer times exhibit a power law dependence on the strength of the non-adiabatic couplings, as measured by U_{eff}. Reproduced from ref. 127 with permission from the PCCP Owner Societies.

is observed as a function of U_{eff} and a linear dependence on the current, while the system appears largely insensitive to the degree of surface pre-coverage. At negative biases, the larger number of available states per palladium atom reduces the non-adiabatic excitation efficiency and, consequently, the rates are smaller than at positive biases.

Figure 8 shows the evolution of the hydrogen probability density as a function of time for a hydrogen atom initially placed in the second subsurface octahedral site, labelled O_{sb}. The other relevant sites are labelled O_{ss}, O_b, and S for the bulk-like state, the subsurface octahedral cavity, and the fcc surface site, respectively. The reaction coordinate is the minimal energy path connecting the different local minima, exhibiting a stepwise behavior for the resurfacing process. The four panels report the dynamics for different hydrogen pre-coverages, ranging from the clean surface to the fully covered case with two types of surface patterning, hcp and fcc. The influence of non-adiabatic couplings can be strongly felt throughout the simulations, as no nodal structure is ever observed in each of the local minima, meaning that the system remains locally in thermal equilibrium at all times ($T = 0$ K was chosen for simplicity). The non-adiabatic coupling to the impinging STM electrons, on the other hand, is felt by all systems strongly. Within a few tens of nanoseconds, the transfer dynamics is all but over, leading to the simple conclusion that the system is driven towards its global minimum at longer times. For all unsaturated surfaces (panels a–b), this happens to be the surface site, and the second subsurface octahedral minimum is preferred for pre-covered surfaces (panels c–d). The main conclusion is that, provided non-adiabatic couplings induced by incoming STM electrons can excite a fraction of the hydrogen population to energetically higher local vibrational states, the hydrogens will resurface and cluster at the metal interface. It appears to be the hydrogen atoms that causes the protuberances observed experimentally after STM excitation.

4 Conclusion

In this chapter, it was shown that non-adiabaticity plays an important role in many aspects of chemical dynamics at metallic surfaces. Within the weak coupling regime, it was demonstrated how the effects of electron–hole pairs creation/annihilation in the metallic environment surrounding an adsorbate can be included seamlessly in a purely quantum mechanical treatment of vibrational dynamics. The importance of surface-induced non-adiabaticity on the vibrational relaxation of selected adsorbates was investigated for hydrogen on palladium(111) and CO on copper(100), revealing that care must be taken to properly account for anharmonicity and intermode coupling to obtain a reliable microscopic description of dissipative dynamics. These can have a large influence on the simulation outcome for other systems subject to strong intramolecular vibrational energy redistribution and for reaction dynamics involving large amplitude motions.

The study of infrared laser excitation in D_2 at a ruthenium(0001) surface shed light on the importance of timescales for controlling efficiency

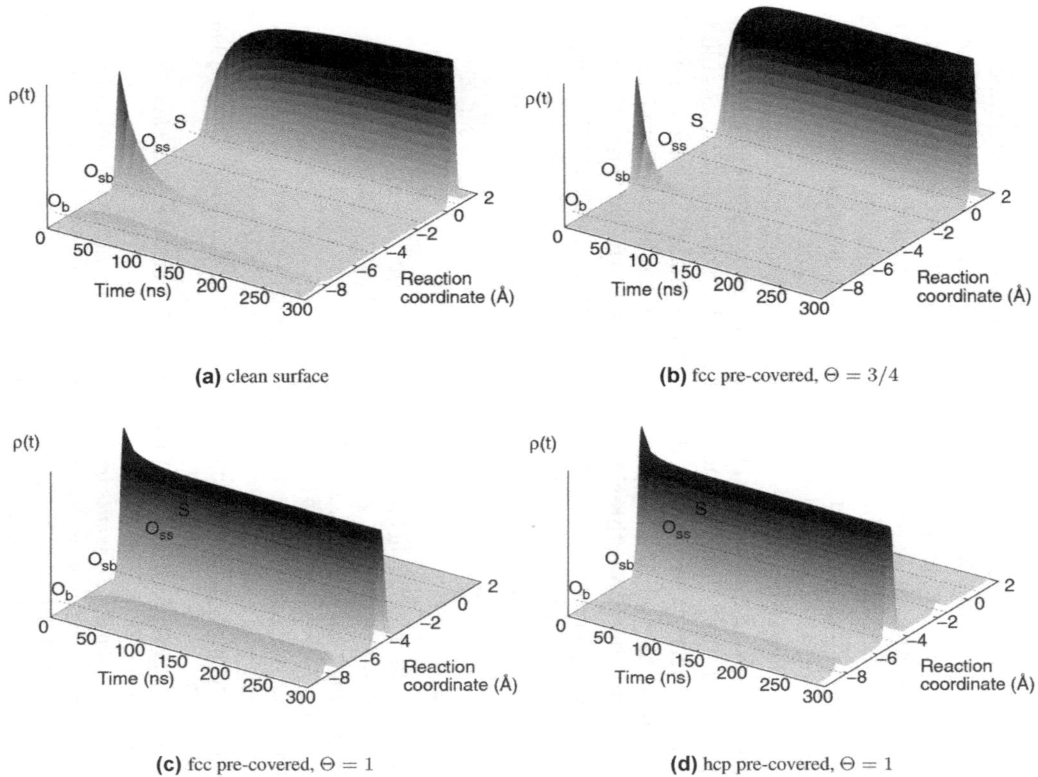

Fig. 8 Time evolution of the density for the STM excitation of a hydrogen impurity in the second subsurface octahedral of a palladium(111) surface at various coverages. The inelastic current contribution from the STM due to non-adiabatic coupling initially excites the adsorbates, which can subsequently migrate to other stable minima. Reproduced from ref. 127 with permission from the PCCP Owner Societies.

and selectivity. Pulses much shorter than the typical system lifetimes proved more efficient at depopulating the initial ground vibrational state, with great mode selectivity. A poor state selectivity was observed for short pulses due to stronger dynamical broadening, as it should be expected for similar systems where the vibrations do not depart strongly from harmonicity. Longer pulses revealed the nefarious effects of non-adiabatic coupling in the form of a loss of both efficiency and selectivity. Laser-induced excitation in the presence of fast dissipation, as is the case for adsorbates at metallic surfaces, can thus be simplified to finding the right balance between the relaxation and dynamical broadening timescales.

Finally, the inelastic contribution to the scanning tunnelling current was revisited from the perspective of an adsorbate, whose dynamics is subjected to electron–hole pair coupling with the impinging electrons. It was argued that, within the perturbative regime, the STM-induced transfer rates reduce to an enhancement factor for the non-adiabatic coupling leading to a skewed thermal distribution. The transfer rates follow a power law dependence with the applied potential bias strength, in agreement with experiment. The time-resolved dynamics is observed to happen on a much longer timescale than the non-adiabatic coupling rates, which merely ensure that the system remains in quasi-thermal distribution at all times. The atom transfer is solely mediated by the tunnelling and the above threshold dynamics of the vibrational states. The latter being computed as solution to the stationary Schrödinger equation in the electronic ground state, they contain implicitly all information about the non-local properties of the adsorbate vibrations.

References

1 M. Born and R. Oppenheimer, *Ann. Phys.*, 1927, **20**, 30.
2 H. Eyring and M. Polanyi, *Z. Phys. Chem., Abt. B*, 1927, **12**, 279.
3 J. C. Polanyi, *Science*, 1987, **236**, 680.
4 H. Eyring, *J. Chem. Phys.*, 1935, **3**, 107.
5 M. G. Evans and M. Polanyi, *Trans. Faraday Soc.*, 1935, **31**, 875.
6 M. G. Evans and M. Polanyi, *J. Phys. Chem.*, 1983, **87**, 2657.
7 D. G. Thrular, B. C. Garret and S. J. Klippenstein, *J. Phys. Chem.*, 1996, **100**, 12771.
8 X. Huang, L. Bytautas, J. M. Bowman and A. J. C. Varandas, *Adv. Phys. Chem.*, 2012, **679869**, 2012.
9 A. Nitzan, *Chemical Dynamics in Condensed Phases: Relaxation, Transfer, and Reactions in Condensed Molecular Systems*, Oxford University Press, 2006.
10 V. May and O. Kühn, *Charge and Energy Transfer Dynamics in Molecular Systems*, Wiley-VCH, Weinheim, 3rd edn, 2011.
11 H.-P. Breuer and F. Petruccione, *The Theory of Open Quantum Systems*, Oxford University Press, Oxford, 2006.
12 R. B. Murphy, D. M. Phillip and R. A. Friesner, *J. Comput. Chem.*, 2000, **21**, 1442.
13 R. A. Friesner and V. Guallar, *Annu. Rev. Phys. Chem.*, 2005, **56**, 389.
14 H. Lin and D. G. Truhlar, *Theor. Chem. Acc.*, 2007, **117**, 185.
15 H. M. Senn and W. Thiel, *Angew. Chem., Int. Ed.*, 2009, **48**, 1198.
16 R. G. Parr and W. Yang, *Density-Functional Theory of Atoms and Molecules*, Oxford University Press, USA, 1989.

17 T. Klüner, N. Govind, Y. A. Wang and E. A. Carter, *J. Chem. Phys.*, 2002, **116**, 42.
18 C. M. Goringe, D. R. Bowler and E. Hernández, *Rep. Prog. Phys.*, 1997, **60**, 1447.
19 M. Elstner, D. Porezag, G. Jungnickel, J. Elsner, M. Haugk, Th. Frauenheim, S. Suhai and G. Seifert, *Phys. Rev. B: Condens. Matter Mater. Phys.*, 1998, **58**, 7260.
20 M. Elstner, *Theor. Chem. Acc.*, 2006, **60**, 316.
21 G. Seifert, *J. Phys. Chem. A*, 2007, **111**, 5609.
22 M. Elstner, *J. Phys. Chem. A*, 2007, **111**, 5614.
23 Q. Cui, M. Gauss and M. Elstner, *J. Chem. Theory Comput.*, 2011, **7**, 931.
24 G. Seifert and J. O. Joswig, *Wiley Interdiscip. Rev.: Comput. Mol. Sci.*, 2012, **2**, 456.
25 G. J. Kroes, *Phys. Chem. Chem. Phys.*, 2012, **14**, 14966.
26 J. K. Nørskov, F. Abild-Pedersena, F. Studta and T. Bligaard, *Proc. Natl. Acad. Sci. U. S. A.*, 2011, **108**, 937.
27 J. C. Tully, *Annu. Rev. Phys. Chem.*, 2000, **51**, 153.
28 P. Saalfrank, *Chem. Rev.*, 2006, **106**, 4116.
29 P. Nieto, E. Pijper, D. Barredo, G. Laurent, R. A. Olsen, E.-J. Baerends, G.-J. Kroes and D. Farías, *Science*, 2006, **312**, 86–89.
30 A. M. Wodtke, D. Matsiev and D. J. Auerbach, *Prog. Surf. Sci.*, 2008, **83**, 167.
31 G.-J. Kroes, C. Diaz, E. Pijpera, R. A. Olsena and D. J. Auerbach, *Proc. Natl. Acad. Sci. U. S. A.*, 2010, **107**, 20881.
32 H. Arnolds, *Prog. Surf. Sci.*, 2011, **86**, 1.
33 P. Saalfrank, G. Füchsel, S. Monturet, J. C. Tremblay and T. Klamroth, in *Dynamics of Gas-Surface Interactions*, ed. R. Díez Muiño and H. F. Busnengo, Springer Series in Surface Sciences, Springer, Berlin Heidelberg, 2013, vol. 50, pp. 323–348.
34 M. Born and K. Huang, *Dynamical Theory of Crystal Lattices*, Oxford University Press, Oxford, 1954.
35 M. Persson and B. Hellsing, *Phys. Rev. Lett.*, 1982, **49**, 662.
36 B. Hellsing and M. Persson, *Phys. Scr.*, 1984, **29**, 360.
37 M. Head-Gordon and J. C. Tully, *J. Chem. Phys.*, 1992, **96**, 3939.
38 J. C. Tully, M. Gomez and M. Head-Gordon, *J. Vac. Sci. Technol., A*, 1993, **11**, 1914.
39 H. Winter, J. I. Juaristi, I. Nagy, A. Arnau and P. M. Echenique, *Phys. Rev. B: Condens. Matter Mater. Phys.*, 2003, **67**, 245401.
40 V. Krishna and J. C. Tully, *J. Chem. Phys.*, 2006, **125**, 054706.
41 E. V. Chulkov, A. G. Borisov, J. P. Gauyacq, D. Sánchez-Portal, V. M. Silkin, V. P. Zhukov and P. M. Echenique, *Chem. Rev.*, 2006, **106**, 4160.
42 J. I. Juaristi, M. Alducin, R. Díez Muiño, H. F. Busnengo and A. Salin, *Phys. Rev. Lett.*, 2008, **100**, 116102.
43 C. Frischkorn and M. Wolf, *Chem. Rev.*, 2006, **106**, 4207.
44 P. A. M. Dirac, *Proc. R. Soc. London, Ser. A*, 1927, **114**, 243.
45 E. Fermi, *Nuclear Physics*, University of Chicago Press, 1950.
46 S. Monturet and P. Saalfrank, *Phys. Rev. B: Condens. Matter Mater. Phys.*, 2010, **82**, 075404.
47 J. C. Tremblay, S. Monturet and P. Saalfrank, *Phys. Rev. B: Condens. Matter Mater. Phys.*, 2010, **81**, 125408.
48 J. C. Tremblay, S. Monturet and P. Saalfrank, *J. Phys. Chem. A*, 2011, **115**, 10698.
49 M. Morin, N. J. Levinos and A. L. Harris, *J. Chem. Phys.*, 1992, **96**, 3950.

50 A. Peremans, A. Tadjeddine and P. Guyot-Sionnest, *Chem. Phys. Lett.*, 1995, **247**, 243.
51 A. Peremans, A. Tadjeddine, W. Q. Zheng, A. LeRille, P. Guyot-Sionnest and P. A. Thiry, *Surf. Sci.*, 1996, **368**, 384.
52 C. Matranga and P. Guyot-Sionnest, *J. Chem. Phys.*, 2000, **112**, 7615.
53 C. Matranga, B. L. Wehrenberg and P. Guyot-Sionnest, *J. Phys. Chem. B*, 2002, **106**, 8172.
54 J. C. Tully, *J. Chem. Phys.*, 1990, **93**, 1061.
55 U. Müller and G. Stock, *J. Chem. Phys.*, 1997, **107**, 6230.
56 L. J. Butler, *Annu. Rev. Phys. Chem.*, 1998, **49**, 125.
57 E. G. d'Agliano, P. Kumar, W. Schaich and H. Suhl, *Phys. Rev. B: Solid State*, 1975, **11**, 2122.
58 A. Nourtier, *J. Phys.*, 1977, **38**, 479.
59 R. Festa and E. Galleani d'Agliano, *Phys. A*, 1978, **90**, 229.
60 K. Schönhammer and O. Gunnarsson, *Z. Phys. B: Condens. Matter*, 1980, **38**, 127.
61 P. M. Echenique, R. M. Nieminen and R. H. Ritchie, *Solid State Commun.*, 1981, **37**, 779.
62 M. J. Puska and R. M. Nieminen, *Phys. Rev. B: Condens. Matter Mater. Phys.*, 1983, **27**, 6121.
63 K. Schönhammer and O. Gunnarsson, *Phys. Rev. B: Condens. Matter Mater. Phys.*, 1983, **27**, 5113.
64 K. Schönhammer, O. Gunnarsson and L. Bönig, *Prog. Surf. Sci.*, 1987, **26**, 1.
65 J. R. Trail, M. C. Graham, D. M. Bird, M. Persson and S. Holloway, *Phys. Rev. Lett.*, 2002, **88**, 166802.
66 A. C. Luntz and M. Persson, *J. Chem. Phys.*, 2005, **123**, 074704.
67 M. S. Mizielinski, D. M. Bird, M. Persson and S. Holloway, *J. Chem. Phys.*, 2005, **122**, 084710.
68 D. M. Bird, M. S. Mizielinskia, M. Lindenblattb and E. Pehlkeb, *Surf. Sci.*, 2008, **602**, 1212.
69 M. Timmer and P. Kratzer, *Phys. Rev. B: Condens. Matter Mater. Phys.*, 2009, **79**, 165407.
70 M. Richter, A. Carmele, S. Butscher, N. Bücking, F. Milde, P. Kratzer, M. Scheffler and A. Knorr, *J. Appl. Phys.*, 2009, **105**, 122409.
71 J. Meyer and K. Reuter, *New J. Phys.*, 2011, **13**, 085010.
72 K. Blum, *Density Matrix Theory and Applications*, Plenum Press, New York, 1996.
73 S. Nakajima, *Prog. Theor. Phys.*, 1958, **20**, 948.
74 R. Zwanzig, *J. Chem. Phys.*, 1960, **33**, 1338.
75 C. Meier and D. J. Tannor, *J. Chem. Phys.*, 1999, **111**, 3365.
76 A. Ishizaki and Y. Tanimura, *J. Phys. Soc. Jpn.*, 2005, **74**, 3131.
77 K. H. Hughes, C. D. Christ and I. Burghardt, *J. Chem. Phys.*, 2009, **131**, 024109.
78 K. H. Hughes, C. D. Christ and I. Burghardt, *J. Chem. Phys.*, 2009, **131**, 024108.
79 V. Gorini, A. Kossakowski and E. C. G. Sudarshan, *J. Math. Phys.*, 1976, **17**, 821.
80 V. Gorini and A. Kossakowski, *J. Math. Phys.*, 1976, **17**, 1298.
81 G. Lindblad, *Commun. Math. Phys.*, 1976, **48**, 119.
82 M. J. Bramley and T. Carrington Jr., *J. Chem. Phys.*, 1993, **99**, 8519.
83 M. J. Bramley and N. C. Handy, *J. Chem. Phys.*, 1993, **98**, 1378.
84 M. J. Bramley, J. W. Tromp, T. Carrington Jr. and G. C. Corey, *J. Chem. Phys.*, 1994, **100**, 6175.
85 M. J. Bramley and T. Carrington Jr., *J. Chem. Phys.*, 1994, **101**, 8494.

86 X. Wang and T. Carrington Jr., *J. Chem. Phys.*, 2002, **117**, 6923.
87 X. Wang and T. Carrington Jr., *J. Chem. Phys.*, 2003, **119**, 101.
88 X. Wang and T. Carrington Jr., *Int. J. Quantum Chem.*, 2004, **99**, 556.
89 X. Wang and T. Carrington Jr., *J. Chem. Phys.*, 2004, **121**, 2937.
90 J. C. Tremblay and T. Carrington Jr., *J. Chem. Phys.*, 2006, **125**, 094311.
91 G. Avila and T. Carrington Jr., *J. Chem. Phys.*, 2009, **131**, 174103.
92 U. Manthe, H.-D. Meyer and L. S. Cederbaum, *J. Chem. Phys.*, 1992, **97**, 3199.
93 U. Manthe, H.-D. Meyer and L. S. Cederbaum, *J. Chem. Phys.*, 1992, **97**, 9062.
94 I. Burghardt, K. Giri and G. Worth, *J. Chem. Phys.*, 2008, **129**, 174104.
95 G. Worth, H.-D. Meyer, H. Koeppel, L. Cederbaum and I. Burghardt, *Int. Rev. Phys. Chem.*, 2008, **27**, 569.
96 A. G. Redfield, *IBM J. Res. Dev.*, 1957, **1**, 19.
97 L. Uranga-Pi na and J. C. Tremblay, *J. Chem. Phys.*, 2014, **141**, 074703.
98 L. Uranga-Pi na and J. C. Tremblay, *Rev. Cubana Fis.*, 2014, **31**, 41.
99 J. C. Tremblay and P. Saalfrank, *J. Chem. Phys.*, 2009, **131**, 084716.
100 O. M. Løvvik and R. A. Olsen, *Phys. Rev. B: Condens. Matter Mater. Phys.*, 1998, **58**, 10890.
101 N. Ozawa, T. A. Roman, H. Nakanishi, H. Kasai, N. B. Arboleda Jr. and W. A. Diño, *J. Appl. Phys.*, 2007, **101**, 123530.
102 R. Marquardt, F. Cuvelier, R. A. Olsen, E. J. Baerends, J. C. Tremblay and P. Saalfrank, *J. Chem. Phys.*, 2010, **132**, 074108.
103 J. C. Tremblay and T. Carrington Jr., *J. Chem. Phys.*, 2004, **121**, 11535.
104 J. C. Tremblay, G. Füchsel and P. Saalfrank, *Phys. Rev. B: Condens. Matter Mater. Phys.*, 2012, **86**, 045438.
105 G. Füchsel, J. C. Tremblay, T. Klamroth and P. Saalfrank, *Isr. J. Chem.*, 2012, **52**, 438.
106 *Laser Spectroscopy and Photochemistry on Metal Surfaces, Part 1*, *Advanced Series in Physical Chemistry*, ed. H.-L. Dai and W. Ho, World Scientific Pub Co Inc, 1995, vol. 5.
107 B. W. Shore, *Manipulating Quantum Structures Using Laser Pulses*, Cambridge University Press, Cambridge, UK, 2011.
108 M. Shapiro and P. Brumer, *Principles of the Quantum Control of Molecular Processes*, Wiley, Hoboken, 2003.
109 S. Sharma, H. Singh and G. Balint-Kurti, *J. Chem. Phys.*, 2010, **132**, 064108.
110 R. Kosloff, S. A. Rice, P. Gaspard, S. Tersigni and D. J. Tannor, *Chem. Phys.*, 1989, **139**, 201.
111 Y. Ohtsuki, W. Zu and H. Rabitz, *J. Chem. Phys.*, 1999, **110**, 9825.
112 Y. Ohtsuki, K. Nakagami, W. Zhu and H. Rabitz, *Chem. Phys.*, 2003, **287**, 197.
113 *Scanning Probe Microscopy and Spectroscopy: Theory, Techniques, and Applications*, ed. D. A. Bonnel, Wiley, New York, 2001.
114 C. J. Chenl, *Introduction to Scanning Tunneling Microscopy*, Oxford University Press, New York, 1993.
115 B. N. J. Persson and J. E. Demuth, *Solid State Commun.*, 1986, **57**, 769.
116 S. Gao, M. Persson and I. Lundqvist, *Solid State Commun.*, 1992, **84**, 271.
117 H. OuYang, B. Källbering and R. A. Marcus, *J. Chem. Phys.*, 1993, **98**, 7565.
118 S. Gao, M. Persson and I. Lundqvist, *J. Electron Spectrosc. Relat. Phenom.*, 1993, **64**, 665.
119 S. Gao, *Phys. Rev. B: Condens. Matter Mater. Phys.*, 1997, **55**, 1876.
120 S. Gao, M. Persson and I. Lundqvist, *Phys. Rev. B: Condens. Matter Mater. Phys.*, 1997, **55**, 4825.
121 J. C. Tremblay, *J. Chem. Phys.*, 2013, **138**, 244106.

122 E. C. H. Sykes, L. C. Fernández-Torres, S. U. Nanayakkara, B. A. Mantooth, R. M. Nevin and P. S. Weiss, *Proc. Natl. Acad. Sci. U. S. A.*, 2005, **102**, 17907.
123 D. M. Eigler, C. P. Lutz and W. E. Rudge, *Nature*, 1991, **352**, 600.
124 T. Mitsui, M. K. Rose, E. Fomin, D. F. Ogletree and M. Salmeron, *Nature*, 2003, **422**, 705.
125 M. Blanco-Rey, M. Alducin, J. I. Juaristi and P. L. de Andres, *Phys. Rev. Lett.*, 2012, **108**, 115902.
126 M. Blanco-Rey and J. C. Tremblay, *J. Chem. Phys.*, 2015, **17**, 13973.
127 J. C. Tremblay and M. Blanco-Rey, *Phys. Chem. Chem. Phys.*, 2015, **142**, 154704.

Theoretical studies of supercapacitors

Mathieu Salanne[a,b]

DOI: 10.1039/9781782622703-00119

1 Introduction

Although electricity can be produced in multiple ways, it remains very difficult to store it efficiently. At the moment most of our storage capacity relies on pumped-storage hydroelectricity, in which the energy is stored in the form of gravitational potential energy of water. A huge amount of efforts are currently devoted to the development of electrochemical storage devices. There are two important characteristics which have to be optimized: the specific energy and specific power. In practice, when used in an electric vehicle, the former measures how far one can go on a single charge, while the latter shows how fast one can go. The two main families of electrochemical energy storage devices are supercapacitors and batteries. Their respective ranges of performance are shown on Fig. 1. It appears clearly that from batteries one can obtain much larger specific energies, with the best performances corresponding to approximately 500 Wh kg^{-1} for the most recent Li-ion technologies. In comparison, supercapacitors can only reach 5 Wh kg^{-1}, but they show much better specific powers, *i.e.* 10 kW kg^{-1}. This difference is better understood by comparing the characteristic charge/discharge times, which are given by the diagonal lines on Fig. 1: Typical values of 1 h and 1 s are respectively obtained for batteries and supercapacitors. They have therefore different applications, but they may also complement one another in a system which needs both high power and energy density. While batteries are more and more used as the primary energy storage for electric cars, supercapacitors are used in their stop & start systems. Nevertheless supercapacitors are also used alone in new applications, such as trolley buses. Indeed, such vehicles can be charged when they board/unboard passengers, with a typical charging time of 1 minute, and the needed autonomy is the distance between two bus stops, *i.e.* less than 1 km.

The origin of the difference in the performances between batteries and supercapacitors lies in their energy storage mechanisms.[1,2] In batteries, it occurs via chemical redox reactions occuring in the bulk of a material. The power limitations are due to the slow reaction kinetics and mass transport. In particular, Li^+ ions need to migrate into solid-state electrodes with relatively low ionic conductivities. On the contrary, supercapacitors are based on the physical adsorption of the ions from the electrolyte on the surface of the electrodes. The lower specific energy comes from the fact that only the surface of the material is used, but

[a]*Sorbonne Universités, UPMC Univ Paris 06, CNRS, UMR 8234, PHENIX, F-75005 Paris, France. E-mail: mathieu.salanne@upmc.fr*

[b]*Maison de la Simulation, USR 3441, CEA – CNRS – INRIA – Université Paris-Sud – Université de Versailles, F-91191 Gif-sur-Yvette, France*

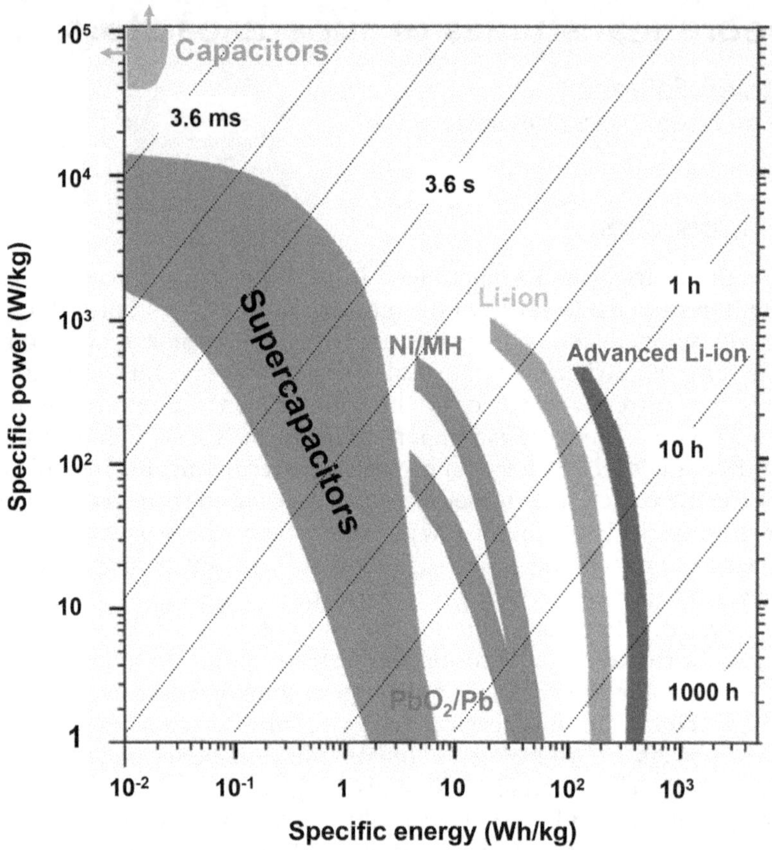

Fig. 1 Ragone plot, showing the specific power *versus* the specific energy stored in a series of electrical energy storage devices. The diagonal lines are the characteristic time constants of the devices.

there is no specific power limitations except the one due to the ionic conductivity of the electrolyte. Another consequence of the different mechanisms is the much better long-term stability of supercapacitors, which have lifetimes of more than one million of charge/discharge cycles while batteries generally sustain one thousand of cycles only.

Here the focus will be on the simulation studies of supercapacitors. These devices are also called electrical double-layer capacitors (EDLCs) due to the formation of a charged layer of electrolyte on the electrode surface, even if other mechanisms may occur, *i.e.* redox reactions at the surface in the case of pseudocapacitors.[3–7] In conventional EDLCs, the electrodes are usually made of porous carbon materials, due to their low price, good stability and good electronic conductivity.[8,9] Many options are available for both the carbon[10–16] and the electrolyte. Aqueous solvents mixed with H_2SO_4 or K_2SO_4 salts[17,18] or organic liquids (acetonitrile[19] or propylene carbonate[20]) with organic ions such as tetraethylammonium tetrafluoroborate salt are generally used, but in the recent years pure ionic liquids have appeared as a promising alternative.[21–25] For many years, the improvements of the devices have mainly focused on the increase of the specific surface of the electrodes, which can only be done by using porous materials. It was implicitly assumed that the energy storage capacity was directly proportional to the surface accessible to the ions. The latter were supposed to remain solvated inside the electrodes, so that the target range of size for the pores was between 2 and 10 nanometres, *i.e.* the mesopore range.

Were this assumption be fulfilled, there would indeed be very few ions adsorbed inside nanoporous carbons, which have a pore size distribution ranging over values smaller than 2 nanometres. The top panels of Fig. 2 schematize the corresponding structure of the electrolyte inside the pores. Nevertheless, starting from 2006, many studies have shown that the specific capacitance, which measures the amount of energy stored by surface area, could be strongly increased when using nanoporous carbons, with typical pore size distributions centered below 1 nm, *i.e.* equalling the size of the ions.[26,27] These results have been interpreted by a mechanism based on the desolvation of the ions, which allows them to enter the pores of smaller sizes. Since then, additional studies performed

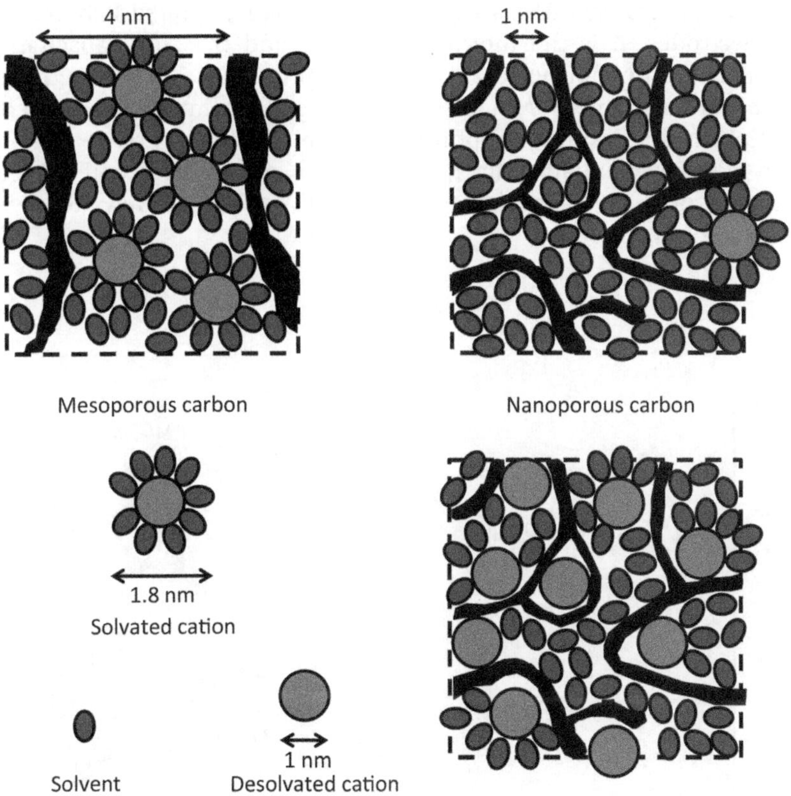

Fig. 2 Schemes illustrating possible adsorption mechanisms for the ions inside mesoporous carbons (left) and nanoporous ones (right). For the latter, the top scheme shows a situation in which the ions would always remain solvated, they would not be able to enter the smaller pores. The bottom scheme shows the adsorption of (partially) desolvated ions. In that case the amount of electricity stored at the surface of the carbon is strongly enhanced.

under many different conditions have confirmed this result.[28–30] Typical results obtained with an ionic liquid electrolyte and nanoporous carbon electrodes with varying average pore size are shown on Fig. 3. The bottom panel of Fig. 2 shows a scheme of the adsorption of the ions when their partial desolvation is allowed.

These results have brought to light the importance of the interactions between the ions, the solvent and the carbon electrode at the molecular scale. In particular, theoretical studies of supercapacitors had so far been limited to mesoscale approaches. The quantities of interest for determining the maximal energy $E_{\max}$ and power $P_{\max}$ which can be delivered by a supercapacitor can be extracted from the following relations:

$$E_{\max} = \frac{1}{2}C(\Delta\Psi)^2 \tag{1}$$

$$P_{\max} = \frac{1}{4}\frac{(\Delta\Psi)^2}{R} \tag{2}$$

In these expressions, $\Delta\Psi$ is the operating voltage, applied between the two electrodes, R is the internal resistance and C is the capacitance. Although the two former are mostly *bulk* properties of the electrolyte, which are easily measured, the capacitance is an interfacial property. Most of the studies aim at optimizing the electrolyte/electrode combination in order to enhance this quantity. For a given interface, if the total charges stored on each side are $+Q$ and $-Q$ respectively, then the capacitance is

$$C = \frac{Q}{\Delta\Psi} \tag{3}$$

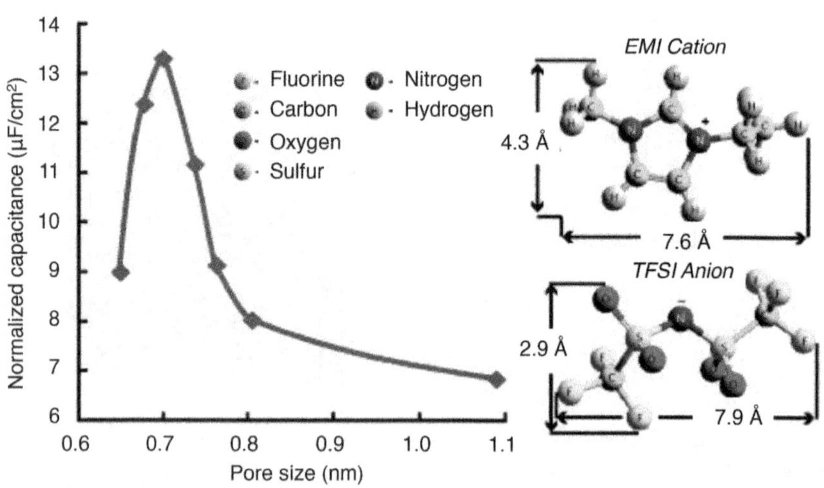

Fig. 3 "Anomalous" increase of the capacitance in nanoporous carbons with an ethyl-methylimidazolium-bis(trifluoromethane-sulfonyl)imide ionic liquid electrolyte. Reprinted with permission from ref. 28. Copyright 2008 American Chemical Society.

Although in the case of a conventional capacitor, which is composed of a dielectric fluid (with a dielectric constant ε_r) separating two parallel-plate electrodes of surface A, it is given by

$$C = \varepsilon_r \varepsilon_0 \frac{A}{d} \tag{4}$$

where ε_0 is the vacuum permittivity and d the separation between the plates, this expression becomes unpractical for EDLCs. Indeed, these devices are characterized by two electrode/electrolyte interfaces, and for each of them the charges on one side of the interface is an adsorbed layer of ions. Under these conditions, ε_r, A and d are ill-defined quantities. In addition, the porous electrodes can no longer be considered as infinite plates. However this expression remains used by fitting or estimating the various parameters, and it has also been extended to more complex electrode geometries.[31,32] Nevertheless this approach remains empirical, and it is based on a mesoscale picture of the interface.

In order to allow for a fundamental understanding of the mechanisms at play during the functioning of nanoporous carbon-based supercapacitors, it is necessary to model them at the molecular scale. A large array of techniques, ranging from mean-field theory to classical density functional theory, molecular dynamics or Monte-Carlo approaches, have been used in the recent years. Nevertheless, most of the works are performed by treating the electrodes as walls with constant charge (*i.e.* each atom or surface element wears an identical charge). Realistic electrodes have metallic electronic properties, *i.e.* the local charge should fluctuate in the presence of the solvent, and it is the electric potential which is constant.[33] Here the focus will be on methods which account for the latter effects; they are described in the next section. The main results obtained from these works, which concern both equilibrium and dynamic properties of the systems, will then be summarized – the methods with which they have been obtained will not necessary be detailed because most of the simulations result agree qualitatively with each other. In parallel to this theoretical work, a great deal of experimental studies involving spectroscopic methods have also been performed to probe the adsorption of the ions inside the pores. Whenever it is possible, a comparison of the results from simulations and experiments will be provided.

2 Methods

2.1 Mean field theory

A mean field theory (MFT) has been derived by Kondrat and Kornyshev.[34] The system they consider consists in point charges of densities $\rho_\pm$ and valencies $Z_\pm$ confined in a single slit-like pore, as shown on Fig. 4. For simplicity, the latter is made of two metallic plates infinitely extended in the lateral directions, and the boundary effects at the frontier between the pore and the bulk electrolyte are neglected. Their method consists in minimizing the free energy of the system with respect to the ion densities. The total charge Q accumulated over a surface A in the pore can

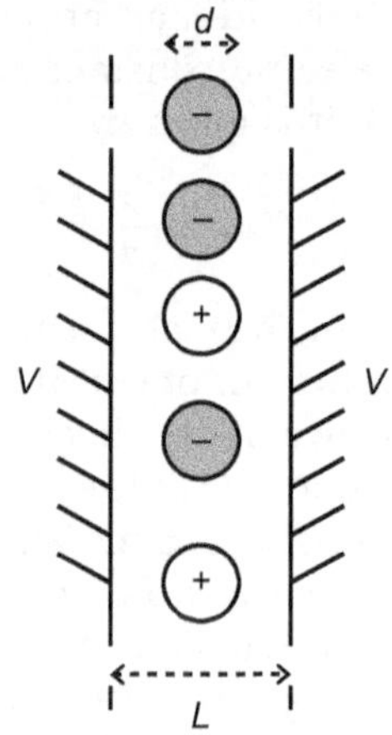

Fig. 4 Scheme of a typical system simulated using MFT. It shows the cross-section of a slit-like pore. The walls are held at a voltage V (with respect to a reference electrode located in the bulk electrolyte), the distance between them is noted L. d is the diameter of the ions.

then be calculated for each voltage V. V is fixed with respect to a reference electrode located in the bulk electrolyte, outside the pore (so that $V^+ - V^- = \Delta\Psi$). From this data it is straightforward to extract the capacitance using eqn (3), as well as the differential capacitance C_{diff}, which is defined per surface area as

$$C_{\text{diff}} = \frac{1}{A}\frac{\partial Q}{\partial V} \tag{5}$$

The free energy (which is calculated per surface area) is given by

$$F = U + E_S + \sum_{\alpha=\pm} \mu_\alpha \rho_\alpha - TS; \tag{6}$$

it contains four contributions. The first one is the electrostatic part of the internal energy. The main originality of this MFT is that this term differs from the usual Coulomb potential in order to account for the screening of the interactions due to the presence of the metallic plates,[34] separated by a distance L. This term is then given by

$$\beta U(\rho, c) = \frac{\beta eVc}{4\pi\varepsilon_0} + 4c^2(\pi\rho)^{-1/2} L_B \sum_{m=1}^{\infty} \frac{\sin^2(\pi m/2)}{m} \times K_1\left(\pi m(\pi\rho)^{-1/2}/L\right) \tag{7}$$

where $\beta = 1/k_B T$, with k_B being the Boltzmann constant and T the temperature and where $L_B = \beta e^2/(4\pi\varepsilon_0\varepsilon_r)$ is the Bjerrum length associated to the ionic liquid. c and ρ are the charge density and total density of the ions in the pore. $K_1(x)$ is the modified Bessel function of the second kind of order 1. This screening may strongly reduce the energy penalty for packing ions of similar charges in a pore, as well as the affinity between cations and anions. In such cases a strong impact is expected on the capacitance.

The second term is the contribution of the ion-pore interaction to the internal energy; it contains a "resolvation" free energy δE_α associated to the changes of the solvation shell when an ion (of type α) enters into the pore and the change in the self-energy of the ions due to confinement:

$$\beta E_S(\rho_\pm) = \sum_{\alpha=\pm} \left(\beta\delta E_\alpha - L_B/Lf(1/2)Z_\alpha^2\right)\rho_\alpha \tag{8}$$

where

$$f(z) = \int_0^\infty \left(\frac{1}{2} - \frac{\sinh(Q(1-z))\sinh(Qz)}{\sinh(Q)}\right) \mathrm{d}Q \tag{9}$$

Finally, $\mu_\pm$ are the chemical potentials of the cations and anions, which are set to the values of the "bulk" ionic liquid, and S is the entropy of the system, given by

$$S = -k_\mathrm{B}\left[\sum_{\alpha=\pm} \rho_\alpha \ln\left(\frac{\rho_\alpha v_0}{L}\right) + \frac{L}{v_0}\left(1 - \frac{\rho v_0}{L}\right) \times \ln\left(1 - \frac{\rho v_0}{L}\right)\right] \tag{10}$$

It is also possible to study the dynamics of charging of supercapacitors using the MFT,[35] starting from a set of fixed initial conditions. In practice, it is assumed that the evolution of the ion density is governed by the continuity equation

$$\frac{\partial \rho_\pm}{\partial t} = -\nabla\left(-D\rho_\pm \nabla \frac{\beta\delta F}{\delta\rho_\pm}\right) \tag{11}$$

where the term inside the parenthesis is the current of $\pm$ ions, D is the "diffusion constant" setting up the time scale, $\tau = d^2/D$ where d is the ion diameter (which is taken the same for both cations and anions).

2.2 Monte-Carlo simulations

In Monte-Carlo (MC) simulations, the ions are now taken into account explicitly. Two main models have been used to simulate supercapacitors with slit-pore type electrodes using this technique. The first one was developed by Kiyohara *et al.*[36–40] Their approach consists in sampling the constant-voltage grand-canonical ensemble for a simulation cell consisting of the two electrodes, represented by infinite plates, and an electrolyte region. There are three possible moves: translation of ions, charge transfer between the anode and the cathode, and insertion/deletion of ions. The probability of acceptance for a move is written as

$$f(\mathbf{r}, \sigma^+) \propto \frac{\gamma^N}{N_+!N_-!} \exp\left[-\frac{U(\mathbf{r}, \boldsymbol{\sigma}) - A\Delta\Psi\sigma_+}{T}\right] \tag{12}$$

where U is the potential energy of the system, which depends on the positions of all the ions of the systems contained in the vector $\mathbf{r}$ and the surface charges of the four planes (noted S_i, $i \in [1,4]$) forming the electrodes, which are contained in the vector $\boldsymbol{\sigma}$. σ_+ is the total charge on the positive electrode, *i.e.* the sum of the surface charge of the two corresponding planes. A, γ, N_+ and N_- are the surface of the electrode, the activity of a pair of ions and the number of cations and anions, respectively.

The potential energy of the system is a sum of three terms, corresponding to the ion–ion U_{II}, ion–surface U_{IS} and surface–surface U_{SS} interactions. The primitive model is used for representing the ions. The ion–ion term is summed over all the pairs of atoms:

$$U_{\mathrm{II}}(\mathbf{r}) = \sum_{i<j} u_{\mathrm{I}_i\mathrm{I}_j}(r_{ij}) \tag{13}$$

where r_{ij} is the distance between the two atoms of the pair and

$$u_{\mathrm{I}_i\mathrm{I}_j}(r_{ij}) = \frac{Z_i Z_j \mathrm{e}^2}{4\pi\varepsilon_0\varepsilon_r r_{ij}} \quad \left(\text{if } r_{ij} > \frac{d_i + d_j}{2}\right) \tag{14}$$

$$= \infty \quad \text{(otherwise)} \tag{15}$$

Z_i is the valence of ion I_i (+1 for cations and −1 for anions) and d_i is its mean diameter. The dielectric constant accounts for the presence of the solvent and the potential is infinite at short distances due to the overlap repulsion effects. In practice, the Leckner–Sperb method[41,42] is used in order to take into account the long range nature of the interaction.

The ion-surface term is

$$U_{\mathrm{IS}}(\mathbf{r}, \boldsymbol{\sigma}) = \sum_i \sum_j u_{\mathrm{I}_i\mathrm{S}_j}(z_i, \sigma_j), \tag{16}$$

it depends on the charge of the plates σ_i and the coordinate z_i of each ion (the plates are taken perpendicular to the z direction of the simulation cell by convention, their coordinate is written z_{S_i}), and we have

$$u_{\mathrm{I}_i\mathrm{S}_j}(z_i, \sigma_j) = -\frac{\sigma_j Z_i \mathrm{e}}{2\varepsilon_0\varepsilon_r}|z_i - z_{\mathrm{S}_j}| + \text{constant} \quad \left(\text{if } |z_i - z_{\mathrm{S}_j}| > d_i/2\right) \tag{17}$$

$$= \infty \quad \text{(otherwise)} \tag{18}$$

Finally the surface–surface term is

$$U_{\mathrm{SS}}(\boldsymbol{\sigma}) = \sum_{i<j} u_{\mathrm{S}_i\mathrm{S}_j}(\sigma_i, \sigma_j), \tag{19}$$

where

$$u_{\mathrm{S}_i\mathrm{S}_j}(\sigma_i, \sigma_j) = -\frac{\sigma_i\sigma_j}{2\varepsilon_0\varepsilon_r}|z_{\mathrm{S}_i} - z_{\mathrm{S}_j}| + \text{constant} \tag{20}$$

The constant applied voltage condition is enforced through eqn (12), which allows the surface charges of the plate to fluctuate. Nevertheless, the corresponding charge is distributed uniformly over the whole plate, which is a simplified representation of realistic systems. Recently, Punnathanam has extended Kiyohara *et al.*'s method for simulating single electrode systems.[43]

A different grand canonical MC approach was proposed by Kondrat and Kornyshev to model ionic liquids in a slit-like metallic nanopore.[44] Their approach is similar in spirit to their MFT approach, which consists in modifying the electrostatic interactions between pairs of ions by introducing an effective screening by the metallic wall. They also simulate a single-pore, and the allowed MC moves are: translation, insertion/deletion, ion/counter-ion switch and molecule-type swap. The potential energy is given by a sum of ion–ion interactions only. In the case of slit pores, the electrostatic interaction between two ions of respective positions z_i and z_j, charges q_i (which equals Z_ie) and q_j, separated by a lateral distance R, is then given by

$$U_{ij}^{c} = \frac{q_i q_j}{\pi \varepsilon_0 \varepsilon_r L} \sum_{n=1}^{\infty} \sin(\pi n z_i/L) \sin(\pi n z_j/L) K_0(\pi n R/L) \quad \left(\text{if } r_{ij} > \frac{d_i + d_j}{2}\right) \tag{21}$$

$$= \infty \quad (\text{otherwise}) \tag{22}$$

where ε_r is the dielectric constant of the medium between the plates, L is the pore size and $K_0(x)$ is the zero order modified Bessel function of the second kind. This expression was extended by Rochester *et al.* to allow finite electric field penetration into the pore walls.[45]

In addition, they account for the work needed to create a charge in the field created by itself. This self-energy term is

$$U_i^s(z_i) = -\frac{q_i^2}{4\pi \varepsilon_0 \varepsilon_r L} f(z_i/L) \tag{23}$$

where $f(x)$ is the function defined in eqn (9).

In this method the applied voltage V is introduced by adding a shift in the chemical potential of the ions. Note that this approach is limited to pore geometries for which the screening function can be evaluated analytically.

2.3 Molecular dynamics

In molecular dynamics (MD) simulations, the sampling of the thermodynamic ensemble is performed in a different way.[46] In this method, the trajectory of the molecules is gathered by solving iteratively the Newton's equation of motion. The configurations are therefore produced according to appropriate Boltzmann probabilities.

In MD all the molecules are accounted explicitly, even the solvent if needed. It has therefore been the method of choice in recent years for studying the electric double layer forming at the interface between ionic liquids and metallic electrodes.[47–64] The interaction potential is generally composed of bonded and non-bonded terms.[65] When all the atoms of the molecules are included in the simulations, it takes the form

$$U_{\text{total}} = U_{\text{bonds}} + U_{\text{angles}} + U_{\text{dihedrals}} + U_{\text{non-bonded}} \tag{24}$$

where the bonded terms U_{bonds}, U_{angles} and $U_{\text{dihedrals}}$ take the following forms:

$$U_{\text{bonds}} = \sum_{ij}^{\text{bonds}} \frac{k_{ij}^r}{2} \left(r_{ij} - r_{ij}^0\right)^2 \tag{25}$$

$$U_{\text{angles}} = \sum_{ijk}^{\text{angles}} \frac{k_{ijk}^{\theta}}{2} \left(\theta_{ijk} - \theta_{ijk}^{0}\right)^2 \tag{26}$$

$$U_{\text{dihedrals}} = \sum_{ijkl}^{\text{dihedrals}} \sum_{m=1}^{4} \frac{V_{ijkl}^{m}}{2} \left[1 + (-1)^m \cos m\phi_{ijkl}\right] \tag{27}$$

The sums respectively run over the full set of instantaneous bonds r_{ij}, angles θ_{ijk} and dihedrals ϕ_{ijkl} which must be defined prior to the simulations (*i.e.* the bonds are not allowed to break or form during the simulation). The quantities k_{ij}^{r}, r_{ij}^{0}, k_{ijk}^{θ}, θ_{ijk}^{0} and V_{ijkl}^{m} are parameters of the force field.

The non-bonded term is in principle composed of three contributions

$$U_{\text{non-bonded}} = U_{\text{repulsion}} + U_{\text{dispersion}} + U_{\text{electrostatic}} \tag{28}$$

but in most of the generic force fields the repulsion and dispersion terms are gathered into a Lennard-Jones potential:

$$\begin{aligned} U_{\text{Lennard-Jones}} &= U_{\text{repulsion}} + U_{\text{dispersion}} \\ &= \sum_{i} \sum_{j>i} 4\varepsilon_{ij} \left[\left(\frac{\sigma_{ij}}{r_{ij}}\right)^{12} - \left(\frac{\sigma_{ij}}{r_{ij}}\right)^{6} \right] \end{aligned} \tag{29}$$

Finally, the electrostatic interactions are calculated with the usual Coulomb potential:

$$U_{\text{electrostatic}} = U_{\text{Coulomb}} = \frac{1}{4\pi\varepsilon_0} \sum_{i} \sum_{j>i} \frac{q_i q_j}{r_{ij}} \tag{30}$$

Each atom type i is therefore characterized by three parameters only, ε_i, σ_i and q_i; the parameters of the Lennard-Jones potential are then derived using the Lorentz-Berthelot mixing rules:

$$\varepsilon_{ij} = \sqrt{\varepsilon_i \varepsilon_j} \tag{31}$$

$$\sigma_{ij} = \frac{1}{2}(\sigma_i + \sigma_j) \tag{32}$$

Nevertheless, supercapacitors necessitate very large simulation cells with explicit electrodes. This limits the number and the sizes of the systems that can be studied. A method which is often used to partially overcome this difficulty is to use coarse-grained force fields, *i.e.* to group several atoms together into a common interacting site, called grain.[66–69] In the following we will label this method as CGMD. The possibilities to describe a molecule using a coarse-grained approach are very large. Prior to the choice of the functional form for the non-bonded terms (and to the determination of the corresponding parameters), it is necessary to establish which atoms will be put together in a grain and whether the "bonds" should be held rigid or treated with bonded terms similar to

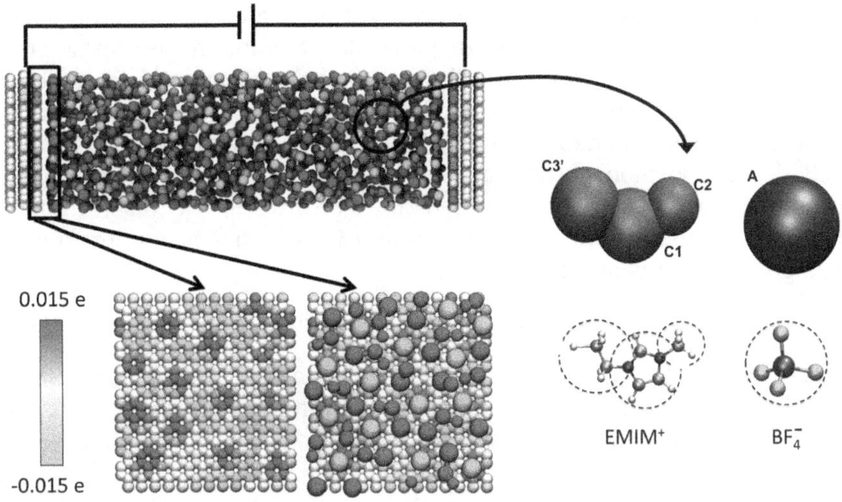

Fig. 5 The top left snapshot shows a typical supercapacitor simulation cell. In this example the electrolyte is a EMIM-BF_4 ionic liquid (red: $EMIM^+$, green: BF_4^-), which is represented using a coarse-grained model as illustrated on the right panel. The electrodes, which are made of graphite, are held at constant potential by allowing the charges of the carbon atoms to fluctuate. An illustration of the instantaneous charges is shown on the bottom left snapshot, where green and red colors are respectively employed for negatively and positively charged carbon atoms.

those of eqn (25)–(27). The right panel of Fig. 5 shows an example for the definition of grains for the 1-ethyl-3-methyl imidazolium tetrafluoroborate (EMIM-BF_4) ionic liquid electrolyte. The anion is represented by a single grain, while three grains corresponding to the imidazolium ring and the two alkyl chains are used for the cation. Similarly to the case of molecular simulations with explicit atoms, the non-bonded interaction are calculated using Lennard-Jones (eqn (29)) and Coulomb (eqn (30)) potentials.

In the special case of supercapacitors, the most difficult part is the treatment of the electrostatic problem, *i.e.* the presence of the charged and conducting electrode, in the effective interactions. Contrarily to the general case, where the charge on each atom is fixed at the beginning of the simulation and does not change after, the electrodes may be treated as ideal conductors. In macroscopic theory the charge $\rho_{\text{ind}}(\mathbf{r})$ induced on the electrode by the applied potential and the charges of the ions in the electrolyte is obtained from the Poisson equation under the condition that, for $\mathbf{r}$ inside the metal,

$$\Psi(\mathbf{r}) = \int \mathrm{d}\mathbf{r}' \frac{\rho(\mathbf{r}')}{|\mathbf{r}' - \mathbf{r}|} = \Psi_0 \tag{33}$$

where $\rho(\mathbf{r}) = \rho_{\text{ind}}(\mathbf{r}) + \rho_{\text{electrolyte}}(\mathbf{r})$ and $\rho_{\text{electrolyte}}(\mathbf{r})$ is the charge density due to the electrolyte (represented by a set of point charges located on the atomic sites). Ψ_0 is the applied potential, which should be uniform inside a conducting electrode. To solve this equation self-consistently, which must be done "on-the-fly" as the ions move, it is necessary to include specific developments in the simulation codes.

Several methods have been developed in order to perform such *constant potential* simulations. In particular, Siepmann and Sprik proposed a model for studying the adsorption of water molecules at a metal surface and on the tip of a model scanning tunneling microscope probe,[70] which was adapted by Reed *et al.* to the case of electrochemical cells.[33] In this model, the electrode consists of atoms which can be arranged with a crystalline order or in a disordered way; each atom j carries a Gaussian charge distribution $\rho_j(\mathbf{r})$ which has an integrated charge of q_j and is of fixed width η:

$$\rho_j(\mathbf{r}) = q_j A \exp(-|\mathbf{r} - \mathbf{r}_j|^2 \eta^2), \tag{34}$$

where $A = \eta^3 \pi^{3/2}$ is a normalization constant. In a supercapacitor, the full system consists of two electrodes, which are infinite in the x and y direction, enclosing a set of melt ions. The electrostatic energy, given by,

$$U_{\text{electrostatic}} = \frac{1}{2} \iint \mathrm{d}\mathbf{r}\mathrm{d}\mathbf{r}' \frac{\rho(\mathbf{r})\rho(\mathbf{r}')}{|\mathbf{r}' - \mathbf{r}|}, \tag{35}$$

has to be expressed through a two-dimensional Ewald summation, for which the correct expressions are provided in ref. 71 (note that the use of three-dimensional Ewald summation may lead to artifacts, although no systematic comparison has been undertaken up to now). By combining eqn (33) and (35), we immediately see that the potential experienced by any charge is obtained from the partial derivative of this expression with respect to that charge

$$\Psi_j = \left[\frac{\partial U_{\text{electrostatic}}}{\partial q_j} \right]_{\{q_i\}_{i \neq j}}. \tag{36}$$

The value of the charge on each electrode atom at each time step in a MD procedure is obtained by requiring that the potential experienced by each charge j in that electrode be equal to the preset electrode potential value, Ψ_0, thus satisfying the constant potential condition. This condition is achieved by adding a constraint term to the energy,

$$U_{\text{constraint}} = -\sum_{j=1}^{M} \Psi_0 q_j \tag{37}$$

where M is the number of electrode atoms, and by minimizing the total potential energy with respect to all the variable charges simultaneously (note that the two electrodes will have different preset potential values Ψ_0^{left} and Ψ_0^{right}, which will be noted Ψ^+ and Ψ^- in the following).

A second approach has also been used in the ionic liquid/electrode context, in which the electrode surface is modeled as an equi-potential smooth surface with a net charge.[72] Here, the entire electrode surface is effectively modeled as a metal foil separating the ions and a set of fixed charges located inside the electrode. The constant electrical potential is enforced on numerical grid points lying over the electrode surface by solving an auxiliary Laplace equation, following a method proposed in another context by Raghunathan and Aluru.[73] This method differs from the Reed *et al.* approach by treating the surface as smooth and by replacing the self-consistent calculation of the charges by solving Poisson's equation on a grid.

More recently, other computationally efficient alternative method were proposed by Petersen *et al.*[74] and Arnold *et al.*[75] In the former, for example, the polarization of the electrode by the electrolyte is accounted for by introducing explicit image charges together with a fluctuating uniform electrode charge located on the electrode surface. The constant potential condition is then enforced by adding a constant uniform charge.[74] The main advantage is that it avoids the need of a self-consistent calculation at each time step, but it presents the drawback of being *a priori* limited to the case of electrodes with planar geometries. Another approach based on the extension of reactive force field to electrochemical systems has been proposed in another context,[76] but it was not applied yet to the study of supercapacitors.

3 Results

3.1 Simple electrodes

Most of the first simulation studies on EDLCs have been devoted to the understanding of systems consisting of room-temperature ionic liquid electrolytes in contact with electrodes that have simple geometries. Indeed, even in the case of a planar electrode, the structure adopted by concentrated electrolytes is far from the trivial picture provided by simple theories such as the one of Gouy and Chapman.[77] Even at null applied voltage, far from adopting a structure in which their densities varies monotonously with respect to the distance to the electrode, the ionic species form successive layers over a few nanometres. This is true not only for molten salts[78–80] or ionic liquids,[81–84] but also for electrolytes composed of ionic species dissolved in an organic solvent such as acetonitrile.[85] For the latter, typical adsorption profiles obtained using CGMD are shown on Fig. 6, together with some representative snapshots of the system. When a non-null voltage is applied, the layered structure persists but the relative concentrations of cations and anions in the layers change in order to allow the charging of the electrode. An important characteristic of the interface is that owing to the correlation between ions of opposite charge, the first adsorbed layer carries a larger charge than the electrode, which is counterbalanced in the second layer. This phenomenon extends up to several layers leading to an effect called overscreening,[86–88] which is more pronounced in concentrated electrolytes such as ionic liquids.

Many experimental studies have confirmed the existence of a layered structure at various interfaces involving ionic liquids, using a vast array of techniques such as high-energy X-ray reflectivity,[62,89–91] atomic force microscopy[92–94] or surface force apparatus.[95–98] The changes in the composition of the layers with voltage may be at the origin of a reported change of the lubricant properties of the liquid on a metal surface when the potential of the latter is modified.[99]

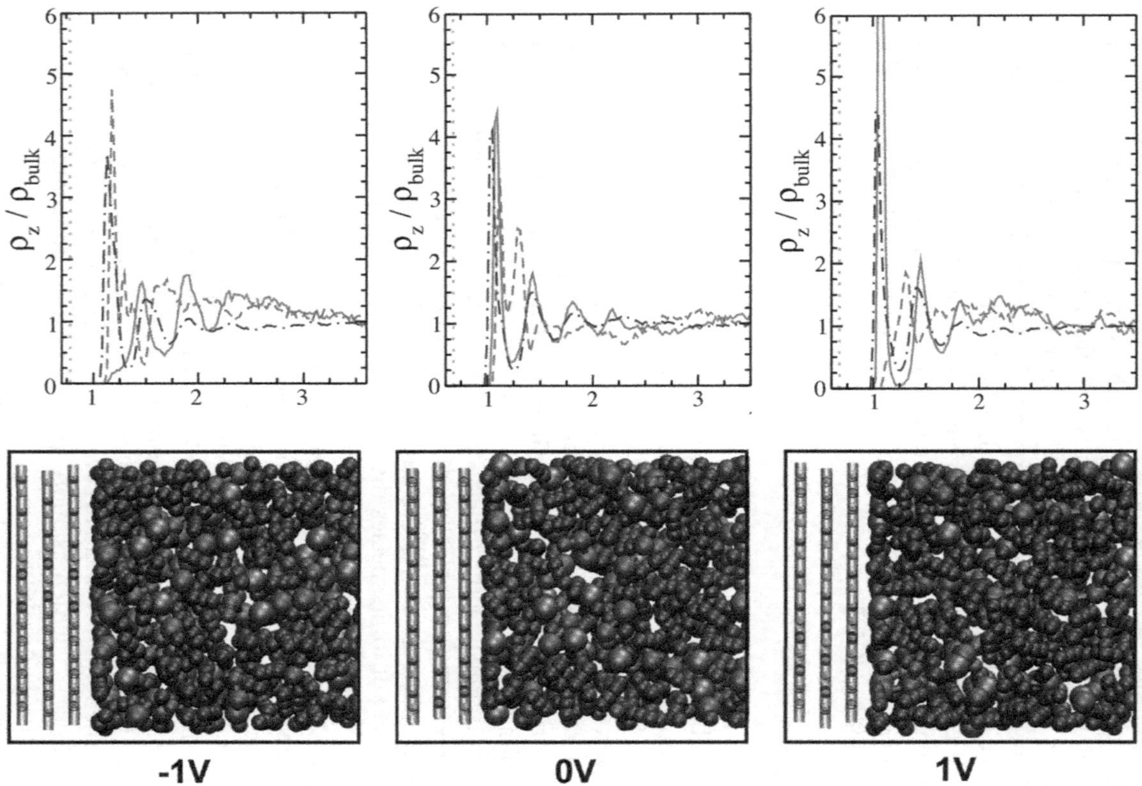

Fig. 6 Top: Molecular densities of the center of mass of the different species of an electrolyte composed of butyl-methylimidazolium (red line) and tetrafluoroborate (green line) ions dissolved in acetonitrile (blue line), with respect to their distance from a graphite electrode, at applied voltages of −1 V (left), 0 V (middle) and +1 V (right). ρ_{bulk} is the density of the bulk electrolyte. Bottom: Typical snapshots of the system at the corresponding conditions. Adapted with permission from ref. 85.

For planar electrodes, it is straightforward to extract the electrostatic potential across the simulation cell from these density profiles, using Poisson equation,

$$\Psi(z) = \Psi(z_0) - \frac{1}{\varepsilon_0}\int_{z_0}^{z} dz' \int_{-\infty}^{z'} dz'' \rho(z'') \quad (38)$$

where $\Psi(z_0)$ is an arbitrary integration constant which is fixed by the boundary condition $\Psi(z_0) = \Psi^+$ inside the positive electrode. A typical electrostatic potential profile determined for a pure ionic liquid is shown on Fig. 7. Note that compared to the MFT formulation, several potential drops can be defined: the two electrode/electrolytes are noted $\Delta\Psi^+$ and $\Delta\Psi^-$, they correspond to the voltage V defined in MFT. $\Delta\Psi$ is the potential drop inside the full simulation cell, *i.e.* between the two electrodes. In principle, the simpler way to determine the differential capacitance is thus to perform simulations of a system for various applied potentials. The average surfacic charge $\langle\sigma\rangle = \langle Q\rangle/A$ of the electrode is sampled and $C_{\rm diff}$ can be extracted from a fit of the $\langle\sigma\rangle = f(\Delta\Psi)$ plot. Here also, differential capacitances at both the positive and the negative electrodes ($C^+_{\rm diff}$ and $C^-_{\rm diff}$), as well as the total capacitance of the supercapacitor $C_{\rm diff}$, can be defined by using the corresponding potential drops. In many cases, this function is simple and this method is adequate.[85,101] Nevertheless, when $\langle Q\rangle(\Delta\Psi)$ has a non-trivial shape, different fits may lead to different results for the differential capacitance.

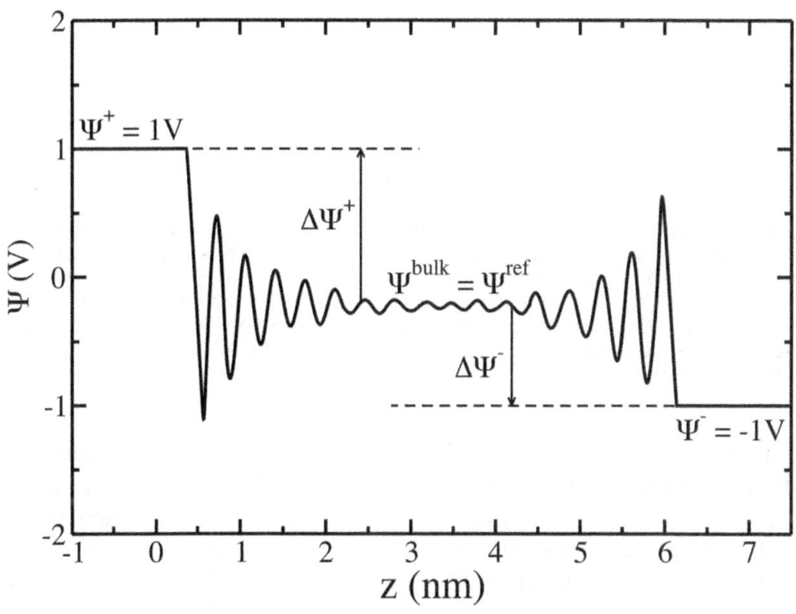

Fig. 7 Typical electrostatic potential between two planar graphite electrodes separated by a concentrated electrolyte. The left electrode has a fixed potential of $\Psi^+ = 1$ V, whereas the right one is at $\Psi^- = -1$ V. The potential difference between the electrodes is of $\Delta\Psi = 2$ V. When the potential reaches a plateau far from the surfaces, the corresponding value of the potential is noted $\Psi^{\rm bulk}$, which can be seen as a reference electrode. The potential drops at each electrodes are then $\Delta\Psi^+$ and $\Delta\Psi^-$.

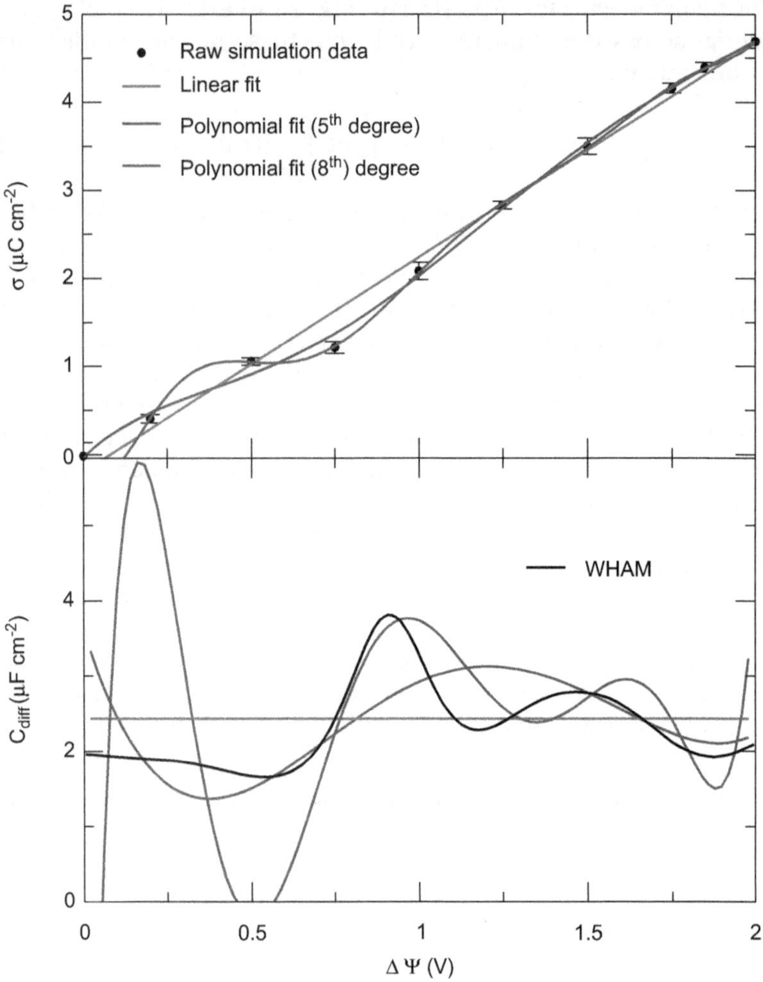

Fig. 8 Top: variation of the average surface charge $\langle\sigma\rangle = \langle Q\rangle/A$ with potential, for a supercapacitor composed of a 1-butyl-3-methylimidazolium hexafluorophosphate ionic liquid electrolyte and graphite electrodes. The points are raw data extracted from CGMD simulations while the lines are different polynomial fits of the data. Bottom: Surfacic differential capacitance, which is either calculated by differentiating $\sigma = f(\Delta\Psi)$ (the colors match with the top panel plots), or from the fluctuations of the charge, using importance sampling methods (WHAM).[100]

Such an example is shown on Fig. 8, which corresponds to the case of a butyl-methylimidazolium hexafluorophosphate ionic liquid (with graphite electrodes). The polynomial fits lead to differential capacitances which vary very diffently with potential from one case to another. Several peaks are observed, but their positions change markedly depending on the fit.

To overcome this difficulty, it is possible to determine the differential capacitance from the variance of the total charge distribution[101]

$$C_{\text{diff}} = \frac{A}{k_B T}\left\langle(\delta\sigma)^2\right\rangle \tag{39}$$

with $\delta\sigma = \sigma - \langle\sigma\rangle$. This fluctuation-dissipation relation is known in electronics as the Johnson–Nyquist relation.[102,103] In a recent work,[100] we have used importance sampling methods (weighted histogram analysis method, WHAM) to estimate this quantity with a good accuracy for the butyl-methylimidazolium hexafluorophosphate ionic liquid/graphite supercapacitor, the corresponding results are compared to the ones obtained using polynomial fits of the data on Fig. 8. It clearly shows that for such cases, the fitting approach is not appropriate. In addition, we could show that the anomalous response of the liquid, indicated by the presence of the peaks of the differential capacitance observed in Fig. 8, was due to structural changes induced in the interfacial region of the ionic liquid. In particular, an abrupt local ordering of the ions adsorbed in the plane of the electrode surface was observed during our CGMD simulations,[100] further underlining the complexity of this apparently simple interface.

However, the electrodes which are used in supercapacitors have far more complex topologies than planar surfaces. The nanoporous carbons, such as the nanotubes,[104] cluster-assembled nanostructured carbons,[105] nano-onions[22,106] or carbide-derived nanoporous carbons[107] have surfaces which are mostly curved, and can also contain many topological defects. It is important to understand how this impacts the structure of the adsorbed electrolytes and the resulting capacitance of the interface, even in the absence of confinement effects.

To this end, the case of rough surfaces has been studied by Vatamanu *et al.*[49] In their study, the carbon electrode consisted in several graphite layers stacked perpendicularly to the interface. They used an ABAB stacking sequence, which leads to a surface with a regular indentation of 1.43 Å due to the relative positions of the atoms at the edges of the A and B layers. The electrolyte was an ionic liquid consisting of 1-ethyl-3-methylimidazolium cations and bisfluorosulfonylimide anions. They observed a large increase of the capacitance per unit area compared to a planar graphene-like surface. The shape of the differential capacitance *vs.* potential curve is also different, since a peak is obtained for the roughened interface but not for the planar one. This means that the shape of the capacitance *vs.* potential plot can vary dramatically depending on the surface which is facing the liquid, a point which should be kept in mind when comparing simulations with experimental data. For the latter, it was also shown using broadband electrochemical impedance spectroscopy on "planar" gold electrodes that the surface roughness has a strong impact on the double layer formation.[108]

3.2 Porous electrodes

3.2.1 Slit pores and nanotubes. The simplest model of a porous electrode is a slit-like pore similar to the one shown on Fig. 4. The main difference with planar electrodes is that the multi-layered structure cannot be formed anymore. Instead, the ions are surrounded on both sides by the electrified walls. As discussed in the methods, Kondrat *et al.* have shown that it results in an exponential screening of the electrostatic interactions. By using either MFT or Monte-Carlo simulations,[34,44] they have shown that accounting for these effects allows for a qualitative reproduction of the anomalous capacitance increase observed in experiments. In practice, the charging mechanism is based on the exchange of the ions between the pores and the bulk electrolytes. When a potential is applied, the screening of the electrostatic interactions inside the pores increases. For example, in the positive electrode, the attraction between cation and anion decreases while the repulsion between anions becomes weaker. Under these conditions, it becomes favorable to pack more and more densely ions of similar charge. The structure of the liquid inside electrified slit pores has therefore been named "superionic state" by Kondrat *et al.*[34,44]

The adsorption of ions inside slit pores and carbon nanotubes has also been studied by using molecular dynamics.[109–113] Although not all of these studies did use the constant potential methods introduced in the methods section (instead, constant and similar partial charges were often attributed to all the carbons of each electrode), they provide results in general agreement with one another. Note also that classical density functional theory simulations have been performed using constant charge walls, leading to similar results.[114,115] For small pores, a peak in the capacitance when the pore size matches the ion size has been observed, in qualitative agreement with the experimental results shown on Fig. 3. An advantage of the slit-pores simulations over experiments is that a single pore size can be investigated for the former. This feature was exploited further by several groups who studied pore sizes of more than 1 nm.[111,112,114] They could show that a second peak in the capacitance *vs.* pore size was observed when the electrolyte was able to form favorable structures. It is then possible to address the effect of pore size dispersity by injecting the experimental pore size distributions, as was made by Kondrat *et al.* for examining the conditions leading to a maximal energy density storage.[116]

The dynamics of charging of slit-pore based supercapacitors has also been investigated using both MFT and MD.[35,117] The two methods agreed for predicting that the charging of the pores is governed by a simple diffusion process. As a result, the accumulated charge grows as a square root of time. In the transient régime, an overfilling of the pores is observed, which is followed by a de-filling. In the MD simulations, it was possible to show that the diffusion coefficients of the ions may change drastically during charging, due to the compositional changes inside the electrode.[117] At null charge, all the ions are surrounded by oppositely charged species. They occupy the sites of a two-dimensional lattice and cations/anions are interlocked with each other. As a result, the diffusion coefficients are two orders of magnitude smaller than in the bulk ionic liquid. When the charging of the pore starts, the exchange of ions with the bulk breaks this arrangement, and ions diffuse more freely. The effect on dynamics is spectacular, since at the maximum they even reach diffusion coefficients larger than in the bulk by one order of magnitude. Then, when the electrode is almost fully charged, the counter-ions form a quasi-Wigner crystal in which the small number of remaining co-ions play the roles of impurities, so that the diffusion coefficients substantially decrease.

3.2.2 Carbide-derived carbons. The commercial supercapacitors employ activated carbons, which include many defects, a wide pore size distribution, *etc.*, which cannot be easily characterized. Studying them using molecular-scale simulations seems for the moment out-of-reach. Nevertheless, many experimental studies use carbide-derived carbons (CDCs). These amorphous carbons are synthesized at high temperature by chlorinating crystalline carbides such as TiC, SiC and ZrC.[118] Their main advantage is that they have a narrow pore size distribution, and the average pore size can be tuned by changing the crystalline precursor and the chlorination temperature.[107] For example, the experimental results shown on Fig. 3 were obtained using such carbons.

By using quenched molecular dynamics simulations, Palmer *et al.* have been able to generate realistric structure of CDCs.[119] Their method consisted in simulating liquid carbon, using a reactive force field which allows the formation and breaking of carbon bonds during the simulations, namely the Reactive Summation State.[120] By quenching this liquid, they were able to form nanoporous carbon structures with varying features. The accessible surface area and the pore size distribution of these structures were calculated.[121] They compare very well with the experimental ones for CDC synthesized from TiC precursor at several temperatures: Similarly to the experiments it is indeed possible to tune the carbon properties by changing the quenching rate. They also simulated high resolution transmission electron microscopy images, showing that the structural ordering of the simulated carbons was also in agreement with the experimental images.

These structures contain roughly 4000 carbon atoms, and it was thus possible to build some molecular dynamics simulation cells with two such electrodes and an ionic liquid as an electrolyte.[122,123] A coarse-grained model was used for the latter in order to minize the computational time (which is several orders of magnitudes larger than in studies involving slit-pores due to the size of the system). A snapshot extracted from such a simulation is shown on the top panel of Fig. 9. The integral capacitance calculated with CDCs having a pore size distribution centered around a value slightly larger than the ions (1-butyl-3-methylimidazolium, BMI^+ and hexafluorophosphate, PF_6^-) diameters reach values ranging between 87 and 125 F g^{-1}. Such values are in qualitative agreement with the experimental data,[26,28,124] and much larger than the one obtained with graphite electrodes using the same model (30 F g^{-1}).

An advantage of simulations is that they are able to provide a microscopic insight on the molecular mechanism at play. Although experiments had shown that the increase of capacitance in CDCs was not due to an increase of the accessible surface (indeed, the observed increase in the experimental data on Fig. 3 is per unit of surface), it was not sure whether there was an artefact due to the surface estimation. This quantity is not measurable; it is usually extracted from the adsorption isotherms obtained from gas adsorption experiments. In MD simulations, the

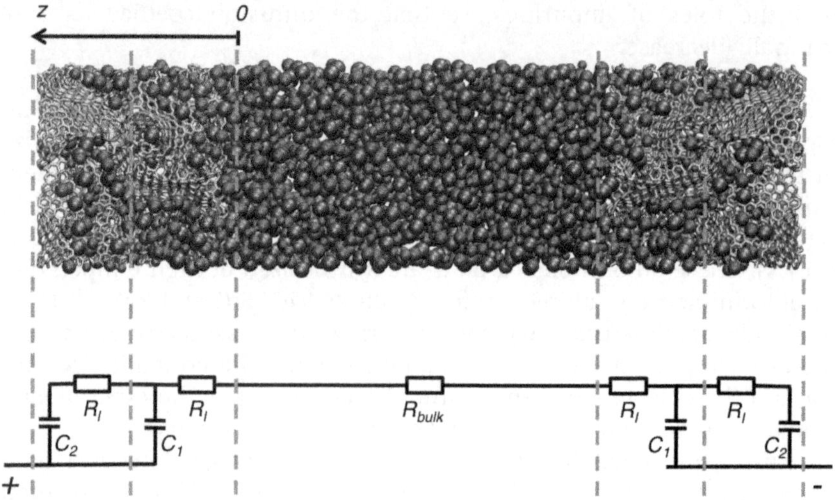

Fig. 9 Top: The simulation cell consists of a BMI-PF_6 ionic liquid electrolyte surrounded by two CDC electrodes held at constant potentials (blue: C atoms; red, the three sites of BMI^+ cations; and green, PF_6^- anions). Bottom: Equivalent electric circuit. R_{bulk} is the resistance of the electrolyte in the bulk region, R_1 is the resistance of the electrolyte adsorbed in an electrode slice, C_1 and C_2 are the capacitances of the two slices. Reprinted with permission from ref. 135. Copyright 2014 American Chemical Society.

coordinates of all the atoms are known so it is possible to use geometric definitions. In porous materials, the accessible surface area is generally defined as the surface between the accessible volume and the excluded volume. To calculate the accessible surface area, one has to choose a probe sphere which determines the minimum distance allowed between ions and wall atoms. In the simulation works on CDCs,[122] argon was taken as a probe sphere, which allowed an easier comparison of the results with experiments. However, doing the analysis with nitrogen as a probe sphere leads to equivalent conclusions (Fig. 10).

The indicator function $\chi(\mathbf{r})=1$ if accessible, 0 otherwise, defining the accessible volume, is estimated on a grid. The accessible volume is then given by

$$V_{\text{acc}} = \int \chi(\mathbf{r})\text{d}\mathbf{r} \tag{40}$$

The accessible surface is the region where the gradient $\|\nabla\chi\| \neq 0$ and the corresponding total area is

$$S_{\text{acc}} = \int \|\nabla\chi(\mathbf{r})\| \ \text{d}\mathbf{r} \tag{41}$$

In practice the gradient is computed numerically by a finite element method. Finally, the local unit vector normal to the surface, which is used to compute the density profiles is given by:

$$n = \frac{\nabla\chi(\mathbf{r})}{\|\nabla\chi(\mathbf{r})\|} \tag{42}$$

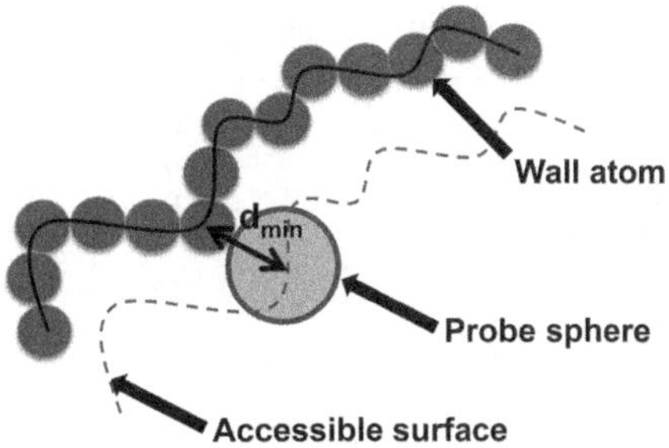

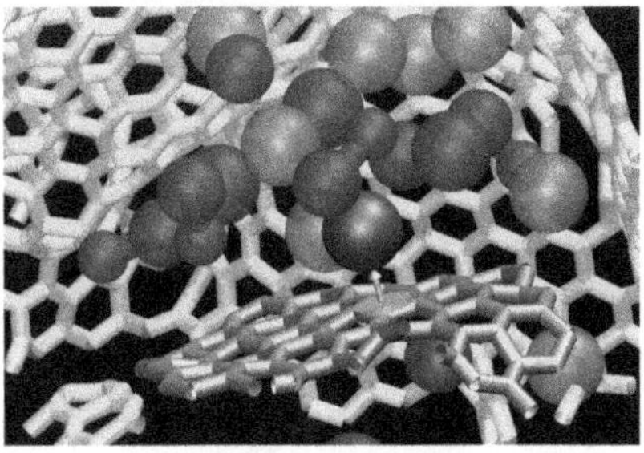

Fig. 10 The accessible surface area depends on the radius of the probe sphere. Following the experimental procedure, argon is used as the probe sphere. The snapshot shows the local structure of a BMI-PF_6 ionic liquid near a positive surface (+0.5 V) inside a CDC electron. Blue: C–C bonds, red: BMI^+ and green: PF_6^-. The ionic local densities are calculated with respect to the normal to the local surface (represented as a yellow arrow on the snapshot). Reprinted with permission from ref. 122.

Using these definitions, it was shown that the accessible surface of CDCs is comparable to the one of graphite, and that the calculated capacitances are much larger for the former, even when they are normalized by the surface area.[122] This is even more clear when examining the ionic density profiles. Indeed, the number of ions which are adsorbed at the surface of porous carbons is smaller than in the case of graphite electrode. This counterintuitive result shows that a different charging mechanism is at play. In fact, for planar electrodes, the overscreening effect, which arises from ionic correlations, couples the polarization of the first adsorbed layer to the next layers. As a consequence, only a fraction of the adsorbed ions are effectively used for storing electricity at the surface of the electrode. In CDCs, for steric reasons, the layered structure cannot form and there is only one adsorbed layer, so that each charge on the liquid side is exactly compensated on the carbon surface, resulting in a much better efficiency. In addition, because the attraction of the ions in the adsorbed layer is not balanced by that of the additional layers, the ions are able to approach the surface more closely than in the planar case, by approximately 0.7 Å.

Another characteristic of CDC electrodes is that their structure is strongly disordered. In order to understand further the effect of the local structure, a degree of confinement can also be defined for the ions[125] as the percentage of the solid angle around them which is occupied by the carbon atoms.[126] The distributions of the degrees of confinement experienced by each ionic molecule adsorbed in CDC electrodes held at negative, null and positive potentials are shown in Fig. 11. The values can differ markedly, ranging from 0 to 70%, and several broad peaks are present in the distribution. This indicates that the molecules have some preferred adsorption modes. By examining the trajectories, it is possible to define four different adsorption modes for the molecules.[125] These were classified following their corresponding degrees of confinement as (1) edge sites, *i.e.* carbon surfaces with a concave curvature, (2) planes, which will have the local structure of a graphene sheet, (3) hollow sites, *i.e.* a carbon surface with a convex curvature and (4) pockets, when they are inside pores with a cylinder-like shape.

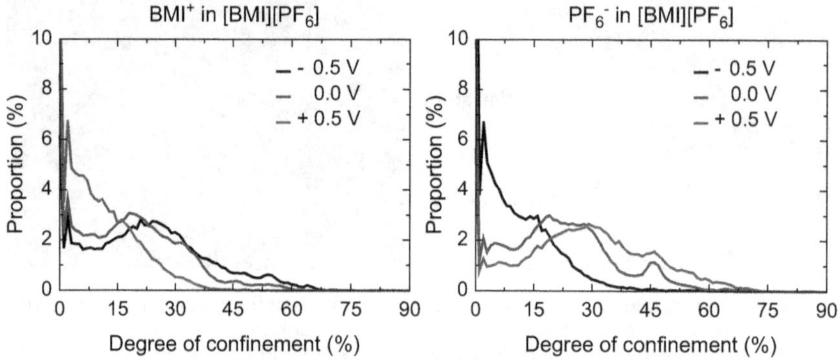

Fig. 11 Distributions of the degree of confinement experienced by ions for pure BMI-PF6 with different applied potentials (left: BMI^+ cations, right: PF_6^- anions).

At null potential, most of the ions are located in edge and plane sites. A minority of them are adsorbed in hollow ones, and pocket sites are almost empty. Anions show a stronger affinity for more confined sites than cations, which is consistent with their relative sizes. When a 1 V potential difference is applied between the electrodes, the number of ions in the electrodes and their degree of confinement change drastically.[122] Fig. 11 shows the distributions of the degrees of confinement experienced by the ionic species inside electrodes at −0.5 V, 0 V and +0.5 V respectively. When passing from 0 V to +0.5 V, anions are dragged toward the surface and they become less coordinated by cations. This permits them to enter into more confined pores. Approximately 25% of them are now adsorbed in the initially almost ions-free hollow and pocket sites. On the contrary, when the potential is decreased to −0.5 V the anions which remain inside the electrode mostly occupy edge sites. A symmetrical behavior is observed for the cations. A striking point is that despite their relatively large size some of them are able to occupy pocket sites. This clearly shows that in addition to the already reported exchange of ions with the bulk,[122] the charging mechanism of supercapacitors involves strong rearrangements in the distribution of ions inside the electrode, with a higher propensity for the ions charged oppositely to the electrode to adsorb in highly-confined sites. Note that in the presence of a solvent (acetonitrile), similar results are obtained despite the smaller ionic concentrations in the electrolyte.[125]

Although in experiments it is difficult to probe the structure very precisely inside the pores, NMR and electrochemical quartz microbalance seems to be the most promising experimental technique for studying supercapacitor electrodes.[127–133] In particular, a setup consisting of tetraethylammonium (TEA^+) and tetrafluoroborate (BF_4^-) ions dissolved in acetonitrile and various CDC electrodes was studied using *in situ* NMR.[127,128] The ^{11}B and ^{19}F chemical shifts were measured in order to understand the adsorption of the anions. In qualitative agreement with the MD simulations, this work identified the presence of different environments for the confined BF_4^- ions. These environments were assigned by increasing progressively the quantity of the electrolyte inside the CDC. For low loadings, only one peak is observed, which is thus assigned to ions which are the most strongly adsorbed inside the nanopores. As the quantity of electrolyte increases, the intensity of this peak reaches a maximal value and a second peak is observed. The latter was assigned to anions which are in a confined liquid, *i.e.* located in spaces between the primary particles of CDCs. Higher resolution may be possible in the future leading to specific sites being distinguished, which would allow a quantitative comparison with simulations.

In the case of acetonitrile-based electrolytes, it is also possible to probe the solvation of the ions, and how it changes with the corresponding adsorption site. Although it was expected from experimental studies[29,124] that the adsorption of ions in confined environments leads to their partial desolvation, it is not easy to quantify exactly how much they are desolvated. In simulations, the solvation number is easily defined as the number of solvent molecules that are located at a distance shorter than a given cut-off from a central ion. The only parameter is the spherical cut-off distance, which is defined using the first minimum of the corresponding radial distribution function.[125] The desolvation is clearly visible on the distributions of the acetonitrile solvent molecule coordination number around each ion, as shown on Fig. 12 for PF_6^- (similar results are obtained for the cations[125]). In the bulk electrolyte, this distribution is centered around 9. Nevertheless, occurrences of coordination numbers down to 4 and up to 14 are also observed, which is a sign that the solvation sphere of the ions is loosely bonded. When the ions are confined, they become less solvated, with distributions centered around 8, 6–7 and 5–6 for edge, planar and hollow sites respectively. In the case of pocket

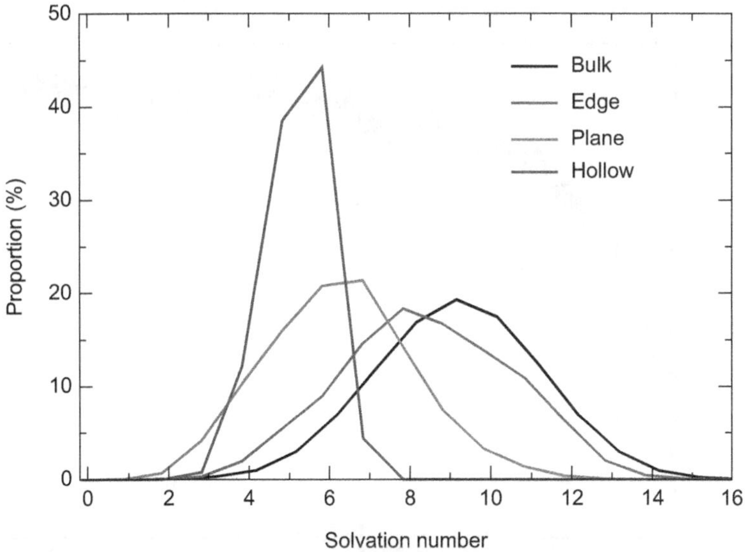

Fig. 12 Distributions of the solvation number of the PF_6^- anions by acetonitrile molecules depending on their environment, *i.e.* either in the bulk (for a concentration of BMI-PF_6 of 1.5 mol L^{-1}) or confined inside CDCs, for various degrees of confinement (edge, plane or hollow).

sites, it was not possible to extract such a distribution due to the lack of statistics. Such a large desolvation effect has already been observed experimentally for aqueous solutions by Ohkubo *et al.* who have shown that Rb^+ and Br^- hydration numbers decrease in slit-shape carbon nanospace by using extended X-ray absorption fine structure spectroscopy.[134]

Finally the dynamics of charging of CDC-based supercapacitors has also been investigated using MD simulations.[135] To this end, after an equilibration of the system at 0 V, an electric potential difference of 1 V is suddenly imposed between the electrodes. Figure 13 shows the time evolution of the total charge (normalized by the carbon mass) inside the positive electrode. For this CDC electrode, it takes more than 6 ns for reaching a plateau, but this result cannot be generalized: by changing the shape of the carbon structure, for example by decreasing the average pore size, the charging time becomes quite slower.[135]

The interpretation of experiments on supercapacitors usually relies on the use of equivalent electric circuits.[108,136–138] The MD simulations allow to test the relevance of such macroscopic models. The most commonly used model in the case of nanoporous electrodes is that of a transmission line, in which the charge penetrates progressively into the electrode.[136,137] This model is based on an infinite succession of slices connected in parallel through the constant potential condition. Each of them consists of a capacitor, which accounts for the possibility to store the charge by adsorption of the ions at the surface of the pores, and of a resistance which describes the transport of the electrolyte inside the pores.

In order to test this model for the simulated supercapacitors, given the small size of the simulated system, the only possibility was to divide each

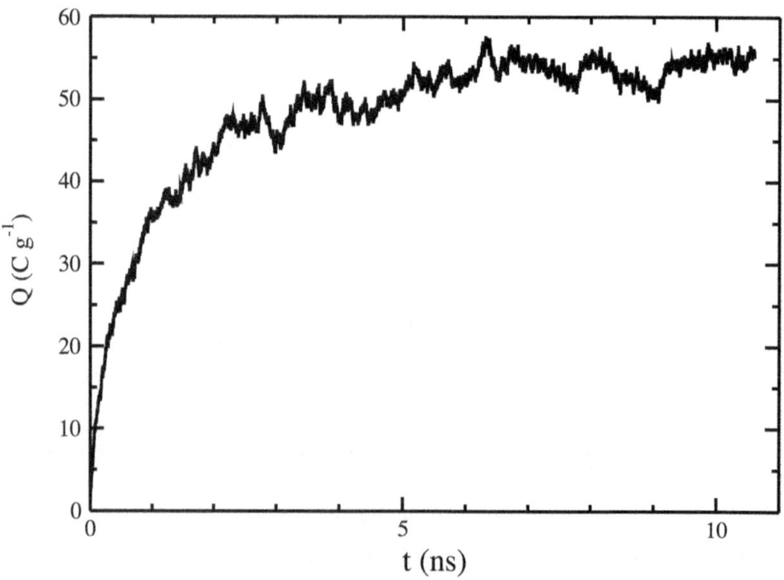

Fig. 13 Evolution of the total charge stored on the carbon electrodes (normalized by its mass) when the applied potential is suddenly switched from 0 V to 1 V.

electrode into two slices of equal thickness l, resulting in the equivalent electric circuit illustrated on Fig. 9. On the figure, R_{bulk} is the resistance of the electrolyte in the bulk region, R_1 is the resistance of each electrode slice, C_1 and C_2 are the capacitances of the two slices. Among these quantities, $R_{bulk} = 1.0 \times 10^8\ \Omega$ is known through the electrical conductivity of the bulk ionic liquid, and the capacitances ($C_1 = 4.3 \times 10^{-18}$ F and $C_2 = 2.7 \times 10^{-18}$ F) are a direct output from the MD simulations (through the distributions of charge inside the electrodes). The only unknown is R_1, which can be determined by fitting the total electrode charge evolution with time. The latter is given by

$$Q(t) = Q_{max}\left[1 - A_1 \exp\left(-\frac{t}{\tau_1}\right) - A_2 \exp\left(-\frac{t}{\tau_2}\right)\right] \quad (43)$$

where the expressions for Q_{max}, A_1, A_2, τ_1 and τ_2 are

$$Q_{max} = \frac{c}{b}V_0 = \frac{C_1 + C_2}{2}\Psi^+$$

$$\tau_1 = \frac{2}{a + \sqrt{a^2 - 4b}}$$

$$\tau_2 = \frac{2}{a - \sqrt{a^2 - 4b}}$$

$$A_1 = \frac{1}{2}\left[1 + \frac{2bd - ac}{2c\sqrt{a^2 - 4b}}\right]$$

$$A_2 = \frac{1}{2}\left[1 - \frac{2bd - ac}{2c\sqrt{a^2 - 4b}}\right].$$

with

$$a = \frac{(R_{bulk} + 2R_l)C_1 + (R_{bulk} + 4R_l)C_2}{R_l(R_{bulk} + 2R_l)C_1C_2}$$

$$b = \frac{2}{R_l(R_{bulk} + 2R_l)C_1C_2}$$

$$c = \frac{C_1 + C_2}{R_l(R_{bulk} + 2R_l)C_1C_2}$$

$$d = \frac{1}{R_{bulk} + 2R_l}.$$

The fit yields a value of $R_1 = 1.2 \times 10^8\ \Omega$ for the supercapacitor shown on Fig. 9. Given the relative widths of the electrolyte (50 Å for each half-cell) and electrode regions ($2l = 42$ Å for each electrode), the fact that R_{bulk} is of the same order of magnitude as R_1 shows that the transport of the ions is only slightly affected inside the pores. The low resistivity of the electrolyte inside the electrodes may arise from the existence of collective transport effects, as suggested from the results obtained for slit-pores based supercapacitors by Kondrat *et al.*,[117] in which the diffusion coefficients of the ions from the ionic liquid adsorbed inside the pore were shown to be strongly dependent on the charged state of the electrode.

From this effective macroscopic model, it is possible to estimate the characteristic time constant for charging as a function of the electrode width. In the full transmission line model, the diffusive charging occurs over the thickness l_{macro} of the electrodes with a characteristic time scale RCl^2_{macro}, with R and C being the resistance and capacitance *per unit length*, respectively. This time can be determined from our simulation results as:

$$\tau(l_{\text{macro}}) = R_1 \times C_{\text{tot}} \times \left(\frac{l_{\text{macro}}}{l_{\text{sim}}}\right)^2, \tag{44}$$

where C_{tot} is given by the total capacitance of the electrodes and l_{sim} is their thickness in the MD simulations. In a typical commercial supercapacitor, the electrode thickness is $l_{\text{macro}} \sim 100$ μm. Applying eqn (44) using the parameters extracted from the simulations yields a charging time of half a second.[135] This agrees well with the experimental data (albeit obtained with different electrolytes).[26] The estimate does not take into account the thickness of the electrolyte region and the resistance of other components of the supercapacitor such as the separator, but these should not impact the charging time significantly.

4 Conclusion and perspectives

The numerous theoretical studies of supercapacitors which have been performed in the past ten years have substantially contributed to our better understanding of these electricity storage devices. The simulation studies have focused on carbon-based supercapacitors, in which the charge is stored by reversible adsorption of the ions on the surface of the electrodes, which are generally made of carbon. Their success is partly due to the absence of chemical reaction, unlike pseudocapacitors or batteries which are much more difficult to simulate using classical techniques. Nevertheless, conventional techniques such as the algorithms currently implemented in molecular dynamics codes are not sufficient to account for the polarization of the electrodes. A special treatment of the electrostatic problem must indeed be included in order to fix their potential, in order to simulate systems in a realistic way.

Most of the simulations have up to now been performed on simple electrode geometries. For example, not much was known on the structure of ionic liquids at electrochemical interfaces,[139] this gap has now mostly been filled.[140–143] Simple porous structures, such as slit-pores or nanotubes, have also been widely studied, which has allowed for a better understanding of the molecular mechanism at the origin of the enhancement of the capacitance in nanoporous carbons with pore size distribution centered close to the size of the ions from the electrolytes.

Finally, realistic carbon structures have also been investigated. Due to their intrinsic disorder and to the relatively large size of the electrodes, it was not possible up to now to sample a wide range of electrolytes (which is in that case represented using a coarse-grained force field), but these simulations have shown that the microstructure of the carbon and the degree of confinement of the ions played major roles on the performances of the supercapacitor. Recent studies have shown that the transport of the ions remains efficient even in sub-nanometre pores, which explains why nanoporous-carbon based supercapacitors keep good performances in terms of specific power.

Now that supercapacitors are better understood, the next objective for simulations consists in proposing optimized electrode – electrolyte combinations or new storage concepts. To this end, high-throughput screening seems like a promising way.[144] Indeed, in the spirit of what is currently done in materials science for photocatalysts[145] and for battery electrodes[146,147] or electrolytes.[148] First steps towards this direction have been made by Balducci *et al.*[149] in order to predict via computation and to test experimentally the best electrolytes for supercapacitors.

New concepts may also emerge from simulations. A typical example is the proposition by Kondrat *et al.* to use "ionophobic" pores in supercapacitors.[117] Such porous materials, which would be initially empty at null voltage, would require to impose a potential for allowing the entrance of the ions. The main advantage is that the resulting dynamics of charging would be much faster since it would not require co-ions to be expelled concomitantly with the entrance of the counter-ions in the pore. Altough the experimental realisation is probably not straightforward, this is an interesting example on how theory can provide new solutions for energy storage.

References

1 J. R. Miller and P. Simon, *Science*, 2008, **321**, 651–652.
2 P. Simon, Y. Gogotsi and B. Dunn, *Science*, 2014, **343**, 1210–1211.
3 M. Toupin, T. Brousse and D. Bélanger, *Chem. Mater.*, 2004, **16**, 3184–3190.
4 T. Brousse, M. Toupin, R. Dugas, L. Athouel, O. Crosnier and D. Belanger, *J. Electrochem. Soc.*, 2006, **153**, A2171–A2180.
5 O. Ghodbane, M. Louro, L. Coustan, A. Patru and F. Favier, *J. Electrochem. Soc.*, 2014, **160**, A2315–A2321.
6 V. Augustyn, J. Come, M. A. Lowe, J. W. Kim, P.-L. Taberna, S. H. Tolbert, H. D. Abruna, P. Simon and B. Dunn, *Nat. Mater.*, 2013, **12**, 518–522.
7 T. Brousse, D. Belanger and J. W. Long, *J. Electrochem. Soc.*, 2015, **162**, A5185–A5189.
8 P. Simon and Y. Gogotsi, *Nat. Mater.*, 2008, **7**, 845–854.
9 E. Frackowiak, Q. Abbas and F. Béguin, *J. Energy Chem.*, 2013, **22**, 226–240.
10 L.-F. Chen, X.-D. Zhang, H.-W. Liang, M. Kong, Q.-F. Guan, P. Chen, Z.-Y. Wu and S.-H. Yu, *ACS Nano*, 2012, **6**, 7092–7102.
11 Z.-S. Wu, Y. Sun, Y.-Z. Tan, S. Yang, X. Feng and K. Mullen, *J. Am. Chem. Soc.*, 2012, **134**, 19532–19535.
12 Z.-S. Wu, K. Parvez, X. Feng and K. Müllen, *Nat. Commun.*, 2013, **4**, 2487.

13 G.-P. Hao, A.-H. Lu, W. Dong, Z.-Y. Jin, X.-Q. Zhang, J.-T. Zhang and W.-C. Li, *Adv. Energy Mater.*, 2013, **3**, 1421–1427.
14 X. Yang, C. Cheng, Y. Wang, L. Qiu and D. Li, *Science*, 2013, **341**, 534–537.
15 C. R. DeBlase, K. E. Silberstein, T.-T. Truong, H. D. Abruna and W. R. Dichtel, *J. Am. Chem. Soc.*, 2013, **135**, 16821–16824.
16 C. Ramirez-Castro, C. Schutter, S. Passerini and A. Balducci, *J. Power Sources*, 2014, **270**, 379–385.
17 K. Kierzek, E. Frackowiak, G. Lota, G. Gryglewicz and J. Machnikowski, *Electrochim. Acta*, 2004, **49**, 515–523.
18 Q. Abbas, P. Ratajczak, P. Babuchowska, A. Le Comte, D. Bélanger, T. Brousse and F. Béguin, *J. Electrochem. Soc.*, 2015, **162**, A5148–A5157.
19 J. Gamby, P.-L. Taberna, P. Simon, J.-F. Fauvarque and M. Chesneau, *J. Power Sources*, 2001, **101**, 109–116.
20 S. Pohlmann, T. Olyschlager, P. Goodrich, J. A. Vicente, J. Jacquemin and A. Balducci, *Electrochim. Acta*, 2015, **153**, 426–432.
21 M. Armand, F. Endres, D. R. MacFarlane, H. Ohno and B. Scrosati, *Nat. Mater.*, 2009, **8**, 621–629.
22 R. Y. Lin, P.-L. Taberna, S. Fantini, V. Presser, C. R. Perez, F. Malbosc, N. L. Rupesinghe, K. B. K. Teo, Y. Gogotsi and P. Simon, *J. Phys. Chem. Lett.*, 2011, **2**, 2396–2401.
23 D. R. MacFarlane, N. Tachikawa, M. Forsyth, J. M. Pringle, P. C. Howlett, G. D. Elliott, J. H. Davis, Jr., M. Watanabe, P. Simon and C. A. Angell, *Energy Environ. Sci.*, 2014, **7**, 232–250.
24 F. Béguin, V. Presser, A. Balducci and E. Frackowiak, *Adv. Mater.*, 2014, **26**, 2219–2251.
25 S. Pohlmann, T. Olyschlager, P. Goodrich, J. A. Vicente, J. Jacquemin and A. Balducci, *J. Power Sources*, 2015, **273**, 931–936.
26 J. Chmiola, G. Yushin, Y. Gogotsi, C. Portet, P. Simon and P. L. Taberna, *Science*, 2006, **313**, 1760–1763.
27 E. Raymundo-Piñero, K. Kierzek, J. Machnikowski and F. Béguin, *Carbon*, 2006, **44**, 2498–2507.
28 C. Largeot, C. Portet, J. Chmiola, P. L. Taberna, Y. Gogotsi and P. Simon, *J. Am. Chem. Soc.*, 2008, **130**, 2730–2731.
29 C. O. Ania, J. Pernak, F. Stefaniak, E. Raymundo-Piñero and F. Béguin, *Carbon*, 2009, **47**, 3158–3166.
30 C. R. Perez, S.-H. Yeon, J. Segalini, V. Presser, P.-L. Taberna, P. Simon and Y. Gogotsi, *Adv. Funct. Mater.*, 2013, **23**, 1081–1089.
31 J. Huang, B. G. Sumpter and V. Meunier, *Angew. Chem., Int. Ed.*, 2008, **47**, 520–524.
32 J. Huang, B. G. Sumpter and V. Meunier, *Chem. – Eur. J.*, 2008, **14**, 6614–6626.
33 S. K. Reed, O. J. Lanning and P. A. Madden, *J. Chem. Phys.*, 2007, **126**, 084704.
34 S. Kondrat and A. A. Kornyshev, *J. Phys.: Condens. Matter*, 2011, **23**, 022201.
35 S. Kondrat and A. A. Kornyshev, *J. Phys. Chem. C*, 2013, **117**, 12399–12406.
36 K. Kiyohara and K. Asaka, *J. Phys. Chem.*, 2007, **111**, 15903–15909.
37 K. Kiyohara, T. Sugino and K. Asaka, *J. Chem. Phys.*, 2010, **132**, 144705.
38 K. Kiyohara, T. Sugino and K. Asaka, *J. Chem. Phys.*, 2011, **134**, 154710.
39 K. Kiyohara, H. Shioyama, T. Sugino and K. Asaka, *J. Chem. Phys.*, 2012, **136**, 094701.
40 K. Kiyohara, H. Shioyama, T. Sugino, K. Asaka, Y. Soneda, K. Imoto and M. Kodama, *J. Chem. Phys.*, 2013, **138**, 234704.
41 J. Leckner, *Phys. A*, 1991, **176**, 485.

42 R. Sperb, *Mol. Simul.*, 1998, **20**, 179.
43 S. N. Punnathanam, *J. Chem. Phys.*, 2014, **140**, 174110.
44 S. Kondrat, N. Georgi, M. V. Fedorov and A. A. Kornyshev, *Phys. Chem. Chem. Phys.*, 2011, **13**, 11359–11366.
45 C. C. Rochester, A. A. Lee, G. Pruessner and A. A. Kornyshev, *ChemPhysChem*, 2013, **14**, 4121–4125.
46 D. Frenkel and B. Smit, *Understanding Molecular Dynamics*, Academic Press, 2nd edn, 2002.
47 J. Vatamanu, O. Borodin and G. D. Smith, *J. Am. Chem. Soc.*, 2010, **132**, 14825–14833.
48 J. Vatamanu, O. Borodin and G. D. Smith, *J. Phys. Chem. B*, 2011, **115**, 3073–3084.
49 J. Vatamanu, L. Cao, O. Borodin, D. Bedrov and G. D. Smith, *J. Phys. Chem. Lett.*, 2011, **2**, 2267–2272.
50 N. N. Rajput, J. D. Monk, R. Singh and F. R. Hung, *J. Phys. Chem. C*, 2012, **116**, 5170–5182.
51 X. Si, S. Li, Y. Wang, S. Ye and T. Yan, *ChemPhysChem*, 2012, **13**, 1671–1676.
52 J. Vatamanu, Z. Hu, D. Bedrov, C. Perez and Y. Gogotsi, *J. Phys. Chem. Lett.*, 2013, **4**, 2829.
53 X. Jiang, J. Huang, B. G. Sumpter and R. Qiao, *J. Phys. Chem. Lett.*, 2013, **4**, 3120–3126.
54 Z. Hu, J. Vatamanu, O. Borodin and D. Bedrov, *Phys. Chem. Chem. Phys.*, 2013, **15**, 14234–14247.
55 Z. Hu, J. Vatamanu, O. Borodin and D. Bedrov, *Electrochim. Acta*, 2014, **145**, 40–52.
56 N. N. Rajput, J. D. Monk and F. R. Hung, *J. Phys. Chem. C*, 2014, **118**, 1540–1553.
57 S. Li, G. Feng and P. T. Cummings, *J. Phys.: Condens. Matter*, 2014, **26**, 284106.
58 J. Vatamanu, L. Xing, W. Li and D. Bedrov, *Phys. Chem. Chem. Phys.*, 2014, **16**, 5174–5182.
59 K. Breitsprecher, P. Kosovan and C. Holm, *J. Phys.: Condens. Matter*, 2014, **26**, 284108.
60 K. Breitsprecher, P. Kosovan and C. Holm, *J. Phys.: Condens. Matter*, 2014, **26**, 284114.
61 X. Liu, Y. Han and T. Yan, *ChemPhysChem*, 2014, **15**, 2503–2509.
62 A. Uysal, H. Zhou, G. Feng, S. S. Lee, S. Li, P. T. Cummings, P. F. Fulvio, S. Dai, J. K. McDonough, Y. Gogotsi and P. Fenter, *J. Phys.: Condens. Matter*, 2015, **27**, 032101.
63 Y. He, J. Huang, B. G. Sumpter, A. A. Kornyshev and R. Qiao, *J. Phys. Chem. Lett.*, 2015, **6**, 22–30.
64 E. Paek, A. J. Pak and G. S. Hwang, *J. Chem. Phys.*, 2015, **142**, 024701.
65 M. Salanne, *Phys. Chem. Chem. Phys.*, 2015, **17**, 14270–14279.
66 Y. T. Wang, S. L. Feng and G. A. Voth, *J. Chem. Theory Comput.*, 2009, **5**, 1091–1098.
67 D. Roy, N. Patel, S. Conte and M. Maroncelli, *J. Phys. Chem. B*, 2009, **114**, 8410–8424.
68 D. Roy and M. Maroncelli, *J. Phys. Chem. B*, 2010, **114**, 12629–12631.
69 Z. Cao, J. F. Dama, L. Lu and G. A. Voth, *J. Chem. Theory Comput.*, 2013, **9**, 172–178.
70 J. I. Siepmann and M. Sprik, *J. Chem. Phys.*, 1995, **102**, 511–524.
71 T. R. Gingrich and M. Wilson, *Chem. Phys. Lett.*, 2010, **500**, 178–183.
72 P. Wu, J. Huang, V. Meunier, B. G. Sumpter and R. Qiao, *J. Phys. Chem. Lett.*, 2012, **3**, 1732–1737.

73 A. V. Raghunathan and N. R. Aluru, *Phys. Rev. E: Stat., Nonlinear, Soft Matter Phys.*, 2007, **76**, 011202.
74 M. K. Petersen, R. Kumar, H. S. White and G. A. Voth, *J. Phys. Chem. C*, 2012, **116**, 4903–4912.
75 A. Arnold, K. Breitsprecher, F. Fahrenberger, S. Kesselheim, O. Lenz and C. Holm, *Entropy*, 2013, **15**, 4569–4588.
76 N. Onofrio, D. Guzman and A. Strachan, *Nat. Mater.*, 2015, **14**, 440–446.
77 S. Perkin, M. Salanne, P. Madden and R. Lynden-Bell, *Proc. Natl. Acad. Sci. U. S. A.*, 2013, **110**, E4121.
78 D. M. Heyes and J. H. R. Clarke, *J. Chem. Soc., Faraday Trans. 2*, 1981, **77**, 1089–1100.
79 M. Pounds, PhD thesis, University of Edinburgh, 2009.
80 S. Tazi, M. Salanne, C. Simon, P. Turq, M. Pounds and P. A. Madden, *J. Phys. Chem. B*, 2010, **114**, 8453–8459.
81 G. Feng, J. S. Zhang and R. Qiao, *J. Phys. Chem. C*, 2009, **113**, 4549–4559.
82 S. Kislenko, I. Samoylov and R. Amirov, *Phys. Chem. Chem. Phys.*, 2009, **11**, 5584–5590.
83 C. Merlet, M. Salanne, B. Rotenberg and P. A. Madden, *J. Phys. Chem. C*, 2011, **115**, 16613–16618.
84 C. Merlet, M. Salanne and B. Rotenberg, *J. Phys. Chem. C*, 2012, **116**, 7687–7693.
85 C. Merlet, M. Salanne, B. Rotenberg and P. A. Madden, *Electrochim. Acta*, 2013, **101**, 262–271.
86 M. V. Fedorov and A. A. Kornyshev, *J. Phys. Chem. B*, 2008, **112**, 11868–11872.
87 G. Feng, J. Huang, B. G. Sumpter, V. Meunier and R. Qiao, *Phys. Chem. Chem. Phys.*, 2011, **13**, 14723–14734.
88 M. Z. Bazant, B. D. Storey and A. A. Kornyshev, *Phys. Rev. Lett.*, 2011, **106**, 046102.
89 M. Mezger, H. Schröder, H. Reichert, S. Schramm, J. S. Okasinski, S. Schröder, V. Honkimäki, M. Deutsch, B. M. Ocko, J. Ralston, M. Rohwerder, M. Stratmann and H. Dosch, *Science*, 2008, **322**, 424–428.
90 A. Uysal, H. Zhou, G. Feng, S. S. Lee, S. Li, P. Fenter, P. T. Cummings, P. F. Fulvio, S. Dai, J. K. McDonough and Y. Gogotsi, *J. Phys. Chem. C*, 2014, **118**, 569–574.
91 M. Mezger, R. Roth, H. Schröder, P. Reichert, D. Pontoni and H. Reichert, *J. Chem. Phys.*, 2015, **124**, 164707.
92 R. Atkin and G. Warr, *J. Phys. Chem. C*, 2007, **111**, 5162–5168.
93 E. Hayes, G. G. Warr and R. Atkin, *Phys. Chem. Chem. Phys.*, 2010, **12**, 1709–1723.
94 R. Atkin, N. Borisenko, M. Drüschler, S. Zein El Abedin, F. Endres, R. Hayes, B. Huber and B. Roling, *Phys. Chem. Chem. Phys.*, 2011, **13**, 6849–6857.
95 S. Perkin, T. Albrecht and J. Klein, *Phys. Chem. Chem. Phys.*, 2010, **12**, 1243–1247.
96 S. Perkin, L. Crowhurst, H. Niedermeyer, T. Welton, A. M. Smith and N. N. Gosvami, *Chem. Commun.*, 2011, **47**, 6572–6574.
97 S. Perkin, *Phys. Chem. Chem. Phys.*, 2012, **14**, 5052–5062.
98 A. M. Smith, K. R. J. Lovelock, N. N. Gosvami, P. Licence, A. Dolan, T. Welton and S. Perkin, *J. Phys. Chem. Lett.*, 2013, **4**, 378–382.
99 J. Sweeney, F. Hausen, R. Hayes, G. B. Webber, F. Endres, M. W. Rutland, R. Bennewitz and R. Atkin, *Phys. Rev. Lett.*, 2012, **109**, 155502.
100 C. Merlet, D. T. Limmer, M. Salanne, R. van Roij, P. A. Madden, D. Chandler and B. Rotenberg, *J. Phys. Chem. C*, 2014, **118**, 18291–18298.

101 D. T. Limmer, C. Merlet, M. Salanne, D. Chandler, P. A. Madden, R. van Roij and B. Rotenberg, *Phys. Rev. Lett.*, 2013, **111**, 106102.
102 J. Johnson, *Phys. Rev.*, 1928, **32**, 97.
103 H. Nyquist, *Phys. Rev.*, 1928, **32**, 110.
104 A. Al-Zubaidi, T. Inoue, T. Matsuhita, Y. Ishii and S. Kawasaki, *Phys. Chem. Chem. Phys.*, 2012, **14**, 16055–16061.
105 L. G. Bettini, M. Galluzzi, A. Podestá, P. Milani and P. Piseri, *Carbon*, 2013, **59**, 212–220.
106 D. Pech, M. Brunet, H. Durou, P. H. Huang, V. Mochalin, Y. Gogotsi, P.-L. Taberna and P. Simon, *Nat. Nanotechnol.*, 2010, **5**, 651–654.
107 Y. Gogotsi, A. Nikitin, H. Ye, W. Zhou, J. E. Fischer, B. Yi, H. C. Foley and M. W. Barsoum, *Nat. Mater.*, 2003, **2**, 591–594.
108 T. Jänsch, J. Wallauer and B. Roling, *J. Phys. Chem. C*, 2015, **119**, 4620–4626.
109 L. Yang, B. H. Fishbine, A. Migliori and L. R. Pratt, *J. Am. Chem. Soc.*, 2009, **131**, 12373–12376.
110 Y. Shim and H. J. Kim, *ACS Nano*, 2010, **4**, 2345–2355.
111 P. Wu, J. S. Huang, V. Meunier, B. G. Sumpter and R. Qiao, *ACS Nano*, 2011, **5**, 9044–9051.
112 G. Feng and P. T. Cummings, *J. Phys. Chem. Lett.*, 2011, **2**, 2859–2864.
113 L. Xing, J. Vatamanu, O. Borodin and D. Bedrov, *J. Phys. Chem. Lett.*, 2013, **4**, 132–140.
114 D.-E. Jiang, Z. Jin and J. Wu, *Nano Lett.*, 2011, **11**, 5373–5377.
115 O. Pizio, S. Sokolowski and Z. Sokolowska, *J. Chem. Phys.*, 2012, **137**, 234705.
116 S. Kondrat, C. R. Perez, V. Presser, Y. Gogotsi and A. A. Kornyshev, *Energy Environ. Sci.*, 2012, **5**, 6474–6479.
117 S. Kondrat, P. Wu, R. Qiao and A. A. Kornyshev, *Nat. Mater.*, 2014, **13**, 387–393.
118 R. Dash, J. Chmiola, G. Yushin, Y. Gogotsi, G. Laudisio, J. Singer, J. E. Fisher and S. Kucheyev, *Carbon*, 2006, **44**, 2489–2497.
119 J. C. Palmer, A. Llobet, S.-H. Yeon, J. E. Fisher, Y. Shi, Y. Gogotsi and K. E. Gubbins, *Carbon*, 2010, **48**, 1116–1123.
120 Y. Shi, *J. Chem. Phys.*, 2008, **128**, 234707.
121 L. Gelb and K. E. Gubbins, *Langmuir*, 1999, **15**, 305–308.
122 C. Merlet, B. Rotenberg, P. A. Madden, P.-L. Taberna, P. Simon, Y. Gogotsi and M. Salanne, *Nat. Mater.*, 2012, **11**, 306–310.
123 C. Pean, B. Daffos, C. Merlet, B. Rotenberg, P.-L. Taberna, P. Simon and M. Salanne, *J. Electrochem. Soc.*, 2015, **162**, A5091–A5095.
124 J. Chmiola, C. Largeot, P.-L. Taberna, P. Simon and Y. Gogotsi, *Angew. Chem., Int. Ed.*, 2008, **47**, 3392–3395.
125 C. Merlet, C. Péan, B. Rotenberg, P. A. Madden, B. Daffos, P.-L. Taberna, P. Simon and M. Salanne, *Nat. Commun.*, 2013, **4**, 2701.
126 J. A. van Meel, L. Filion, C. Valeriani and D. Frenkel, *J. Chem. Phys.*, 2012, **136**, 234107.
127 H. Wang, T. K. J. Koster, N. M. Trease, J. Segalini, P. L. Taberna, P. Simon, Y. Gogotsi and C. P. Grey, *J. Am. Chem. Soc.*, 2011, **133**, 19270–19273.
128 A. C. Forse, J. M. Griffin, H. Wang, N. M. Trease, V. Presser, Y. Gogotsi, P. Simon and C. Grey, *Phys. Chem. Chem. Phys.*, 2013, **15**, 7722–7730.
129 H. Wang, A. C. Forse, J. M. Griffin, N. M. Trease, L. Trognko, P.-L. Taberna, P. Simon and C. Grey, *J. Am. Chem. Soc.*, 2013, **135**, 18968–18980.
130 M. Deschamps, E. Gilbert, P. Azais, E. Raymundo-Piñero, M. R. Ammar, P. Simon, D. Massiot and F. Béguin, *Nat. Mater.*, 2013, **12**, 351–358.
131 A. J. Ilott, N. M. Trease, C. P. Grey and A. Jerschow, *Nat. Commun.*, 2014, **5**, 4536.

132 A. C. Forse, J. M. Griffin, C. Merlet, P. M. Bayley, H. Wang, P. Simon and C. P. Grey, *J. Am. Chem. Soc.*, 2015, **137**, 7231–7242.
133 W.-Y. Tsai, P.-L. Taberna and P. Simon, *J. Am. Chem. Soc.*, 2014, **136**, 8722–8728.
134 T. Ohkubo, T. Konishi, Y. Hattori, H. Kanoh, T. Fujikawa and K. Kaneko, *J. Am. Chem. Soc.*, 2002, **124**, 11860–11861.
135 C. Péan, C. Merlet, B. Rotenberg, P. A. Madden, P.-L. Taberna, B. Daffos, M. Salanne and P. Simon, *ACS Nano*, 2014, **8**, 1576–1583.
136 B. E. Conway, *Electrochemical Supercapacitors: Scientific Fundamentals and Technological Applications*, Kluwer, 1999.
137 P.-L. Taberna, P. Simon and J.-F. Fauvarque, *J. Electrochem. Soc.*, 2003, **150**, A292–A300.
138 B. Roling, M. Drüschler and B. Huber, *Faraday Discuss.*, 2012, **154**, 303–311.
139 A. A. Kornyshev, *J. Phys. Chem. B*, 2007, **111**, 5545–5557.
140 M. V. Fedorov and A. A. Kornyshev, *Chem. Rev.*, 2014, **114**, 2978–3036.
141 C. Merlet, B. Rotenberg, P. A. Madden and M. Salanne, *Phys. Chem. Chem. Phys.*, 2013, **15**, 15781–15792.
142 R. Burt, G. Birkett and X. S. Zhao, *Phys. Chem. Chem. Phys.*, 2014, **16**, 6519–6538.
143 R. Hayes, G. G. Warr and R. Atkin, *Chem. Rev.*, 2015, **115**, 6357–6426.
144 S. Curtarolo, G. L. W. Hart, M. B. Nardelli, N. Mingo, S. Sanvito and O. Levy, *Nat. Mater.*, 2013, **12**, 191–201.
145 Y. Wu, P. Lazic, G. Hautier, K. Persson and G. Ceder, *Energy Environ. Sci.*, 2013, **6**, 314–319.
146 G. Hautier, A. Jain, S. P. Ong, B. Kang, C. Moore, R. Doe and G. Ceder, *Chem. Mater.*, 2011, **23**, 3495–3508.
147 A. Jain, G. Hautier, S. P. Ong, S. Dacek and G. Ceder, *Phys. Chem. Chem. Phys.*, 2015, **17**, 5942–5953.
148 L. Cheng, R. S. Assary, X. Qu, A. Jain, S. P. Ong, N. N. Rajput, K. Persson and L. A. Curtiss, *J. Phys. Chem. Lett.*, 2015, **6**, 283–291.
149 C. Schütter, T. Husch, M. Korth and A. Balducci, *J. Phys. Chem. C*, 2015, **119**, 13413–13424.

Nanotubes with well-defined structure: imogolites

Luciana Guimarães,*[a] Maicon P. Lourenço[b] and Hélio A. Duarte[b]
DOI: 10.1039/9781782622703-00151

1 Introduction

The progress achieved in nanoscience and nanotechnology has opened new frontiers in environmental sciences, biology, physics, chemistry and engineering. New materials are being designed with enhanced chemical and physical properties, which are determined not only by their chemical bonds and composition but also by morphology and size. Nanotubes (NTs) are important building blocks for the design of materials at nanometric scale. Their unusual properties are assigned to their chirality, radii, size and morphology. Notwithstanding, it remains a challenge in the synthesis process to control the dimensions and produce monodisperse nanostructures. Reaching the perfect control of nanoparticles morphology is the daily work of many researchers around the world. Among the wide family of NTs, imogolite is an example of NT with well-defined structure in which the diameter control is feasible.

Imogolite is an aggregation of single-walled (SW) aluminosilicate NTs which are found in soils of volcanic origin and have the general formula $(OH)_3Al_2O_3SiOH$ (Fig. 1). The character of mineral imogolite was established by Yoshinaga *et al.*[1] in 1962, who identified it as fibrous component of acid-dispersible clay. The imogolite local structure was proposed by Cradwick *et al.*[2] in 1972. The authors described the mineral as a tubular structure where the external part consists of a curved gibbsite sheet $Al(OH)_3$ and the inner hydroxyls are replaced by $SiO_3(OH)$ groups. Naturally occurring NTs have an external diameter of around 2.0 nm, internal diameter equal to 1.0 nm and length around few hundred nanometers.[3] Few years after Cradwick's structure elucidation, Farmer *et al.*[4] demonstrated a synthetic route to produce imogolite and ever since, a variety of procedures has been described.[3,5–9] The lattice misfit caused by the binding of orthosilicate group at gibbsite hexagonal ring, *i.e.*, the shortening of O–O distances, leads to its curvature and formation of imogolite NT. Independently of the synthesis procedure (different anionic ligands and temperature)[6,9] the obtained imogolite NTs tend to be monodisperse with well-defined diameter.

The monodisperse imogolite NTs attracted interest due to their vast range of potential applications such as gas storage,[10–13] additive of

[a]*Departament of Natural Science, Universidade Federal de São João del Rei, São João del Rei, Minas Gerais, Brazil. E-mail: lucianaguimaraes@ufsj.edu.br*
[b]*Departament of Chemistry, Instituto de Ciências Exatas, Universidade Federal de Minas Gerais, Belo Horizonte, Minas Gerais, Brazil*

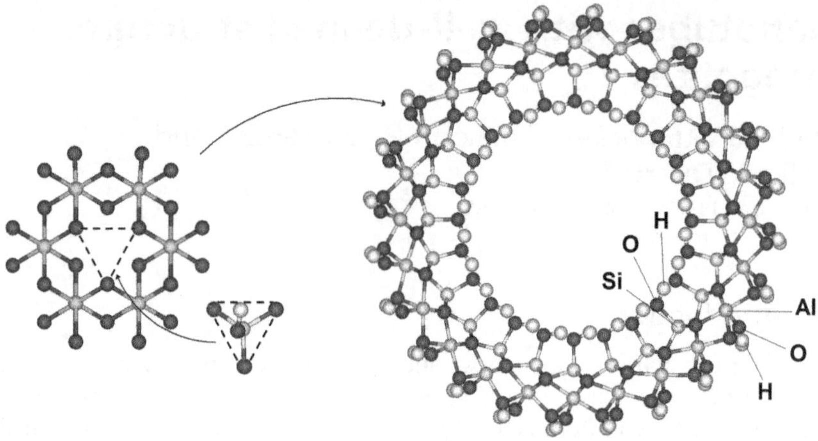

Fig. 1 Gibbsite hexagonal ring where the silanol group is bonded (left) and the cross section view of imogolite unit cell (right). White atoms, H; red, O; blue, Al and yellow, Si. Adapted with permission from ref. 29. Copyright 2007 American Chemical Society.

transparent polymers,[14] catalyst support,[15] base for organic–inorganic nanohybrides,[16–23] and confinement of molecules and ions in the channel allowing interactions with species inside the NT inner wall.[24–28] Besides, imogolite structure has also attracted interest since it can be synthesized using low-temperature solution-phase process without any catalyst and with inexpensive reactants.

In the last ten years, much progress has been reached in the synthesis, characterization and functionalization of imogolite, and only one more example of monodisperse SW NT could be produced so far. The aluminogermanate imogolite-like analogue has been synthesized replacing the silicon atoms by germanium atoms.[3,30] The substitution of Si by Ge results in NTs with shorter length (around 20 nm in place of few hundred nm) and larger diameter than imogolite, since the larger size of GeO_4 tetrahedra decreases the stress in the gibbsite hexagonal unit. Besides, the control over the synthesis conditions defines the formation of SW, double-walled (DW) aluminogermanate or even hollow spheres and nanoscrolls.[31,32] In this context, the striking imogolites monodispersity has motivated investigations aiming to elucidate the NTs formation mechanism[8,33,34] and controlling the shape of NTs.[9] The term imogolites refers to the family of both aluminosilicate and aluminogermanate NTs. Hereafter, img-Ge will denote aluminogermanate NTs and img-Si refer to imogolite.

Theoretical methods have been used since 2002 to investigate the electronic, structural and mechanical properties of imogolites aiming to contribute to the understanding of these systems at atomic level. Most of the theoretical investigations at that time were based on force fields specially developed for aluminosilicate or aluminogermanate systems[35–38]. The large size of NT unit cell is still a limitation for the use of more sophisticated quantum mechanical methods.

The need for elucidating the electronic structure of large scale systems has been outstanding by the development of approximated and scalable methods that provide ways of simulating materials. The following quantum methods are examples of these achievements and have been extensively used to study imogolites and similar large scale systems: the SCC-DFTB approach,[39,40] the DFT method approximated by employing numerical basis sets as implemented at SIESTA[41] and at DMol3 codes[42,43] and, finally, the DFT applying helical boundary symmetry as available in the CRYSTAL software.[44] As it will be presented here, the outcomes of quantum mechanical methods allowed the elucidation of imogolite electronic and structural properties.

The present chapter is structured as follows: we will discuss the synthesis and formation mechanism of imogolites at Section 2, whereas Section 3 features the computational aspects of the approximate method Self-Consistent-Charge Density-Functional Tight-binding and its parameterization steps. The theoretical analysis for NT structures and stability will be presented at Section 4, followed by the electronic and mechanical properties at Section 5. Final remarks will be presented at Section 6.

2 Synthesis and formation mechanism

2.1 Aluminosilicate nanotubes

Unlike carbon and other inorganic NTs, imogolite can be synthesized in aqueous solution in mild conditions. The first synthesis was performed by Farmer *et al.*[4] over thirty years ago, employing aluminium chloride and sodium silicate solutions in millimolar concentration at a temperature of 95 °C. Ever since, the most explored sol–gel route is prepared by employing an aluminium salt (chloride or perchlorate) and tetraethoxysilane (TEOS) at millimolar concentrations.[3] TEOS is preferred instead of sodium silicate due to its slower condensation kinetics.[23] The initial solution is stirred and aged for 18 h at 25 °C. After a dilution, the temperature is increased to 95 °C and kept constant under stirring for 4 days (96 h).[8] Imogolite synthesis process is highly dependent on the medium pH and the control during the initial steps is crucial to obtain tubular structures in place of amorphous structures or allophane.[45]

Although the synthesis process is well established, theoretical and experimental studies have not been able to show the mechanism of self-assembling that leads to the formation of well-defined NTs. Earlier works[46,47] have proposed that imogolite formation takes place *via* an intermediate named protoimogolite, which presents a sheet-like structure. The enclosure of protoimogolite would occur in the beginning of the synthesis process and would serve as nuclei for the imogolite NT growth. However, the protoimogolite composition has not been proven experimentally. As mentioned by Yang *et al.*,[48] the progressive nucleation and sequential growth describe a kinetic based approach albeit additional evidence is needed. Notwithstanding, based on dynamic light scattering (DLS) and transmission electron microscopy (TEM), Mukherjee *et al.*[3] have shown that imogolite NT average lengths maintain unchanged at the synthesis' later phase and have proposed imogolite growth may occurs by a thermodynamic self-assembly process. On the other hand, Yang *et al.*,[48] observing individual imogolite NTs by TEM, have proposed a controlled kinetic formation process, although in none of the cases the structures of the precursors and intermediate species were identified.

Nevertheless, in order to understand the molecular structure of precursors and intermediates, Yucelen *et al.*[8] have reported a detailed molecular-level study concerning the imogolite formation in the liquid phase, identifying the structure, composition and shapes of the proto-imogolites. This achievement was taken by a detailed analysis of nuclear magnetic resonance (NMR), electrospray ionization mass spectrometry (ESI-MS) and DLS, with the support of quantum chemistry calculations for the key structures. From NMR analysis, it was possible to follow the evolution of local environment around the Al atoms. In particular, a broad peak around 6 ppm was observed during the aging and heating and assigned to six-coordinated Al species with a chemical environment very similar to that found in the NT structure. A sharp peak near 0 ppm is assigned to octahedral monomer aluminate units and a peak around 63.3 ppm is attributed to a four-fold coordinated Al environment and both disappear during the heating step. Thus, while the monomer peak decreases the NT like peak increases as the equilibrium is established. It was not observed any significant change in the chemical shift related to the octahedral peak at 6 ppm during the heating step. This result indicates that no significant new Al coordination environment is formed after the aging step, *i.e.*, the Al octahedral species formed during the aging step are taking part in the NT formation. After increasing the temperature from 25 °C to 95 °C, the condensation and rearrangement processes occur.

From ESI-MS data, the authors have shown that NT formation is based on aluminate speciation and the main species formed in the solution do not change along the aging process. The signals for silicate species were not detected. In the aluminium speciation, several aluminate species were detected, containing from 1 to 13 aluminium atoms. The gibbsite basic unit, an Al_6 ring, is detected as complete or incomplete ring in several isomeric organizations of perchlorate anions, water molecules and hydroxyl groups on the edges of the structures. The authors have observed that almost all the aluminate species have formed complexes with perchlorate ions, suggesting this anion may stabilize species whose basic unit is an Al_6 ring and may also avoid further condensation of species larger than Al_{13}. Besides, according to the ESI study on aluminium speciation in aqueous $AlCl_3$ solution,[49] the perchlorate anion may also help to bridge two aluminium atoms. Hence, Yucelen *et al.*[8] have suggested the aluminosilicate species complexation with perchlorate ion is important for understanding the NT formation. The distribution of aluminosilicate species, as in aluminate solution, presents structures with aluminium atoms ranging from 1 to 13 and the differences are related to the number of attached silicon tetrahedra and/or perchlorate anion, as represented by the formula $[Al_1Si_x(ClO_4)_y\text{–}Al_{13}Si_x(ClO_4)_y]$. The complexes formed are the first hint at the protoimogolite formation in solution and some points may be highlighted concerning this stage, as follow: (i) the NT formation is based on aluminate speciation; (ii) the octahedral aluminium atoms are linked by μ_2-OH groups; (iii) the tetrahedral silicon groups are attached to three Al atoms in a Q^3(6Al) coordination; (iv) aluminosilicate complexes

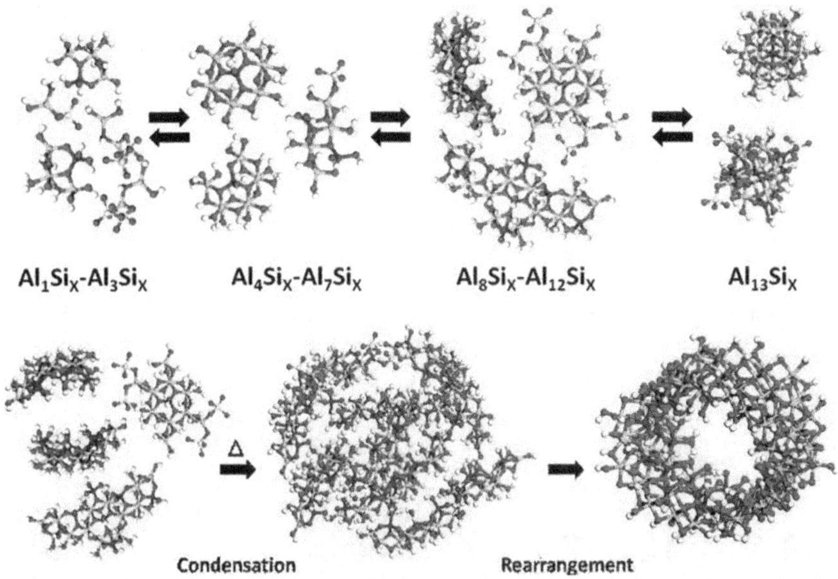

Fig. 2 Fomation mechanism of aluminosilicate nanotubes. (Top) Equilibrium established between aluminosilicate species during the aging stage at 25 °C. (Bottom) Condensation and rearrangement of Al_8Si_x–$Al_{12}Si_x$ species upon heating. Reprinted with permission from ref. 8. Copyright 2011 American Chemical Society.

with Al_6 ring and Si atoms at Q^3(6Al) configuration are present; (v) the perchlorate ions are bonded at the edges of aluminosilicate complexes; (vi) tetrahedral aluminium sites are in the edges of aluminosilicate species. According to the authors, the conversion and rearrangement from tetrahedral aluminium to octahedral aluminium take place during the assembly process *via* condensation of protoimogolite complexes. Thus, from ESI-MS and NMR results, the authors were able to propose a molecular-level mechanism for imogolite formation, as outlined in Fig. 2.

The anions in solution, *i.e.*, chlorate, perchlorate, are found complexing the initial aluminosilicate precursors. The intermediate species have an intrinsically curved structure, which are formed within 1 h of aging and remain unaltered during this period. Around 7 hours after the beginning of the synthesis, a dynamic equilibrium is settled between "small", "medium" and "large" precursors. After heating the aluminosilicate solution, the intermediate species, represented by "Al_8Si_x–$Al_{13}Si_x$", disappear from ESI-MS spectra since they condense into larger species, and then rearrange into the NT structure. The rearrangement occurs by the conversion of tetra-coordinated end groups into octahedral configuration in the moment that the precursors condense. Finally, the growing in length occurs by a set of processes.

Currently, it is possible to control imogolite shape, size, composition and structure from the modification of a number of synthesis conditions such as kinetics, temperature, ionic strength, pH and reactants concentration.[9] Yucelen *et al.*[9] have shown that it is possible to control the imogolite diameter at the Ångström level *via* the control of the precursors shape as it is proposed in Fig. 3. According to them, the binding of

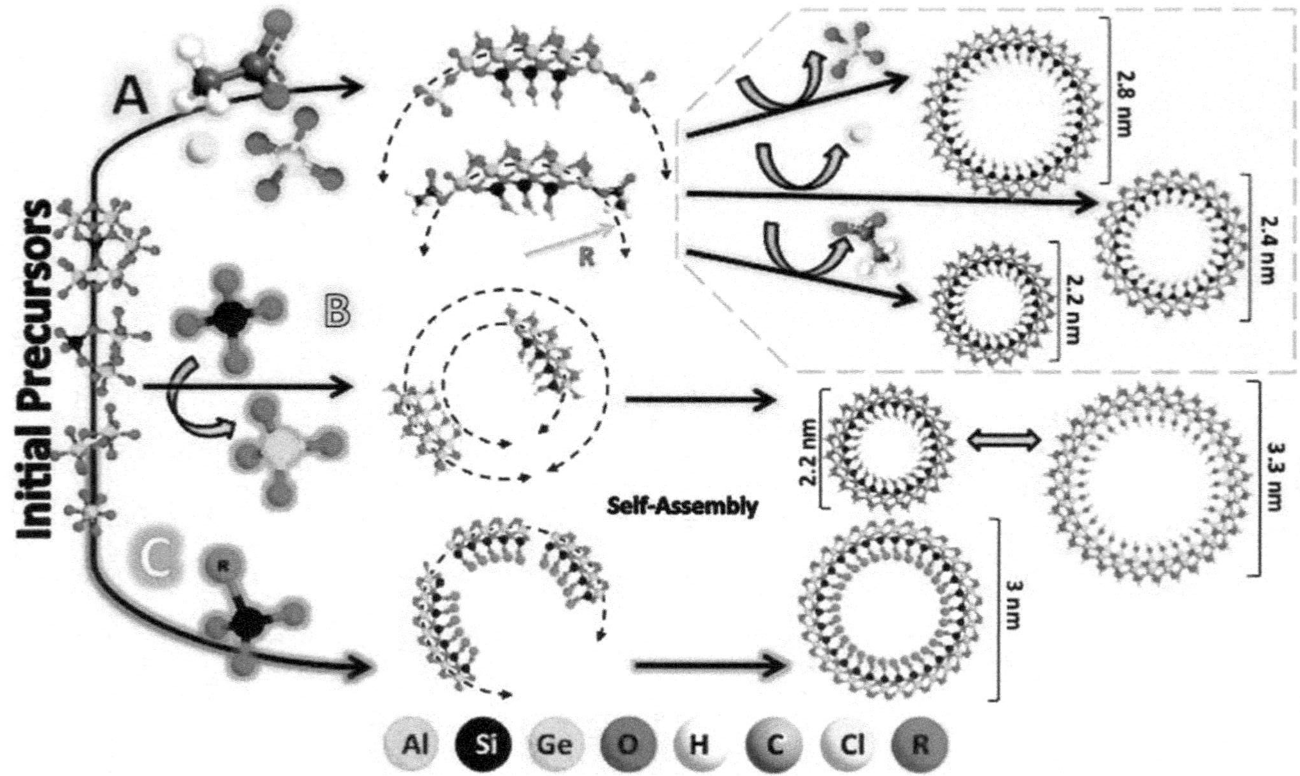

Fig. 3 Shape control of metal oxide nanotubes. (A) Anion complexation: use of different anions (*e.g.*, acetate, chloride, perchlorate) that complex with the precursors and alter their curvature. The anions are released during condensation of the precursors, leaving the nanotube structure unaffected but allowing control of curvature, as shown in this work. Other methods for altering the nanotube structure (not investigated here) may also follow similar but yet unexplained mechanisms, such as (B) substitution. The silicate tetrahedra can be substituted with other tetrahedra (*e.g.*, germanate) or (C) by organosilicate (Si–R) tetrahedra; in both cases leading to a change in diameter as well as composition. Reprinted with permission from ref. 9. Copyright 2012, American Chemical Society.

different anions, such as acetate, chloride or perchlorate, to the precursors can be used to modify their curvature and produce NTs with controlled diameters but with the same structure. From XRD analysis, it has been shown that the low-angle peak for the synthesized NTs shifts as a function of the anion used in the synthesis, suggesting a systematic change in the NT diameter. For instance, the smallest external diameter for imogolite synthesis is achieved using acetate anions whereas the largest external diameter is obtained with perchlorate anions. In order to demonstrate this control over the NT diameter, high-resolution cryo-electron microscopy images have been obtained for individual imogolite NTs. The analyses have shown remarkably sharp diameter distributions in all NTs. The average diameters are equal to 2.2 ± 0.2 nm for NTs synthesized with acetate anions; 2.8 ± 0.2 nm for perchlorate; 2.5 ± 0.2 nm for 50% perchlorate and 50% chloride and 2.4 ± 0.2 nm for chloride. Once the experimental NT diameters have been determined, the number of hexagonal repeat units around the circumference (hereafter Nu) could be assigned by using a theoretical geometrical relation previously obtained.[50] The use of perchlorate anions led to the largest diameters, with $Nu = 15$, while the use of chloride and acetate resulted to NTs with $Nu = 12$ and $Nu = 11$, respectively, as depicted in Fig. 4.

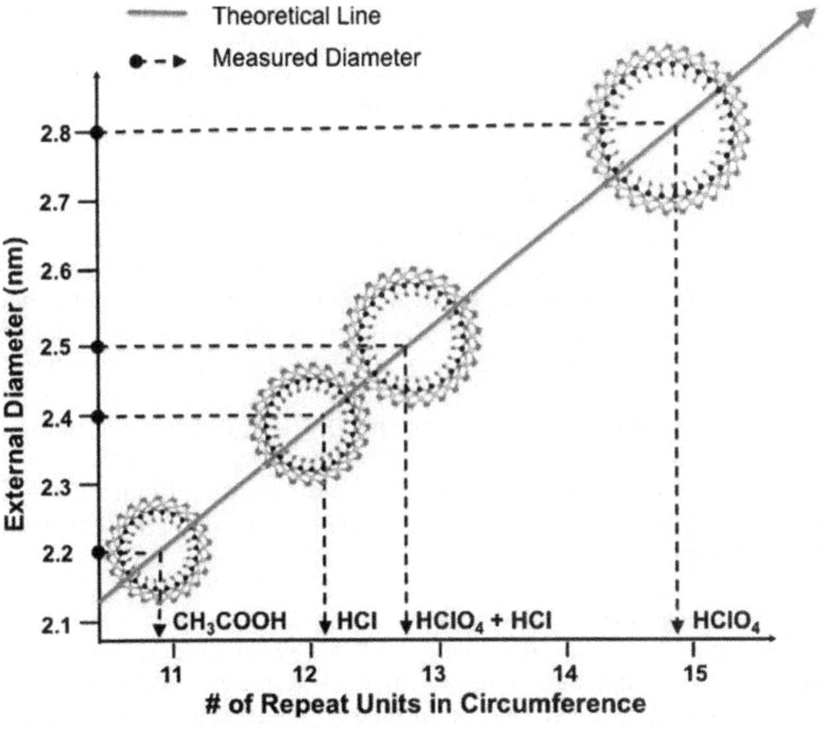

Fig. 4 Average external nanotube diameters measured from cryo-EM micrographs and the corresponding number of aluminosilicate repeat units in the circumference for each type of anion used in the synthesis. The structures of the nanotubes are also shown. The theoretical line is obtained from geometry-optimized nanotube models and is used to find the number of repeating units in the nanotube circumference using the experimental cryo-EM data. Reprinted with permission from ref. 9. Copyright 2012 American Chemical Society.

Moreover, Yucelen *et al.*[9] have shown that the precursor's intrinsic curvature are dependent on the anion type in the solution. Several experiments indicated that aluminate species are affected by the change of the electrolyte used to control the ionic strength of the Al(III) solution. The perchlorate and acetate anions formed complexes with Al^{3+} by replacing the pendant –OH groups and in the monodentate coordination, while chloride ions are directly bonded to Al^{3+}. The chloride coordination ability leads to the formation of brucite-like clusters. All solutions containing the different anions have presented a limited set of nanometer sized precursors containing $[Al_1Si_x–Al_{14}Si_x]$ anions. After heating all these solutions, the nanometer precursors disappear from ESI-MS spectra in all cases and in the meantime, nanostructures of around 5 nm are detected in the DLS spectra. Consequently, these phenomena are crucial to the NT nucleation process that occurs *via* the precursor's temperature-induced condensation into nanoclusters. The next step is the rearrangement of the nanoparticles into a nanotubular structure. The driving force for the nucleation is the same for all the solutions; however the NT curvature is related to the complex formed between the anion and the nanoscale precursors. DFT calculations have been performed for several selected aluminosilicate intermediate species found in ESI-MS analysis in acidic solutions of perchloric acid, acetic acid and hydrochloric acid. In all species the aluminium atoms at the edges are complexed with different anions (perchlorate, chloride or acetate). All the studied species presented inherent curvatures. The curvature was obtained by squares fitting of a cylindrical radius and the molecular curvature (κ) was defined as $\kappa = 1/R$, where R is the radius of the NT. As it has been already discussed in the literature, the curvature occurs due to the mismatch between Al–O bond length (1.9 Å) and Si–O bond length (1.6 Å). However, as demonstrated by Yucelen *et al.*,[9] the curvature can be controlled by the binding of the anion to the precursor. The complexation of acetate anions lead to the intermediates with the largest curvature, followed by chloride and finally perchlorate with the smallest curvature. Consequently, NT diameters present the inverse trend. Therefore, by using a combination of different anions in solution, XRD, NMR, ESI-MS, cryo-EM experiments and DFT calculations, the authors were able to demonstrate the relationship between the structures of the precursor complexes in solution and the final diameter of the NTs. Then, the diameter control of the synthesized NTs has been demonstrated.

2.2 Aluminogermanate nanotubes

The single-walled aluminogermanate NT is a synthetic derivative of the aluminosilicate imogolite NT, in which Si atoms are replaced by Ge atoms. The first synthesis was performed by Wada and Wada[30] in 1982, employing aluminium chloride and $GeCl_4$ solutions in millimolar concentration at a temperature of 95–100 °C for 120 hours. Following the same procedure used by Farmer *et al.*[4] for imogolite synthesis, the solution pH was set to 5.0 by addition of NaOH and then reacidified to pH equal to 4.5. The obtained NTs were similar to imogolite in their tubular morphology, as observed by electron diffraction patterns and differential thermal analysis curves. However, the external diameter of 3.3 nm is much larger than the imogolite diameter, due to the larger Ge–O equilibrium distance. Nowadays, the Ge source used is tetraethyl orthogermanate (TEOG).

Img-Si synthesis occurs at millimolar concentration of reactants since attempts to synthesize it with more concentrated solutions at atmospheric pressure have failed, leading to allophane. Notwithstanding, img-Ge NTs can be obtained in large amount with an increase in the reactant concentration to decimolar, *i.e.*, 100 times more concentrated, with a similar formation kinetic (5 days) comparable to imogolite synthesis at millimolar concentration.[51] According to Levard *et al.*,[45] a possible explanation concerning the large amount of img-Ge NT produced due to the higher solubility of GeO_2 species at solution in comparison to SiO_2 species, which precipitates at higher concentration (decimolar).

On the other hand, Maillet *et al.*[52] have shown that by changing the initial Al concentrations, 0.25 M and 0.5 M, it is possible to obtain DW and SW img-Ge NTs, respectively. In the synthesis process, the aluminium and germanium were slowly hydrolysed by the addition of NaOH solution up to reach the hydrolysis ratio ([OH]/[Al]) of 2. As always, the solution was heated at 95 °C for 5 days. Based on small-angle X-ray scattering (SAXS) and cryo-TEM images, the authors have shown the diameter of the external DW NTs (4.0 ± 0.1 nm) is larger than the SW NTs (3.5 ± 0.15 nm), while the internal NT diameter is around 2.4 ± 0.1 nm. In addition, they have reported that the distance between the Ge layer of the external tube and the Al layer of the internal tube is 2.7 Å, suggesting that no covalent bonds are present between the inner and outer DW NTs walls. Such distance corresponds to the typical value of one water layer. In a sequential work, Maillet *et al.*[53] have shown the length distributions for SW and DW NTs, assessed through AFM measurements, are much broader than the diameter distribution. For instance, the DW NTs length is equal to 20 ± 15 nm while the average length for SW NTs is significantly longer (50 ± 30 nm). Thereafter, Amara *et al.*[54] have prepared micron-long (up to 2.5 μm) DW img-Ge NTs at high concentrations by using a novel synthesis route.

In the quest for understanding the conditions allowing the formation of SW and DW img-Ge NTs, Thill *et al.*[33] have explored the physico-chemical conditions controlling the NTs formation. They have analysed the effect of the hydrolysis ratio ([OH]/[Al]) and reactant concentration in the SW or DW structure once both enable the control over the final NT structure. On the one hand, it is possible to go from SW to DW NT by changing the hydrolysis ratio and with any tested concentration. For instance, SW NTs are obtained with a low hydrolysis ratio and DW NTs with a higher hydrolysis ratio. On the other hand, at a given hydrolysis ratio, the proportion of SW or DW NT can be modified by the change of the reactant concentration.

In order to progress in the understanding of SW and DW NT formation mechanisms, Thill *et al.*[33] have proposed a simple model by taking into account a balance between attractive interaction energy and internal

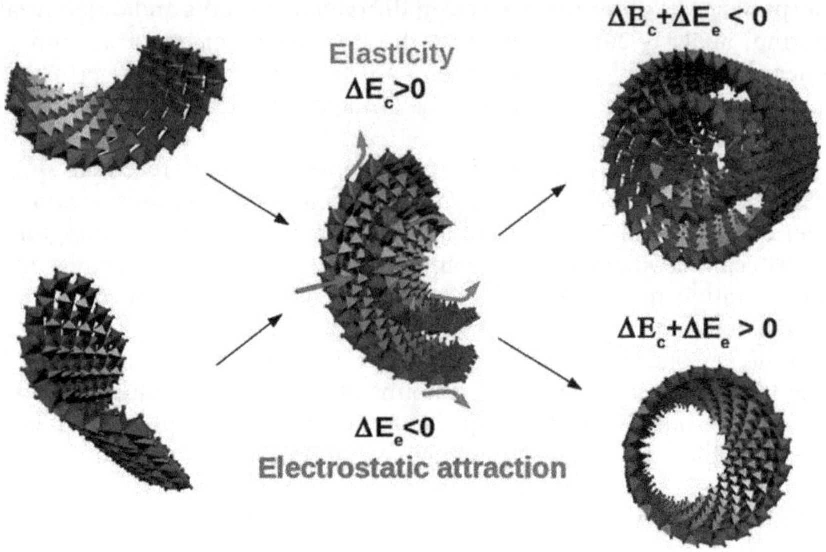

Fig. 5 Qualitative schematic scenario for the protoimogolite transformation into SW or DW imogolite nanotubes. Reprinted with permission from ref. 33. Copyright 2012 American Chemical Society.

curvature energy, as depicted in Fig. 5. The protoimogolite structure is described by them as a flexible sheet in which the natural curvature is controlled by the internal bonded atoms (Si or Ge) and the quantity of adsorbed tetrahedra. The protoimogolite surfaces are dominated by either positive charges due to Al–OH and Al_2–OH sites in one side or dominated by negative charges because of SiOH sites in the other side, leading to electrostatic interaction between protoimogolites. This characteristic is in accordance to experimental findings that protoimogolite spieces tend to form aggregates, *i.e.*, two protoimogolite structures attract each other due to the electrostatic attraction between the positive charged external wall of one and the negative charged inner wall of the other. Besides, the hydrogen bonds between protoimogolites and van der Waals interactions contribute to the electrostatic attractions. The flexibility of the two protoimogolite species leads to the increase or decrease of their curvature at the expense of elastic energy, ΔE_c.

As schematized in Fig. 5, there are two possible ways of forming SW or DW NTs. In the first one, the loss of attractive energy is more than the acquisition of elastic energy, leading to DW NTs. In the second one, the loss of attractive energy is less than the gain in elastic energy, so the two curved structures detach and SW NTs are formed. It is possible to better understand the two scenarios by analysing the effect of hydrolysis ratio and reactants concentrations. When the hydrolysis ratio ([OH]/[Al]) is decreased, defects and lacunas appear at the surface of imogolite. The consequences are: (i) the increase of the AlOH surface site modifies the quantity of surface charge as a function of pH, as Al_2OH and AlOH present different pK_a. The change in the surface charge leads to loss of electrostatic attraction and decreases the interaction between two protoimogolite species. As a result, SW NTs are formed. (ii) The defects and gaps can also modify the imogolite mechanical properties. However, in the variation of reactants concentration, the ionic strength is modified. The electrostatic interaction between protoimogolite species is controlled by the magnitude of ionic strength. For instance, at high reactants concentration, *i.e.*, high ionic strength, the interaction of protoimogolites is reduced and the curvature energy will not be sustained while the species are coupled. As consequence, SW NTs will be formed. Counterwise, at low reactants concentration, the strength of protoimogolite interaction is higher and the protoimogolite species can maintain the increased curvature energy, leading to DW NTs. Furthermore, according to Thill *et al.*,[33] among the explored synthesis conditions, none allowed the formation of triple-walled or even multi-walled NTs.

Computer simulations describing the mechanism of formation would be useful for its fully understanding at a molecular level. However, at the present state-of-the-art of the theoretical chemistry these systems are much complex to be treated. The kinetics are very slow taking some days to occur; the water solvent is very important and participates in the hydrolysis reactions, hence, has to be treated explicitly; the reaction free energies have to be described at tenths of kJ mol^{-1} of precision in order to correctly describe the acidic constants of the species and the reactions involved. Therefore, theoretical calculations are limited to some species identified by ESI-MS and NMR, evidencing their curvature and stability.[8,9] Notwithstanding, the stability and the electronic, structural and mechanical properties of the imogolite NTs have been subject to many theoretical investigations providing important insights in this compelling class of materials.

3 Computational aspects of the self-consistent-charge density-functional tight-binding (SCC-DFTB) method

Imogolites and other clay mineral NTs (*e.g.* halloysite, chrysotile) have been investigated by different theoretical methods. Force field methods have been used to calculate the geometry and to perform molecular dynamics of imogolites in solution. These methods are classical approaches that use empirical equations for describing the potential energy with respect to bond distances, angles and dihedrals. Density functional theory (DFT) has also been applied to elucidate the geometries and the electronic structure of a limited number of NTs. DFT is probably the most efficient, accurate and affordable approach to describe the electronic structure of such systems. However, imogolites and other clay mineral NTs are large systems and the computational cost prevents one for performing more extended calculations. The self-consistent-charge density-functional tight-binding (SCC-DFTB) method[39,55] is the most used for electronic structure calculations of imogolites and other clay mineral NTs. This is an approximation of the density functional theory (DFT) that uses a minimal basis set and the tight-binding approach (TB) to the Kohn–Sham hamiltonian. The basic concepts of the SCC-DFTB approach, its implementation and how it is parameterized will be presented. For a review of the SCC-DFTB method we refer the readers to the references.[39,56–58]

The SCC-DFTB total energy for a closed shell system is described by the eqn (1):

$$E = E_{\mathrm{bnd}} + E_{\mathrm{rep}} + E_{\mathrm{SCC}}. \tag{1}$$

The first term, the so-called band energy (E_{bnd}) is calculated by the sum of the one-electron energies accordingly to the eqn (2):

$$E_{\mathrm{bnd}} = 2\sum_{i=1}^{M/2} \varepsilon_i, \tag{2}$$

where M is the number of electrons and ε_i is the i-th occupied molecular orbital energy. In order to calculate E_{bnd}, one has to build the tight-binding Kohn–Sham like hamiltonian[56] for the system. Then, for a closed shell finite system, the hamiltonian is described according to eqn (3).

$$H^0_{\mu\nu} = \begin{cases} \varepsilon_\mu^{\text{free atom}}, & \mu = \nu, \\ \langle \phi_\mu | T + v_{KS}[\rho_0^A + \rho_0^B] | \phi_\nu \rangle, & \mu \in \{A\}, \nu \in \{B\},\ A \neq B \\ 0, & \mu \neq \nu \in \{A\} \text{ or } \in \{B\}. \end{cases} \tag{3}$$

μ and ν are the indexes of the atomic basis function centered on the atoms A and B. With this groundwork, one can build the whole $\boldsymbol{H}^0$ and $\boldsymbol{S}$ matrices, whose dimension is $K \times K$, where K is the size of all atomic basis. The matrix elements $H^0_{\mu\nu}$ and $S_{\mu\nu}$ are all collected from the Slater–Koster Files (SLAKOs) for certain interatomic distances of the atoms A and B – which are already computed and tabulated in an interatomic distance grid. Based on the spatial position of the atoms, a set of rotations applied in the matrix elements, called the Slater–Koster transformations, are performed.[57] After the matrices are built from eqn (3), generalized eigenvalue problem $\boldsymbol{H}^0\boldsymbol{C} = \boldsymbol{S}\boldsymbol{C}\varepsilon$ is solved[56] and the eigenvectors ($\boldsymbol{C}$) as well as the eigenvalues (ε) are obtained and, finally, E_{bnd} is calculated. It is important to note that the tight-binding approach neglects the three and four center integrals permitting that the integrals of eqn (3) be tabulated in a pairwise manner within the SLAKO files. The set of equations and procedures discussed above form the foundation of the so called *standard* DFTB.

From the eigenvectors $\boldsymbol{C}$, one can calculate the density matrix $\boldsymbol{P}$ according to eqn (4):

$$P_{\mu\nu} = 2\sum_{i}^{M/2} C_{\mu i} C^*_{\nu i}. \tag{4}$$

From $\boldsymbol{P}$ the Mulliken population analysis of the atom A is obtained:

$$q_A = \sum_{\mu \in A}^{K_A} (\boldsymbol{PS})_{\mu\mu}, \tag{5}$$

where K_A is the size of atomic basis of atom A, $\boldsymbol{PS}$ is the matrix multiplication of the density and overlap matrices. The net charge is

$$\Delta_{q_A} = q_A^0 - q_A, \tag{6}$$

where q_A^0 is the valence number of electrons of the isolated atom A.

The E_{SCC} contribution to the energy comes from the self-consistent charge (SCC) correction[39] and it is calculated according to the eqn (7):

$$E_{\text{SCC}} = \frac{1}{2}\sum_{A,B}^{N} \gamma_{AB}\Delta_{q_A}\Delta_{q_B}, \tag{7}$$

where N is the number of atoms in the system and γ_{AB} is the gamma matrix which is obtained by analytic functions which depend on the Hubbard parameter of the atoms U_A and U_B, for instance. The Hubbard parameter is related to the hardness of the atoms and drives how the electron density is distributed between the atoms.

The SCC contribution to the hamiltonian is given by eqn (8):

$$H_{\mu\nu}^1 = -\frac{1}{2}S_{\mu\nu}\sum_{\xi}^{N}(\gamma_{A\xi} + \gamma_{B\xi})\Delta_{q\xi}. \tag{8}$$

The $S_{\mu\nu}$ element is the overlap matrix of the confined atomic orbitals. Finally, the SCC-DFTB hamiltonian is the sum of the two contributions according to eqn (9):

$$H_{\mu\nu} = H_{\mu\nu}^0 + H_{\mu\nu}^1\,;\ \forall\mu \in \{A\},\ \nu \in \{B\}. \tag{9}$$

The secular equation $\boldsymbol{HC} = \boldsymbol{SC}\boldsymbol{\varepsilon}$ is solved by a simplified general eigenvalue problem[56] for obtaining the molecular orbitals as well as their eigenstates. One can observe that the hamiltonian (eqn (9)) depends on the charges which are calculated using the molecular orbital coefficients $\boldsymbol{C}$. Therefore, a self-consistent procedure is necessary for solving the problem. The new ε, $\boldsymbol{C}$, $\boldsymbol{P}$ and Δ_{qA} are used to compute the total energy, eqn (1). The net charge criteria $net_{\text{criteria}} = \sqrt{\sum_{\xi}^{N}\frac{1}{N}\left(\Delta_{q_\xi}^{I} - \Delta_{q_\xi}^{I-1}\right)^2}$ (I is the current iteration and $I-1$ is the previous one) as well as the energy criteria ($|E^I - E^{I-1}|$) have to be achieved in order to have the final converged energy solution. If the convergence is not satisfied, a new $H_{\mu\nu}^1$ is constructed (eqn (8)) and another diagonalization ($\boldsymbol{HC} = \boldsymbol{SC}\varepsilon$) is performed, new ε, $\boldsymbol{C}$, $\boldsymbol{P}$ and Δ_{qA} are computed and used to calculate the new total energy. This iterative scheme continues until the net charge and total energy convergences are settled.

The repulsion energy (E_{rep}) term corrects the total energy ($E_{\text{bnd}} + E_{\text{SCC}}$) with respect to the reference calculated at the DFT level of theory. Therefore a set of adequate molecular models might be used as reference. The $E_{\text{rep}}(R_{\text{AB}})$ is calculated according to eqn (10) and fitted to a polynomial:

$$E_{\text{rep}}(R_{AB}) = E_{\text{DFT}}(R_{AB}) - (E_{\text{bnd}}(R_{\text{AB}}) + E_{\text{SCC}}(R_{\text{AB}})). \tag{10}$$

Actually, the E_{rep} corrects the tight-binding approximation of neglecting the three-center integrals as exemplified in Fig. 6. The

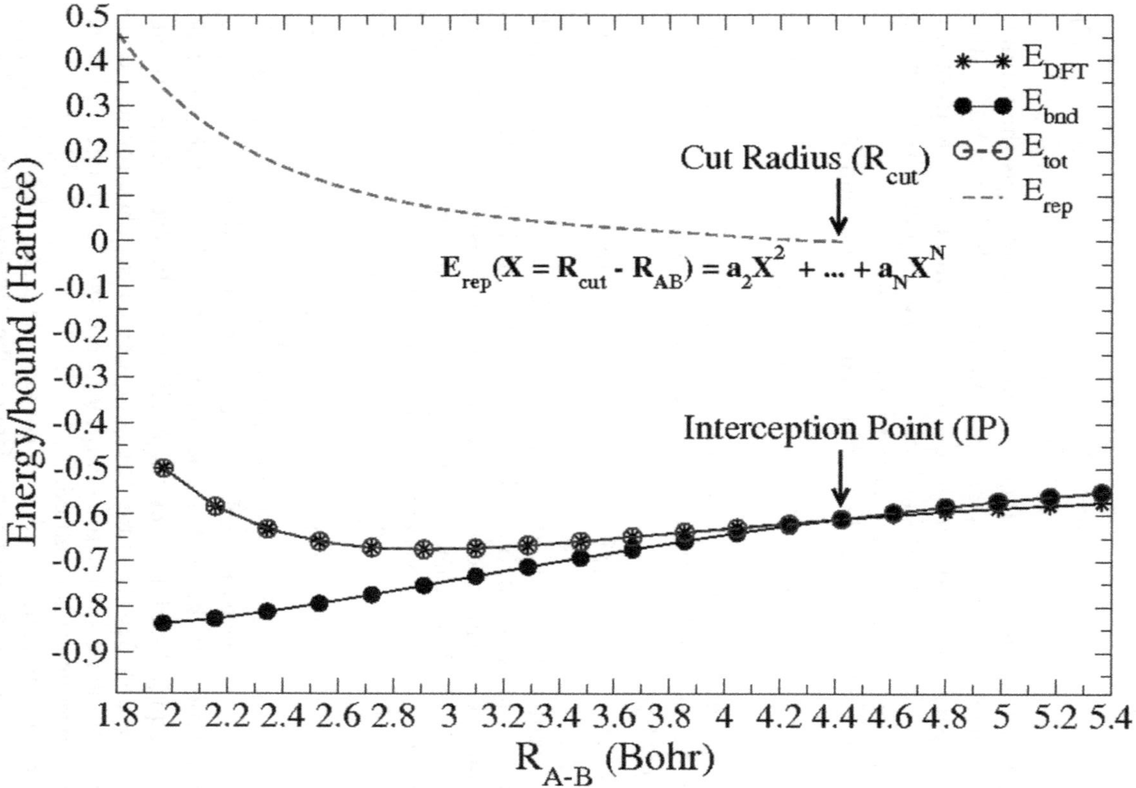

Fig. 6 DFT (E_{DFT}) and SCC-DFTB (E_{bnd}) potential curves and the coefficients of the polynomial ($P_{rep} = E_{rep}$) which describe the repulsion energy (E_{rep}), all data are expressed as a function of the interatomic distance (R_{AB}).

interception point and the cut radius as indicated in Fig. 6 are variables that one can handle to improve the parameterization. The quality of this parameterization will ensure the accuracy of the SCC-DFTB calculations.

Finally, the forces on atom A are calculated straightforward by applying vector $\boldsymbol{F}_A = -\boldsymbol{\nabla}_A E$ on the total SCC-DFTB energy (eqn (1)) to obtain the eqn (11):

$$\boldsymbol{F}_A = -2\sum_{i}^{M/2}\sum_{\mu\nu}^{K} C_{\mu i}C_{\nu i}\left(\frac{\partial H^0_{\mu\nu}}{\partial \vec{R}_A} - \left(\varepsilon_i - \frac{H^1_{\mu\nu}}{S_{\mu\nu}}\right)\frac{\partial S_{\mu\nu}}{\partial \vec{R}_A}\right) - \Delta_{q_A}\sum_{\xi}^{N}\frac{\partial \gamma_{A\xi}}{\partial \vec{R}_A}\Delta_{q_\xi} - \frac{\partial E_{\text{rep}}}{\partial \vec{R}_A}. \tag{11}$$

3.1 The structure of the SLAKO files

The Slater–Koster files (SLAKO) have the electronic part which includes all integrals evaluated in a grid of interatomic distances and the polynomial describing the E_{rep}.

In the parameterization procedure, the atomic calculation is performed by both atoms (A and B, for instance) according to eqn (12) with a confinement perturbation:

$$\left[\hat{T} + v_{\text{eff}}\left[\rho_A^0\right] + \left(\frac{\vec{r}}{r_0}\right)^2\right]\phi_\nu(\vec{r}) = \varepsilon_\nu\phi_\nu(\vec{r}). \tag{12}$$

The perturbed term $(\vec{r}/r_o)^2$ added in the one electron hamiltonian contracts the atomic orbitals $\phi_\nu(\vec{r})$ (the basis set), which results in an appropriate basis set to study molecules and periodic systems within the tight-binding approximation. The value of the confinement radius r_0 is usually set between 1.85 r_{cov} and 2.00 r_{cov} (r_{cov} is the covalent radius of atom A). Eqn (12) is solved for the atoms A and B, separately. Then, it is possible to evaluate the integrals for A–A, A–B and B–B atomic pairs. Recently, Wahiduzzaman *et al.*[59] have shown that the confinement radius could be tuned to reproduce the band electronic structure of many solids.

The eigenvalues obtained from atomic calculations for the free atom $\left(\varepsilon_\mu^{\text{free atom}}\right)$ are tabulated into the A–A and B–B SLAKO files and, in the SCC-DFTB calculation, used to build the diagonal of the hamiltonian matrix $\boldsymbol{H}^0$ (eqn (3)).

The molecular integrals for the atomic pairs are evaluated and tabulated as a function of the interatomic distances. Once the confined $\rho_A^0(\vec{r})$ for atom A and $\rho_B^0(\vec{r})$ for atom B are obtained, the off diagonal elements of $\boldsymbol{S}$ and $\boldsymbol{H}^0$ are tabulated as a function of the interatomic distances. To accomplish so, the Kohn Sham like potential for the A–B pair is constructed by the superposition of the effective atomic potentials as expressed in eqn (13):

$$v_{\text{KS}} = v_{\text{eff}}\left[\rho_A^0\right] + v_{\text{eff}}\left[\rho_B^0\right]. \tag{13}$$

Wahiduzzaman *et al.*[59] have demonstrated that superposition of the effective atomic potentials presents, in general, superior results for the band energies and their curvatures. Finally, after getting v_{KS}, the off diagonal matrix elements of $\boldsymbol{H}^0$ are built by solving the molecular integrals of eqn (14):

$$\langle \phi_\mu | T + v_{KS}[\rho_0^A + \rho_0^B] | \phi_\nu \rangle, \quad \mu \in \{A\}, \nu \in \{B\}, A \neq B. \tag{14}$$

The integrals are stored, for every angular momentum of the basis (ϕ_u and ϕ_v) in the *A*–*B* SLAKO as a function of the *A*–*B* distance in a grid of, usually, 0.05 bohr. Following the same procedure for *A*–*A* and *B*–*B*, the full set of homonuclear and heteronuclear SLAKOs used for simulating any system containing *A* and *B* atoms is obtained. The homonuclear SLAKO files have the tabulated atomic energies of the valence orbitals ($\varepsilon_\mu^{\text{free atom}}$) and the integrals for several angular momenta in a grid of distances. The heteronuclear SLAKO files do not hold the atomic energies of the valence orbitals ($\varepsilon_\mu^{\text{free atom}}$).

Therefore, during the SCC-DFTB calculations the hamiltonian is built based on the integrals previously tabulated in the SLAKOs and no explicit calculations of the integrals are required. SLAKO files are available in the website (http://www.dftb.org) for many different systems and pairs of atoms. However, the transferability of the SLAKO files for different classes of systems is limited and their use might be done with caution. Tests ought to be done before using them for calculating different systems.

4 Structures and stability

Since 2002, theoretical works[29,35,50,60–64] have been reported in describing the structure and the stability of imogolites. Depending on the used theoretical method, structures with different diameters and number of gibbsite hexagonal units around the circumference (*N*u) were found as a minimum. However, the driving force in all cases is attributed to lattice mismatch caused by shortening of O–O distances of orthosilicate group when bonded to gibbsite hexagonal ring. The difference in bond distances Al–O (1.9 Å) and Si–O (1.6 Å) imposes a curvature which leads to the minimum strain in the structure.

Although experimental works[6,8,9,33] have shown the imogolite NTs shape can be controlled by changing experimental synthesis protocol (as anionic ligands, temperature, pH), the synthesized NTs at the same conditions are monodisperse in diameter. Consequently, the imogolite monodispersity has motivated several studies in order to explain this unique behaviour at a molecular level.

The stability of NTs is estimated by calculating the strain energy, which is defined as the difference of the total energies between the NT and the planar layer. The strain energy reflects the energy required to curl a planar sheet onto a cylinder. In the imogolites case, the strain energy presents a minimum that reflects the monodispersity and well-defined structure. In order to model imogolite, usually the tubes are built starting from the slab, which is easy and allow the identification of symmetry. The roll-up process can lead to different symmetries as follow: *armchair* (n,n), *zigzag* $(n,0)$ and "chiral" (n,m), with $n \neq m$.

Imogolites present supercells with hundreds of atoms and, for this reason, most of the theoretical studies that performed a scan over different structures are based on empirical force fields or DFT tight-binding methods.[29,35,37,50,63] Due to the large computational effort associated with first-principles approach, as DFT, several works[38,65–68] limited their investigations to values of *N*u = 10 for natural imogolite[2] or *N*u = 12 to the synthetic one.[4] In this way, the results from different theoretical works are dependent on the applied method, *e.g.* molecular dynamics simulation, density functional theory or the SCC-DFTB. A comparison among the experimental and calculated diameters as well as the number of gibbsite hexagonal units around the circumference is shown in Table 1. One has to note that the theoretical simulations deal with idealized structures without any defect while real structures present defects. Besides, each work has calculated NT diameters in a different manner, *e.g.*, distance between the outmost O or H atoms, center-to-center separation; for this reason the diameters can change up to 4 Å for the same structure.

The first theoretical scan over *zigzag* and *armchair* NTs has been performed by Guimaraes *et al.*[29] on the basis of SCC-DFTB method.

Table 1 Experimental and calculated diameter and number of gibbsite hexagonal units for SW and DW imogolites.

Nanotube	Method	*N*u[a]	Diameter/nm	Ref.
SW img-Si	exprtl	10	2.0	2
		12	2.3	3
		11 to 15[b]	2.2 to 2.8[b]	9
	MD	16	2.9	35
	MD/CLAYFF	12	2.26	50
	DFT/PW91	12	2.48	60
	DFT/PBE	9	1.75	69
	DFT/B3LYP	10	1.45	61
	DFT/PBE	8	1.69	70
	SCC-DFTB	12	1.97	29
	SCC-DFTB	11	2.1	63
DW img-Si	SCC-DFTB	9/18	1.85/3.2[d]	63
SW img-Ge	exptl	18	3.3	30
		18	3.3	3
		18	3.03 ± 0.11	51
		18	3.5 ± 0.15	52
		—[c]	2.5 ± 0.5	53
		—[c]	3.8	
	SCC-DFTB	14	2.64	63
DW img-Ge	exptl	—[c]	2.4 ± 0.1/4.0 ± 0.1[d]	52
		—[c]	3.0 ± 0.5 (external)	53
		—[c]	2.6/4.3[d]	33
	SCC-DFTB	12/21	2.3/3.7[d]	63

[a] *N*u = Number of gibbsite hexagonal repeat units at the circumference.
[b] Dependent on the anion used in the synthesis.
[c] The value was not available.
[d] First value corresponds to internal NT while second value corresponds to external NT.

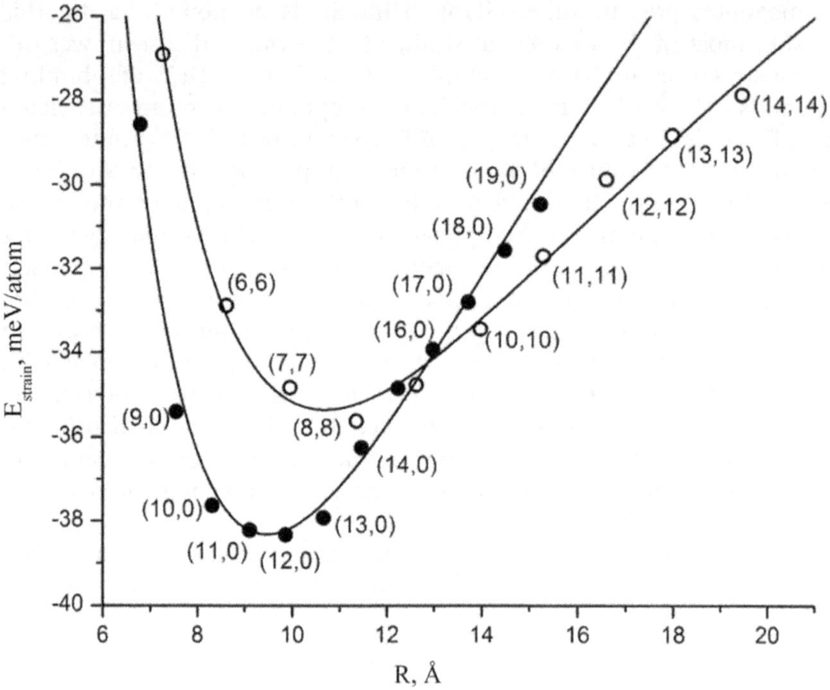

Fig. 7 Calculated strain energies E_{str} as a function of the radius R for zigzag (closed circles) and armchair (open circles) imogolite NTs. Reprinted with permission from ref. 29. Copyright 2007 American Chemical Society.

The calculated strain energy per atom for both chiralities have shown the same behaviour, although *zigzag* NTs are more stable than *armchair* ones and have a minimum at $Nu = 12$, *i.e.*, (12,0) (Fig. 7).

The NT stability can also be explained in the framework of a model based on the classical theory of elasticity. For several NTs, including C, BN, MoS_2, TiO_2[71–73] the NT strain energy E_{str} per atom can be related to the elastic modulus Y, the thickness h of monolayer and by the tube radius R:

$$E_{str} = \frac{a}{R^2} \sim \frac{Yh^3}{R^2} \tag{15}$$

The strain energy per atom follows the general trend $1/R^2$ for all known NTs except for imogolite. When the tube is formed by a symmetric layer, eqn (15) is valid. Imogolite is composed of nonsymmetrical aluminosilicate layer and a difference in the surface tensions $\Delta\sigma$ of outer and inner tube surfaces must be taken into account. As a result, an additional contribution is included to strain energy as can be seen in eqn (16) and (17)

$$E_{str} = \frac{a}{R^2} + \frac{b}{R} \sim \frac{Yh^3}{R^2} + \frac{\Delta\sigma \cdot h}{R}, \tag{16}$$

$$E_{str} = \frac{5.2}{R^2} - \frac{1.1}{R}, \tag{17}$$

in which E_{str} is given in eV atom^{-1}, R in Å, a in eV atom^{-1} Å^2, and b in eV atom^{-1} Å. The surface energy $\Delta\sigma$ supports a negative curvature, which decreases the strain energy and introduces a minimum into the $E_{str}(R)$ curve. The fit of the obtained E_{str} and R values for imogolite NTs using eqn (16) describes the change of the strain energy in the wide range of radii quite well (Fig. 7).

In a sequential work using SCC-DFTB, Lourenço *et al.*[63] have evaluated the influence of the hydrogen bonding network from the inner side of the imogolite tube over the structure stabilization. The inner hydroxyl orientations appear to influence the strength of the hydrogen bonding network, as suggested elsewhere.[61,70] Hence, Lourenço *et al.*[63] have used three different orientations for the inner hydroxyl groups as follows: clockwise (img-Si-1), anticlockwise (img-Si-2) and both (img-Si-3) directions (Fig. 8). The study has been performed for *zigzag* and *armchair* img-Si and img-Ge NTs. The results for img-Ge will be presented later. The imogolite clockwise orientation (img-Si-1) is the most stable one, being around 0.71 meV atom^{-1} lower in energy than img-Si-2 (anticlockwise orientation). The authors also performed SCC-DFTB calculations for the hypothetical DW aluminosilicate NTs. The optimized NTs were used to construct the $(p,0)@(q,0)$ DW NTs for $p = 7$–13 and $q - p = 8$, 9, 10, 11. The strain energy curve for the DW NTs as a function of the outermost radii is presented in Fig. 9. A bilayer was used as reference for the strain energy calculation. The curve was fitted using eqn (16). Each point presented in Fig. 9 corresponds to the minimum of the $(p,0)@(q,0)$ NTs. The most stable imogolite DW NT is the (9,0)@(18,0), in which the best combination is $q - p = 9$. One can notice that for the smaller radii, the strain energy contribution of the inner NT increases drastically. For the most stable DW NT, the inner tube presents large strain energy compared to the most stable SW img-Si NT and consequently, it does not favor the DW formation. Moreover, the intertubular interaction is only −1.74 meV per atom. As it will be shown later, for DW img-Ge NT, the inner NT presents a relatively small increase for the strain energy and the intertubular interaction is higher, rationalizing its existence.

The first theoretical assessment on the img-Si stability was carried out by Tamura and Kawamura[35] by means of molecular dynamics simulations. In the study, a specific interatomic potential, including

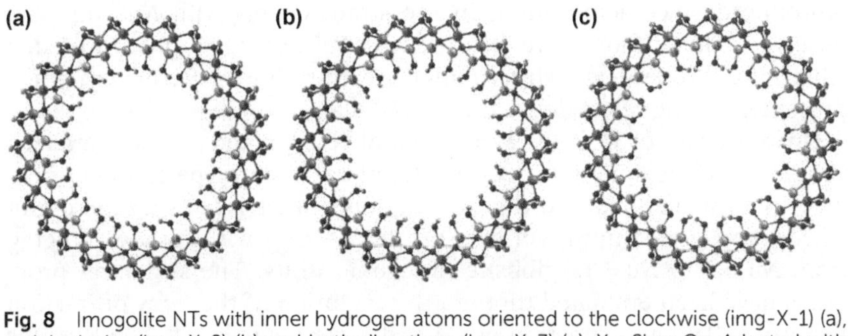

Fig. 8 Imogolite NTs with inner hydrogen atoms oriented to the clockwise (img-X-1) (a), anticlockwise (img-X-2) (b) and both directions (img-X-3) (c). X = Si or Ge. Adapted with permission from ref. 63. Copyright 2014 American Chemical Society.

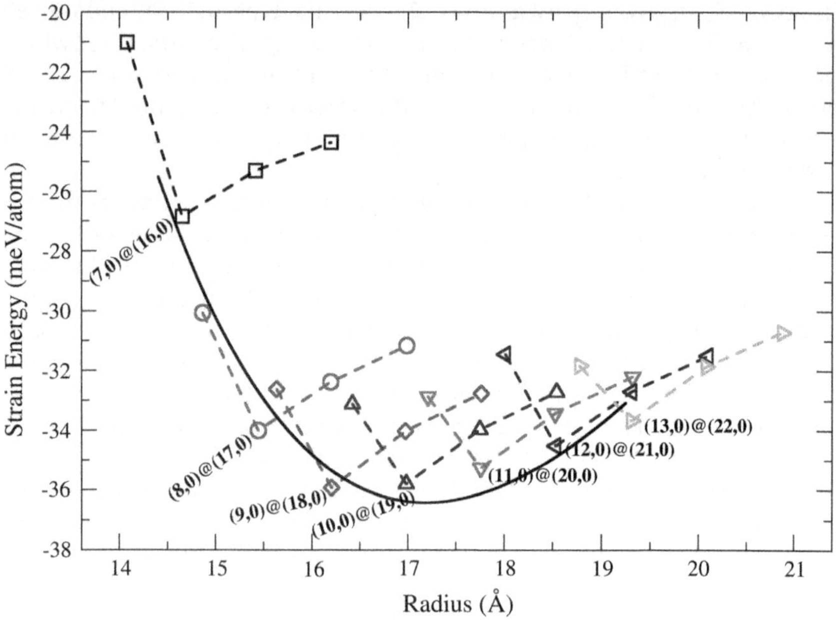

Fig. 9 Strain energy per atom as a function of the outer radius for DW (p,0)@(q,0) img-Si NTs taking into account different sizes and walls interactions: $q-p=8$, 9, 10 and 11. Dashed lines are related to the same inner NT but different outer NT. The reference used to calculate the double-walled NTs strain energies has been the bilayer. Reprinted with permission from ref. 63. Copyright 2014 American Chemical Society.

three-body and two-body terms, has been developed specially to treat img-Si and gibbsite NTs. The study has been performed for NTs from 10 to 24 gibbsite hexagonal units around the circumference (Nu). The obtained total energy has the minimum around diameter 2.6–2.9 nm, *i.e.*, Nu = 14–16. The authors have also simulated the vibrational spectra of Si–O bond which were in good accordance to experimental infrared spectra. Konduri *et al.*[50] have studied the origin of img-Si strain energy minimum by using a harmonic force-constant model and molecular dynamics simulations. The CLAYFF force field[36] has been employed, since it has been suitable to reproduce the properties of clay minerals including gibbsite. The total energy was obtained for NTs with Nu ranging from 9 to 24. The experimental findings of Farmer *et al.*[5] have been reproduced since a minimum in the strain energy with Nu = 12 were observed. The authors have also used a strain energy model to predict vibrational properties such as radial breathing mode frequency which presented strong dependence on the NT diameter.

On the basis of GGA-PW91 functional (GGA stands for Generalized Gradient Approximation), Alvarez-Ramirez[60] have performed first-principle calculations in order to study imogolites NTs as a function of silicon and germanium content (X = Si/(Si + Ge)) with models ranging from Nu = 9 to Nu = 13 gibbsite hexagonal units. The structural properties have been evaluated through the evolution of the X-ray diffraction pattern (XRDP) as a function of Nu and X for a single NT. The XRDP is barely affected by silicon content X and mostly affected by Nu, in such a way that the two main peaks in the low 2θ region are shifted to lower values as *N*u is increased. Thus, a linear correlation between the NT radius and the position of the two first low peaks on the XRD was found. Besides, the simulations of XRDP for bundles of img-Si were in good agreement with Mukherjee *et al.*[3] experimental data, with $Nu = 12$ and tube–tube distance separation of 2.48 nm. Lastly, the performed frequency analysis has shown two main regions, one characterized by collective movements of all the atoms in the fiber (phonons) and another with localized modes associated to O–H vibrations.

First principle calculations have been performed by Zhao *et al.*[69] and Lee *et al.*[70] to study imogolite NTs using the DFT method, as implemented in the SIESTA code.[41] In both studies, DZP numerical basis sets were adopted for the valence electrons and the GGA-PBE functional were used. Firstly, Zhao *et al.*[69] have studied the energetics of imogolite NT with structures ranging from $Nu = 7$–14. The strain energy curve presents a minimum for the NT with $Nu = 9$ and a local energy minimum for $Nu = 12$. The structure with $Nu = 9$ is 0.14 kJ mol^{-1} more stable than the NT with $Nu = 12$, as depicted in Fig. 10. According to the authors,[69] the two energetic minimum structures correlate to the natural and synthetic img-Si NTs.

Few years later, Lee *et al.*[70] studied the energetics and structure of *zigzag* and *armchair* img-Si NTs in order to understand the physicochemistry governing the origin of the strain energy minimum. The authors have shown that the organization of inner and outer hydroxyl

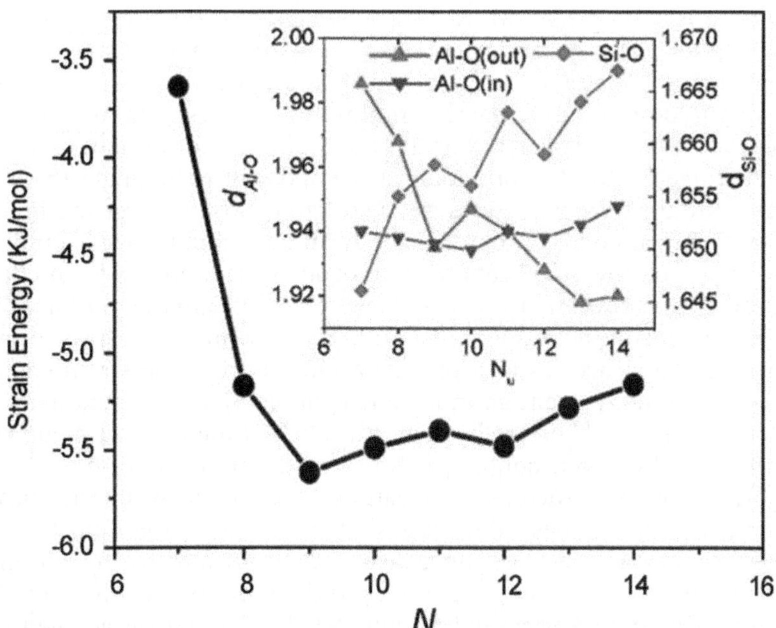

Fig. 10 Variation of the strain energy of imogolite nanotubes relative to imogolite sheet as a function of *N*u. The error bars (<0.003 KJ mol^{-1}) are smaller than symbols. The bond length evolution of Al–O and Si–O bonds is plotted in the inset of this figure. Reprinted with permission from ref. 69. Copyright 2009 American Chemical Society.

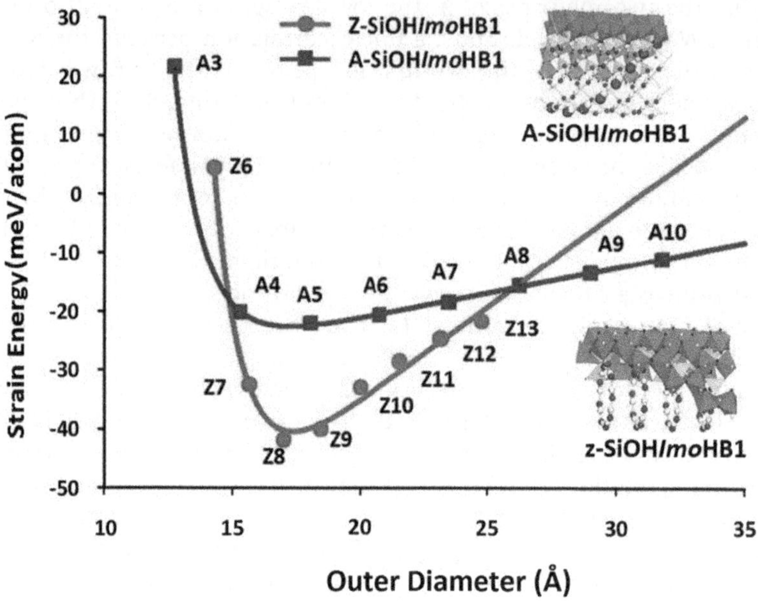

Fig. 11 Calculated strain energies as a function of the outer diameter for Z-SiOHlmoHB1 (red circles) and A-SiOHlmoHB1 (blue squares). Adapted with permission from ref. 70. Copyright 2011 American Chemical Society.

(OH) groups, *i.e.* hydrogen bonding networks, are important in the formation of imogolite NTs. The high monodispersity in diameter, with a minimum at *zigzag* NT with Nu = 8, is assigned to the inner hydrogen bonding networks. According to Lee *et al.*,[70] the unique disposition of inner silanol groups and their hydrogen bonding networks are reason for the origin of the strain energy minimum, as presented in Fig. 11. Nevertheless, the mismatches at Al–O and Si–O bond distances are the driving force for the NT formation. It is important to highlight that the authors are the only ones to assign the origin of strain energy minimum to the hydrogen bonds. The inner side hydrogen bond network certainly contributes to the stabilization of imogolites. However, a theoretical analysis[74] has shown that this is not enough to explain the minimum in the strain energy curve. For instance, the arsenic and phosphorous imogolite-like structures also present a minimum in the strain energy curve, but do not present an inner side hydrogen bonding network.

Demichelis *et al.*,[61] on the basis of GGA-B3LYP functional, also pointed out that the hydrogen bonding network in the inner imogolite wall is important for the structure stabilization. In contrast to the previously presented DFT studies, Demichelis *et al.*[61] have used all-electron Gaussian type basis sets and the helical boundary symmetry to perform a quantum mechanical study. According to the authors, the difference in stability of *zigzag* and *armchair* imogolite NTs is related to the geometrical organization of the inner wall. The *zigzag* tubes have SiO_4 tetrahedra closer and, consequently, the distance between neighboring oxygen atoms is shorter than in *armchair* tubes, as shown in Fig. 12. For instance, the O···O distance between two SiO_4 group is 3.09 Å for NT with

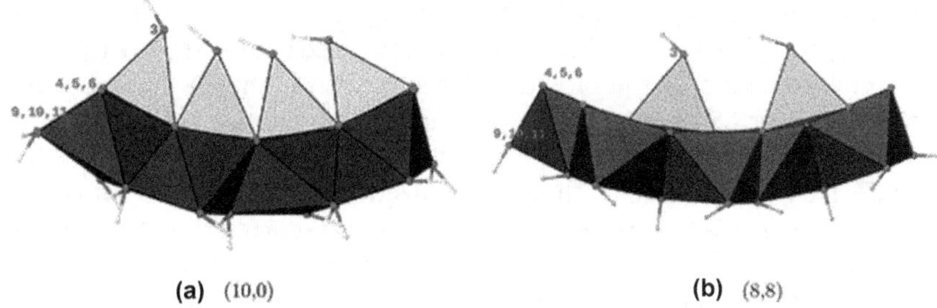

Fig. 12 View along the tube axis for (10,0) and (8,8) imogolite tubes. The distance between (10,0) apical oxygens are much shorter than in (8,8), because SiO_4 are much closer. Adapted from ref. 61 with permission from The Royal Society of Chemistry.

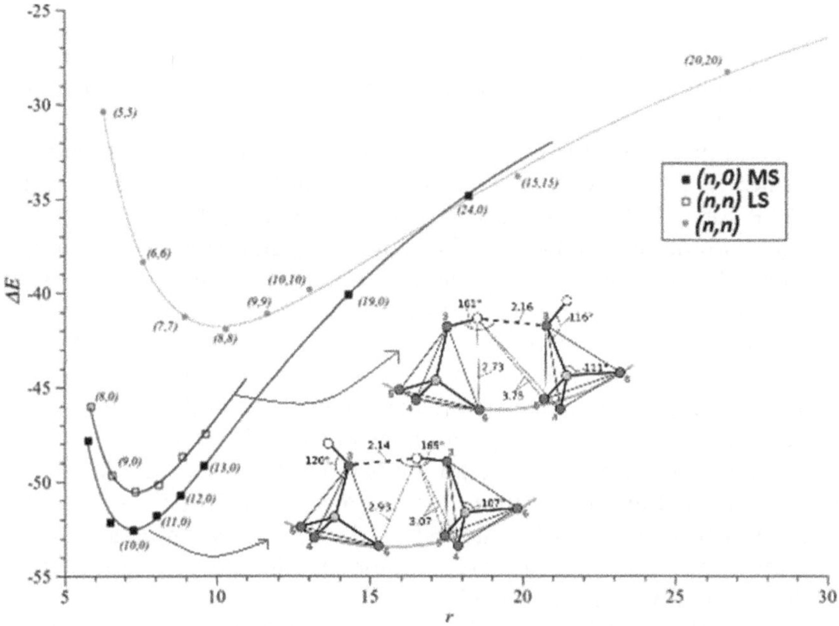

Fig. 13 Relative energy with respect to the flat slab (ΔE, kJ mol^{-1} per formula unit – fu) *vs.* tube radius measured between the basal O6 oxygen (r, Å). The curves with solid and empty squares correspond to the most (MS) and least (LS) stable (*n*, 0) *zigzag* tubes, respectively; circles correspond to (*n*, *n*) *armchair* tubes. Adapted from ref. 61 with permission from The Royal Society of Chemistry.

*N*u = 10 and 3.98 Å for NT with *N*u = 8. The stronger hydrogen bonds in the *zigzag* structure are responsible for the stability difference over the *armchair*. Besides, the presence of a structure with a minimum in both chiralities indicates that the imogolite natural curling is due to the misfit between Al–O and Si–O bond distances rather than hydrogen bonds.

Moreover, the obtained total energy curve presents a well-defined minimum at *N*u = 10 for *zigzag* NTs and *N*u = 8 for *armchair*. For *zigzag*, two close parallel curves were found, and the energy difference between the minima is 2.01 kJ mol^{-1} only, as presented in Fig. 13. The different stability between the most stable (MS) and least stable (LS) tubes is related to minor modification in the structural parameters associated with the apical O–H group, as highlighted in Fig. 13. The H···O interaction (involving an apical O) is slightly weaker in the LS structure, due to a slightly stronger interaction of the H with a basal oxygen atom (O_b) of the same tetrahedron (for LS the H···O_b distance is 2.73 Å while for MS is 3.07 Å).

In order to closely compare the results with the ones obtained by Zhao *et al.*,[69] Demichelis *et al.*[61] have optimized the most stable imogolite structures (*N*u = 8–13) using the PBE exchange–correlation functional. The total energy curve presents a minimum at *N*u = 9, in contrast to *N*u = 10 from B3LYP, although the absolute energy difference is only 0.4 kJ mol^{-1} per formula unit (fu). The position of the minimum tends to be slightly dependent on the method due to the small difference among tubes with *N*u around the minimum. Moreover, Demichelis *et al.*[61] have pointed out two hypotheses for the not explicable local and global minimum found by Zhao *et al.*[69] The first one is the NTs with *N*u = 7, 8, 9, 12, 13, 14 could belong to a MS and *N*u = 10, 11 to a LS structure; the second one is *N*u = 7–11 corresponding to the LS and *N*u = 12–14 to the MS structure.

In attempt to describe the dynamics of the img-Si NT rolling process, González *et al.*[64] have performed molecular dynamics simulation with CLAYFF force field using the large-scale atomic/molecular massively parallel simulator (LAMMPS). The simulation has started with a flat aluminosilicate sheet and has evolved freely until reaching its minimum energy configuration, ending up as a NT or a related structure, over a temperature range from 10 K up to 368 K. The role played by the precursors as shown by Yucelen *et al.*[8,75] was neglected. Depending on the temperature, the minimal structures adopted by the system are tubular, scrolled, more complex configuration, as presented in Fig. 14.

The minimum tubular configuration is monodisperse in diameter and dependent on the temperature. For instance, at 150 K, the NT at the minimum has *N*u = 11 while at 368 K (the synthesis temperature) the minimum has *N*u = 12. Moreover, a tendency in the direction of NTs agglomeration was also observed.

On the other hand, concerning img-Ge NTs, Lourenço *et al.*[63] have studied for the first time the stability and structural properties of SW and DW NTs. The *zigzag* and *armchair* configurations have been investigated for SW NTs while only *zigzag* chirality was investigated for DW NTs since it is significantly favoured on energetic grounds. As it has been presented for img-Si NTs, the influence of the hydrogen bonding network from the img-Ge tube inner side over structure stabilization have also been evaluated. The clockwise orientation (Fig. 8) is also the most stable one, being around 0.84 and 2.07 meV per atom lower in energy than both directions and anticlockwise direction, respectively. The strain energy curves for all studied structures were also fitted using eqn (16). The *zigzag* configuration is the most stable in the whole range of radii investigated and the minimum occurs at *N*u = 14 with diameter around 2.6 nm (Fig. 15). As expected, the diameter increases with the replacement of silicon by germanium, as can be seen in Table 1. The DW NTs selected for the study follow the combination ($p < q$) (p, 0)@(q, 0), where $q - p = 8, 9, 10, 11$. The most stable configuration concerning all studied NTs is the one with

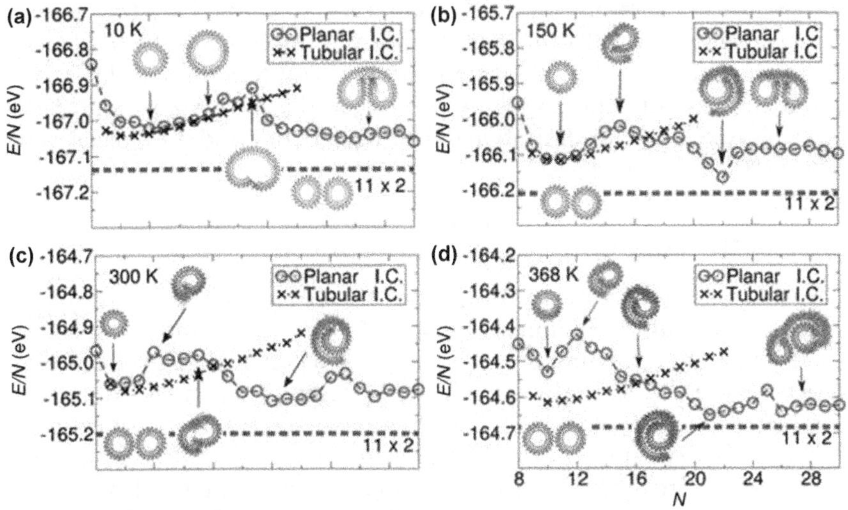

Fig. 14 Thermodynamic average of E/N for the temperatures 10, 150, 300, and 368 K. The black x correspond to tubular (TIC), and the red circles to planar (PIC), initial conditions. Snapshots of the scrolled conformations, obtained for the minimum energy, are provided for representative N values. The dashed horizontal line corresponds to the total energy of two interacting $N=11$ NTs at the same temperature. Reprinted with permission from ref. 64. Copyright 2014 American Chemical Society. The N in this figure corresponds to the Nu notations along the text, *i.e.*, number of gibbsite hexagonal repeat units in the circumference.

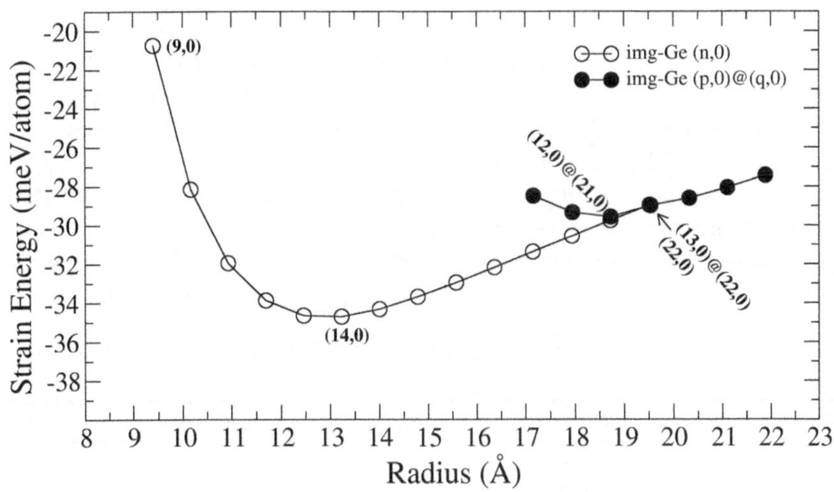

Fig. 15 Strain energy curves for the *zigzag* SW and DW img-Ge NTs. Adapted with permission from ref. 63. Copyright 2014 American Chemical Society.

$q\text{-}p=9$, which are plotted in Fig. 15. A bilayer was used as reference to calculate the strain energy. The SW and DW curves get close to each other for larger diameter. In the most stable DW NT, (12, 0)@(21, 0), the inner NT presents relatively small increase in the strain energy when compared to the most stable (14, 0) SW NT. One can notice that the most stable SW NT is not involved in the DW formation. This is expected since the img-Ge formation is thermodynamically controlled leading to the most stable structure. The distance between the walls (d_{wall}) (12, 0)@(21, 0) is around 1.43 Å, favoring the stabilization of DW NTs through hydrogen bonds. In this way, the DW with $q–p = 9$ are the most stable structures probably due to the maximization of hydrogen bonds and the smaller sum of the NTs strain energies.

5 Electronic and mechanical properties

The electronic structure of imogolites has been studied by means of different quantum mechanical methods,[29,60,61,63,67,69] providing insights about the band structure, density of states (DOS), local density of states (LDOS), band gap and the charge of the atoms. Thus, one can infer the conductivity degree of the NTs and their chemical reactivity, since there is a lack of experimental information on this subject. Essentially, the SCC-DFTB approximation,[40] the DFT applying the numerical basis sets approach (as implemented in the SIESTA[41] and DMol3 codes[43]) and the DFT using helical boundary symmetry (as implemented in the CRYSTAL program[44]) have been successfully used for studying imogolites and other NTs. The calculated band gaps for SW and DW imogolites are presented in Table 2.

In 2007, the imogolite electronic structure was reported for the first time on the basis of two different quantum mechanical methods.[29,60] Alvarez-Ramirez[60] has performed DFT/PW91 calculations using $\vec{\Gamma}$ point and different $\vec{k}$ points mesh and basis set size – double numeric (DN-$\vec{k}$) and double numeric with polarization (DNP-$\vec{k}$). In all cases, the DOS presented the same qualitative behavior and shape, suggesting the electronic behavior of the imogolite NTs is independent of the basis sets and number of $\vec{k}$ points. For instance, the DOS using DNP and several $\vec{k}$ points presented a band gap of 4.94 eV, only 0.2 eV higher than the other cases. Moreover, in the case of NTs with a changeable silicon–germaninum content, Alvarez-Ramirez[60] has pointed out that larger Ge concentration leads to smaller band gaps. Nevertheless, the results

Table 2 Calculated band gap energies (eV) of the SW and DW imogolites by different theoretical methods.

Nanotube	Method	*Nu*[a]	Band gap/eV	Ref.
SW img-Si	DFT/PW91	10	4.9	60
	DFT/PBE	9, 12	5.2, 5.3	69
	DFT/PBE	12	3.67	67
	DFT/B3LYP	10	7.2	61
	SCC-DFTB	12	10.3	29
	SCC-DFTB	11	10.3	63
DW img-Si	SCC-DFTB	9/18	8.0	63
SW img-Ge	DFT/PW91	10	~4.6	60
	SCC-DFTB	14	9.5	63
DW img-Ge	SCC-DFTB	12/21	8.5	63

[a] *Nu* = Number of gibbsite hexagonal repeat units in the circumference.

suggest that imogolites have an insulator behavior. The LDOS results show that the main contributions to the valence band come from the O layers. Hirshfeld charge population analysis has been used to determine the SW img-Si and img-Ge surface charges. The inner and outer H presented the same positive charge $\sim$0.13*e*. The O average Hirshfeld charges for O atoms going from inner to outer NT wall, are −0.273*e*, −0.324*e* and −0.305*e*, respectively. The Al average charge is 0.506*e*. Comparing the Si and Ge atoms in the NT structure, Si presents a charge of 0.616*e* and Ge of 0.670*e*. On the other hand, Guimaraes *et al.*[29] have studied imogolite electronic structure in the framework of SCC-DFTB. The most stable zigzag imogolite NT presented a band gap around 10.3 eV. As it is well known, DFT underestimates the band gap of insulators while SCC-DFTB overestimates it.[29] Based on SCC-DFTB Mulliken population analysis, the charge transfer from Al (around +0.6*e*) to outer O (around −0.45*e*) and from Si (around +0.8*e*) to inner O (around −0.54*e*) were determined. Furthermore, the authors[29] have also estimated the (12, 0) img-Si electrostatic field which express the charge distribution within the NT: outer surface is mostly positive while the inner surface is negative.

On the basis of DFT/PBE calculations, Zhao *et al.*[69] have obtained a direct band gap of 5.2–5.3 eV for the most stable img-Si NTs with $Nu = 9$ and 12. Both of them are insulators. The maximum of valence band and the minimum of conduction band are at $\vec{\Gamma}$ point, being composed by O atoms (connected to Si) and outside hydroxyls, respectively. The calculated charges are in concordance to SCC-DFTB method: a set of positive charges at the outer surface and negative ones at the inner surface.

Demichelis *et al.*[61] have used DFT/B3LYP method with helical boundary symmetry to evaluate imogolite electronic properties. The obtained band gap using the hybrid functional B3LYP is 7.2 eV for all rolling directions. The previous GGA calculations predicted band gaps in the range of 3.7 to 5.3 eV, while the SCC-DFTB values are much larger (around 10 eV). It is well known that pure DFT functionals tend to underestimate band gap whereas SCC-DFTB tends to overestimate it. The obtained value from hybrid functional falls in between GGA and LDA (Local Density Approximation) extremes. The Mulliken population analysis presented the following charges: +2.3*e* for Si, +2.0*e* for Al, −1.2*e* and −0.80*e* for the two types of O and +0.35*e* for H.

In the framework of SCC-DFTB calculations, Lourenço *et al.*[63] have calculated the electronic properties for SW and DW img-Si and img-Ge NTs. The band gap for the SW img-Si and img-Ge NTs are 10.3 eV and 9.5 eV, respectively, while the DW ones are 8.0 and 8.5 eV, respectively. It can be seen that the band gap decreases when the Si atoms are substituted by Ge atoms, as it was observed by Alvarez-Ramirez on the basis of DFT/PW91 method. The PDOS of img-Si and img-Ge showed that the valence band is governed by oxygen states and the largest contribution to the conduction band arises from the silicon and germanium atoms. The Al and Ge atoms in the img-Ge NTs present charge around +0.35*e* and +0.61*e*, respectively. Moreover, the O atoms bound to Al and Ge have the net charge of −0.32*e* and −0.39*e*, respectively. In comparison to the

DW (12, 0)@(21, 0) img-Ge NT, the Al and Ge charges from (14, 0) img-Ge SWNT do not change significantly.

Mechanical properties give insights about the strength of the materials and are important to envisage applications. The Young's modulus is a central property which is accessed by experiments or theoretical estimates that characterizes the NTs stiffness. As far as we know, only theoretical works have obtained imogolite Young's moduli (Y) so far. The Y has been calculated as described in the literature[29,63,76] by performing a series of relaxation calculations for different values of the cell length in the direction of the tube periodic axis, and thus imposing either tensile or compressive strain on the NT. The Y have been estimated using eqn (18), in which $\varepsilon = (a - a_0)/a_0$, where ε is the strain, a and a_0 are the cell parameter along the tube outside and in the equilibrium, respectively. The volume of the tube is estimated by the cylinder equation: $V_0 = \pi C_0 (R_{out}^2 - R_{in}^2)$, C_0 is the cell length along the periodic axis, R_{out} is the outer radius and R_{in} is the inner one. The Young's moduli can be calculated by performing numerical derivatives of the energy (E) expressed as a function of the strain ε obtained by varying a, yielding $E(\varepsilon)$. More details about optimum volume and Y calculation, can be found in ref. 76.

$$Y = \frac{1}{V_0}\left(\frac{\partial^2 E}{\partial \varepsilon^2}\right)_{\varepsilon=0} \tag{18}$$

Guimarães *et al.*[29] evaluated the SW img-Si NTs Young's Moduli by means of the SCC-DFTB method. The results exhibited a range of 175–479 GPa for the *zigzag* NTs and of 280–416 GPa for the *armchair* ones. In comparison to other NTs, the authors have concluded that imogolite NTs are less stiff than carbon NTs, and in the same magnitude order of MoS_2 (~230 GPa),[73] GaS (~270 GPa),[77] halloysite (~230–240 GPa)[78] and chrysotile NTs (~261–323 GPa).[79,80]

DW img-Ge NT have been calculated by Lourenço *et al.*[63] also on the basis of SCC-DFTB method. For the SW img-Ge, the range is 320–350 GPa for the zigzag and 332–370 GPa for the armchair NTs. For the most stable img-Ge DW NT, the (12, 0)@(21, 0), the Young's modulus is 328 GPa. Img-Si and Img-Ge NTs Young's moduli are in the same order of magnitude of other inorganic NTs, as previously discussed.

6 Final remarks

Imogolites are NTs with extraordinary structural and mechanical properties that can be found in nature or be synthesized in mild conditions. They are usually uncapped, single-walled with well-defined diameter and length, hence, a perfect target for developing new materials with enhanced properties. Much progress has been achieved in the last decade concerning the synthesis, controlling the NT diameter to some extent and, for the case of the aluminogermanate, the synthesis of single- or double-walled NTs. Although the synthesis is relatively simple, the establishing of the adequate parameters such as temperature, ionic strength, pH and reactants concentration is delicate. Much effort has been made for understanding the imogolites formation mechanism. However, the mechanism of this complex process at a molecular level is still incomplete.

The molecular modelling and computer simulation of imogolites have been applied for understanding this important class of materials. The computational chemistry techniques have contributed for understanding the stability and monodispersity of the material. In fact, a minimum in the strain energy explains why imogolites have very well-defined structures with respect to other clay mineral NTs such as chrysotile and halloysites.[78,79] The preferred $(n, 0)$ zigzag chirality is a consensus with respect to the theoretical investigations reported in the literature. The role of the internal hydrogen bonding to imogolites stability has been discussed in details. The question of why the double-walled img-Ge was successfully synthesized while the img-Si seems to be impossible was explained based on molecular modelling. Insights about their electronic structure and mechanical properties have also been provided. The imogolites are considered insulator with large band gaps and the stiffness are in the same range as other inorganic NTs and comparable to steel.

Different theoretical chemistry methodologies have been used to investigate the imogolites. The Self-Consistent-Charge Density-Functional Tight-Binding method (SCC-DFTB) is one of the most used for investigating this system and has been briefly described in the present chapter. It is important to note that molecular modelling has also been applied to explore the properties of other possible imogolite-like NTs either than img-Si and img-Ge. It is really disconcerting the fact that other imogolite-like NTs have not been synthesized either than img-Si and img-Ge, in spite of the effort that has been employed for the development of the field. Indeed, it has been shown that many other imogolite-like NTs replacing the silicon by elements such as C, Si, Ge, Sn, P, As are stable enough to exist.[74,81] However, the synthesis in aqueous solution involves a complex chemical speciation and self-organizing process that requires finding the adequate experimental conditions to occur. Furthermore, its complex mechanism precludes the fully simulation by the state-of-the-art of theoretical chemistry and more methodological developments are still necessary.

It is worthwhile to mention that the functionalization of the NTs inner and outer surface has been successfully accomplished, broadening, then, the frontiers of imogolites application.[18,82] They are envisaged for the development of nanoreactors, controlled release devices, ion conductors for batteries, gas storage and separation systems. Computer modelling of such functionalized systems have already been reported providing insights about their structural, electronic and mechanical properties.[76]

Finally, in spite of the progress realized in the development of new materials based on imogolites, the synthesis of new imogolite-like structures is an important issue and deserves more attention. Actually, imogolites have not received much attention commensurate with their potential for technological application.

Acknowledgements

We would like to thank the Brazilian agencies Conselho Nacional para o Desenvolvimento Científico e Tecnológico (CNPq), Coordenação de Aperfeiçoamento de Pessoal de Nível Superior (CAPES) and Fundação de Amparo a Pesquisa do Estado de Minas Gerais (FAPEMIG) for the long term support of our research projects. The National Institute of Science and Technology for Mineral Resources, Water and Biodiversity (INCT-ACQUA) is gratefully acknowledged.

References

1 N. Yoshinaga and S. Aomine, *Soil Sci. Plant Nutr.*, 1962, **8**, 22–29.
2 P. D. Cradwick, K. Wada, J. D. Russell, N. Yoshinag, C. R. Masson and V. C. Farmer, *Nature, Phys. Sci.*, 1972, **240**, 187–189.
3 S. Mukherjee, V. A. Bartlow and S. Nair, *Chem. Mater.*, 2005, **17**, 4900–4909.
4 V. C. Farmer, A. R. Fraser and J. M. Tait, *J. Chem. Soc., Chem. Commun.*, 1977, 462–463.
5 V. C. Farmer, M. J. Adams, A. R. Fraser and F. Palmieri, *Clay Miner.*, 1983, **18**, 459–472.
6 H. Lee, Y. Jeon, Y. Lee, S. U. Lee, A. Takahara and D. Sohn, *J. Phys. Chem. C*, 2014, **118**, 8148–8152.
7 C. Levard, A. Masion, J. Rose, E. Doelsch, D. Borschneck, C. Dominici, F. Ziarelli and J.-Y. Bottero, *J. Am. Chem. Soc.*, 2009, **131**, 17080–17081.
8 G. I. Yucelen, R. P. Choudhury, A. Vyalikh, U. Scheler, H. W. Beckham and S. Nair, *J. Am. Chem. Soc.*, 2011, **133**, 5397–5412.
9 G. I. Yucelen, D.-Y. Kang, R. C. Guerrero-Ferreira, E. R. Wright, H. W. Beckham and S. Nair, *Nano Lett.*, 2012, **12**, 827–832.
10 I. Bottero, B. Bonelli, S. E. Ashbrook, P. A. Wright, W. Zhou, M. Tagliabue, M. Armandi and E. Garrone, *Phys. Chem. Chem. Phys.*, 2011, **13**, 744–750.
11 F. Ohashi, S. Tomura, K. Akaku, S. Hayashi and S. I. Wada, *J. Mater. Sci.*, 2004, **39**, 1799–1801.
12 C. Zanzottera, M. Armandi, S. Esposito, E. Garrone and B. Bonelli, *J. Phys. Chem. C*, 2012, **116**, 20417–20425.
13 D.-Y. Kang, N. A. Brunelli, G. I. Yucelen, A. Venkatasubramanian, J. Zang, J. Leisen, P. J. Hesketh, C. W. Jones and S. Nair, *Nat. Commun.*, 2014, **5**, 1–9.
14 K. Yamamoto, H. Otsuka, S. Wada, D. Sohn and A. Takahara, *Polymer*, 2005, **46**, 12386–12392.
15 S. Nakagaki and F. Wypych, *J. Colloid Interface Sci.*, 2007, **315**, 142–157.
16 D. Geraldo, N. Arancibia-Miranda, N. Villagra, G. Mora and R. Arratia-Perez, *J. Nanopart. Res.*, 2012, **14**, 1–9.
17 T. Hongo, J. Sugiyama, A. Yamazaki and A. Yamasaki, *Ind. Eng. Chem. Res.*, 2013, **52**, 2111–2115.
18 D. Y. Kang, J. Zang, C. W. Jones and S. Nair, *J. Phys. Chem. C*, 2011, **115**, 7676–7685.
19 Y. Kuroda, K. Fukumoto and K. Kuroda, *Appl. Clay Sci.*, 2012, **55**, 10–17.
20 W. Ma, H. Otsuka and A. Takahara, *Chem. Commun.*, 2011, **47**, 5813–5815.
21 W. Ma, W. O. Yah, H. Otsuka and A. Takahara, *Beilstein J. Nanotechnol.*, 2012, **3**, 82–100.
22 H. Otsuka and A. Takahara, in *Inorganic and Metallic Nanotubular Materials: Recent Technologies and Applications*, ed. T. Kijima, Springer, 2010, vol. 117, pp. 169–190.

23 B. Thomas, T. Coradin, G. Laurent, R. Valentin, Z. Mouloungui, F. Babonneau and N. Baccile, *RSC Adv.*, 2012, **2**, 426–435.
24 B. Bonelli, M. Armandi and E. Garrone, *Phys. Chem. Chem. Phys.*, 2013, **15**, 13381–13390.
25 E. Belorizky, P. H. Fries, A. Guillermo and O. Poncelet, *ChemPhysChem*, 2010, **11**, 2021–2026.
26 Y. Kuroda and K. Kuroda, *Chem. Lett.*, 2011, **40**, 46–48.
27 J. Zang, S. Konduri, S. Nair and D. S. Sholl, *Acs Nano*, 2009, **3**, 1548–1556.
28 M. Dvoyashkin, J. Zang, G. I. Yucelen, A. Katihar, S. Nair, D. S. Sholl, C. R. Bowers and S. Vasenkov, *J. Phys. Chem. C*, 2012, **116**, 21350–21355.
29 L. Guimarães, A. N. Enyashin, J. Frenzel, T. Heine, H. A. Duarte and G. Seifert, *Acs Nano*, 2007, **1**, 362–368.
30 S. Wada and K. Wada, *Clays Clay Miner.*, 1982, **30**, 123–128.
31 B. H. Bac, Y. Song, M. H. Kim, Y.-B. Lee and I. M. Kang, *Chem. Commun.*, 2009, 5740–5742.
32 A. Thill, B. Guiose, M. Bacia-Verloop, V. Geertsen and L. Belloni, *J. Phys. Chem. C*, 2012, **116**, 26841–26849.
33 A. Thill, P. Maillet, B. Guiose, O. Spalla, L. Belloni, P. Chaurand, M. Auffan, L. Olivi and J. Rose, *J. Am. Chem. Soc.*, 2012, **134**, 3780–3786.
34 G. I. Yucelen, D.-Y. Kang, I. Schmidt-Krey, H. W. Beckham and S. Nair, *Chem. Eng. Sci.*, 2013, **90**, 200–212.
35 K. Tamura and K. Kawamura, *J. Phys. Chem. B*, 2002, **106**, 271–278.
36 R. T. Cygan, J.-J. Liang and A. G. Kalinichev, *J. Phys. Chem. B*, 2004, **108**, 1255–1266.
37 S. Konduri, S. Mukherjee and S. Nair, *Acs Nano*, 2007, **1**, 393–402.
38 B. Creton, D. Bougeard, K. S. Smirnov, J. Guilment and O. Poncelet, *J. Phys. Chem. C*, 2008, **112**, 10013–10020.
39 M. Elstner, D. Porezag, G. Jungnickel, J. Elsner, M. Haugk, T. Frauenheim, S. Suhai and G. Seifert, *Phys. Rev. B*, 1998, **58**, 7260–7268.
40 B. Aradi, B. Hourahine and T. Frauenheim, *J. Phys. Chem. A*, 2007, **111**, 5678–5684.
41 E. Artacho, D. Sánchez-Portal, P. Ordejón, A. García and J. M. Soler, *Phys. Status Solidi B*, 1999, **215**, 809–817.
42 B. Delley, *J. Chem. Phys.*, 1990, **92**, 508–517.
43 B. Delley, *J. Chem. Phys.*, 2000, **113**, 7756–7764.
44 R. Dovesi, V. R. Saunders, C. Roetti, R. Orlando, C. M. Zicovich-Wilson, F. Pascale, B. Civalleri, K. Doll, N. M. Harrison, I. J. Bush, P. D'Arco and M. Llunell, *CRYSTAL09*, University of Torino, Torino, 2009.
45 C. Levard, A. Masion, J. Rose, E. Doelsch, D. Borschneck, L. Olivi, P. Chaurand, C. Dominici, F. Ziarelli, A. Thill, P. Maillet and J. Y. Bottero, *Phys. Chem. Chem. Phys.*, 2011, **13**, 14516–14522.
46 V. C. Farmer and A. R. Fraser, *Int. Clay Conf.*, 1979, 547–553.
47 M. A. Wilson, G. S. H. Lee and R. C. Taylor, *J. Non-Cryst. Solids*, 2001, **296**, 172–181.
48 H. Yang, C. Wang and Z. Su, *Chem. Mater.*, 2008, **20**, 4484–4488.
49 T. Urabe, T. Tsugoshi and M. Tanaka, *J. Mass Spectrom.*, 2009, **44**, 193–202.
50 S. Konduri, S. Mukherjee and S. Nair, *Phys. Rev. B*, 2006, **74**, 033401–033404.
51 C. Levard, J. Rose, A. Masion, E. Doelsch, D. Borschneck, L. Olivi, C. Dominici, O. Grauby, J. C. Woicik and J.-Y. Bottero, *J. Am. Chem. Soc.*, 2008, **130**, 5862–5863.

52 P. Maillet, C. Levard, E. Larquet, C. Mariet, O. Spalla, N. Menguy, A. Masion, E. Doelsch, J. Rose and A. Thill, *J. Am. Chem. Soc.*, 2010, **132**, 1208–1209.
53 P. Maillet, C. Levard, O. Spalla, A. Masion, J. Rose and A. Thill, *Phys. Chem. Chem. Phys.*, 2011, **13**, 2682–2689.
54 M.-S. Amara, E. Paineau, M. Bacia-Verloop, M.-E. M. Krapf, P. Davidson, L. Belloni, C. Levard, J. Rose, P. Launois and A. Thill, *Chem. Commun.*, 2013, **49**, 11284–11286.
55 D. Porezag, T. Frauenheim, T. Kohler, G. Seifert and R. Kaschner, *Phys. Rev. B*, 1995, **51**, 12947–12957.
56 T. Frauenheim, G. Seifert, M. Elsterner, Z. Hajnal, G. Jungnickel, D. Porezag, S. Suhai and R. Scholz, *Phys. Status Solidi B*, 2000, **217**, 41–62.
57 P. Koskinen and V. Mäkinen, *Comput. Mater. Sci.*, 2009, **47**, 237–253.
58 A. F. Oliveira, G. Seifert, T. Heine and H. A. Duarte, *J. Braz. Chem. Soc.*, 2009, **20**, 1193–1205.
59 M. Wahiduzzaman, A. F. Oliveira, P. Philipsen, L. Zhechkov, E. van Lenthe, H. A. Witek and T. Heine, *J. Chem. Theory Comput.*, 2013, **9**, 4006–4017.
60 F. Alvarez-Ramirez, *Phys. Rev. B: Condens. Matter Mater. Phys.*, 2007, **76**, 125421–125434.
61 R. Demichelis, Y. Noel, P. D'Arco, L. Maschio, R. Orlando and R. Dovesi, *J. Mater. Chem.*, 2010, **20**, 10417–10425.
62 A. Kuc and T. Heine, *Adv. Mater.*, 2009, **21**, 4353–4356.
63 M. P. Lourenço, L. Guimaraes, M. C. da Silva, C. de Oliveira, T. Heine and H. A. Duarte, *J. Phys. Chem. C*, 2014, **118**, 5945–5953.
64 R. I. González, R. Ramirez, J. Rogan, J. Alejandro Valdivia, F. Munoz, F. Valencia, M. Ramirez and M. Kiwi, *J. Phys. Chem. C*, 2014, **118**, 28227–28233.
65 B. Creton, D. Bougeard, K. S. Smirnov, J. Guilment and O. Poncelet, *Phys. Chem. Chem. Phys.*, 2008, **10**, 4879–4888.
66 G. Teobaldi, N. S. Beglitis, A. J. Fisher, F. Zerbetto and A. A. Hofer, *J. Phys.: Condens. Matter*, 2009, **21**, 195301–195309.
67 L. Li, Y. Xia, M. Zhao, C. Song, J. Li and X. Liu, *Nanotechnology*, 2008, **19**, 175702–175711.
68 K. S. Smirnov and D. Bougeard, *J. Phys.: Condens. Matter*, 2010, **22**, 284115–284121.
69 M. Zhao, Y. Xia and L. Mei, *J. Phys. Chem. C*, 2009, **113**, 14834–14837.
70 S. U. Lee, Y. C. Choi, S. G. Youm and D. Sohn, *J. Phys. Chem. C*, 2011, **115**, 5226–5231.
71 A. N. Enyashin and G. Seifert, *Phys. Status Solidi B*, 2005, **242**, 1361–1370.
72 E. Hernandez, C. Goze, P. Bernier and A. Rubio, *Phys. Rev. Lett.*, 1998, **80**, 4502–4505.
73 G. Seifert, H. Terrones, M. Terrones, G. Jungnickel and T. Frauenheim, *Phys. Rev. Lett.*, 2000, **85**, 146–149.
74 L. Guimaraes, Y. N. Pinto, M. P. Lourenco and H. A. Duarte, *Phys. Chem. Chem. Phys.*, 2013, **15**, 4303–4309.
75 G. I. Yucelen, R. P. Choudhury, J. Leisen, S. Nair and H. W. Beckham, *J. Phys. Chem. C*, 2012, **116**, 17149–17157.
76 M. C. da Silva, E. C. dos Santos, M. P. Lourenco, M. P. Gouvea and H. A. Duarte, *Front. Mater.*, 2015, **2**, 1–10.
77 T. Köhler, T. Frauenheim, Z. Hajnal and G. Seifert, *Phys. Rev. B*, 2004, **69**, 193403–193406.
78 L. Guimaraes, A. N. Enyashin, G. Seifert and H. A. Duarte, *J. Phys. Chem. C*, 2010, **114**, 11358–11363.

79 M. P. Lourenço, C. de Oliveira, A. F. Oliveira, L. Guimarães and H. A. Duarte, *J. Phys. Chem. C*, 2012, **116**, 9405–9411.
80 S. Piperno, I. Kaplan-Ashiri, S. R. Cohen, R. Popovitz-Biro, H. D. Wagner, R. Tenne, E. Foresti, I. G. Lesci and N. Roveri, *Adv. Funct. Mater.*, 2007, **17**, 3332–3338.
81 F. Alvarez-Ramirez, *J. Comput. Theor. Nanosci.*, 2009, **6**, 1120–1124.
82 D. Y. Kang, J. Zang, E. R. Wright, A. L. McCanna, C. W. Jones and S. Nair, *Acs Nano*, 2010, **4**, 4897–4907.

Application of DFT modeling in Fischer–Tropsch synthesis over Co-based catalysts

Xin-Chao Xu,[a] Pengfei Tian,[a] Yong Cao,[b] Jing Xu*[a] and Yi-Fan Han*[a]

DOI: 10.1039/9781782622703-00184

1 Introduction

Fischer–Tropsch Synthesis (FTS) is regaining interest as a platform process for the production of long-chain hydrocarbons from syngas ($CO + H_2$), which is produced from natural/shale gas, coal and renewable biomass because FTS fuels meet stringent environmental requirements, and is almost free of sulfur, aromatics and heavy metals. Numerous efforts have been dedicated to FTS since it was discovered by Fischer and Tropsch in the 1920s.[1–3] The metals in group VIII exhibit catalytic activity for CO hydrogenation, while Fe- and Co-based catalysts have already been applied commercially. In particular, Co-based catalysts have been attracting much attention because of their good compromise of activity, selectivity, stability and low water-gas-shift activity in FTS.[4,5]

Over the past decades, first principles-based modeling has rapidly gained importance as a new tool in the study of catalytic reactions.[6] Structure changes in catalysts and the formation of intermediates in the typical catalytic reactions are too rapid to be detected experimentally. First principles modeling, on the other hand, is best placed to study molecular level effects and validates our chemical intuition on a reaction. Breakthrough advances in heterogeneous catalysis will only occur if catalytic systems are understood and well controlled. As Somorjai[7] pointed out, the ultimate goal for catalysis science in the 21st century is the "design" of the active sites to provide 100% selectivity for catalyst-based chemical processes. In this aspect, molecular modeling can provide new insights into reaction pathways and collate information from different experimental techniques into a coherent picture. In particular, in combination with careful experimental work on model catalysts, rational design of active sites, as "a dreamful methodology", is becoming real.

The reaction mechanism of FTS is fairly complex, composing a network of the elementary bond-breaking and bond-formation steps, which include the activation of CO and H_2 as well as hydrogenation and chain growth/termination over metal catalysts. FTS is assumed to follow a chain growth mechanism *via* the addition of CH_2 groups; while alternative mechanisms, such as CO insertion *via* carbide, have also been proposed. A better understanding of the reaction mechanism, the key reaction

[a] *State Key Laboratory of Chemical Engineering, East China University of Science and Technology, Shanghai, China. E-mail: xujing@ecust.edu.cn; yifanhan@ecust.edu.cn*
[b] *Department of Chemistry, Fudan University, Shanghai, China*

intermediates, steps and the nature of the active sites, is crucial to further improve the process.

The optimal metals for FTS have several prerequisites that can promote CO activation yet provide a balanced degree of surface hydrogenation and C–C coupling, in order to produce longer chain hydrocarbons. Co shows excellent properties thus allowing to dissociate adsorbed CO and subsequently undergo both hydrogenation and coupling reactions with similar rates. The molecular transformations, which determine the mechanism, the kinetics and the influence of the atomic scale features of Co-based catalysts on the catalytic activity, however, are not very well understood. A more detailed understanding of the catalyst surface structure, the elementary steps involved and deactivation mechanism will provide a deep insight into the mechanism, reliable kinetic models and the structure-reactivity relationships. Therefore, those studies at the molecular level are very crucial for developing efficient Co-based FTS catalysts. This review will focus on the theoretical studies on following three aspects: (1) adsorption of CO and H_2 on the Co surface, and the nature of the surface intermediates, as well as the surface reconstruction during the reaction, induced by the effects of adsorbates, temperature and pressure, (2) the reaction mechanisms involved, and (3) the catalyst deactivation mechanism.

2 Surface chemistry of Co

Pioneered by Langmuir, surface chemistry has long been regarded as a primary part of heterogeneous catalysis since the concept "catalysis" was introduced.[8] According to the two prevailing catalytic mechanisms, the Langmuir–Hinshelwood (LH) mechanism and the Eley–Rideal (ER) mechanism,[9] solid catalysts accelerate the chemical reactions by altering the reaction paths that reactants contact each other on catalysts' surfaces instead of interacting directly. Catalytic reactions are considered to take place on the active sites of catalysts, and the surface reconstruction determining the active sites eventually is responsible for the catalyst performance. Accordingly, the surface science is of vital importance for the study of heterogeneous catalysis.

To establish the relationship between catalytic performance and the catalyst structure and unravel the reaction mechanism, it is necessary to investigate various properties relating to the surface nature of Co catalysts, including the adsorption of reactants, the nature of the surface intermediates, active sites, and the dynamics of geometric structures of the surfaces during reactions. Langmuir and Taylor[10,11] have developed general theories and technologies to investigate surfaces of solid catalysts. Modern characterization techniques, such as temperature-programmed desorption (TPD), infrared spectroscopy (IR), X-rays photon electron spectroscopy (XPS), scanning transmission microscopy (STM) and low energy electron diffraction (LEED), and so on, have been widely used to investigate the surface catalysis.[8] In last decades, many experiments have been conducted to explore the surface chemistry concerned in FTS on Co catalysts. However, due to the difficulties in the preparation

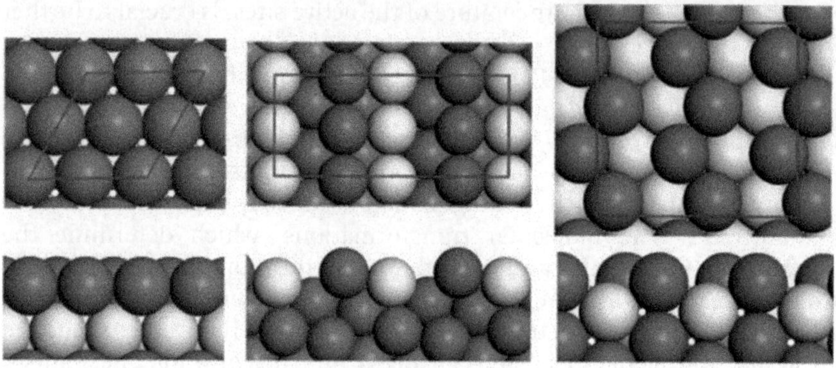

Fig. 1 Models of 3 represented surfaces (Co(0001), Co(10 − 12), Co(11 − 20)). Atoms in the second layer are lighter. Reprinted from ref. 29 with permission from Elsevier.

and stabilization of Co single crystals and the limitations of experimental conditions, the study of Co surfaces in model catalysts is less studied compared to other industrially important metals.[12,13] Thus, it's necessary for us to find alternative tools to gain deeper insights into the surface chemistry of Co.

It is well accepted that theoretical calculations can provide valuable information undiscovered by experimental methods.[6] These works bring new insight into surface chemistry of FTS on Co catalysts. In this part, the advances on the computational study of cobalt surface chemistry are reviewed. It is noted that this topic is restricted to the FTS synthesis by preliminarily divided into two sections: (i) adsorption of CO, H_2, and key intermediates on Co surfaces; (ii) surface reconstruction during the reaction, induced by the effects of adsorbates, temperature and pressure.

Co catalysts employed in theoretical studies were described as slab models and cluster models.[14,15] The former is suitable to simulate bulk Co catalysts. Based on the two different crystallographic structures, the hexagonal close-packed (hcp) Co and the face centered-cubic (fcc Co, slab models were used to calculate on various hcp surfaces, such as Co(0001), Co(10 − 10), Co(10 − 11), Co(10 − 12), Co(11 − 20), Co(11 − 21) and Co(11 − 24) and fcc surfaces as Co(111), Co(100), Co(110), Co(211) and Co(311), respectively. The latter is usually used to simulate some nanoparticles without specific crystallographic structures. Three typical surfaces as displayed in Fig. 1 will be discussed in detail.

2.1 The thermochemistry of surface species on Co

In FTS, being consistent with the experiment work and the Monte Carlo simulation, the quantum chemical calculations indicate that CO molecularly adsorbs on Co surfaces, while H_2 directly dissociates into two H atoms.[16] Here we firstly discuss the adsorption of CO, and then the other intermediates as H, C, O, and CH_x ($x = 1$–4).

2.1.1 Adsorption of CO. The adsorption of CO on Co surfaces has found to be dependent on the electronic features of CO molecule and the ensembles of Co surface atoms.[16–19] According to the molecular orbital theory, the interaction between the orbitals of Co atoms in the surface and the highest occupied 5σ and lowest unoccupied 2π* CO orbitals results in the chemisorption of CO (Fig. 2).[20,21] The stability of adsorbed CO is determined by the distribution of the electrons over the bonding and *anti*-bonding orbital fragments formed between CO and Co surfaces.[20] The bond-order overlap-population densities are widely used to analyze the bonding nature of the orbital fragments between the chemisorbed molecule and surface metal atoms. The early investigations by this method can be dated back to 1990.[18] Using quantum chemical modeling methods, van Santen and co-workers investigated the chemisorption of CO on several transition-metals including Co.

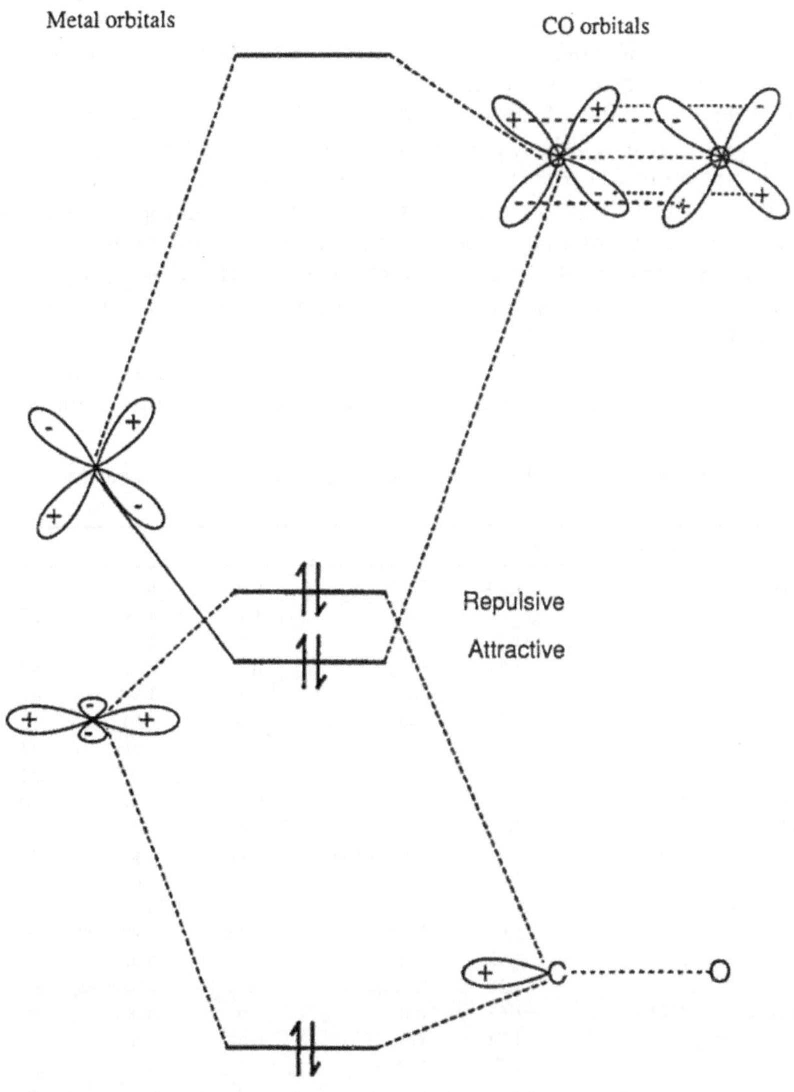

Fig. 2 Molecular orbital view of CO 5σ orbital interacting with metal d-valence electrons, Reprinted from ref. 18 with permission from Springer.

Those results indicate that the CO 5σ orbital interacts with d-valence electrons of Co, leading to a strong σ type interaction with highly occupied d-valence orbitals, thus resulting in a favorable configuration of atop bonded CO.

A good catalyst should adsorb reactants in a moderate degree by Sabatier principle.[22] Therefore, a quantitative study of adsorption energies, adsorption sites, and bond lengths of CO on Co surfaces is necessary for FTS. Here, adsorption energies, binding sites and bond lengths of CO on various Co surfaces were summarized in Table 1.

The adsorption of CO on Co surfaces has proved mainly influenced by two factors: (i) The interaction between the surface Co atoms and CO molecule, which determines the adsorption site preference and the adsorption properties of adsorbed CO molecule on Co surfaces; (ii) The interactions among the adsorbates. The first effect is decided by the models, while the second one is up to the coverage of adsorbates and the co-adsorption with other species.

Co(0001) is a kind of hexagonal close packed surface, being a model system since it is relatively stable and simple.[8,13] The typical adsorption configuration of CO adsorption on Co(0001) is shown in Fig. 3. As listed in Table 1, the adsorption of CO on Co(0001) prefers on top sites with an adsorption energy of 1.60 ~ 1.90 eV, which is 0.3 ~ 0.6 eV higher than the experimental results of 1.35 eV and 1.32 ± 0.13 eV. This divergence comes from the inaccurate description of the chemical bond of adsorbed CO molecule by DFT methods.

Abild-Pedersen and Andersson[23] have proposed a simple empirical correction formula based on the empirical correction scheme reported by

Table 1 Summary of adsorption energies, binding sites, and bond lengths of CO on various Co surfaces.

Surfaces	E_s (meV Å^{-2})	E_{ads} (eV)	Sites	d_{C-O} (Å)	d_{C-Co} (Å)	Ref.
Co(0001)	131	−1.64	Top	n.a.	n.a.	29, 30
		−1.62	Top	1.17	1.76	60
		−1.36	Top	1.17	1.75	26
		−1.78	hcp	n.a.	n.a.	40
		−1.77	hcp	1.19	1.98	41
		−1.88	n.a.	n.a.	2.01	38
		−1.66	hcp	1.19	1.50	27
		−4.47	Top	1.09	n.a.	25
Co(10−10)	140	−1.70	n.a.	n.a.	n.a.	30
Co(10−11)	149	−1.85	n.a.	n.a.	n.a.	30
Co(10−12)	156	−1.77	4-Fold hollow	n.a.	n.a.	29, 30
		−1.71	Bridge	1.19	1.86/1.99	60
Co(11−20)	155	−1.65	Short bridge	n.a.	n.a.	29, 30
		−1.65	Bridge	1.18	1.90/1.92	60
Co(11−21)	163	−1.82	n.a.	n.a.	n.a.	30
Co(11−24)	n.a.	−1.70	Bridge	1.19	1.87/1.96	60
Co(111)	127	−1.61	n.a.	n.a.	n.a.	30
Co(100)	154	−1.71	n.a.	n.a.	n.a.	30
Co(211)	n.a.	−1.69	C-3F,O-4F	1.29	n.a.	33
Co(311)	156	−1.71	n.a.	n.a.	n.a.	30
Co(110)	151	−1.61	n.a.	n.a.	n.a.	30

Fig. 3 The adsorption configuration of CO on Co (0001). Reprinted from ref. 29 with permission from Elsevier.

Mason *et al.*[104] Following this formula, Ma *et al.*[24] calculated the CO chemisorption on Co(0001) with improved adsorption energy on the most favorable top sites. These results indicate that the correction is somehow useful for FTS on Co(0001). It is noted that an empirical correction formula was used only if we are entirely familiar with its primary principle.

Except for the divergence of the adsorption energy from the experimental values, the results reported (Table 1) are not consistent with each other because of the different calculation methods and parameters involved in calculations. Légaré *et al.*[25] utilized a version of Anderson's atom superposition and electron delocalization (ASED)–MO theory and obtained the chemisorption energy of CO on Co(0001) as high as −4.47 eV, even much higher than experimental results.

For DFT calculations, the chemisorption energy of CO has found to be dependent on the functional employed. Ma *et al.*[24,26] adopted four different GGA exchange and correlation functionals, PW91, PBE, non-self-consistent PKZB (*Meta*-GGA), and revised Perdew–Burke–Ernzerhof (RPBE), to examine the effects of functional on CO chemisorption on Co(0001). The most stable binding sites calculated with PW91, PBE, PKZB, and RPBE are top sites with adsorption energies in the range from −1.36 to −1.68 eV. The bond length of C–O is 1.16–1.18 Å, and the bond lengths of C–Co are 1.86 Å. Among different functionals, the results from RPBE are more consistent with experiment values.

Apart from the flat Co(0001) surface, modified Co(0001) surfaces have also attracted much attention. Hu *et al.*[27] investigated a stepped Co(0001) surface (Fig. 4), which involves near-edge-hcp, edge-bridge and step-corner sites. The most stable adsorption sites for CO is step-corner sites with an adsorption energy of 1.90 eV, higher than that on fcc hollow sites by 0.24 eV. CO shows the stronger adsorption on the stepped Co(0001) surface than on the flat Co(0001) surface (1.66 eV).

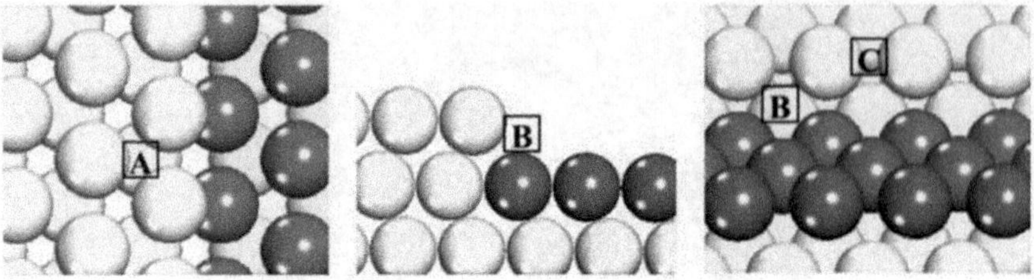

Fig. 4 The top view (left), the side view (middle) and another side view (right) of the stepped Co(0 0 0 1). The Co atoms on the terrace below the step edge are lighter, and the other Co atoms are darker. A labels the near-edge-hcp site, B the step corner site and C the edge-bridge site. Reprinted from ref. 27 with permission from Elsevier.

Jiao *et al.*[28] investigated a double-stepped Co(0001) model by removing three or two neighboring rows of Co atoms on the top and the second layers from a four-layer $p(5 \times 2)$ slab (Fig. 5) with the utilization of PW91 and RPBE. The optimized configurations of CO on Co(0001) (Figs. 5–6) have seven different adsorption sites on the modified Co(0001) surface. The most stable chemisorption sites to CO is the e2-t site with a binding energy of −1.52 eV (with RPBE functional). By switching the functional from RPBE to PW91, the most stable adsorption site remains unchanged; while the differences in CO adsorption energies between the edge-top site and other high-coordination binding sites are quite small, which is inconsistent with the experimental results. Therefore, RPBE is more reliable to describe the chemisorption of CO on Co surfaces.

Similar to the stepped Co(0001), more open surfaces, such as Co(10 − 10), Co(10 − 11), Co(10 − 12), Co(11 − 20), Co(11 − 21), Co(11 − 24), have been studied. The adsorption of CO on those surfaces showed not much difference compared to that on Co(0001),[29,30] because CO is an electron saturated molecule and its adsorption is not sensitive to the surface structure.

Li *et al.*[30] also reported an average adsorption energy (−1.66 ± 0.058 eV) of CO over different fcc Co facets (Co(111), Co(100), Co(110), and Co(311)), which is lower than that on hcp Co facets (−1.74 ± 0.088 eV). It indicates that the average binding on hcp Co is stronger than that on fcc Co.

It is found that the fcc-Co is the dominant crystal phase when the size of Co particles is less than 100 nm.[31,32] In particular, Co(211) (Fig. 7) consists of a small Co(111) terrace and a B_5 step site (consisting of 5 atoms in both a triangular (bottom 3F) and a square (4F) arrangement). van Helden *et al.*[33] investigated the adsorption of CO on the fcc Co(211) stepped surface (Fig. 8). CO is found to locate in a 5-fold orientation (B_5 site) with C and O in the bottom 3F and 4F sites, respectively. The adsorption energy and bond length of C–O are 1.69 eV and 1.29 Å, respectively.

Except for the slab models, a few works on cluster models have been also performed. Miyamoto *et al.*[34] built a Co_6 cluster model for FTS with a CO adsorption energy of 1.13 eV, which is much lower than those on the most slab models. Swang *et al.*[35] studied FTS on a larger cluster, a Co_{20} model. The adsorption energy and the length of C–Co bond are −1.69 eV and 1.78 Å, respectively, being close to those on slab models. This

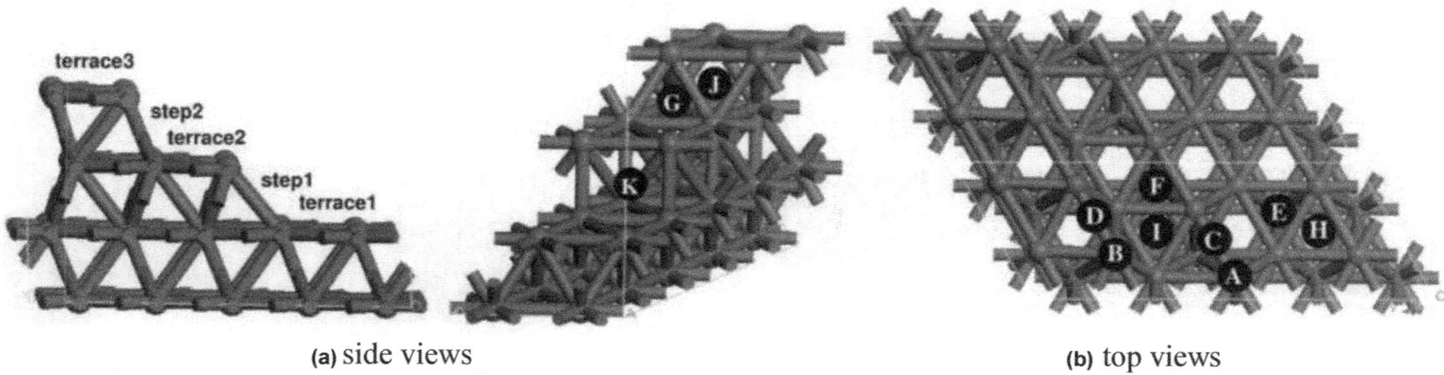

Fig. 5 Side and top views of the double-stepped Co(0001) surface: edge-top site (A–B), edge bridge site (C–D), terrace-hcp site (E–F), step-hcp site (G), terrace-fcc site (H–I), step-fcc site (J), and step-4-fold site (K). Reprinted with permission from ref. 28. Copyright 2008 American Chemical Society.

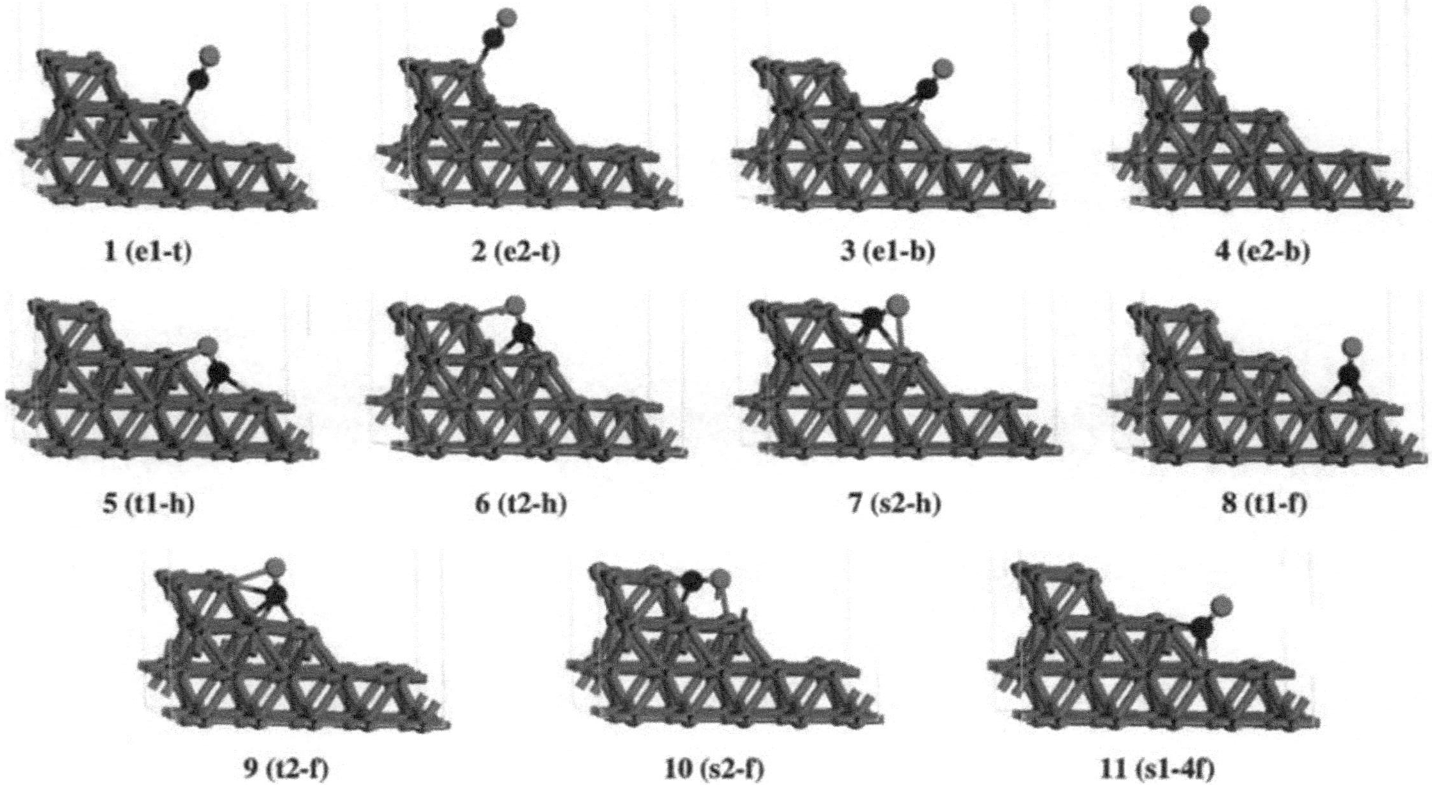

Fig. 6 The adsorption structures of CO on Co(0001). Reprinted with permission from ref. 28. Copyright 2008 American Chemical Society.

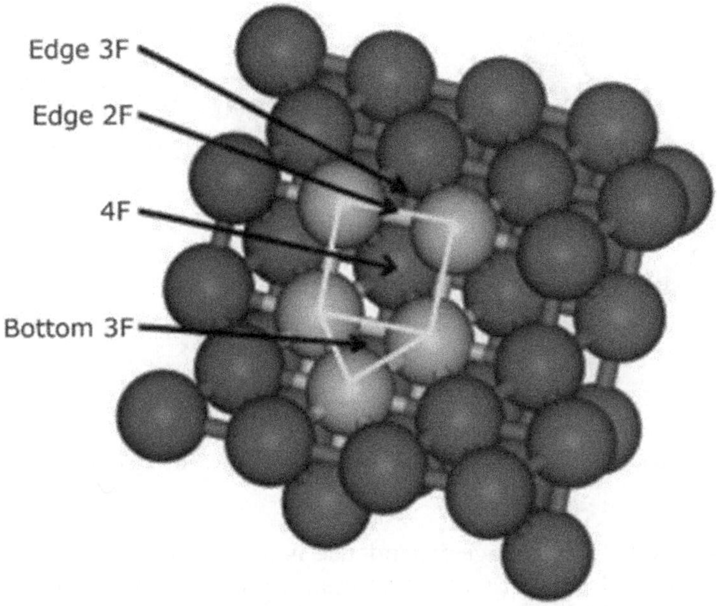

Fig. 7 Configuration of the B_5 site on the Co(211) surface. Local adsorption sites are indicated. Reproduced from ref. 33 by permission of The Royal Society of Chemistry.

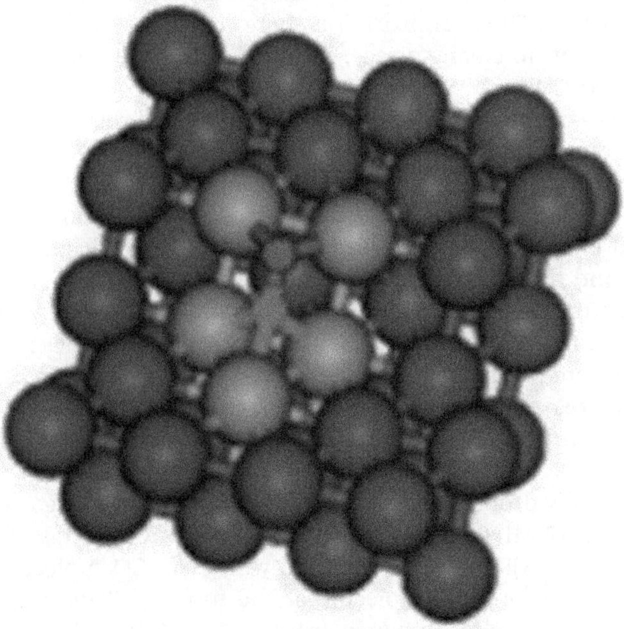

Fig. 8 The adsorption of CO on Co(211). Reproduced from ref. 33 by permission of The Royal Society of Chemistry.

indicates that clusters with more atoms (larger particle size) possess more characteristic of bulk crystals. The structures of Co_6 and Co_{20} clusters are illustrated in Fig. 9.

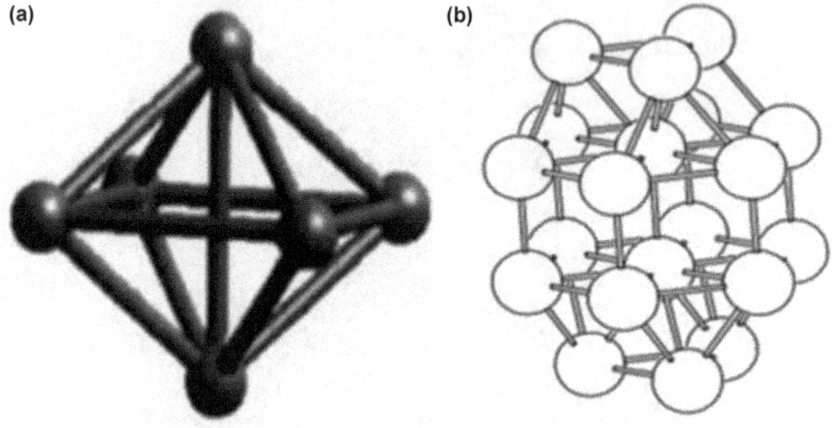

Fig. 9 Co_6 and Co_{20} cluster models. Reprinted from ref. 34 and 35 with permission from Elsevier.

It is noted that coverage-dependence is also an important factor relating to CO adsorption on Co surfaces. In line with the result of LEED patterns,[36,37] CO molecules found to be adsorbed at top sites on Co(0001), and the coverage increased gradually to 0.33 ML until a $(\sqrt{3} \times \sqrt{3})R30^\circ$ structure was formed. Iglesia *et al.*[38] compared the adsorption energy of CO on Co(0001) between the clean surface and the 0.5 ML CO coverage. The chemisorption energy of CO on the latter is 1.10 eV, lower than that on clean surface, indicating that the bonding energy of the adsorbed CO become weaker with an increase in the coverage.

The Saeys group[39] conducted a thorough study of the Co(0001) surface with different CO coverages. At low CO partial pressure (7×10^{-9} mbar), the adsorption energy decrease slightly from −1.35 eV (1/9 ML) to −1.32 eV (1/3 ML). The stable CO coverage is 1/3 ML under FT conditions (500 K, 20 bar). The $(\sqrt{3} \times \sqrt{3})R30^\circ$ – CO structure will form at 1/3 ML coverage, and the adsorption of additional CO molecules is unfavorable on such a surface. Only high pressure (100 mbar) of CO can convert the $(\sqrt{3} \times \sqrt{3})R30^\circ$ – CO structure to a $(2\sqrt{3} \times 2\sqrt{3})R30^\circ$-7CO structure. It should be noted that the adsorption energy does not decrease linearly with an increase in the coverage during this process. The absolute value of adsorption energy first decreases to a local minimum as −0.97 eV at 4/9 ML coverage, and then increases to −1.06 eV at 5/9 ML coverage, finally drops down to −0.86 eV at 2/3 ML coverage. The stable structure at 7/12 ML CO coverage with an adsorption energy of −1.08 eV is the $(2\sqrt{3} \times 2\sqrt{3})R30^\circ$-7CO structure, where six CO molecules occupy bridge sites with one CO molecule at the top site. This indicates that there is a sudden phase change from $(\sqrt{3} \times \sqrt{3})R30^\circ$-CO structure to $(2\sqrt{3} \times 2\sqrt{3})R30^\circ$-7CO structure around 5/9 ML CO coverage.

The effect of the coverage-dependence on the adsorption of CO is the repulsive interaction of the adjacent co-adsorbed CO molecules. Similarly, the co-adsorption of CO with other surface species involved in FTS also affects CO adsorption. Inderwildi *et al.*[40] found that the chemisorption energy was 0.02 eV higher with the co-adsorption of H, indicative of a slightly attractive lateral interaction between CO and H. By investigating the CO adsorption on an oxygen pre-covered Co(0001) surface, Jiao *et al.*[41] reported that the adsorption energy of CO was decreased by 0.35 eV. There are few theoretical calculations focusing on the effects of co-adsorption, especially, it cannot be negligible in real industrial FTS under harsh reaction conditions.

2.1.2 Adsorption of H, C, O.

Compared to CO, H_2 is only adsorbed dissociatively on the Co surfaces. The configuration is determined by the electronic and geometrical nature of H_2 and Co surfaces. In addition, the dissociation of CO is regarded as an important reaction of FTS, the adsorption of C and O mono atoms from CO dissociation is of vital importance for the reaction followed. We summarized the main results of atomic H, C, and O adsorption on Co surfaces in Table 2.

Similar to CO, the effects of models, coverage, and co-adsorption with other species on the adsorption of H, C and O will be discussed in detail.

van Helden *et al.*[42] suggested the adsorption of H on Co(111) was rather similar to that on the Co(0001) surface. The hollow sites are favorable for H adsorption on Co(111) and Co(100), while both the bridge and hollow sites are possible for H adsorption on Co(100). Compared to the flat surface, stepped sites allow for a variety of adsorption sites with different coordination for H adsorption on Co(211). The adsorption energy on Co(211) is slightly higher than that for Co(100), indicating that defect sites facilitate H adsorption. Clearly, H adsorption does exhibit a moderate dependence on the Co surface structures.

As for CO, the hydrogen binding enthalpy decreases gradually with an increase in the hydrogen coverage due to the repulsive lateral interactions with neighboring H adsorbates. In addition, the co-adsorption of other adsorbates also affects the adsorption of H. Saeys *et al.*[39] investigated the adsorption of H on a $(\sqrt{3} \times \sqrt{3})R30^\circ$ – CO surface and found

Table 2 Adsorption energies and binding sites of H, C, and O atoms on various Co surfaces.

	H		C		O		
Surfaces	E_{ads}/eV	Sites	E_{ads}/eV	Sites	E_{ads}/eV	Sites	Ref.
Co(0001)	−2.78	fcc	−6.83	hcp	−5.65	hcp	29
	−0.51	hcp	n.a.	n.a.	−2.48	hcp	40
	n.a.	n.a.	n.a.	n.a.	−5.39	hcp	41
	n.a.	n.a.	−6.62	hcp	−5.34	hcp	28
	−2.88	n.a.	−6.71	n.a.	−5.43	n.a	38
	−0.62	n.a.	−1.68	n.a.	−1.20	n.a.	103
	−2.94	fcc	−6.62	hcp	n.a.	n.a.	45
Co(0001)-1ST	−2.74	e-hcp	−7.44	st-corner	−5.59	Edge-hcp	27
	−2.90	Near-hcp	−7.53	Corner	n.a.	n.a.	45
Co(0001)-2ST	n.a.	n.a.	−7.84	B_5	−4.38	Step 1-4 fold	28
Co(10−12)	−2.73	4-Fold	−7.85	4-Fold	−6.04	4-Fold	29
Co(11−20)	−2.66	Short br	−7.22	Hollow	−5.63	Hollow	29
Co(111)	−0.48	fcc	n.a	n.a.	n.a.	n.a.	42
Co(100)	−0.46	Hollow	n.a.	n.a.	n.a.	n.a.	42
Co(211)	−0.50	Edge-hcp	n.a.	n.a.	n.a.	n.a.	42
Co(221)	−0.55	Edge-fcc	n.a.	n.a.	n.a.	n.a.	42

that the hydrogen adsorption enthalpy was decreased by 0.49 eV with the pre-adsorption of CO. Despite the slight attractive lateral interaction between CO and H at low coverages, the repulsive lateral interaction dominates at high coverages.

Atomic C and O are yielded by the dissociation of CO in FTS, the C adsorption on Co surfaces is crucial for this reaction. As listed in Table 2, the adsorption of C on most Co models is much stronger than the adsorption of H and O. Weststrate *et al.*[43] made a systematic investigation on the C adsorption on Co surfaces. The experimental results indicated that atomic carbon weakened the adsorption of CO and H_2, whereas a saturated (reconstructed) atomic carbon-covered surface can still adsorb ~60% of the CO and H compared to the clean surface. Thus, a high coverage of atomic carbon may not be a strong poison to FT cobalt catalysts. In other words, FTS on Co catalysts is feasible in spite of the strong adsorption of C atoms on Co surfaces.

The adsorption of O is stronger than that of H on Co surfaces. The coverage dependence of O on Co(0001) is different from that of CO. Ma *et al.*[44] reported the adsorption energy increased constantly with the rise of the coverage to 0.25 ML; while the repulsive interaction becomes to prevail with continuous increase in O coverage. Since each surface species is competitively adsorbed on Co surfaces, the existence of O will affect the nature of other surface species. In addition, O would participate in certain reactions that produce by-products as CH_3OH, CH_3CH_2OH and so on. Thus, the adsorption of O should be taken into account when calculating FTS on Co catalysts.

2.1.3 CH_x ($x=1$–4). Key intermediates (CH, CH_2, CH_3, CH_4) are produced during FTS, and the detailed results of the adsorption of these intermediates on Co surfaces are listed in Table 3.

It was found that the adsorption energies of CH_x ($x=1$–4) decreased gradually with an increase in the number of C–H bonds.[29,45] On the Co(0001) surface, the adsorption of CH_x ($x=1$–3) prefers adsorbing on hcp hollow sites, while CH_4 prefers on top sites. The adsorption energy of CH_4 is much lower than those of electron unsaturated CH_x ($x=1$–3). The interaction between CH_4 and the surface is very weak, in other words, CH_4 surface species is very easy to desorb from the surface.

2.2 Surface reconstruction during the reaction

The variation of surface environments during the reaction, taking into account the factors such as adsorbates, temperature and pressure, can frequently cause restructuring of catalyst surfaces, ranging from small atomic relaxations and reconstructions to macroscopic shape modifications. The occurrence of such surface reconstruction may alter the reaction process significantly. Thus, in order to obtain a thorough understanding of the Co surface chemistry during the FT process, it is of great importance to investigate the reconstruction process.

Currently, theoretical studies on this process are mainly focused on the adsorbate-induced reconstruction. Ciobîcă *et al.*[46] have studied the reconstruction process systematically. One interesting phenomenon

Table 3 Adsorption energies and binding sites of CH, CH_2, CH_3, CH_4 on various Co surfaces.

	Co(0001)		Co(10 − 12)		Co(11 − 20)		
Species	E_{ads}/eV	Sites	E_{ads}/eV	Sites	E_{ads}/eV	Sites	Ref.
CH	−6.30	hcp	−6.84	4-Fold hollow	−6.47	Hollow	29
	−6.14	n.a.	n.a.	n.a.	n.a.	n.a.	41
	−6.31	n.a.	n.a.	n.a.	n.a.	n.a.	38
	−5.99	hcp	n.a.	n.a.	n.a.	n.a.	28
	−5.99	hcp	n.a.	n.a.	n.a.	n.a.	45
CH_2	−3.95	hcp	−4.25	4-Fold hollow	−3.97	Hollow	29
	−3.91	n.a.	n.a.	n.a.	n.a.	n.a.	41
	−3.86	n.a.	n.a.	n.a.	n.a.	n.a.	38
	−3.85	hcp	n.a.	n.a.	n.a.	n.a.	28
			n.a.	n.a.	n.a.	n.a.	45
	−3.85	hcp					
CH_3	−1.89	hcp	−2.08	Step bridge	−1.97	Short bridge	29
	−1.89	hcp	n.a.	n.a.	n.a.	n.a.	45
CH_4	−0.02	Top	−0.02	Top	−0.02	Top	29

predicted by their calculations is that, instead of converting a more open and more reactive surface to a more compact surface that is less reactive, adsorption induced surface reconstruction actually converts compact surfaces to more open surfaces. This phenomenon may result from the larger adsorption energy on the more open surface compared to the adsorption energy on the more compact surface, which compensates for the higher surface energy of the former. Actually, among a series of adsorbates during the reaction (O, CO, CH, CH_2 and C), only carbon is capable of inducing a surface reconstruction (fcc-Co(111) to fcc-Co(100)). The unique behavior of carbon appears to be related to its unique stability when adsorbed in the so-called B5 sites (a fourfold hollow site with five neighboring metal atoms). However, this result is somewhat in contradiction with previous experimental evidences, which showed that other adsorbates, such as CO, could also induce the surface reconstruction. Further studies on this area as well as the influence of temperature and pressure are necessary for achieving a more comprehensive understanding.

2.3 Summary

Most theoretical studies on surface chemistry of Co catalysts for FTS are mainly contributed to the adsorption of CO, H_2 and other intermediates. Both CO and H_2 can be adsorbed on Co surfaces favorably. The computational work has proved that CO was adsorbed molecularly, whereas H_2 dissociated into two H atoms during adsorption process. The adsorption of surface species is affected by two factors: (i) The interactions between the surface atoms and adsorbates, (ii) and the interactions among the adsorbates. The former is much stronger than the later. As a result, binding energies of molecules, as CO and CH_4, are lower than that of atoms as H, C, O, and $CH_x(x = 1–3)$. The adsorption energies of specific species will be reduced with increasing coverage or the completive adsorption of other species.

Not only surfaces can affect adsorbates, but also can adsorbates alter the surface vice versa. Adsorbates-induced surface reconstruction has been demonstrated by computation, indicating that adsorbates-induced surface reconstruction converts compact surfaces to more open surfaces.

The computational studies on surface chemistry of Co catalysts have offered significant supports to the investigation of FTS on Co catalysts. However, the work is far from decent. As the experimental studies indicated, surface reconstruction and phase transition were certain to take place under practical FTS conditions. The theoretical studies about surface reconstruction and phase transition of Co catalysts, however, are fairly rare. In addition, cluster models were less studied in the previous theoretical work compared to slab models. However, practical catalytic reactions do not always happen as proposed in ideal plane models, nor the active sites distribute homogeneously on the surface. The investigation on cluster models is acting a crucial role in the study of heterogeneous catalysis. Accordingly, more considerations on surface reconstruction, phase transition, and cluster models should be taken into account in future work.

3 FTS mechanisms

3.1 Introduction of the proposed FTS mechanisms

Ever since the discovery of FTS in 1926,[1,2] many efforts have been made to provide a general reaction mechanism for the whole process.[4,16,47–51] However, discovering the FTS reaction mechanism is a thorny issue due to the following two aspects: (i) The reaction network is very complex, which consists of various intermediates and elementary steps; (ii) Cobalt may undergo structural reconstruction or phase transition during the reaction. Experimental studies by measures of kinetics and characterization can only give limited information. Computational studies have recently generated important information regarding reaction intermediates and activation barriers of elementary reaction steps involved in FTS.

The FTS process can be commonly divided into three steps: chain initiation, propagation and termination. The resulting products may also readsorb and undergo a secondary (de)hydrogenation, which makes the evaluation for detailed reaction mechanisms more complex. Generally, two seemingly conflicting reaction mechanisms have been proposed based on the three steps above.

The first one is the **carbide mechanism** (Fig. 10), which was initially advanced by Fischer and Tropsch in 1925 and later further developed by Sachtler, Biloen and others.[52,53] Therefore it is also known as Sachtler–Biloen mechanism. The main steps of this mechanism involve CO cleavage to a surface carbide (CH_x) as the monomer for chain propagation, hydrogenation (eventually to methane), and the formation and coupling of surface hydrocarbyl (C_nH_x) species, from which 1-alkenes are produced by β-elimination.

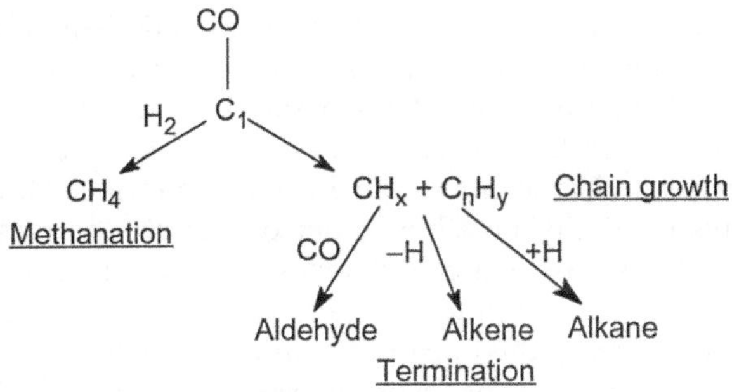

Fig. 10 Schematic diagram of the carbide mechanism. Reprinted from ref. 16 with permission from Elsevier.

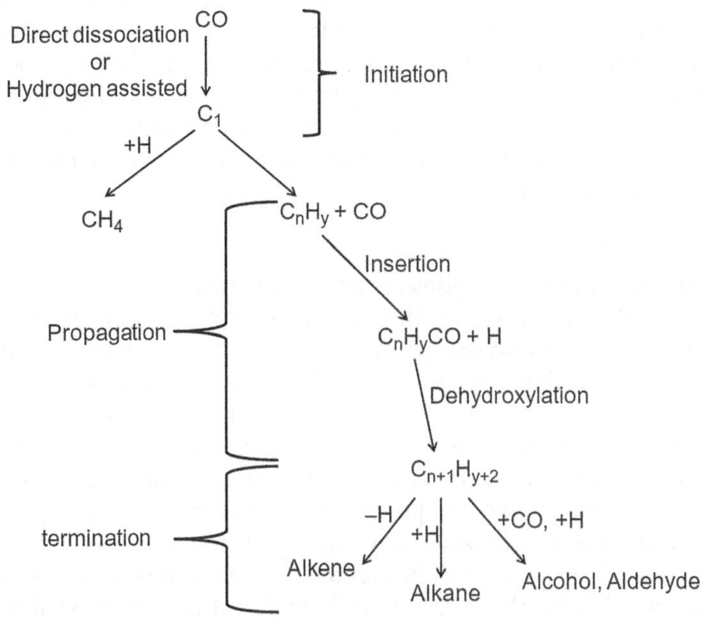

Fig. 11 Schematic diagram of the CO insertion mechanism. Reprinted from ref. 16 with permission from Elsevier.

The carbide mechanism can well explain the formation of long chain hydrocarbons, while fails to give some ideas on the production of oxygenates in the process. Thus, a second one is the **CO insertion mechanism** (Fig. 11), which was introduced from the organometallic catalysis.[54,55] This mechanism holds the view that the monomer for propagation is the associatively activated CO (without initial C–O bond cleavage). There are some experimental findings[56,57] to support this mechanism: The disappearance of IR bands of adsorbed CO during the FTS and during alkene hydroformylation, consistent with CO insertion between metal atoms and alkyl ligands. It is important to realize that to insert CO into metal alkyl groups, the alkyl group has to be bonded to surface metal sites with empty d-orbitals.

In general, both mechanisms are mainly different in the way of CO activation, dissociatively or associatively. Thus, to gain more insights into the FTS mechanism, we have to investigate the relationships between the reaction conditions and the CO activation pathway. Many recent computational results have analyzed the issues as presented in the next section. Different reaction mechanisms account for the formation of different species and it seems that we need to consider all of them in order to explain all the possible Anderson–Schulz–Flory (ASF) distributions of products observed experimentally. Again, the limitations imposed by the characterization techniques, which are often carried out on model systems, increase the uncertainty in this debate.

It is noted that, in a recent review by van Santen *et al.*[16] that rationalizes the main FT reaction mechanisms (the carbide and the CO insertion pathways) using a BEP type calculation, they concluded that chain growth does not occur by CO insertion. Alternatively, recent works suggest the co-existence of two reaction mechanisms. Gaube and Klein[58,59] suggested that the ASF plots obtained from analysis of FT hydrocarbon synthesis reactions arose from the participation of two "incompatible" mechanisms. Both involve surface alkyls, but in one case the chain growth takes place by insertion of CO, while in the other the monomer is a surface carbene (CH_2).

3.2 Density functional studies on FTS mechanisms

3.2.1 Chain initiation. As illustrated by both the carbide mechanism and the CO insertion mechanism, the FTS reaction is initiated through CO activation. The divergence for two mechanisms is that the carbide mechanism requires CO to firstly dissociate, while the CO insertion mechanism proposes CO to be hydrogenated at the initial step. Some recent density functional studies based on periodic catalyst models have evaluated the possibility of these different activation pathways by calculating the activation energy barriers of the involved steps under different reaction conditions (temperatures, pressures and coverages). Here, we will discuss these results in both direct and H-assisted routes.

Direct route. Neurock and coworkers[60] studied the adsorption and direct activation of CO over flat Co(0001), corrugated Co(11 − 20), and stepped Co(10 − 12) and Co(11 − 24) surfaces computationally. The barriers on flat and corrugated surfaces are all larger than 1.97 eV independent of CO coverages, whereas the stepped surfaces are much more active. Direct CO activation shows a strong dependence on the surface structure. The step sites, which contain the so called B5 sites, may act as the reaction center for the direct channel. Hu and coworkers[27] found that the step sites were more favored in activating the direct CO dissociation compared to flat sites.

H-Assisted route. CO dissociation has evidenced experimentally to be favorable in the presence of co-adsorbed hydrogenation. This is also evaluated by the theoretical calculations as well. Saeys *et al.*[61] studied the dissociation of a CO molecule on a flat Co(0001) surface with the activation barrier of 2.28 eV, and the hydrogenation barriers of this molecule to form HCO and H_2CO are 1.51 and 0.62 eV, respectively. Moreover, the C–O bond scission is kinetically favorable for HCO (0.93 eV) and H_2CO

(0.70 eV) compared to CO alone. Ojeda *et al.*[38] also studied the direct and H-assisted CO activation pathways on the Co(0001) surface. As shown in Fig. 12(a), monomer formation *via* the H-assisted CO activation route was found to dominate completely on the Co surface. The key intermediates for C–O bond scission are HCO and HCOH.

H-assisted routes on defected surfaces were also studied by several groups. King *et al.*[40] calculated the energy barriers of both CO dissociation and hydrogenation in flat Co(0001) and corrugated Co(10–12) models. From their results, the activation energy for CO dissociation on the corrugated surface is 1.61 eV while the hydrogenation of CO to form HCO is only 0.45 eV. A similar work was conducted by Ciobîcă *et al.*[33] on a fcc-Co(211) stepped surface, while H-assisted CO dissociation mechanism was compared to direct CO dissociation on the B5 site as exposed. As shown in Fig. 12(b), direct CO dissociation showed an overall barrier of 1.47 eV, while the alternative mechanism involving a HCO intermediate proceeded with a slightly lower overall barrier of 1.27 eV endothermically. This indicates that on stepped Co surfaces, direct and H-assisted route may become comparable. On a double-stepped Co surface, Huo *et al.*[28] demonstrated the role of hydrogen-assistance. But these calculations have not taken into account the effects of CO coverage, which may need a further evaluation in the future.

As a summary, the FT mechanism on Co nanoparticles is likely to mainly initiate on the stepped surface *via* the H-assisted CO dissociation mechanism, rather than the direct dissociation route. However, considering that the overall barriers of the H-assisted routes are still comparable to the direct one, we could not simply rule out the direct route. A noticeable fact is that the detailed mechanism for the chain initiation or CO activation in FTS process may be influenced by the corresponding surface sites. H-assisted route is the most feasible one on flat models while the direct route is favored on corrugated and stepped models. A recent study by Liu *et al.*[30] has verified both the direct and H-assisted CO activation on hcp Co and fcc Co. Hcp Co prefers the direct CO dissociation, whereas fcc Co is in favor of H-assisted route, due to their distinct crystallographic structure and morphology.

3.2.2 Chain propagation. Herein, we review the theoretical achievements based on the formation of C_2 and C_3 products. According to the choice of the propagation monomer, there are two mechanisms: CH_x–CH_y recombination and CO insertion. In the preceding section, the recent DFT investigations on the chain propagation mechanism will be presented in two aspects.

CH_x–CH_y recombination. The typical propagation mechanism is that CH_x monomers are produced from the CO activation steps, while the propagation proceeds *via* CH_x–CH_y recombination. Carbene (CH_2) is usually proposed as the monomer, which is very common in the polymer science. But due to the difficulty in experimental evaluation, an unambiguous conclusion is currently hardly to make. In the following, the recent results on the CH_x–CH_y recombination or C–C coupling process will be presented.

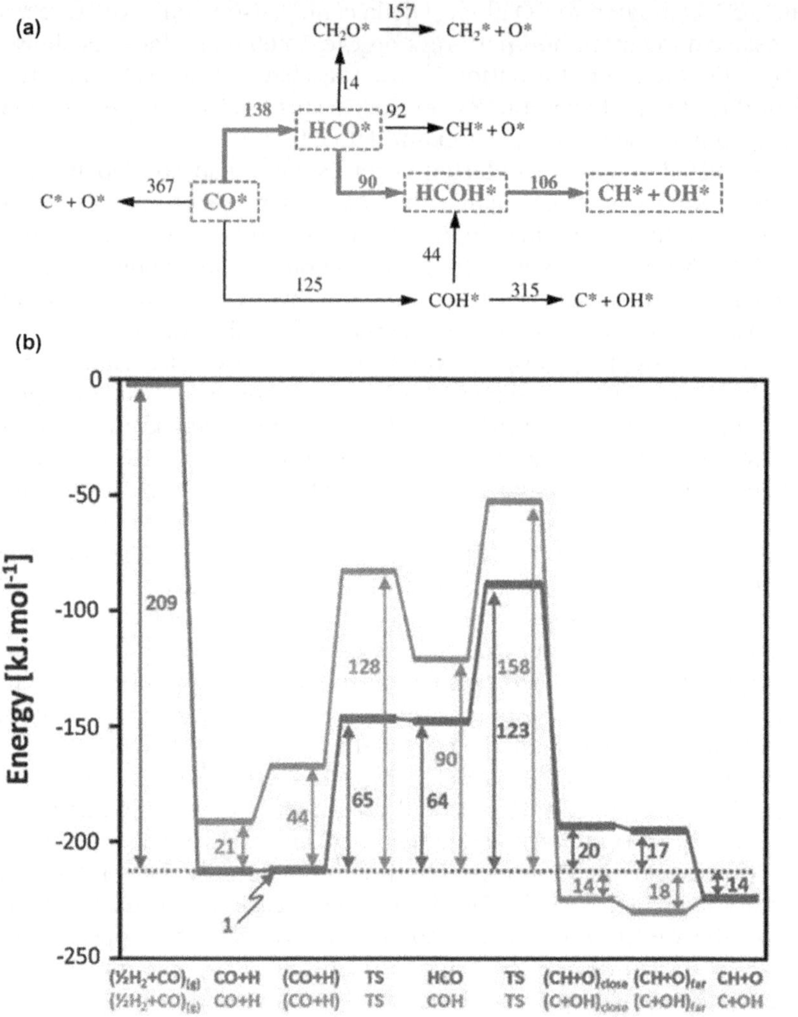

Fig. 12 (a) Schematic diagram of the chain initiation mechanism on Co(0001). Values on the arrows refer to the individual barriers (in kJ mol^{-1}) for each elementary step. The preferred H-assisted CO dissociation path is the lighter colour. Reprinted from ref. 38 with permission from Elsevier; (b) Energy profile of H-assisted CO dissociation pathways on the stepped Co(211) surface: HCO pathway (darker) and COH pathway (lighter). Reproduced from ref. 33 by permission of The Royal Society of Chemistry.

van Santen *et al.*[62] investigated the C–C coupling of $C + CH_x$ ($x = 1–3$) on Co cluster models, which is endothermic and cluster-size dependent. Neurock and coworkers[63] studied the coupling of CH and CH_2, to form CH–CH_2 over the Co(0001) surface. The reaction has proved to be endothermic with an activation barrier of 0.58 eV. However, these studies have not covered all the possible recombination steps. The Hu's group has reported the CH_x–CH_y recombination mechanism more comprehensively.[45,64–66] They extensively calculated hydrogenation reactions and C_1–C_1 coupling reaction on both flat and stepped Co surfaces and compared the reactions on the two surfaces (Fig. 13). CH_3 hydrogenation ($CH_3 + H \rightarrow CH_4$) is the most difficult one among all the elementary

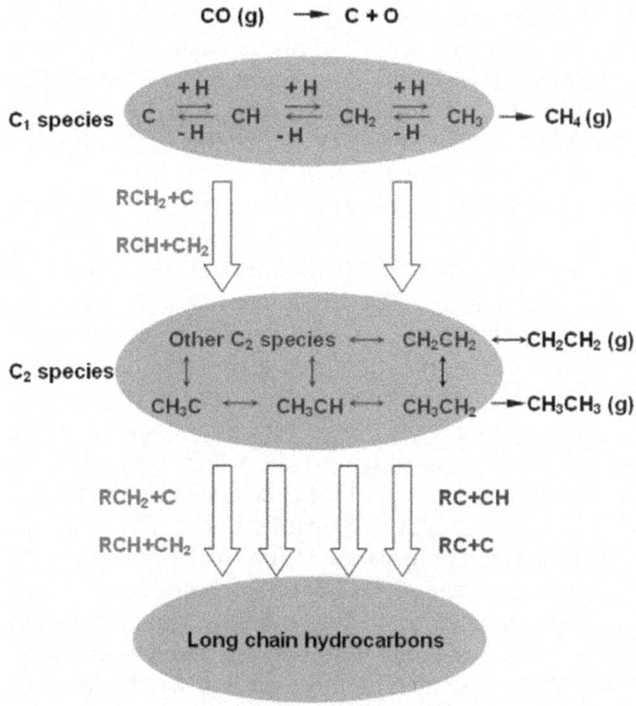

Fig. 13 The CH_x–CH_y coupling mechanism of FT synthesis: (1) CO and H_2 dissociation. (2) Hydrogenation/dehydrogenation of C_1 species (in quasi-equilibrium) and CH_4 desorption. (3) Two pathways of chain growth, the $CH_3 + C$ and $CH_2 + CH_2$ coupling. (4) Hydrogenation or dehydrogenation of C_2 species, and CH_3CH_3 and CH_2CH_2 desorption. (5) The coupling of $RCH_2 + C$, $RCH + CH_2$, $RC + CH$ and $RC + C$. Reprinted from ref. 64 with permission from Elsevier.

reactions on both surfaces, possessing barriers of around 1.0 eV. Another noticeable result is that CH_2 is actually the least stable species among all the CH_x ($x = 1$–3) species on both surfaces. The barriers of each elementary step on both flat and stepped surfaces are similar, indicating that CH_x hydrogenation is a structure-insensitive reaction. As for C–C coupling reactions, it is interesting to note that all coupling reactions on Co(0001) are faster at terraces, by an overall factor of 2, than at step sites except for CH_2–CH_2 and CH_2–CH_3 combinations, respectively. Metallic cobalt is generally oxidized by water during FTS;[4,67] Jiao and coworkers[41] have studied the adsorption and dissociation of CO as well as CH_x coupling and hydrogenation on both the clean and oxygen pre-covered Co(0001) surfaces. The ability for CO adsorption and activation was weakened for the latter, while CO_2 formation was improved by raising the barrier of CO dissociation and CH/CH coupling. Hence, in order to facilitate the CH_x–CH_y recombination, the oxygen removal is crucial for the carbide route.

CO insertion. The CO insertion mechanism, in which CO inserts into R–CH_2 directly, was originally proposed by Pichler and Schulz.[54] Hu[68] and Saeys[61] have calculated the activation barrier for the $CH_2 + CO$ coupling step. The barrier for insertion is about 1.87 eV, being too high to account for the fast chain growth in FTS. Therefore, the other CO insertion step was also considered,[61,69] the calculated barrier for CH + CO coupling proposed by Masters[55] was about 0.83 eV. Subsequent hydrogenation to CH_2CHO is also feasible, while CH_2CHO undergoes C–O bond scission with a barrier of 0.62 eV. Alternatively, C–O bond scission is proposed to occur before hydrogenation with a barrier of 0.73 eV for CHCO. On a low coverage Co(0001) terraces, the calculated rate in turnover frequency (TOF) for the CO insertion mechanism is 100 times faster than for the hydrogen-assisted CO activation mechanism, though still significantly slower than the TOF value measured in experiments. However, the coverage effects were not included in the calculation.

Under FTS conditions, the Co surface is nearly saturated with CO, while lateral interactions between surface species are expected to decrease the adsorption energies.[70] Recently, Saeys and co-workers[39] have studied the effect of CO coverage on the CO-insertion-type FTS mechanism on Co catalysts (Fig. 14). The CO coverage on Co(0001) increased gradually until a $(\sqrt{3} \times \sqrt{3})R30^{\circ}$-CO configuration (1/3 ML) was formed. The 1/3 ML CO coverage reduces the H_2 binding enthalpy from −1.25 to −0.77 eV with a H coverage of <0.3 ML. The bonding energy of reaction intermediates was reduced by 10–30 kJ mol^{-1}. Thus, the overall surface barrier of CO insertion into CH species decreased from 1.81 to 1.15 kJ mol^{-1}. The above results indicate that the higher CO coverage inhibits C–O bond cleavage but enhance oxygenate desorption. The occurrence of actual chain growth after CO insertion depends on the rate ratio of oxygenate desorption to C–O bond cleavage,[71] while CO insertion may not serve as the propagation mechanism under practical conditions.

Zhao *et al.*[72] studied the chain propagation by HCO insertion on Rh(111) and Co(0001) by comparing with those by CO insertion. The barriers for HCO insertion into CH_x ($x = 1$–3) species are rather low and more favorable than the corresponding CO insertion steps. However, HCO insertion may be limited by the high formation barriers of HCO.

3.2.3 Chain termination. After the chain propagation, CH_x–CH_y or R–C(H)O fragments may be formed according to different propagation mechanisms. These intermediates could undergo (de)hydrogenation to terminate the propagation process, resulting in the formation of methane, longer chain alkanes, olefins and oxygenates.[48] Due to this complex product distribution, there is still no theoretical work involving all the above termination routes.

Hu *et al.*[64] established a descriptor to quantify the relationship between methane selectivity and surface properties. Using the descriptor, they found that the fast C–O bond scission led to higher activity, while increased the selectivity to the undesired methane. Thus, the catalytic surfaces should be carefully optimized to achieve a compromise between selectivity and activity.[65] It was also shown that the hydrogenation and dehydrogenation of different CH_xCH_y fragments have similar activation barriers. Studies on the formation of oxygenated products (*e.g.* aldehydes, alcohols) are rare, because they are not the main products of FTS. In general, there is still no mechanism that can elucidate the formation of

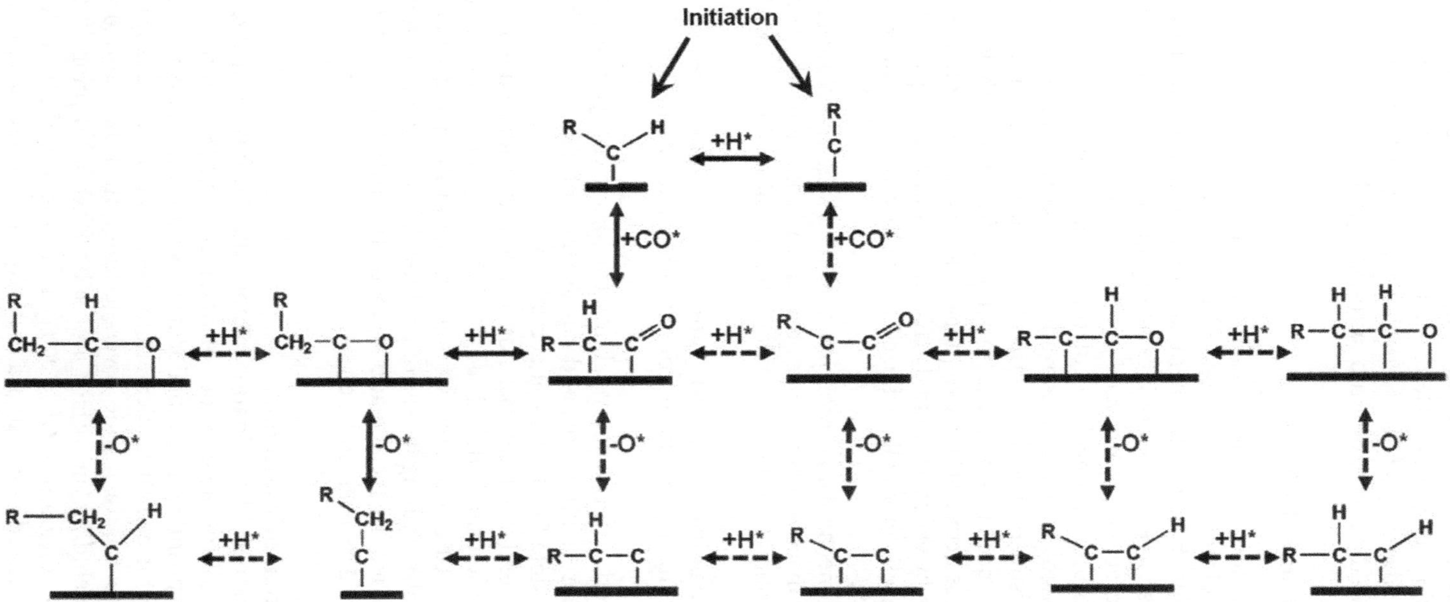

Fig. 14 Possible propagation reaction paths for the CO insertion mechanism. The reaction starts by C–C coupling ("+CO*"), followed by hydrogenation ("+H*") and C–O scission steps ("O*"). The full arrows indicate the dominant reaction path and the dotted arrows the minor reaction paths. R represents hydrogen or an alkyl group. Reprinted from ref. 39 with permission from Elsevier.

all the products. The carbide mechanism is good at the explanation of the formation of hydrocarbons, but fails in that of oxygenates. The situation for CO insertion is reverse. Hence a unified new mechanism which combines the advantages of both carbide route and CO insertion route is necessary to finally address all the untracked problems.

4 Catalyst deactivation

Catalyst deactivation is a major challenge in cobalt-based FTS. However, it is very difficult to trace the origin of catalyst deactivation with experimental methods, because the chemical environment in the FTS reactor encompasses a large number of interacting species which may negatively affect catalytic activity. Fortunately, using well-chosen and sufficiently accurate quantum chemical calculations, scientists are able to gain qualitative insight into the deactivation mechanism. In this part, four main mechanisms of deactivation have been described in the literature:

In the first mechanism, the active phase (metallic cobalt) is transformed into less active or inactive phase due to partial or total oxidation of Co particles by water. Despite the fact of FTS in reducing atmosphere, the active phase is gradually oxidized to CoO by water,[67] which is a by-product of FTS;

The second mechanism[73] proposes that deactivation is caused by the formation of inactive carbon species, or carbidization (*e.g.* graphitic carbon, amorphous carbon, coke) on the surface of catalyst, thereby losing the active surface area;

Thirdly, cobalt-based catalysts are poisoned and deactivated by chemical poisoning (*e.g.* sulfur compounds in most industrial syngas feeds.);[74,75]

Finally, sintering of cobalt crystallites, the loss of catalytic surface area due to ripening or migration and coalescence of the cobalt phase, is mainly responsible for the loss of activity.[73,76]

In short, several mechanisms, not just a single mechanism, are responsible for the overall deactivation of Co catalysts. At the molecular level, normally, DFT has rarely worked on the problems of catalyst sintering and catalyst fouling by wax or attrition. Therefore, in the following section, we review DFT calculations about the first three deactivation mechanisms of FTS over Co catalysts.

4.1 Deactivation by oxidation

As abovementioned, the oxidation of metallic Co has been intensively studied, especially in the presence of water during FTS, and this process has long been postulated as a major deactivation mechanism.[77]

Ma *et al.* performed DFT calculations to investigate the adsorption of oxygen on Co(0001) for a wide coverage range from 0.11 to 2.0 monolayers (ML), including on-surface and subsurface adsorption.[44] A first-principle phase diagram of O/Co(0001) as a function of the O chemical potential was derived. It was found that oxygen preferred on-surface hcp-hollow site at coverages lower than 1.0 ML, then oxygen incorporated into subsurface layers with the completion of the full monolayer oxygen

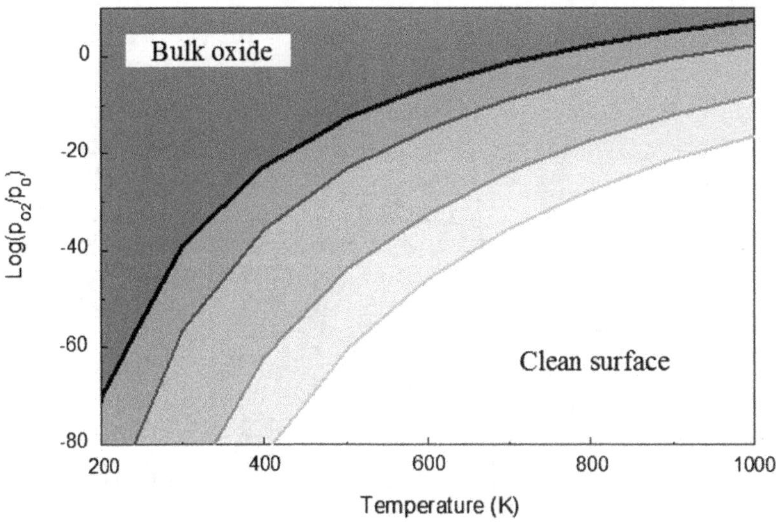

Fig. 15 Surface pressure–temperature (*P*, *T*) phase diagram for O/Co(0001) indicating the stability range of the lowest energy structures. (The successive deeper gray regions denote for *p*(2 × 2)-O, *p*(2 × 2)-2O and *p*(2 × 2)-3O, respectively). Reprinted from ref. 44 with permission from Elsevier.

chemisorbed on Co(0001). Furthermore, a two-dimensional phase diagram (Fig. 15) was derived to show the stability of the *p*(2 × 2)-O overlayer; meanwhile, two bulk cobalt oxides are predicted thermodynamically at a pressure about 10^{-13} atm at $T = 700$ K, and at ~300 K under UHV conditions, respectively.

Nevertheless, it is generally accepted that the oxidation behavior of cobalt catalysts is related to the FTS reaction conditions, especially, to water and cobalt crystallite size. Gong *et al.*[27,78] studied the water formation on stepped and flat Co(0001). The calculations on stepped Co(0001) (Fig. 16) indicated that the O hydrogenation (O + H → OH) was a favored step ($E_a = 0.73$ eV) at low coverages. However, the second step of OH hydrogenation (OH + H → H_2O) is unlikely to occur due to the higher barrier of 1.72 eV. They suggested that water formation could occur readily only at high coverages. Meanwhile, the O hydrogenation at the interface is easier ($E_a = 0.32$ eV in the O–O system, $E_a = 1.10$ eV in the O–OH system) than that on the same terrace ($E_a = 1.49$ eV in the O–O system, $E_a = 1.80$ eV in the O–OH system) for both the O–O and O–OH co-adsorption systems; the OH hydrogenation can occur on the same terrace near the edge. Compared to the OH hydrogenation in the O–OH system ($E_a = 1.46$ eV), the reaction in the OH–OH system ($E_a = 0.64$ eV) is more feasible. A possible reaction route with low barriers for water formation was proposed as: O + O + 3H → O + OH + 2H → 2OH + H → H_2O + OH.

The formation of Co_3O_4 has proved to be thermodynamically unfavorable under FTS conditions. Thermodynamic analysis by van Berge *et al.*[79] revealed that oxidation of metallic Co does not occur during FTS. However, the calculations reported by Steen[67] indicated that the oxidation was thermodynamically possible under realistic FTS conditions (443–513 K and $p_{H_2O}/p_{H_2} = 1-1.5$) only when the size of the reduced Co particles is

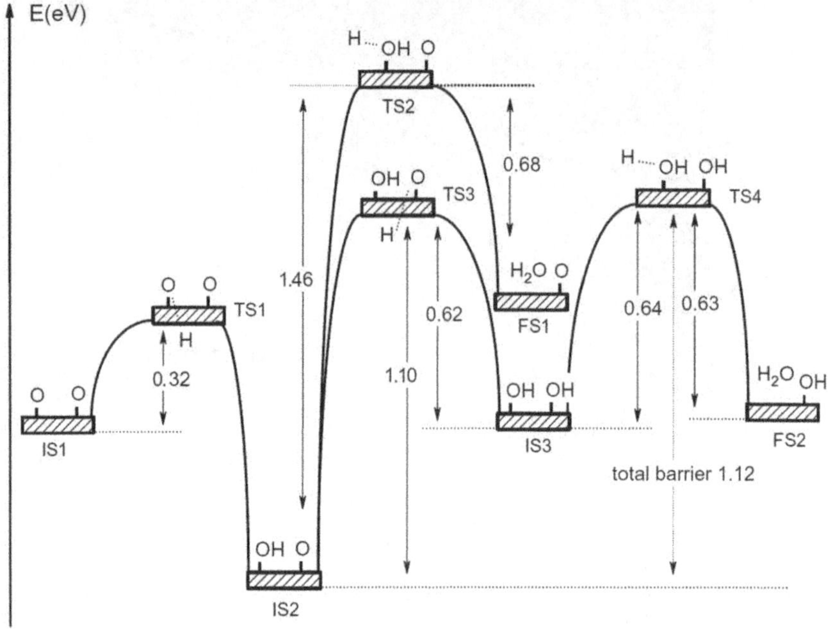

Fig. 16 The energy profile of the O removal on the stepped Co(0001). IS1 is the initial state (IS) of O hydrogenation in the presence of O (O-(near-edge-hcp) -O-(near-edge-bridge) and H-hcp). The H is omitted for clarity. TS1 is the transition state (TS) for O + H → OH at the interface in the presence of O nearby. The H in this state is on the terrace below which is indicated by placing the H underneath the surface. IS2 is the state of O-(near-edge-hcp)-OH-(edge-bridge) and H-hcp. TS2 is the TS for OH + H → H_2O on the same terrace in the presence of O nearby. TS3 is the TS for O + H → OH at the interface in the presence of OH nearby. FS1 is the final state (FS) with water in the gas phase and the O on the near-edge-hcp site. IS3 is the state of OH-(edge-bridge)-O-(edge-bridge) and H-hcp. TS4 is the TS for OH + H → H_2O on the same terrace in the presence of OH nearby. FS2 is the final state with water in the gas phase and the OH on the edge-bridge site. Reprinted from ref. 78 with permission from Taylor & Francis.

smaller than 4–5 nm. This is derived from the contribution of the surface energy of nanosized cobalt crystallites to the overall oxidation process.

More recently, Saib *et al.*[80] reported that the (surface) oxidation under realistic FTS conditions does not occur for cobalt crystallites larger than 2 nm. To get further evidence on surface oxidation, Saib and co-authors performed molecular modelling calculations to study the stability of oxygen from water on the fcc-Co(111) and fcc-Co(100) surfaces as well as on fcc-Co(111) subsurface (Fig. 17). The reaction used for the calculations can be expressed as: $H_2O + Co \leftrightarrow CoO + H_2$. Accordingly, the dissociation of water on Co(100) is exothermic at an oxygen coverage of 25%. However, for higher coverages water dissociation becomes endothermic. Compared to Co(100), water dissociation on fcc-Co(111) is less favorable and the subsurface oxygen resulting from water dissociation is highly endothermic. They argued that water dissociation on the more open (100) surface was more favorable than on the more dense (111) surface. It means that surface roughness or crystallite size may play an important effect in the oxidation process. Thus, surface oxidation *via* water dissociation is thermodynamically favorable for cobalt crystallites.

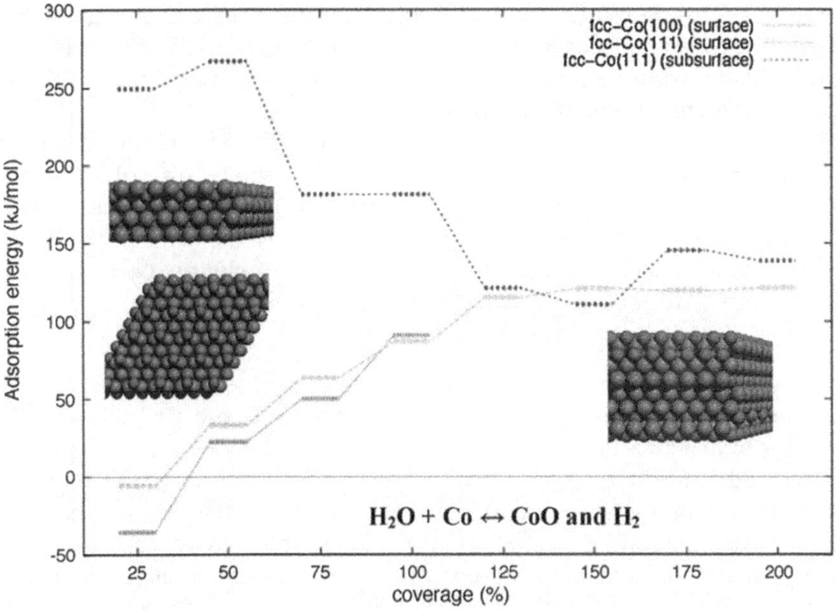

Fig. 17 Molecular modelling results for the dissociation of water on cobalt fcc surface (100) and (111) as well as subsurface oxygen on Co(111). Reprinted from ref. 80 with permission from Elsevier.

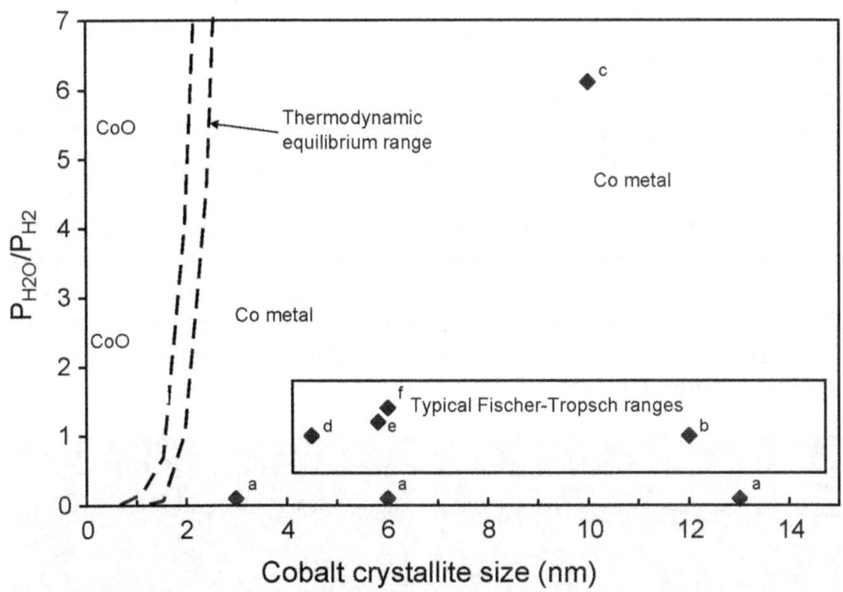

Fig. 18 Reconciliation of the effect of the water/hydrogen reactor ratio as a function of cobalt crystallite size on the oxidation behaviour of cobalt. Reprinted from ref. 80 with permission from Elsevier.

Saib and co-authors also updated the reconciliation with revised thermodynamic calculations (Fig. 18) reported by Swart,[81] who proved that the surface energy for cobalt oxide was previously underestimated.[67]

More recently, Bezemer[82] and Elbashir[83] summarized the effects of the water/hydrogen ratio on the oxidation behaviour of cobalt (Fig. 18) and drew a conclusion that (surface) oxidation under realistic FTS conditions does not occur for cobalt crystallites larger than 2 nm.

In addition, some groups[84–86] also claimed that the rising of water concentration increases the formation of aluminate on cobalt-based catalysts either during FTS or under model conditions in mixtures of H_2/H_2O. The observed deactivation is often ascribed to the formation of aluminate as it is proposed that the irreducible cobalt-support species is generated from/at the expense of active metallic Co.

4.2 Carbon deposition and surface reconstruction

Carbon deposition is of major concern as one of the possible mechanisms of deactivation in FTS.[87] Excessive carbon deposition results from strong adsorption of a graphene sheet on the catalyst surface, thus leading to catalyst encapsulation, pore blockage and catalyst breakage under extreme conditions.[88,89] Due to the exposure to syngas and wide range of carbon products present on the catalyst surfaces, carbon deposition was experimentally observed after FTS.[90–92] The formation of carbon deposits on Co catalysts has received significantly more attention using theoretical calculations because of the difficulty in identifying the nature of the resilient carbon species in FTS.

On cobalt catalysts, carbon is supposed to originate from CO dissociation in FTS, which is considered a key step in FTS.[93] C_1 species (C^*, CH^*, CH_2^* and CH_3^*) are an intermediate, which can be converted *via* hydrogenation and coupling to products. Moreover, C_1 species may also be converted to more stable species which may deteriorate the catalysts.[94] It has been reported that both cobalt carbide and polymeric carbon can negatively impact on FTS activity.[94,95]

To understand the nature of the resilient carbon species on the Co surface, the stability of carbon species at terrace sites and near-step edges under FTS conditions was investigated.[96] So far, there are four types of chemisorbed carbon to be distinguished: (i) On-surface carbon atoms; (ii) A subsurface carbon; (iii) Extended islands of graphene; (iv) *p4g* surface carbide (Fig. 19).

Fig. 19 *P4g* (a space group symmetry) clock surface carbide and graphene strips on a stepped Co surface. Reprinted from ref. 96 with permission from Elsevier.

Thermodynamically, atomic carbon should adsorb preferentially at the hcp hollow site on Co(0001) at low coverages.[16] However, a *p4g* surface carbide produced by carbon diffusion into the step sites is also highly favorable and ultimately grows into small graphene islands. Though carbon adsorption at the step sites is more favorable than a single row of *p4g* surface carbide, a second row of *p4g* surface carbide is likely formed at the step sites near the first row. The calculations further indicate that the *p4g* surface carbide follows a row-by-row growth mechanism. The calculated C_1s core-level binding energies of 284.5 and 283.4 eV further demonstrate the presence of polyaromatic carbon and surface carbide. Extended layers of graphene were the most stable form, followed by a *p4g* surface carbide phase initiating from the step edges. Extended graphene islands are very stable, however, small graphene strips are significantly less stable due to the unsaturated edge sites. Hydrogen termination of the edge carbon atoms enhances the stability of graphene strips.

Based on DFT calculations, Swart *et al.*[97] computed the stability of infinite graphene islands on the Co(111) surface. The lattice constant of graphene of 2.46 Å closely matches the lattice constant of the Co(111) surface of 2.50 Å, and only a <2% expansion of the graphene lattice is required. Infinite graphene islands correspond to a carbon coverage of 2.0 ML. Four high-symmetry orientations, graphene on fcc-Co(111) at ring-top, ring-bridge, ring-fcc and ring-hcp sites, were considered (Fig. 20). The ring-fcc and ring-hcp sites were found to converge at a height of about 2.16 Å while the ring-bridge site is a transition state for surface diffusion at a similar height above the surface.

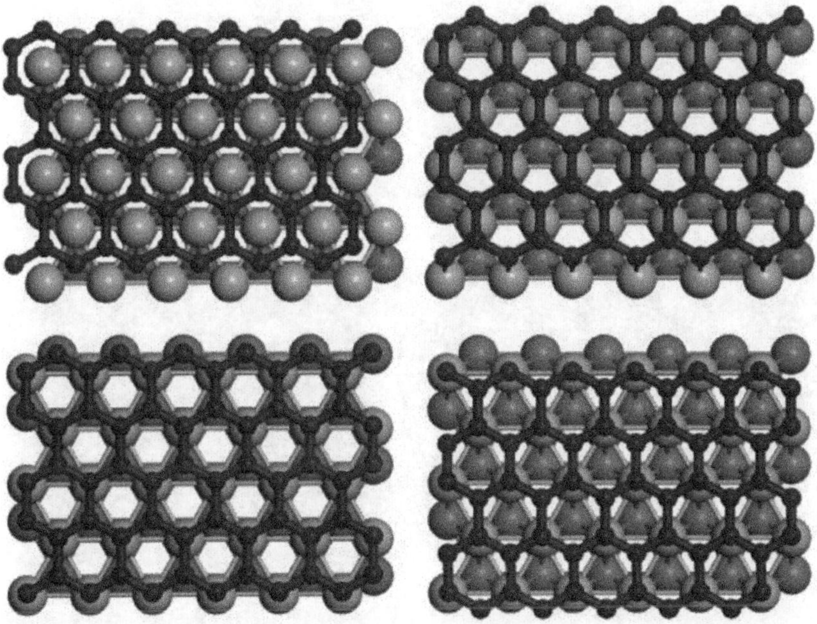

Fig. 20 Adsorption sites for graphene on fcc-Co(111) at ring-top, ring-bridge, ring-fcc and ring-hcp sites (top layer Co atoms:light gray; second layer Co atoms: dark gray; carbon atoms: black). Reproduced from ref. 97 by permission of The Royal Society of Chemistry.

Graphene adsorption on cobalt is a chemisorption process as evidenced by the partial DOS and Bader charge analysis; meanwhile, the charge transfer and the shifts in the Fermi level and electronic levels were evidenced by the vibrational analysis. The vibrational analysis has revealed the vibrational modes within the sheets are weakened while the modes perpendicular to the surface are strengthened.

The Bader charge analysis indicated that adsorption of graphene on the metal resulted in a redistribution of charge density; the surface metal atoms lose 0.144 e while the carbon atoms on top gain 0.151 e. A small change in the charge of the carbon atoms on the hollow site and the metal atoms in the bulk was observed, suggesting a localized interaction between the surface metal atoms and the surface carbon atoms.

Apart from the interaction of graphene with fcc Co(111), Swart and co-authors[98] also investigated the possible intermediates in the formation of graphene on fcc-Co(111) (Fig. 21). At higher coverages (1 ML), carbon clusters are much more stable than atomic carbon atoms, while the linear carbon structures are slightly favored over the branched carbon structures. Further growth of the carbon chains, aromatic clusters become more favorable. The flat graphene sheet on the surface was the most stable of all the aromatic structures under FTS conditions.

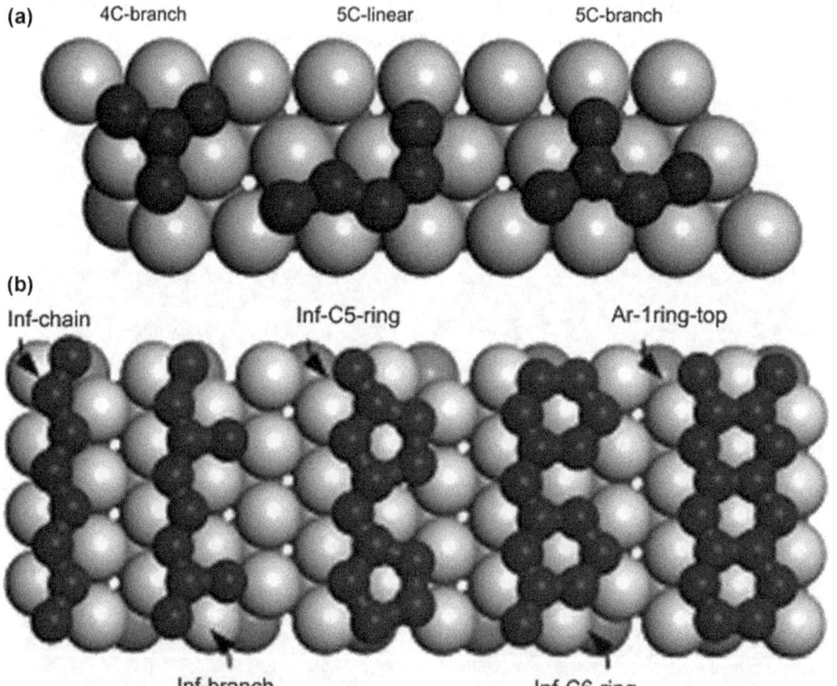

Fig. 21 (a) Larger linear and branched clusters investigated as possible intermediates in the formation of graphene; (b) Infinite structures in one dimension as possible intermediates in the formation of graphene. Reprinted with permission from ref. 98. Copyright 2008 American Chemical Society.

4.3 Deactivation by sulfur

Sulfur compounds are apt to poison most metal catalysts. Sulfur-containing species from coal, natural gas and biomass can be brought into FTS *via* syngas production. The sulfur-containing compounds were found strongly adsorbed on the catalyst surface and deactivate the catalysts. Only few theoretical studies investigated the adsorption of sulfur on Co catalysts.[99,100]

The favored configuration is sulfur atoms locating at the fcc-hollow sites with S–Co distance of 2.19 ± 0.10 Å. The small difference between fcc and hcp adsorption sites can also be interpreted through the changes in the charge density at the surface due to the adsorption of S atoms (Fig. 22). Thus, it seems that the prime interaction occurs between the topmost Co surface atoms and the adsorbed S atoms. S adatoms were

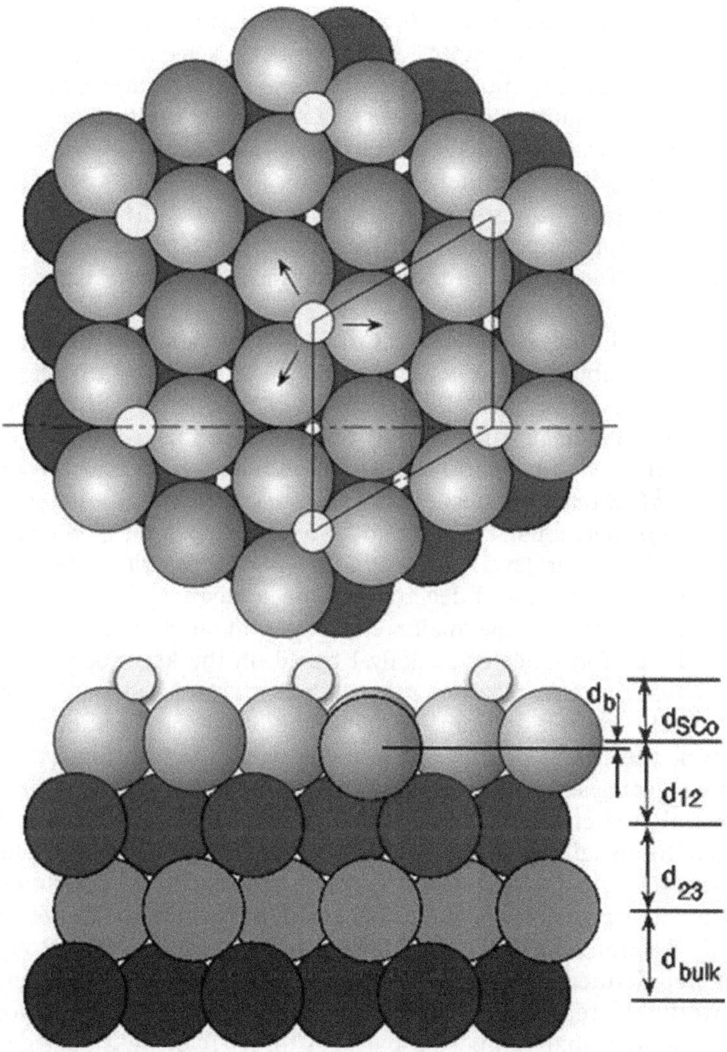

Fig. 22 Top and side views of the optimum (2 × 2)-S adsorption structure on Co(0001). Reprinted from ref. 99 with permission from Elsevier.

found tightly bound to the nearest neighbor Co surface atoms, but only a tiny change in the electronic structure near other Co atoms was identified, based both on the Local DOS and charge density. It can explain that the poisoning effect of S on Co(0001) only is limited to the sites it is located at. This agrees well with experimental results[74] that one S atom blocks effectively 1.2 CO sites, while in the case of hydrogen adsorption,[75] the blocking is complete. From the calculation results on electronic structure, adsorption energy of CO molecule is remarkably reduced upon S adsorbed on Co(0001), it is consistent with the experimental results that the presence of S weakens the bonding of CO to the Co(0001) surface.

5 Conclusions and perspectives

Although FTS has long been studied as early as the 1920s for the production of clean transportation fuels and chemicals, there are still major questions regarding the control of activity and selectivity of the transformation of syngas to produce hydrocarbons or oxygenates. Specifically, the reaction mechanism in detail is still unknown; in particular, the relationships between catalytic performance and catalyst structure are lacking from both experiment and theory. All the statements in this chapter are limited to illustrate the surface structure, the mechanism of initial elementary steps and deactivation over Co-based catalysts during FTS using DFT calculations. Many important additional aspects have been ignored.

Obviously, the development of Co-based catalysts with excellent selectivity and stability is the main target for FTS catalysts. Especially, efforts should be given to the improvement of product selectivity towards diesel fuels, chemicals such as lower olefins and linear alcohols of the desired carbon chain length. The primary challenge lies in the understanding of FTS reaction mechanism. The successful manipulation of the FTS reaction mechanism appears to be a hard task and, obviously, will require a molecular level understanding of the reaction mechanism, nature of active site(s) and deactivation mechanism.

In addition, one of the major challenges in modern catalysis is to predict the performance of a catalyst based on the knowledge of active sites. Nowadays, the reaction rate coefficients and the nature of the intermediate complexes can be simulated by computational tools. Moreover, the use of *in situ*/*operando* spectroscopic techniques, simplified model catalysts and microkinetic modeling are essential to verify the theoretical predictions. This is especially important since first principles calculations usually involve the choice of very approximate models to represent the active sites in heterogeneous catalysis. Indeed, using state-of-the-art first principles calculations, it might be possible to describe a catalytic reaction on a well-defined metal surface, usually a single crystal, under well-defined conditions with reasonable accuracy. However, the "real world" is tremendously more complex, particularly when reactions take place over small supported particles under realistic industrial conditions. The ideal surfaces and those of industrial practice are not only separated by the often cited "pressure gap" but also by "material gaps".

For instance, the dynamic response of small supported particles in different environments is investigated initially, requiring the use of sophisticated *in situ/operando* characterization techniques.

It is clear that a detailed knowledge of the structure of the catalyst under reaction conditions, in particular FTS reaction in harsh conditions (230–300 °C, 2–5 MPa) is required for any theoretical study to be successful described. Recent studies are combining DFT with simple thermodynamics to obtain phase diagrams and predict the thermodynamically favored catalyst structure under different conditions.[101,102] This approach would be appreciable for the elucidation of surface structure of Co-based catalysts for FTS reaction. In addition, Kinetic Monte Carlo methods allow bridging the pressure gap from the vacuum conditions of DFT to real conditions, leading to the understanding, optimization and design of new catalysts.

The complexity of FTS is often quoted as a second challenge that prevents a molecular level understanding of the process. However, recent efforts of building microkinetic models for the reaction indicate that the process can be understood by focusing on a limited number of steps, *i.e.* hydrogenation, C–C coupling and CO activation.[37,103] The complexity of the reaction products, however, makes the kinetics of the process quite hard to be accurately described. Nevertheless, these are beyond this review. The gap between molecular level modeling of simplified catalyst systems and experimental studies of real world catalysts is often quoted as a difficult one to bridge. This will also be one of the main challenges in FTS.

References

1 F. Fischer and H. Tropsch, *Brennst.-Chem.*, 1923, **4**, 276.
2 F. Fischer and H. Tropsch, *Brennst.-Chem.*, 1926, **7**, 79.
3 F. Fischer and H. Tropsch, *Chem. Ber.*, 1926, **59**, 830.
4 E. Iglesia, *Appl. Catal., A*, 1997, **161**, 59.
5 D. J. Moodley, J. van de Loosdrecht, A. M. Saib, M. J. Overett, A. K. Datye and J. W. Niemantsverdriet, *Appl. Catal., A*, 2009, **354**, 102.
6 J. K. Nørskov, F. Abild-Pedersen, F. Studt and T. Bligaard, *Proc. Natl. Acad. Sci. U. S. A.*, 2011, **108**, 937.
7 G. A. Somorjai and K. McCrea, *Appl. Catal., A*, 2001, **222**, 3.
8 G. A. Somorjai and Y. Li, *Introduction to Surface Chemistry and Catalysis*, John Wiley & Sons, Inc., New Jersey, USA, 2010.
9 I. Chorkendorff and J. W. Niemantsverdriet, *Concepts of Modern Catalysis and Kinetics*, Wiley-VCH Verlag GmbH & Co. KGaA, Weinheim, Germany, 2003.
10 I. Langmuir, *J. Am. Chem. Soc.*, 1918, **40**, 1361.
11 E. K. Rideal and H. S. Taylor, *Catalysis in Theory and Practice*, Macmillan, London, UK, 1919.
12 M. E. Bridge, C. M. Comrie and R. M. Lambert, *Surf. Sci.*, 1977, **67**, 393.
13 E. A. Lewis, A. D. Jewell, G. Kyriakou and E. C. H. Sykes, *Phys. Chem. Chem. Phys.*, 2012, **14**, 7215.
14 B. Hammer and J. K. Nørskov, *Adv. Catal.*, 2000, **45**, 71.
15 D. S. Sholl and J. Steckel, *Density Functional Theory : A Practical Introduction*, John Wiley & Sons, Inc., Hoboken, New Jersey, 2009.

16 R. A. van Santen, I. M. Ciobîcă, E. van Steen and M. M. Ghouri, *Adv. Catal.*, 2011, **54**, 127.
17 Z.-J. Wang, Z. Yan, C.-J. Liu and D. W. Goodman, *ChemCatChem*, 2011, **3**, 551.
18 R. A. Van Santen, A. De Koster and T. Koerts, *Catal. Lett.*, 1990, **7**, 1.
19 M. Corral Valero and P. Raybaud, *Catal. Lett.*, 2013, **143**, 1.
20 S. S. Sung and R. Hoffmann, *J. Am. Chem. Soc.*, 1985, **107**, 578.
21 C. Zheng, Y. Apeloig and R. Hoffmann, *J. Am. Chem. Soc.*, 1988, **110**, 749.
22 P. Sabatier, *La Catalyse en Chimie Organique*, Béranger, Paris, France, 1920.
23 F. Abild-Pedersen and M. P. Andersson, *Surf. Sci.*, 2007, **601**, 1747.
24 S.-H. Ma, X.-T. Zu, Z.-Y. Jiao and X.-Z. Zhang, *J. Mol. Struct.*, 2008, **864**, 68.
25 G. F. Cabeza, P. Légaré and N. J. Castellani, *Surf. Sci.*, 2000, **465**, 286.
26 S.-H. Ma, X.-T. Zu and Z.-Y. Jiao, *Solid State Commun.*, 2008, **147**, 152.
27 X.-Q. Gong, R. Raval and P. Hu, *Surf. Sci.*, 2004, **562**, 247.
28 C.-F. Huo, Y.-W. Li, J. Wang and H. Jiao, *J. Phys. Chem. C*, 2008, **112**, 14108.
29 J.-X. Liu, H.-Y. Su and W.-X. Li, *Catal. Today*, 2013, **215**, 36.
30 J.-X. Liu, H.-Y. Su, D.-P. Sun, B.-Y. Zhang and W.-X. Li, *J. Am. Chem. Soc.*, 2013, **135**, 16284.
31 A. M. Saib, M. Claeys and E. Van Steen, *Catal. Today*, 2002, **71**, 395.
32 O. Kitakami, H. Sato, Y. Shimada, F. Sato and M. Tanaka, *Phys. Rev. B: Condens. Matter Mater. Phys.*, 1997, **56**, 13849.
33 P. Van Helden, J.-A. Van den Berg and I. M. Ciobîcă, *Catal. Sci. Technol.*, 2012, **2**, 491.
34 R. V. Belosludov, S. Sakahara, K. Yajima, S. Takami, M. Kubo and A. Miyamoto, *Appl. Surf. Sci.*, 2002, **189**, 245.
35 V. Bakken, E. Rytter and O. Swang, *Surf. Sci.*, 2011, **605**, 513.
36 J. Greeley, J. K. Nørskov and M. Mavrikakis, *Annu. Rev. Phys. Chem.*, 2002, **53**, 319.
37 A. J. Markvoort, R. A. Van Santen, P. A. Hilbers and E. J. Hensen, *Angew. Chem.*, 2012, **51**, 9015.
38 M. Ojeda, R. Nabar, A. U. Nilekar, A. Ishikawa, M. Mavrikakis and E. Iglesia, *J. Catal.*, 2010, **272**, 287.
39 M. Zhuo, A. Borgna and M. Saeys, *J. Catal.*, 2013, **297**, 217.
40 O. R. Inderwildi, S. J. Jenkins and D. A. King, *J. Phys. Chem. C*, 2008, **112**, 1305.
41 C.-F. Huo, Y.-W. Li, J. Wang and H. Jiao, *J. Phys. Chem. C*, 2008, **112**, 14108.
42 P. Van Helden, J.-A. Van den Berg and C. J. Weststrate, *ACS Catal.*, 2012, **2**, 1097.
43 C. J. Weststrate, A. C. Kızılkaya, E. T. R. Rossen, M. W. G. M. Verhoeven, I. M. Ciobîcă, A. M. Saib and J. W. Niemantsverdriet, *J. Phys. Chem. C*, 2012, **116**, 11575.
44 S. H. Ma, Z. Y. Jiao, T. X. Wang and X. Q. Dai, *Surf. Sci.*, 2014, **619**, 90.
45 X.-Q. Gong, R. Raval and P. Hu, *J. Chem. Phys.*, 2005, **122**, 024711.
46 I. M. Ciobîcă, R. A. Van Santen, P. J. Van Berge and J. Van de Loosdrecht, *Surf. Sci.*, 2008, **602**, 17.
47 C. K. Rofer-DePoorter, *Chem. Rev.*, 1981, **81**, 447.
48 R. B. Anderson, Fischer–Tropsch and Related Synthesis, Academic Press, New York, 1983.
49 A. T. Bell, *Catal. Rev.*, 1981, **23**, 203.
50 P. Biloen and W. M. H. Sachtler, *Adv. Catal.*, 1981, **30**, 165.
51 G. J. Hutchings, A. Kiennemann and J. P. Hindermann, *Catal. Rev.*, 1993, **35**, 255.
52 P. Biloen, J. N. Helle and W. M. H. Sachtler, *J. Catal.*, 1979, **58**, 95.

53 R. C. Brady and R. Pettit, *J. Am. Chem. Soc.*, 1981, **103**, 1287.
54 H. Pichler and H. Schulz, *Chem. Ing. Tech.*, 1970, **12**, 1160.
55 C. Masters, *Adv. Organomet. Chem.*, 1979, **17**, 61.
56 G. A. Beitel, A. Laskov, H. Oosterbeek and E. W. Kuipers, *J. Phys. Chem. C*, 1996, **100**, 12494.
57 P. M. Maitlis and V. Zanotti, *Chem. Commun.*, 2009, 1619.
58 J. Gaube and H. F. Klein, *J. Mol. Catal. A: Chem.*, 2008, **283**, 60.
59 J. Gaube and H. F. Klein, *Appl. Catal., A*, 2010, **374**, 120.
60 Q. Ge and M. Neurock, *J. Phys. Chem. B*, 2006, **110**, 15368.
61 M. Zhuo, K. F. Tan, A. Borgna and M. Saeys, *J. Phys. Chem. C*, 2009, **113**, 8357.
62 H. Burghgraef, A. P. J. Jansen and R. A. Van Santen, *J. Chem. Phys.*, 1995, **103**, 6562.
63 Q. Ge, M. Neurock, H. A. Wright and N. Srinivasan, *J. Phys. Chem. B*, 2002, **106**, 2826.
64 J. Cheng, X.-Q. Gong, P. Hu, C. M. Lok, P. Ellis and S. French, *J. Catal.*, 2008, **254**, 285.
65 J. Cheng, P. Hu, P. Ellis, S. French, G. Kelly and C. M. Lok, *J. Catal.*, 2008, **257**, 221.
66 J. Cheng, P. Hu, P. Ellis, S. French, G. Kelly and C. M. Lok, *Top. Catal.*, 2010, **53**, 326.
67 E. Van Steen, M. Claeys, M. E. Dry, J. Van de Loosdrecht, E. L. Viljoen and J. L. Visagie, *J. Phys. Chem. B*, 2005, **109**, 3575.
68 J. Cheng, P. Hu, P. Ellis, S. French, G. Kelly and C. M. Lok, *J. Phys. Chem. C*, 2008, **112**, 9464.
69 O. R. Inderwildi, D. A. King and S. J. Jenkins, *Phys. Chem. Chem. Phys.*, 2009, **11**, 11110.
70 M. C. J. Bradford and M. A. Vannice, *Catal. Rev.*, 1999, **41**, 1.
71 R. A. Van Santen and A. J. Markvoort, *ChemCatChem*, 2013, **5**, 3384.
72 Y.-H. Zhao, K. Sun, X. Ma, J. Liu, D. Sun, H.-Y. Su and W.-X. Li, *Angew. Chem., Int. Ed.*, 2011, **50**, 5335.
73 N. E. Tsakoumis, M. Rønning, Ø. Borg, E. Rytter and A. Holmen, *Catal. Today*, 2010, **154**, 162.
74 K. Habermehl-Ćwirzeń and J. Lahtinen, *Surf. Sci.*, 2004, **573**, 183.
75 K. Habermehl-Ćwirzeń, K. Kauraala and J. Lahtinen, *Phys. Scr.*, 2004, **2004**, 28.
76 A. Y. Khodakov, W. Chu and P. Fongarland, *Chem. Rev.*, 2007, **107**, 1692.
77 C. H. Bartholomew, *Appl. Catal., A*, 2001, **212**, 17.
78 X.-Q. Gong, R. Raval and P. Hu, *Mol. Phys.*, 2004, **102**, 993.
79 P. J. Van Berge, J. Van de Loosdrecht, S. Barradas and A. M. Van der Kraan, *Catal. Today*, 2000, **58**, 321.
80 A. M. Saib, D. J. Moodley, I. M. Ciobîcă, M. M. Hauman, B. H. Sigwebela, C. J. Weststrate, J. W. Niemantsverdriet and J. Van de Loosdrecht, *Catal. Today*, 2010, **154**, 271.
81 J. C. W. Swart, PhD Thesis, University of Cape Town, South Africa, 2008.
82 G. L. Bezemer, J. H. Bitter, H. P. C. E. Kuipers, H. Oosterbeek, J. E. Holewijn, X. Xu, F. Kapteijn, A. J. Van Dillen and K. P. de Jong, *J. Am. Chem. Soc.*, 2006, **128**, 3956.
83 N. O. Elbashir, P. Dutta, A. Manivannan, M. S. Seehra and C. B. Roberts, *Appl. Catal., A*, 2005, **285**, 169.
84 J. Li, X. Zhan, Y. Zhang, G. Jacobs, T. Das and B. H. Davis, *Appl. Catal., A*, 2002, **228**, 203.

85 G. Jacobs, T. K. Das, P. M. Patterson, J. Li, L. Sanchez and B. H. Davis, *Appl. Catal., A*, 2003, **247**, 335.
86 G. Jacobs, P. M. Patterson, Y. Zhang, T. Das, J. Li and B. H. Davis, *Appl. Catal., A*, 2002, **233**, 215.
87 P. G. Menon, *J. Mol. Catal.*, 1990, **59**, 207.
88 J. A. Moulijn, A. E. Van Diepen and F. Kapteijn, *Appl. Catal., A*, 2001, **212**, 3.
89 M. E. Dry, *Catal. Today*, 2002, **71**, 227.
90 G. Kalibaeva, R. Vuilleumier, S. Meloni, A. Alavi, G. Ciccotti and R. Rosei, *J. Phys. Chem. B*, 2006, **110**, 3638.
91 S.-G. Wang, X.-Y. Liao, D.-B. Cao, Y.-W. Li, J. Wang and H. Jiao, *J. Phys. Chem. C*, 2007, **111**, 10894.
92 H. Amara, C. Bichara and F. Ducastelle, *Phys. Rev. B: Condens. Matter Mater. Phys.*, 2006, **73**, 113404.
93 D. J. Moodley, J. Van de Loosdrecht A. M. Saib and J. W. Niemantsverdriet, in *Chemical Industries*, ed. B. H. Davis and M. L. Occelli, CRC Press, Boca Raton, 2009.
94 V. Gruver, R. Young, J. Engman and H. J. Robota, *Prepr. Pap. –Am. Chem. Soc., Div. Pet. Chem.*, 2005, **50**, 164.
95 V. Gruver, X. Zhan, J. Engman and H. J. Robota, *Prepr. Pap. –Am. Chem. Soc., Div. Pet. Chem.*, 2004, **49**, 192.
96 K. F. Tan, J. Xu, J. Chang, A. Borgna and M. Saeys, *J. Catal.*, 2010, **274**, 121.
97 J. C. W. Swart, E. Van Steen, I. M. Ciobica and R. A. Van Santen, *Phys. Chem. Chem. Phys.*, 2009, **11**, 803.
98 J. C. W. Swart, I. M. Ciobîcă, R. A. Van Santen and E. Van Steen, *J. Phys. Chem. C*, 2008, **112**, 12899.
99 J. Lahtinen, P. Kantola, S. Jaatinen, K. Habermehl-Cwirzen, P. Salo, J. Vuorinen, M. Lindroos, K. Pussi and A. P. Seitsonen, *Surf. Sci.*, 2005, **599**, 113.
100 S. H. Ma, Z. Y. Jiao, T. X. Wang and X. T. Zu, *J. Phys. Chem. C*, 2009, **113**, 16210.
101 K. Reuter, D. Frenkel and M. Scheffler, *Phys. Rev. Lett.*, 2004, **93**, 116105.
102 W.-X. Li, C. Stampfl and M. Scheffler, *Phys. Rev. Lett.*, 2003, **90**, 256102.
103 S. Storsæter, D. Chen and A. Holmen, *Surf. Sci.*, 2006, **600**, 2051.
104 S. E. Mason, I. Grinberg and A. M. Rappe, *Phys. Rev. B*, 2004, **69**, 161401.

Structure prediction and its applications in computational materials design

Qiang Zhu,*[a] Artem R. Oganov,[a,b,c] Qingfeng Zeng[c] and Xiangfeng Zhou[d]
DOI: 10.1039/9781782622703-00219

1 Introduction

Structure is the most fundamental characteristics of a material. X-ray crystallography allows one to determine how atoms are arranged in a molecule and how molecules pack into a crystal. However, it requires a high-quality crystal sample, which is time consuming to prepare and often impossible under extreme conditions.

Theory has been playing a significant role in understanding crystal structures. Pauling summarized the rules for crystal structures of ionic solids,[1] however, similar powerful rules are still lacking for metals. First attempts to use computers to predict crystal structures date back to 1980. Although not very successful at the beginning,[2,3] crystal structure prediction (CSP) began to play an important role nowadays, thanks to many progresses in the last decade.[4–13] Indeed, mathematicians have developed algorithms to solve similar problems. Some of them are quite general and thus could be applied to crystal structure prediction. One can refer to a recent book[14] for a discussion of different methods. In this chapter, we will briefly introduce the modern structure prediction techniques, and review the recent developments in the context of the USPEX method, which is based on the evolutionary algorithm (EA), and has been viewed as a revolution in crystallography.[15] Discussions here follow closely the previous literature,[16–18] with primary focus on the most recent developments.

2 Methodology

2.1 Global optimization methods

Several global optimization algorithms have been devised and used with varying degree of success in structure prediction, for instance, simulated annealing,[4,5] metadynamics,[6,7] genetic and evolutionary algorithms,[8,13] random sampling,[9] basin hopping,[10] minima hopping,[11] and data mining.[12] Most of the methods mentioned above are developed to predict inorganic crystals and nano-clusters. However, the same philosophy can be applied to organic crystals[19–23] and proteins[24] as well.

[a]*Department of Geosciences, SUNY Stony Brook, Stony Brook, NY, USA. E-mail: qiang.zhu@stonybrook.edu*
[b]*Moscow Institute of Physics and Technology, Dolgoprudny city, Moscow, Russia*
[c]*International Center for Materials Discovery, School of Materials Sciences and Engineering, Northwestern Polytechnical Univerisity, Xi'an, China*
[d]*School of Physics, Nankai Univeristy, Tianjin, China*

Strictly speaking, all of the above methods (except metadynamics and data mining) are stochastic methods, thus they possess some inherent randomness. It is not guaranteed the one would obtain the same solution even by starting with the same set of parameter values. The divergence depends on the complexity of the landscape.

One either has to start already in a good region of configuration space (so that no effort is wasted on sampling poor regions) or use a "self-improving" method that locates, step by step, the best structures. The first group of methods includes metadynamics, simulated annealing, basin hopping, and minima hopping approaches. The second group essentially includes evolutionary algorithms. Alternatively, data mining approaches use advanced machine learning concepts and predict the structures based on a large database of known crystal structures.[12] Among all these groups of methods, the strength of evolutionary simulations is that they do not require any system-specific knowledge except chemical composition, and are self-improving, *i.e.*, in subsequent generations increasingly good structures are found and used to generate new structures.

2.2 Energy landscape

Before talking about the prediction of the crystal structure, let us first consider the energy landscape that needs to be explored. The dimensionality of the energy landscape is:

$$d = 3N + 3, \tag{1}$$

where $3N - 3$ degrees of freedom are from N atoms, and the remaining six dimensions are defined by the lattice. CSP is an NP-hard problem, *i.e.*, the difficulty increases exponentially with dimensionality. Yet, drastic simplification can be made if structures are relaxed, *i.e.* brought to the nearest local energy minima. Relaxation introduces intrinsic chemical constraints (bond lengths, bond angles, avoidance of unfavorable contacts). Therefore, the intrinsic dimensionality can be reduced:

$$d^* = 3N + 3 - \kappa, \tag{2}$$

where κ is the number of correlated dimensions, which could vary greatly according to the intrinsic chemistry in the system. For example, the dimensionality drops greatly from 99 to 11.6 for $Mg_{16}O_{16}$, while only slightly from 39 to 32.5 for $Mg_4N_4H_4$.[16] Thereby, the reduced difficulty of the problem (*i.e.*, the number of possible structures) is reduced:

$$C^* = \exp(\beta d^*). \tag{3}$$

This implies that any efficient search method must include structure relaxation (**local optimization**). We also note that all global optimization methods rely on the assumption that the reduced energy landscape will have a well-organized overall shape (Fig. 1), which is often true for chemical systems.[28]

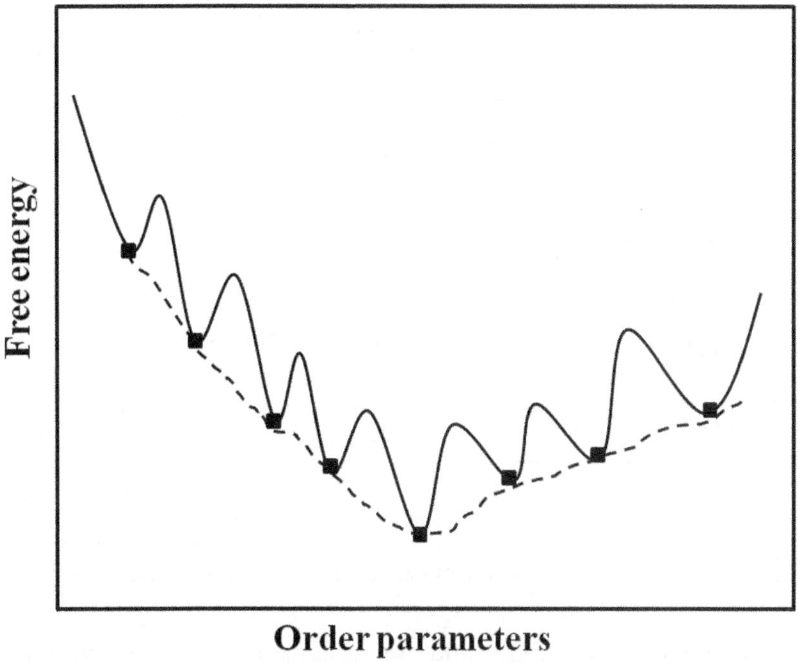

Fig. 1 An illustration of the simplified illustration of energy landscape. The idea of local optimization is to transform the noisy acted landscape (solid line) to a bowl-shaped reduced landscape (dashed line).

2.3 Evolutionary algorithm

Evolutionary algorithms (EA) mimic Darwinian evolution and employ natural selection of survival of the fittest and variation operators including genetic heredity and mutations. It is a stochastic method which is used to solve problems in which there exist many possible solutions for minima. The EA procedure is as shown in Fig. 2:

(1) Initialization of the first generation, that is, a set of structures satisfying the hard constraints are randomly generated;

(2) Perform structural relaxation and determine the quality (fitness) for each member of the population;

(3) Selection of the high-fitness members from the current generation as parents, from which the new generation is created by applying specially designed variation operators;

(4) Repeat steps 2–3 until the halting criteria are achieved;

(5) The above algorithm has been implemented in the USPEX (Universal Structure Predictor: Evolutionary Xtallography) package.[13–17]

2.4 Representation, fitness and variation operators

During evolution, it is important to keep the good structural features from the old generations to the next population. In traditional genetic algorithms, the investigated systems are usually expressed as an array of bits (chromosome), where each bit (gene) represents a different object. This representation behaves like DNA, and is quite convenient for

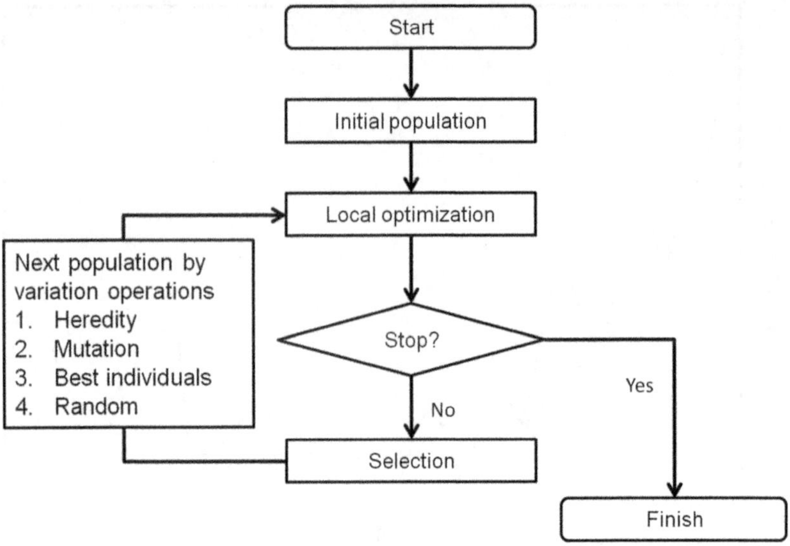

Fig. 2 The EA implemented in the USPEX code for crystal structure prediction.

variation operations (heredity and mutation). However, the disadvantage is that it involves encoding and decoding processes, which make it inconvenient to be applied to chemical systems, and most importantly, structural information is lost rather than transferred from parents to offspring. Deaven and Ho employed a real-space representation, and successfully applied it to the prediction of clusters.[27] The real-space representation in terms of Cartesian or fractional coordinates is more straightforward, and physically more meaningful. And USPEX adopts the **real space representation** as well.

The **Fitness function** mathematically describes the target direction of the global search, which can be either a thermodynamic fitness (to find stable states) or a physical property (to find materials with desired properties).

We rank structures by fitness values, a certain fraction of worst structures are discarded, and the rest are given a chance to be chosen as parents in the **selection** stage. The probability of the survived structures being chosen as a parent increases with the quality of a structure.

An essential step in an EA is to deliver the good genes to the next population, while introducing some variation. In USPEX, it is done *via* **variation operators**.

Heredity is a core part of the EA approach, as it allows communication between different trial solutions or classes of solutions by combining parts from different parents. In USPEX, to generate a child from two parents, the algorithm firstly chooses a plane which is parallel to one lattice plane, and then cuts a slice with a random thickness and random position along the other lattice vector; such slices from two parent structures are then matched to form a child structure. In this process, the number of atoms of each type is adjusted to ensure conservation of chemical composition.

Mutation operators use a single parent to produce a child. *Lattice mutation* applies a stain matrix with zero-mean Gaussian random strains to the lattice vectors; *soft-mode mutation* (which we call *softmutation* for brevity from now) displaces atoms along the softest mode eigenvectors, or a random linear combination of softest eigenvectors; the *permutation* operator swaps chemical identities of atoms in randomly selected pairs of unlike atoms.

A general challenge for global optimization methods is to avoid getting stuck in a local minimum and thus skip the global minimum. To prevent this, the key is to control the diversity of the population, by preventing proliferation of very similar structures and by adding new blood. For the latter purpose, we produce some fraction of each generation with the **random symmetric structures.**

Last, a certain number of best structures in the current generation (**best individuals**) are intentionally transferred to the new generation, and compete with others.

2.5 Fingerprints: a metric of structural similarity

In a global structure search, very similar structures always appear frequently. Duplicate structures do not only create inconvenience in post-processing, but also lead to the situation that the search is 'trapped' in some local minimum but not the ground state. Therefore, a technique to measure the similarities between structures is needed. In USPEX, we use the so-called fingerprint function[28] to describe a crystal structure. It is very similar to pair distribution function (PDF), which for an elemental solid is:

$$\mathrm{PDF}(R) = \sum_{i} \sum_{j \neq i} \frac{1}{4\pi R_{ij}^2 \frac{N}{V} \Delta} \delta(R - R_{ij}) \tag{4}$$

where R_{ij} is the distance between atoms i and j, V is the unit cell volume, N is the number of atoms in the unit cell, and Δ is a bin width (in Å). The index i goes over all atoms in the unit cell and index j goes over all atoms within the cutoff distance from the atom i. The PDF at long distances oscillates around the value $+1$, which is not convenient for our purposes, and we subtract this "background" value for convenience. Generalizing to systems containing more than one atomic type, we introduce fingerprint as a matrix, the components of which are fingerprint functions for A–B type distances:

$$F_{\mathrm{AB}}(R) = \sum_{\mathrm{A}_i,\mathrm{cell}} \sum_{\mathrm{B}_j} \frac{\delta(R - R_{ij})}{4\pi R_{ij}^2 \frac{N_\mathrm{A} N_\mathrm{B}}{V} \Delta} - 1 \tag{5}$$

One can measure the similarity between two structures by calculating the cosine distance between two fingerprint functions,

$$d_{ij} = 0.5 \cdot \left(1 - \frac{F_i F_j}{|F_i||F_j|}\right) \tag{6}$$

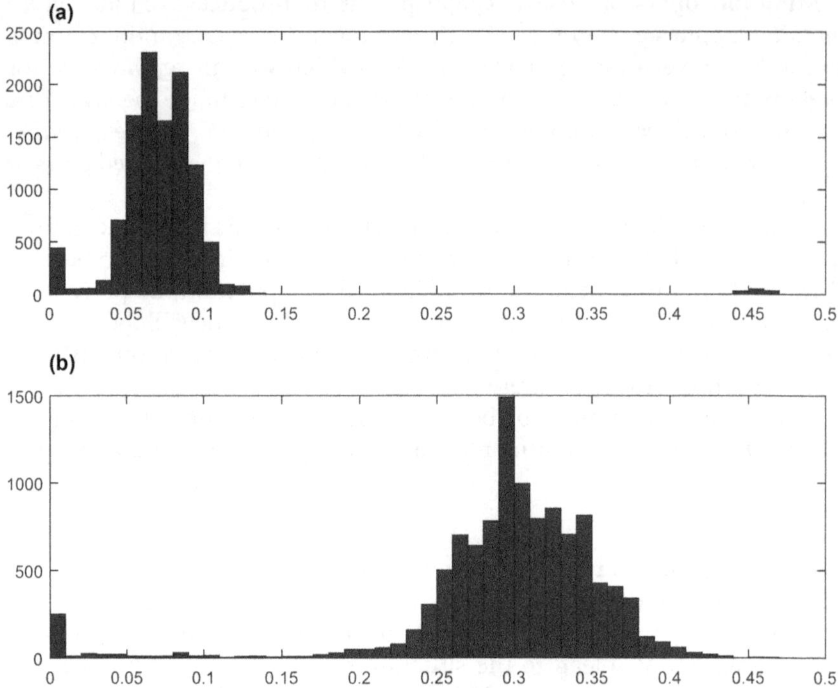

Fig. 3 Fingerprint distance distribution over 1000 structures from a typical USPEX simulation of 4 urea molecule per unit cell, when (a) including all distances (b) excluding intra-molecular distances in the fingerprint calculation for each individual structure.

Using this new crystallographic descriptor, we can improve the selection rules and variation operators above. During the selection process, only one copy of each distinct structure is used, and all its duplicates are killed. Fingerprint theory brings many other benefits (quantification and visualization of energy landscapes, use of ordered fragments of crystal structures, *etc.*).[25] However, it should be noted that the fingerprint function might be revised for different systems, in order to separate structures better. For instance, intramolecular contributions in fingerprint are identical for all different packing of the same molecule and thus decrease the discriminatory power of the fingerprint function (Fig. 3). Therefore, we only consider the intermolecular distances in the computation of the fingerprint function when dealing with crystals made of molecules with the same conformation.[26]

2.6 General choices of EA parameters

In any implementation of EA, the choices of parameters might lead to a different performance of the algorithm. Such parameters include: *population size, number of generations, fractions for each variation operation.* From our experience, a good choice of the population size should be $\sim$2 times of the number of atoms in the studied system, and the percentages should be 50%–70% for heredity, 20% for different types of mutations, and 15%–30% for random structures, respectively. Typically, such settings would be quite efficient. Therefore, we set another parameter (*stopCrit*) to stop the calculation if the best structure does not change for a given number of generations. We usually set 20 for *stopCrit* for system less than 40 atoms. However, larger values would be needed for larger systems.

A more rigorous way is to assign the fractions of the variation operation according to their performances during the calculation. In the current version of USPEX adopts the following strategy.

(1) At the end of each generation i, we discard identical structures, and select structures according to the fitness ranking;

(2) Within the selected structures, we count the origin of each structure, and obtain the fraction f_i*;

(3) Set $f_{i+1} = (f_i + f_i^*)/2$ for the next generation;

(4) In order to make it more robust, we also set the lower bound for those variation operations which have been proved very important from our experience. Here we put the minimum of 20% for heredity, 10% for random, 10% for softmutation.

According to our tests, the new scheme will generally enhance the searching efficiency by up to 2 times.

3 Recent developments

In the past years, we have extended structure prediction techniques to a broad range of systems. In this section, we will discuss the most recent developments. In order to give a complete demonstration, some developments which have been reviewed previously will be also briefly mentioned.

3.1 Choices of fitness functions

Traditional structure prediction is aimed at finding the structure with the lowest energy. Therefore, the fitness function is often defined as the energy or enthalpy. However, one can define the fitness in various ways, based on different applications. For instance, physical properties such as density, hardness and dielectric constants, can be used as the search criterion.[29–32] In these cases, a strategy of hybrid optimization is needed, that is, we search for the global optimum with respect to fitness, considering only structures corresponding to local energy minima. On the other hand, the so-called variable-composition prediction uses a modified energy criterion to evaluate the quality of structures over the whole allowed compositional space. In this case, stability of each individual can be defined as its decomposition energy relative to the easiest decomposition path.[33]

3.2 Low-dimensional systems

Comprehensive extensions of structure prediction in the most recent years are devoted to the low-dimensional systems, which include nanoparticles, two dimensional (2D) crystals, and surface reconstructions. For consistency with most widely used electronic structure codes, we treat

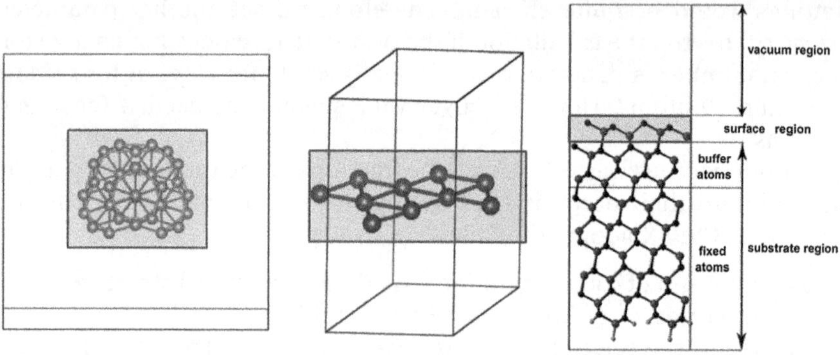

(a) 0D cluster (b) 2D crystal (c) 2D surface

Fig. 4 Two sets of cells in various low-dimensional systems. The small cells (highlighted in shadow) are used for global optimization, while the big cells (the whole structure model) are used for *ab initio* calculation.

the problem with periodic boundary conditions, by adding vacuum to eliminate the interactions with periodic images. Therefore, we have two types of cell representations in EA. Here the small cell represents the structure we want to optimize (excluding vacuum and substrate, this is the cell where variation operators work), and the big cell represents all the structural information (including vacuum and substrate, needed in structure relaxation) (Fig. 4).

3.2.1 Clusters. Several methods have been applied to cluster structure prediction.[11,27] Schönborn *et al.* translated the original version of the USPEX method to predict clusters.[34] However, with some new developments, this can perform even better.[17] The algorithm works as follows. The user gives a list of possible point groups (like, for example, C_2, D_{6h}, *etc.*) and nanoparticles are generated by randomly placing atoms inside the ellipsoid inscribed in the "small cell", and then replicating them using the point group symmetry operators. When the cluster is generated or relaxed, we place it in the center of the cell and rotate it so that principal moment of inertia axis with the highest moment is pointed in the *z*-direction. The "big cell" is then constructed by adding a certain amount of vacuum in all directions. The thickness of the vacuum region around the cluster is a user-defined parameter; more vacuum means more accurate results, but (for some approaches, such as plane-wave methods) greater computational costs. When interfaced with codes based on local basis set methods, the thickness does not strongly affect the *ab initio* calculation; however, it would be still convenient for performing the variation operations (such as heredity and lattice mutations). Before performing a 'cut-and-splice' heredity, the cut plane is randomly rotated around a random axis that goes through the center of mass of the nano-particle. This idea is similar to random 'shifts' for heredity in crystal structure prediction.

3.2.2 2D crystals. In recent years, 2D or quasi-2D materials have attracted great interests for their fascinating properties. Graphene, a single layer of carbon atoms with honeycomb configuration, has been widely studied due to its novel electronic properties (massless Dirac fermions, *etc.*). It has rapidly become a candidate for the next generation of faster and smaller electronic devices. Besides graphene, other 2D-crystals (such as MoS_2) with excellent properties were also discovered.[35] Quite recently, a systematic strategy for searching for flat 2D-crystals based on particle swarm optimization algorithm was proposed and applied to the B–C system.[36] However, it was later found that constraint of flat configuration will miss a number of meaningful systems. Thus, it was later extended to search for both 2D and quasi-2D materials.[37] In USPEX, we allow the non-planar configurations, and describe the system as a slab with a certain thickness. To initialize, the slabs of the 2D-crystal (small cell) are generated by random plane groups, and then the big cell is constructed by adding vacuum along the *c*-axis. After relaxation, we extract the slab from the big cell and apply the variation operators such as heredity and mutation. This allows one to explore more complex structures.

One must keep in mind that a 2D-crystal is always metastable and if allowed, will grow into a 3D-crystal. In other words, the greater the thickness is allowed, the lower-energy structures can be found. Thus, 2D-crystals give an example of constrained optimization, where the final results are determined by the constraint.

3.2.3 Surfaces. In practical calculations, the surface model includes three parts: vacuum, surface and substrate. Vacuum and substrate regions are pre-specified, while the surface region is optimized by the EA.[38] The number of surface atoms varies from zero to a given maximum number. Meanwhile, the cell size is also variable, in order to explore more complex reconstructions involving multiple unit cells. The fitness function needs to indicate the relative stability of structures with various surface stoichiometries and reconstruction cell sizes. We construct the fitness function based on the surface energy.

$$E_{\text{formation}} = E_{\text{total}} - E_{\text{ref}} - \sum_i n_i \mu_i, \tag{7}$$

where E_{total} and E_{ref} are the total energy of the surface under consideration and of the reference cleaved surface; n_i and μ_i are the number of atoms and chemical potentials for each species. The chemical potential is the energy needed to add or remove one atom from the system, assuming there is a reservoir for each species to equilibrate with. For a simple binary compound (AB), if $\mu(\text{A})$ is extremely high, the elemental phase A would condense on the substrate. Therefore, the chemical potentials must satisfy constraints under equilibrium conditions as follows,

$$\begin{aligned} &\mu(\text{A}) \leq \mu(\text{A}_0), \\ &\mu(\text{B}) \leq \mu(\text{B}_0), \\ &\mu(\text{A}) + \mu(\text{B}) = G(\text{AB}). \end{aligned} \tag{8}$$

At 0 K, the Gibbs free energy reduces to the internal energy $E(\mathrm{AB})$. Therefore, the chemical potential is bounded by

$$E(\mathrm{AB}) - \mu(\mathrm{B_0}) \le \mu(\mathrm{A}) \le \mu(\mathrm{A_0}). \tag{9}$$

Thus $E_{\mathrm{formation}}$ can be rewritten as dependent only on $\mu(\mathrm{A})$

$$E_{\mathrm{formation}} = E_{\mathrm{total}} - E_{\mathrm{ref}} - n_{\mathrm{B}}E(\mathrm{AB}) - \mu(\mathrm{A})(n_{\mathrm{A}} - n_{\mathrm{B}}) \tag{10}$$

This method can be employed in different ways (1) fixed number of surface atoms and cell size; (2) fixed number of surface atoms and variable cell sizes; (3) both variable surface atoms and surface unit cells. Here we emphasize the case of variable surface stoichiometry, as illustrated in Fig. 5. First, for two given surface configurations (I and II) which have different numbers of atoms on the same substrate cell, their relative energy differences is a function of the chemical potential $\mu(\mathrm{A})$ according to eqn (7). As shown in Fig. 5, surface I is stable when $\mu_{\mathrm{min}} \le \mu(\mathrm{A}) \le \mu_{\mathrm{eq}}$, while surface II is stable when $\mu_{\mathrm{eq}} \le \mu(\mathrm{A}) \le \mu_{\mathrm{max}}$. For any other unstable configuration, fitness can be viewed as the minimum energy difference compared with the stable configuration. The minimum condition is reached when $\mu(\mathrm{A}) = \mu_{\mathrm{eq}}$. It is useful to express it algebraically. Similar to Qian's work,[39] we define a term E_0 which is invariant to $\mu(\mathrm{A})$:

$$E_0 = E_{\mathrm{total}} - E_{\mathrm{ref}} - n_{\mathrm{B}}E(\mathrm{AB}) \tag{11}$$

Versions (a) and (b) in Fig. 5 contain equivalent information. Stable structures appearing on the phase diagram form a convex hull in energy-composition coordinates. The slope of each section in the convex hull is either the boundary chemical potential, or the equilibrium chemical potential in which stable configurations can coexist. Therefore, we can choose the fitness of a structure to be its distance to the convex hull. The EA search

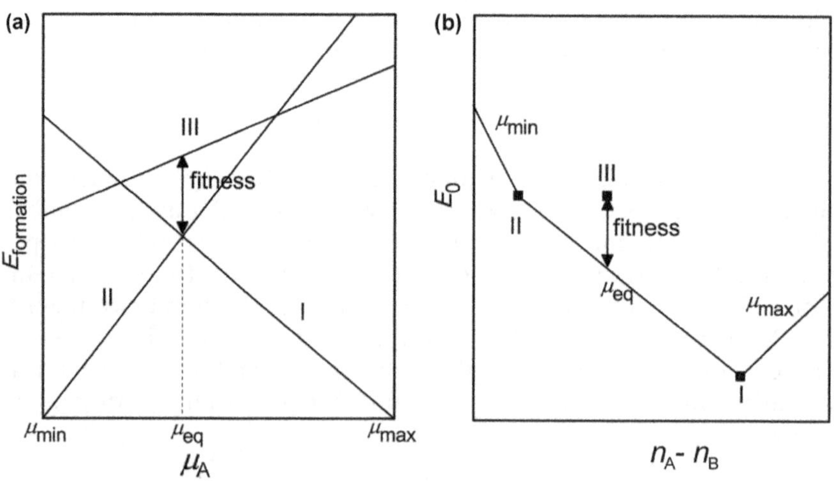

Fig. 5 Illustration of the fitness function used in surface prediction with variable stoichiometry in binary AB system. (a) Phase diagram as a function of $\mu(\mathrm{A})$. (b) Phase diagram as a function of $(n_{\mathrm{A}}-n_{\mathrm{B}})$. The vertices of the convex hull are the stable structures appearing in the phase diagram. The slope of each section is either the boundary chemical potential or the equilibrium chemical potential where stable structures coexist.

then aims to optimize the convex hull. When comparing structures with different surface cell sizes, the energies should be normalized.

3.3 Prediction of molecular crystal structures

The capability to predict molecular crystal structures has been implemented in the USPEX package since 2012.[26] When we adapted EA to organic systems, some concerns need to be addressed.

(1) *Metastablity*. Most of the molecular compounds are thermodynamically less stable than the simple molecular compounds from which they can be obtained (such as H_2O, CO_2, CH_4, *etc.*). This means that a fully unconstrained global optimization approach will produce only a mixture of these simple molecules, instead of the target molecular compounds of interest.

(2) *Weak interactions.* In organic crystals, packing largely depends on the weak inter-molecular interactions, such as hydrogen bonds and van der Waals interactions. These interactions are much weaker and softer than covalent bonds. Therefore, it leads to a very sparse molecular packing and a flat energy distribution in the real space. In this case, a method containing both efficient structural search and accurate energy ranking is needed.

(3) *Symmetry preference.* The distribution of structures over symmetry groups is very uneven. Most organic crystals are found to possess space groups: $P2_1/c$ (36.59%), P-1(16.92%), $P2_12_12_1$(11.00%), $C2/c$(6.95%).[40]

In order to apply EA to organic systems, it is essential to impose constraints, by fixing the bond connectivity and rigid angles; this can be conveniently done when molecular geometry is represented by internal coordinates (bond length, bond angle, torsional angle).[26] Here we introduce two other types of constraints which can be made for different systems.

3.3.1 Linear polymers. In most of the polymeric crystals, the structure can be viewed as packing of polymeric chains. Provided the chain conformation is known, their packing can be described by (1) relative positions of chains; (2) rotational degrees of freedom associated with the lateral groups; (3) the orientation of the chains. For linear polymers, the mutual orientation of the chains can only be either parallel or anti-parallel. As shown in Fig. 6, we assemble the polymeric chains from the monomers by ensuring the neighboring contacts of these bridging atoms are close to the real situation (in terms of bond length and bond angle). Mathematically, the chain's orientation can be determined by the vector between the geometric centers of two connected monomers, C–C′. Thus we can reorient the linear chain in the (001) or (00-1) direction. In the structure initialization stage, we create a 2D primitive cell in the *a*–*b* plane for the geometric centers, according to the randomly assigned plane group symmetry. Then the *c*-axis is defined by the length of the chain, and the monomers are arranged either up and down around the centers in the 3D unit cell. Accordingly, the rotational axis is always fixed to the *c* direction. This linear chain mode has been applied to study the polymorphism of various systems such

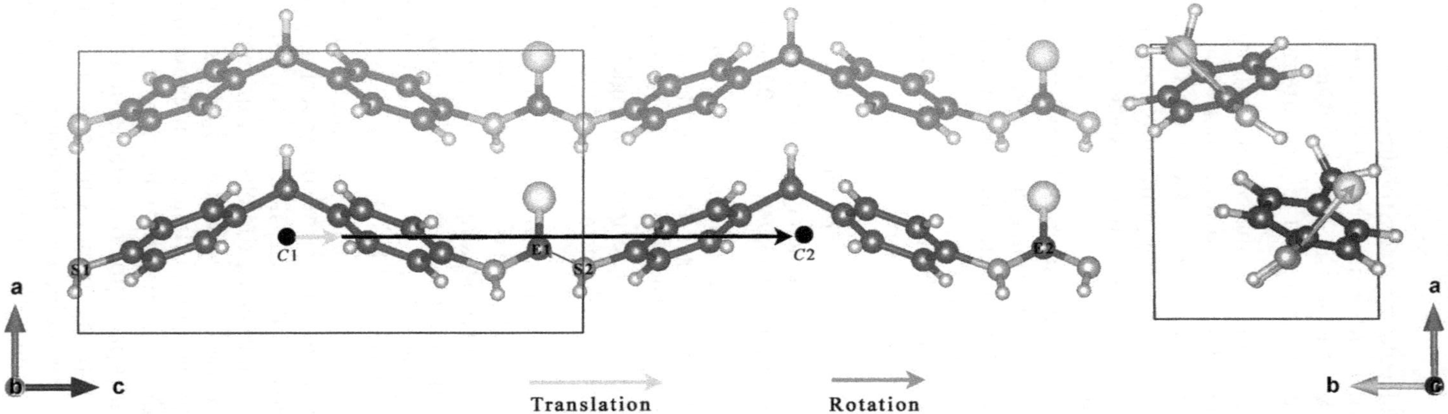

Fig. 6 Structure initialization of linear polymeric crystals. C and C′ are the geometric centers of monomers. The monomers are assembled in such a way that the C–C′ connections are parallel or antiparallel to the *c*-axis of the cell. The operations of translation and rotation will strictly act along *c*-axis.

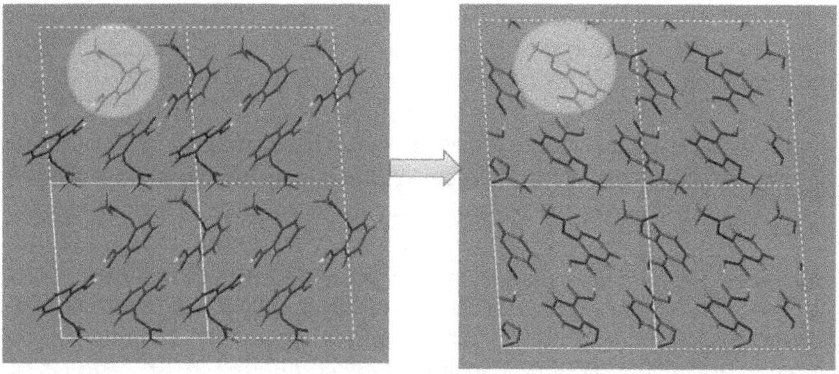

Fig. 7 Illustration of the symmetry-preserving rotational mutation in the case of aspirin. The parent structure has one asymmetric unit ($Z' = 1$) in $P2_1/c$ space group. The child structure is obtained by randomly changing the orientation of one molecule in the parent structure, while the rest molecules are generated according to the symmetry operations.

as poly(vinylidene fluoride) PVDF, and it proved to significantly speed up the searching process.[41]

3.3.2 Symmetry-preserving operations. Based on the fact that the symmetry distribution of organic crystals is very uneven, we design the mutation operators that keep the original symmetry. Since we generate the initial structures with random space groups, we keep the track of the asymmetric units and the corresponding symmetry operations. During mutation, we only perform mutations on those asymmetric units and then reconstruct positions and orientations of the remaining molecules by symmetry operations. As shown in Fig. 7, this operator can efficiently generate structures close to ground state even from a structure with high energy. However, one should use it with caution, as new symmetries are harder to find.

4 Applications

Structure prediction techniques have become increasingly important in materials research. In this section, we will focus on the applications in materials sciences, based on the methodology described above.

In all the calculations described below, global optimizations were done by the USPEX code, and the VASP code[42] was employed for local optimization (*i.e.* structural relaxation), using the PBE exchange-correlation functional[43] and the PAW method.[44] For the soft materials, van der Waals (vdW) dispersion and hydrogen bond are crucial factors in determining the crystal packing. There have been major efforts in improving the accuracy of vdW functional.[45,46] Here we use the Tkatchenko–Scheffler method[47] and optPBE88[48] functional which have proved to give results in satisfactory agreement with experimental data.[49]

4.1 Materials missed in the experiments: CsF_n compounds

Many materials exhibit quite complex chemistry under extreme conditions. Theoretical prediction plays an increasingly important role in this field. A number of new stroichiometric compounds have been predicted[50–56] and even confirmed by experiments.[51] On the other hand, there should be vast opportunities to discover new compounds even under ambient conditions as well. For example, many possible inorganic materials that consist of three or more elements are not studied yet. More surprisingly, chemical space of even a restricted subclass of materials made of two elements is far from being exhausted. Recent studies show a lot of brand new binary compounds which have not been reported yet.[57–60] For instance, our recent work predicted four viable ground-state compounds, with MnB_2, MnB, MnB_4, and another previously never reported MnB_3. Stimulated by the simulation results, the further experiments were able to verify them by annealed samples.[59] Similarly, in the well known Hf–C system, two additional compounds Hf_3C_2 and Hf_6C_5 were predicted to be stable.[57] Therefore, the systematic variable compositional predictions for those materials of interests are in great need.

Here we illustrate the power of the variable-composition prediction by its application to the Cs–F system.[61] Alkali halides *MX* have been viewed as typical ionic compounds, characterized by 1 : 1 ratio necessary for charge balance between M^+ and X^-. It was proposed that group I elements like Cs can be oxidized further under high pressure.[56] We perform a comprehensive study of the CsF–F system at pressures up to 100 GPa, and found extremely versatile chemistry.[61] Our calculation uncovers quite a different scenario (Fig. 8) from Miao's report.[56] A series of

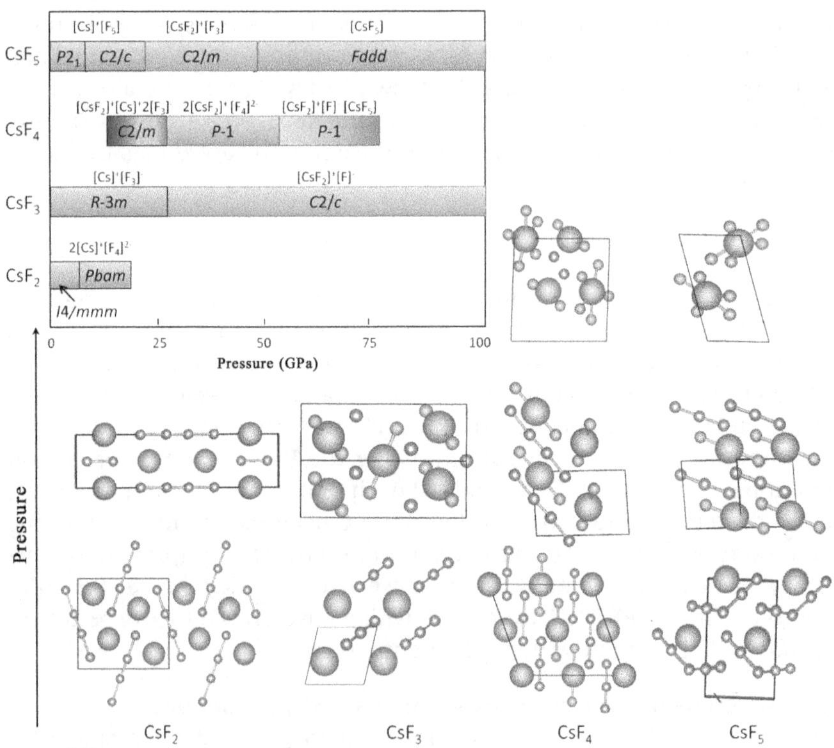

Fig. 8 The pressure-composition phase diagram of the CsF–F system and the corresponding stable crystal structures.

Table 1 Investigated reactions of the CsF–F system at ambient pressure conditions. The wt% gives the weight content of released F_2 gas. $\Delta H^{0\,K}$ and $\Delta H^{300\,K}$ are the calculated enthalpies at $T = 0$ K and 300 K, including the vibrational energies in kJ mol^{-1}. $\Delta S^{300\,K}$ is the corresponding formation entropy in J (K mol)$^{-1}$. T_c is the predicted decomposition temperature at standard atmosphere (1 bar). Note that F_2 is treated as crystalline solid at 0 K.

Reactions	wt%	$\Delta H^{0\,K}$	$\Delta H^{300\,K}$	$\Delta S^{300\,K}$	T_c(°C)
$CsF_2 = CsF + 1/2F_2(g)$	11.1	44.30	37.59	78.25	218
$CsF_3 = CsF + F_2(g)$	20.0	72.24	63.41	152.29	150
$CsF_5 = CsF + 2F_2(g)$	33.3	88.41	76.73	284.96	− 15

CsF_n ($n>1$) compounds are predicted to be stable already at ambient pressure. Under pressure, 5p electrons of Cs atoms become active, with growing tendency to form Cs (III) and (V) valence states at fluorine-rich conditions. Although Cs (II) and (IV) are not energetically favored, the interplay between two mechanisms (polyfluoride anions and polyvalent Cs cations) allows CsF_2 and CsF_4 compounds to be stable under pressure.

Surprisingly, already at ambient pressure several stoichiometric compounds (CsF_2, CsF_3 and CsF_5) are calculated to be thermodynamically stable. The estimated defluorination temperatures of CsF_n compounds at atmospheric pressure (218 °C, 150 °C, −15 °C, respectively), are attractive for fluorine storage applications. Light halogens, fluorine (F) and chlorine (Cl), at normal conditions are highly reactive and toxic gases. For chemical industry and laboratory use, this presents great inconvenience. Their storage in the gaseous form (even as liquefied gases) is very inefficient, and compressed gas tanks may explode, presenting great dangers. At normal conditions, the volume of 22.4 litres (L) of pure fluorine gas weighs just 36 grams (g), illustrating the dismal inefficiency of storage in this form. To the best of our knowledge, no effective and safe fluorine storage materials are known. Both F and Cl have a huge range of industrial applications, which would benefit from such storage materials, especially if they can achieve high storage capacity, stability and reversibility (Table 1).

4.2 Property optimizations – HfO_2–SiO_2

As we discussed above, crystal structure search can be also property-oriented within the framework of hybrid optimization. In this case, the candidate structures should be locally optimized by energy, and globally selected and operated with respect to the target properties. Following this track, researchers have made significant steps towards to the materials design by properties (including density,[29] hardness,[30,62] band gap,[31,63] and so on). In such studies, the fitness function should be properly defined. Here we illustrate it by the example of searching for high-*k* dielectric materials.[32]

High-*k* dielectric materials are important as gate oxides in microelectronics and as potential dielectrics for capacitors. In order to enable computational discovery of novel high-*k* dielectric materials, we propose a fitness model (energy storage density) that includes the dielectric constant (κ) and an intrinsic breakdown field E_{bd}, and expressing the latter through the bandgap (E_g), we obtain the following formula,

$$F_{ED} = \frac{1}{2}\varepsilon_0 k E_{bd}^2 = 8.1882\ \mathrm{J\ cm^{-3}} \times k\left(\frac{E_g}{E_{gc}}\right)^{2\alpha}, \tag{12}$$

where $E_{gc} = 4.0$ eV, the critical bandgap value separating materials into semiconductors and insulators, and ε_0 is the absolute permittivity of the vacuum. With this new fitness descriptor, we can simultaneously account for the dielectric constant, bandgap, and breakdown field during optimization, in a rational and comprehensive way. Remarkably, the same fitness can be used to search for optimal dielectric materials for capacitors and gate oxide materials.

We found a number of high-fitness structures of SiO_2 and HfO_2, some of which correspond to known phases and some of which are new. Our variable-composition searches in the HfO_2–SiO_2 system also uncovered several high-fitness states. The compositional dependences of enthalpy of formation and energy density are illustrated in Fig. 9. The highest FED is shown at each composition. The relationship between compositions and energy density appears to be intriguing (recall that two physical properties E_g and k, display quite different variation with respect to composition). A high concentration of HfO_2 does not necessarily result in high energy storage. As an example of a disordered structure, we take $Hf_{0.9}Si_{0.1}O_2$ (Hf_9SiO_{20}) with a relatively large unit cell containing 30 atoms; its dielectric permittivity is relatively high (22.11), but its low E_g (3.02 eV) results in very low fitness (33.93 $\mathrm{J \cdot cm^3}$). Ordered phases seem to be superior in terms of their fitness. Among the pseudobinary compounds, the best fitness values are seen for $Hf_{0.5}Si_{0.5}O_2$ ($I4_1/amd$) and $Hf_{0.75}Si_{0.25}O_2$ ($I\text{-}42m$); their fitness is three times greater than that of SiO_2 quartz. Clearly, further improvements are possible by considering other systems. The methodology and principles presented here allow a systematic search for such improved materials.

4.3 Nano clusters: B_{36} and LJ_{44}

Baturin *et al.* have applied this method to predict the atomic structure and stability of small silicon nanoclusters passivated by hydrogen.[64] Recently, the B_{36} clusters which was synthesized and theoretically studied by Wang *et al.*[65] Here we applied this method and reproduced the most two stable structures B_{36}–C_{6v} and B_{36}–D_{4h} within 10 generations (30 structures per generation). One can clearly see that B_{36} clusters prefer to adopt high-symmetry structures. For instance, its ground state has six-fold symmetry and a perfect hexagonal vacancy, while the next stable configuration (B_{36}–D_{4h}) has tetragonal symmetry. It is very important to note though, the algorithm does not favor only highly symmetric structures, as shown by tests on the artificial Lenard-Jones systems. LJ_{44} is an example of a cluster with a ground state that has no symmetry. As shown in Fig. 10, the algorithm is still able to identify the ground state even though we start from random symmetric structures. The reason is that

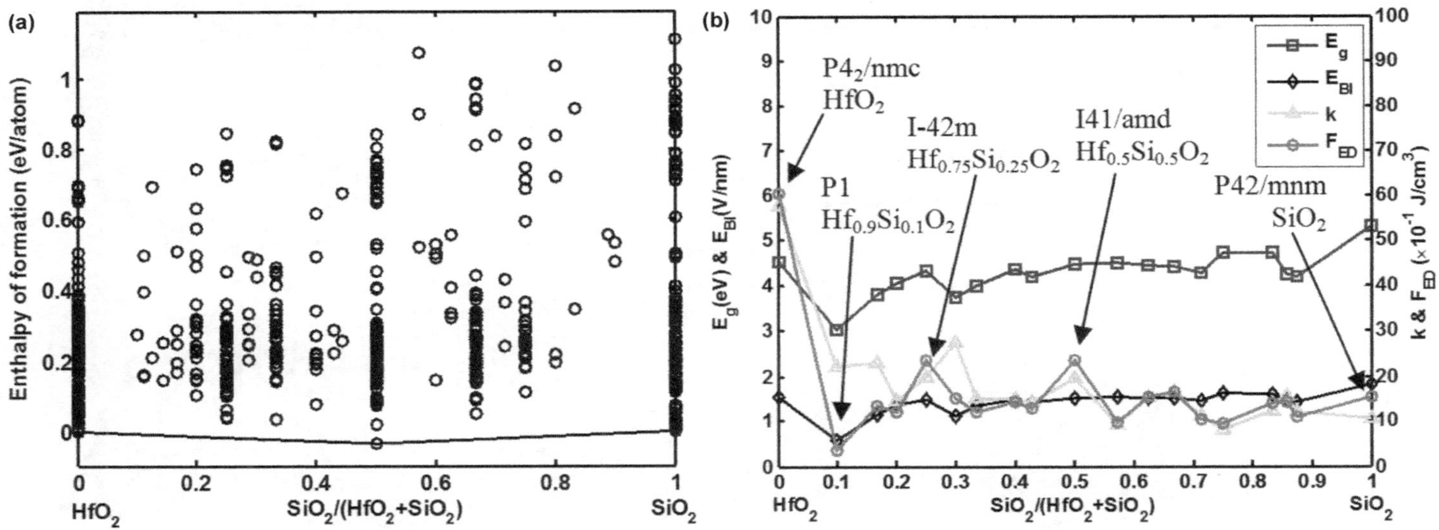

Fig. 9 HfO_2–SiO_2 system: (a) enthalpy of formation, showing stability of hafnon ($HfSiO_4$) and (b) compositional dependence of physical properties.

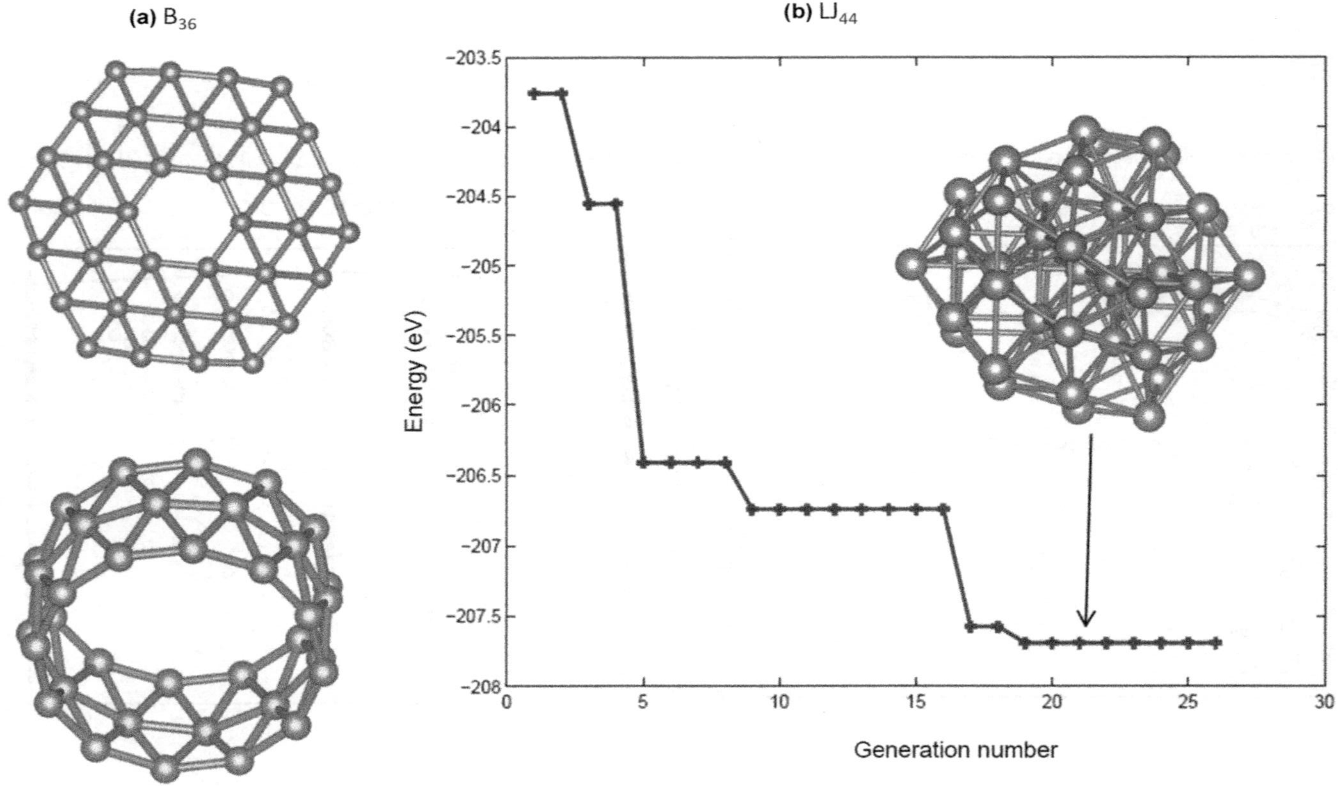

Fig. 10 (a) The best two structures found in B_{36} cluster, C_{6v} and D_{4h}; (b) a typical evolution curve of LJ_{44} cluster in USPEX simulation, showing the lowest energy in each generation.

variation operators in USPEX can break symmetry and enable totally new structures with different symmetry to emerge.

4.4 2D crystals: boron-based Dirac materials

Boron is a fascinating element because of its chemical and structural complexity. Recently, a new class of boron sheets composed of triangular and hexagonal motifs, exemplified by the so-called α-sheet structure, has been identified to be energetically most stable, and argued to be the precursor of B_{80} fullerene.[66] However, the stabilities of both α-sheet and B_{80} were both challenged.[67,68] We explored other potentially stable structures or structures with novel electronic properties. Contrary to the general constructing rules for flat monolayer boron sheets (mixing of triangular and hexagonal patterns),[66] great complexity is uncovered with multilayer structures. The non-flatness of 2D boron sheets enhances its energetic stability and creates novel electronic properties. In particular, we found that a 2D-boron with *Pmmn* symmetry can exhibit anisotropic Dirac cones,[69] after graphene and silicene,[70,71] the third elemental material with massless Dirac fermions. This property may be superior to that of graphene, because transport properties of these Dirac fermions will depend on direction, which gives an additional degree of freedom (with faster-than-graphene and slower-than-graphene directions) for electronic applications (Fig. 11).

4.5 Surfaces

4.5.1 Diamond (100), (111). We firstly studied the known 2×1 reconstructions of diamond (100) and (111) surfaces, which are the two most important surfaces for polycrystalline diamond obtained from chemical vapor deposition (CVD). We tried two and six carbon atoms on a 2×1 surface cell.[72] Our results are in excellent agreement with those reported in previous literature. The cleaved diamond (100) surface, containing one unsaturated carbon atom with two dangling bonds per unit cell, is unstable. Stabilization is achieved *via* a reconstruction with surface atoms forming one π-bonded C–C dimer per 2×1 unit cell. The diamond (111) surface contains two unsaturated carbon atoms with two dangling bonds per 1×1 unit cell. Our search also confirmed the model proposed by Pandey, with surface atoms forming Pandey chains along the [011] direction.[73] From the top view [Fig. 12(b)], the Pandey chains further form an extended two-dimensional (2D) network, having the same period as the unreconstructed (111) surface. From the side view, the surface atoms together with the second layer form an alternating $(5+7)$-ring pattern, which is different from the 6-ring pattern in the bulk, but is similar to the structures of *M*-carbon, a metastable carbon allotrope.[9]

4.5.2 GaN–O (1110). We also studied the semipolar GaN (10-11) surfaces in the presence of oxygen,[38] in which we allowed both variable number of surface atoms and variable cell size (restricted to a 2×2 or

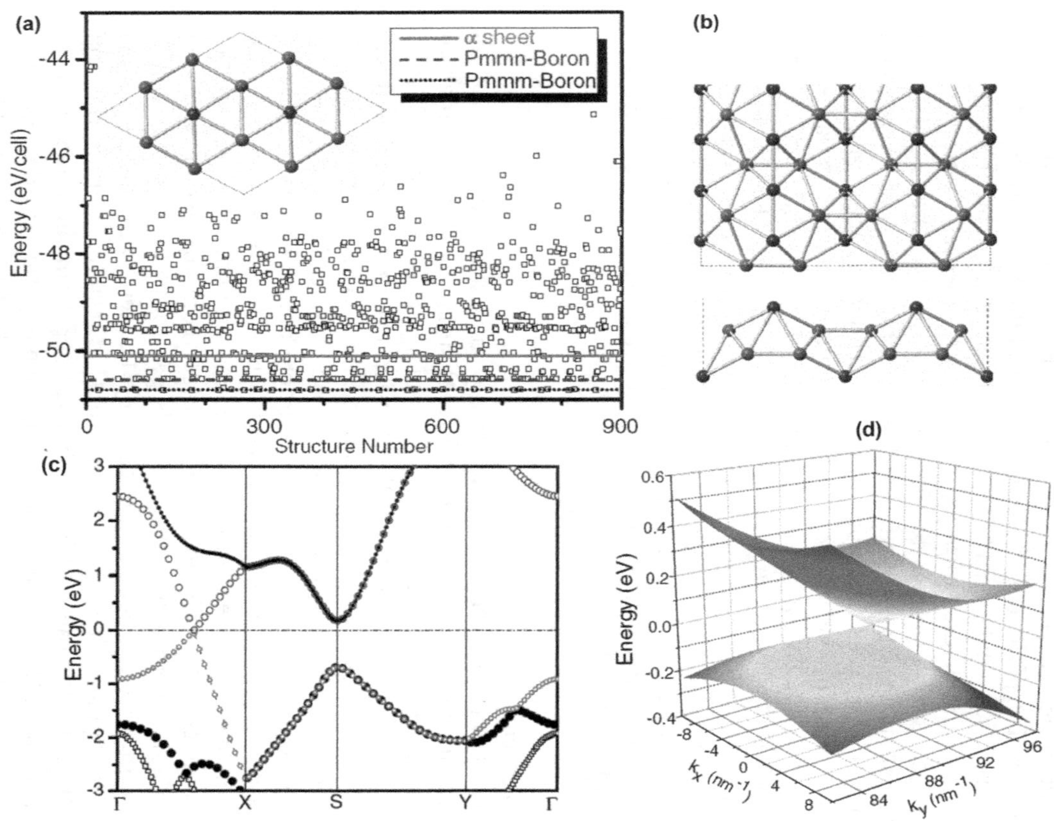

Fig. 11 (a) Enthalpy evolution for an 8 atom 2D boron system during an evolutionary structure search. The insets shows the structure of α-sheet; (b) the top view and side view of *Pmmn*-boron; (c) the band structure of *Pmmn*-boron; (d) the Dirac cone of *Pmmn*-boron.

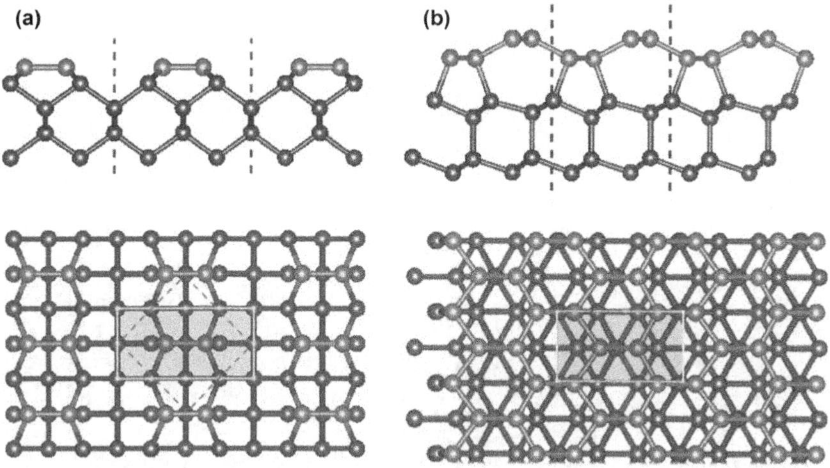

Fig. 12 Reconstructions on diamond (a) 2×1 (100) and (b) 2×1 (111) surface.

smaller surface cell). Figure 13 shows the whole phase diagram as a function of μ(O) and μ(Ga)-μ(N).

Compared to the cleaved surface, structure S1 has two Ga adlayers. Structure S2 has one Ga adlayer. Structure S3 has the top N and half of the second N layer removed. Structure S4 has only the top N layer removed. Structure S5 has an additional N at the bridging position of the two top N atoms. The first four structures were found by Akiyama *et al.*[74] Structure S5 with N_3 trimers is not intuitively obvious, and demonstrates the power of the automated searching by the EA. An analogous Se-trimer has been predicted to be stable on ZnSe (100) reconstructions in Se-rich conditions.[75] Two additional major reconstructions are structures S6 and S7, which appear in presence of oxygen. Compared to the cleaved surface, structure S6 has half of the top N layer removed, and half of the top N and all of the second N layer replaced by O. Structure S7 has the top two N replaced by O. Reconstructions similar to S6 and S7 for the (10-11) surface have been reported.[76]

4.5.3 α-Boron (111) surface. As a neighbor of carbon, boron is in many ways an analog of carbon and its nanostructures. The carbon surface has been thoroughly studied. In contrast, little is known for boron surface due to its exceptional structural complexity. Recently, Amsler *et al.* performed the first study of the reconstruction of the α-boron (111) and predicted several low-energy surface reconstructions by using the minima hopping method. In particular, a metallic reconstructed phase (111)-$I_{R,(a)}$ was predicted to be the most stable configuration, where a conducting boron sheet was adsorbed on a semiconducting substrate, leading to numerous possible applications in nanoelectronics.[77] However, this seems to be in conflict with the general principle the reconstructions usually lower their energies by atomic rearrangement leading to a semiconducting (rather than metallic) surface state.[78] Addressing this contradiction, we found an unexpected surface reconstruction in α-boron (111) using the *ab initio* evolutionary algorithm USPEX

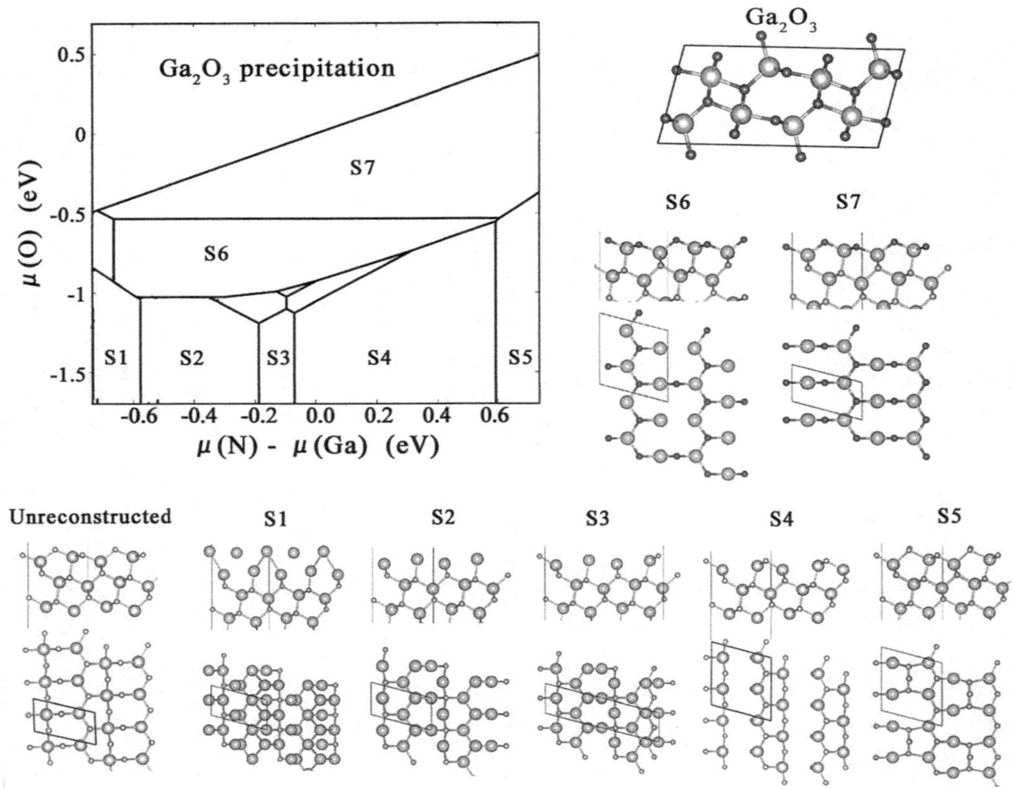

Fig. 13 The phase diagram of GaN (10-11) surface as a function of chemical potentials (μ(O) and μ(N)-μ(Ga)) and the corresponding stable reconstructions at various conditions.

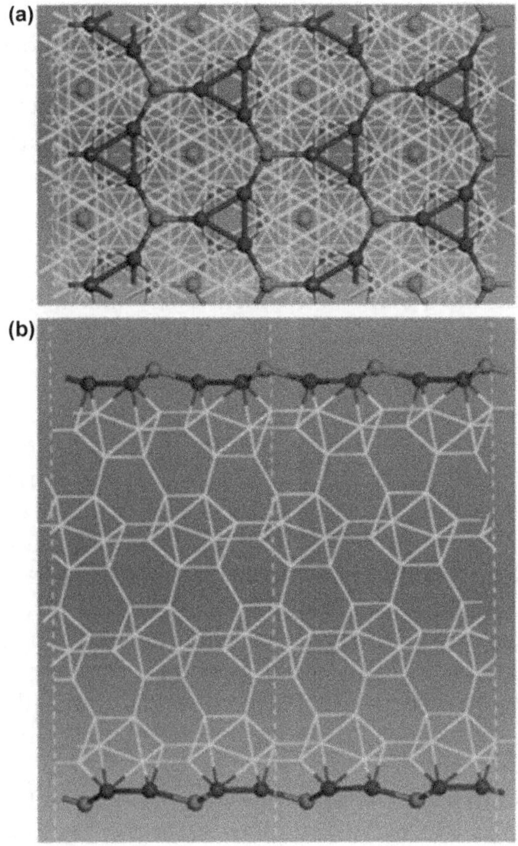

Fig. 14 (a) Projection of the $2 \times 2 \times 1$ supercell of the (111)-$I_{R,(z)}$ structure along the [111] direction. (b) Projection of the $2 \times 2 \times 1$ supercell of the (111)-$I_{R,(z)}$ structure along the [−1 −1 2] direction. The inequivalent surface atoms are shown by different colors.

(see Fig. 14). Our reconstruction has a much lower surface energy and is much simpler than previous predictions.[77] This reconstruction satisfied electron counting rules and is semiconducting.[79]

4.6 Polymers

This capability for prediction of polymeric crystals was developed only quite recently.[41] So far, we have systematically studied ten common polymers. And this module has been used for the design of dielectric polymers.[80] Here we illustrate the application to predict two complex polymers, nylon-6 and cellulose.

4.6.1 Nylon-6. Two crystalline forms of nylon-6 have been experimentally characterized, α and γ. There is a substantial confusion regarding the structure of the α-phase. The earliest reported crystal structure had some incorrect atomic coordinates in Cambridge structural database (CSD entry: LILSUU).[81] Here we used the model suggested in the previous theoretical studies,[82] which is described by the packing of the full-extended chains, possessing eight monomeric units of [-(CH_2)-CO-NH-] per unit cell, while γ-phase has $Z = 4$ based on twisted chains.

We have performed a search with $Z=8$ starting from the full-extended chain, and $Z=4$ starting from the twisted chain, in the hope of finding α- and γ-phase. Indeed, we found that the most stable configuration has a monoclinic symmetry for $Z=4$ (space group $P2_1/c$, $a=4.77$ Å, $b=8.35$ Å, $c=16.88$ Å (fiber axis), $\gamma=121.2°$, in good agreement with experimental results, except that there is a considerable deviation in cell vector b (the direction which largely depends on vdW bonding). In the γ-phase, the antiparallel twisted chains form pleated sheets *via* hydrogen bonds, and the chain directions are opposite in alternating sheets.[41] In our $Z=8$ search, we found the ground state which is very similar to what has been described in the literature.[82] This structure also features nylon sheets joined by H bonds in the antiparallel way.[41]

4.6.2 Cellulose. Cellulose is a polymer with repeating D-glucose units $[-C_6H_{10}O_5-]_n$. Microfibrils of naturally occurring cellulose correspond to two crystalline forms, I_α and I_β.[83,84] I_α has a triclinic unit cell and crystallizes in $P1$ space group. I_α has a simple unit cell and thus is easy to be predicted. Therefore we focused on the more challenging case of I_β. It was found that I_β is the thermodynamically more stable phase. Starting from the D-glucopyranosyl chains ($Z=4$), we indeed identified I_β as the ground state configuration, and the calculated unit cell parameters agree well with previous reports. As shown in Fig. 15(b), cellulose chains are arranged parallel-up and edge to edge, making flat sheets that are held together by H-bonds. Sheets formed by H-bonded D-glucopyranosyl chains are in the *bc*-plane, while there are no strong H-bonds which are perpendicular to the sheets. Most importantly, the complex hydrogen bond network in the flat sheets is also correctly predicted (Fig. 15(b)).

5 Outlook

We have briefly reviewed the principles of evolutionary algorithms and their application to structure prediction. The USPEX method proved to be a powerful tool enabling reliable and efficient prediction of stable crystal structures. In this chapter, we introduced the recent progress in extending the structure prediction technique to a wide range of problems. Despite its huge success in different fields, the current approach is still limited by the followings:

(1) *Energy accuracy*. So far, most applications in structure prediction are to find the most stable structures with the energy as fitness function. The performance is largely limited by the accuracy of todays's *ab initio* simulations, which for some cases is insufficient. For instance, van der Waals systems needs 1 kJ mole^{-1} accuracy to differentiate the crystal packing, which can be only achieved by using extremely expansive quantum-chemistry methods treatment.[85,86] Despite these significant progresses, it is still not feasible for massive structural searches.

(2) *Free energy versus lattice energy*. In organic crystals, the polymorph energy differences are often quite small. Among them, the lattice energy differences are typically very small. However, vibrational energy

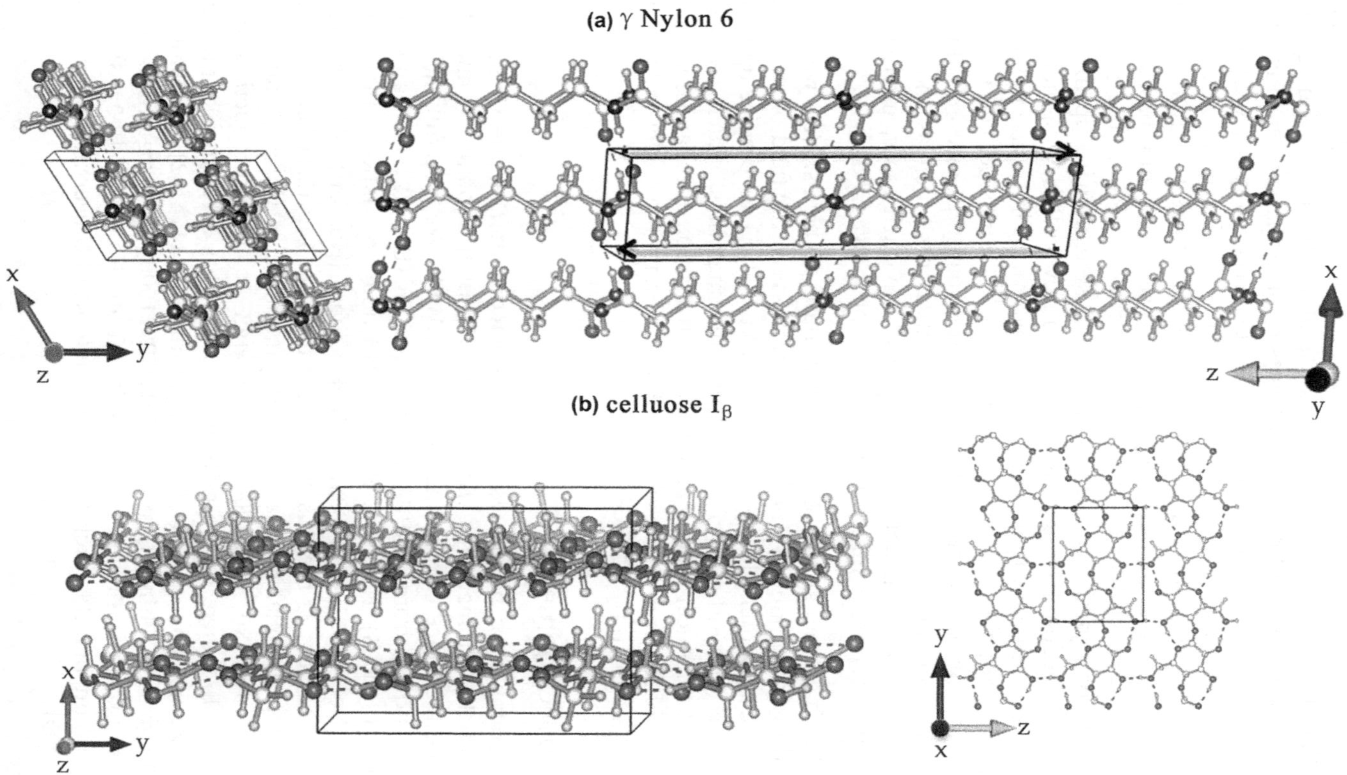

Fig. 15 The crystal structures of (a) γ-nylon 6 and (b) cellulose I_β found by USPEX.

differences are often large enough to cause a re-ranking of polymorph stability at room temperature.[87] Therefore, it is particularly needed for organic crystal structure prediction to evaluate the free energy, instead of lattice energy at 0 K. Enhanced and rare-event molecular dynamics sampling techniques (metadynamics[88] and adiabatic free energy dynamics[89]) provide a solution, but choices of order parameter limit their applications for general purpose. The free-energy sampling techniques would be complementary to crystal structure search technique in practice. Still, it is needed to develop more accurate force fields to make it feasible to evaluate the free energy for the case of structure prediction.

(3) *Structural complexity*. We are suggesting USPEX as the method of choice for crystal structure prediction of systems with up to ∼300 degrees of freedom (∼100 atoms in the primitive cell for non-molecular crystals, and more for molecular crystals), where no information (or just the lattice parameters) is available. Above ∼100 atoms per cell runs become expensive due to the "curse of dimensionality". However, some of them are still tractable by using the constraints (such as molecular geometry, lattice constants, *etc.*). Especially in such cases, interaction with experiment is helpful and should be encouraged.

USPEX has been applied to many important problems. Here we highlighted the methodology and some applications in the field of structure prediction. Another closely related subject is how to predict optimal conditions of synthesis of those predicted materials, which requires studies of chemical reactions and phase transition mechanisms. That direction of research is still wide open and we refer the reader to some of the first steps in it.[90–92]

Acknowledgements

Calculations were performed at the supercomputer of the Centre for Functional Nanomaterials, Brookhaven National Laboratory, and the High Performance Computing Center of NWPU. We gratefully acknowledge funding from DARPA (Grants No. W31P4Q1210008 and No. W31P4Q1310005), NSF (No. EAR-1114313 and No. DMR-1231586), the AFOSR (No. FA9550-13-C-0037), CRDF Global (No. UKE2-7034-KV-11), and Government of the Russian Federation (No. 14.A12.31.0003). Q-F Z thanks the National Natural Science Foundation of China (Grants No. 51372203 and No. 51332004), the Foreign Talents Introduction and Academic Exchange Program (Grant No. B08040). X-F.Z thanks the National Science Foundation of China (Grant No. 11174152), the National 973 Program of China (Grant No. 2012CB921900), the Program for New Century Excellent Talents in University (Grant No. NCET-12-0278), and the Fundamental Research Funds for the Central Universities (Grant No. 65121009).

References

1 L. Pauling, *J. Am. Chem. Soc.*, 1929, **51**, 1010.
2 J. Maddox, *Nature*, 1988, 335.

3 A. Gavezzotti, *Acc. Chem. Res.*, 1993, **27**, 309.
4 J. Pannetier, J. Bassas-Alsina, J. Rodriguez-Carvajal and V. Caignaert, *Nature*, 1990, **346**, 343.
5 J. C. Schon and M. Jansen, *Angew. Chem., Int. Ed. Engl.*, 1996, **35**, 1286.
6 A. Laio and M. Parrinello, *Proc. Natl. Acad. Sci. U. S. A.*, 2002, **99**, 12562.
7 R. Martonak, A. Laio and M. Parrinello, *Phys. Rev. Lett.*, 2003, **90**, 075503.
8 S. M. Woodley, P. D. Battle, J. D. Gale and C. R. A. Catlow, *Phys. Chem. Chem. Phys.*, 1999, **1**, 2535.
9 A. R. Oganov and C. W. Glass, *J. Chem. Phys.*, 2006, **124**, 244704.
10 C. M. Freeman, J. M. Newsam and S. M. Levine, *J. Mater. Chem.*, 1999, **3**, 531.
11 D. J. Wales and J. P. K. Doye, *J. Phys. Chem. A*, 1997, **101**, 5111.
12 S. Goedecker, *J. Chem. Phys.*, 2004, **120**, 9911.
13 S. Curtarolo, D. Morgan, K. Persson, J. Rodgers and G. Ceder, *Phys. Rev. Lett.*, 2003, **91**, 135503.
14 *Modern Methods of Crystal Structure Prediction*, ed. A. R. Oganov, WILEY-VCH, Weinheim, 2010.
15 S. L. Chaplot and K. R. Rao, *Curr. Sci.*, 2006, **91**, 1448.
16 A. R. Oganov, A. O. Lyakhov and M. Valle, *Acc. Chem. Res.*, 2011, **44**, 227.
17 A. O. Lyakhov, A. R. Oganov, H. T. Stokes and Q. Zhu, *Comput. Phys. Commun.*, 2013, **184**, 1172.
18 Q. Zhu, A. R. Oganov and X.-F. Zhou, Crystal Structue Prediction and its applications to Earth and Materials Sciences, *Topics in Current Chemisty*, Springer Verlag, 2014.
19 J. P. M. Lommerse, W. D. S. Motherwell, H. L. Ammon, J. D. Dunitz, A. Gavezzotti, D. W. M. Hofmann, F. J. J. Leusen, W. T. M. Mooij, S. L. Price, B. Schweizer, M. U. Schmidt, B. P. van Eijck, P. Verwer and D. E. Williams, *Acta Crystallogr., Sect. B: Struct. Sci.*, 2000, **56**, 697.
20 W. D. S. Motherwell, H. L. Ammon, J. D. Dunitz, A. Dzyabchenko, P. Erk, A. Gavezzotti, D. W. M. Hofmann, F. J. J. Leusen, J. P. M. Lommerse, W. T. M. Mooij, S. L. Price, H. Scheraga, B. Schweizer, M. U. Schmidt, B. P. van Eijck, P. Verwer and D. E. Williams, *Acta Crystallogr., Sect. B: Struct. Sci.*, 2002, **58**, 647.
21 G. M. Day, W. D. S. Motherwell, H. L. Ammon, S. X. M. Boerrigter, R. G. Della Valle, E. Venuti, A. Dzyabchenko, J. D. Dunitz, B. Schweizer, B. P. van Eijck, P. Erk, J. C. Facelli, V. E. Bazterra, M. B. Ferraro, D. W. M. Hofmann, F. J. J. Leusen, C. Liang, C. C. Pantelides, P. G. Karamertzanis, S. L. Price, T. C. Lewis, H. Nowell, A. Torrisi, H. A. Scheraga, Y. A. Arnautova, M. U. Schmidt and P. Verwer, *Acta Crystallogr., Sect. B: Struct. Sci.*, 2005, **61**, 511.
22 G. M. Day, T. G. Cooper, A. J. Cruz-Cabeza, K. E. Hejczyk, H. L. Ammon, S. X. M. Boerrigter, J. S. Tan, R. G. Della Valle, E. Venuti, J. Jose, S. R. Gadre, G. R. Desiraju, T. S. Thakur, B. P. van Eijck, J. C. Facelli, V. E. Bazterra, M. B. Ferraro, D. W. M. Hofmann, M. A. Neumann, F. J. J. Leusen, J. Kendrick, S. L. Price, A. J. Misquitta, P. G. Karamertzanis, G. W. A. Welch, H. A. Scheraga, Y. A. Arnautova, M. U. Schmidt, J. van de Streek, A. K. Wolf and B. Schweizer, *Acta Crystallogr., Sect. B: Struct. Sci.*, 2009, **65**, 107.
23 D. A. Bardwell, C. S. Adjiman, Y. A. Arnautova, E. Bartashevich, S. X. M. Boerrigter, D. E. Braun, A. J. Cruz-Cabeza, G. M. Day, R. G. Della Valle, G. R. Desiraju, B. P. van Eijck, J. C. Facelli, M. B. Ferraro, D. Grillo, M. Habgood, D. W. M. Hofmann, F. Hofmann, K. V. J. Jose, P. G. Karamertzanis, A. V. Kazantsev, J. Kendrick, L. N. Kuleshova, F. J. J. Leusen, A. V. Maleev, A. J. Misquitta, S. Mohamed, R. J. Needs, M. A. Neumann, D. Nikylov, A. M. Orendt, R. Pal, C. C. Pantelides,

C. J. Pickard, L. S. Price, S. L. Price, H. A. Scheraga, J. van de Streek, T. S. Thakur, S. Tiwari, E. Venuti and I. K. Zhitkov, *Acta Crystallogr., Sect. B: Struct. Sci.*, 2011, **67**, 535.
24 J. Moult, K. Fidelis, A. Kryshtafovych, B. Rost, T. Hubbard and A. Tramontano, *Proteins*, 2007, **69**, 3.
25 A. O. Lyakhov, A. R. Oganov and M. Valle, *Comput. Phys. Commun.*, 2010, **181**, 1623.
26 Q. Zhu, A. R. Oganov, C. W. Glass and H. T. Stokes, *Acta Crystallogr., Sect. B: Struct. Sci.*, 2012, **68**, 215.
27 D. M. Deaven and K. M. Ho, *Phys. Rev. Lett.*, 1995, **75**, 288.
28 A. R. Oganov and M. Valle, *J. Chem. Phys.*, 2009, **130**, 104504.
29 Q. Zhu, A. R. Oganov and M. A. Salvado, *Phys. Rev. B: Condens. Matter Mater. Phys.*, 2011, **83**, 193410.
30 A. O. Lyakhov and A. R. Oganov, *Phys. Rev. B: Condens. Matter Mater. Phys.*, 2011, **84**, 092103.
31 Y. Zhang, W. Gao and S. Chen, *Comput. Mater. Sci.*, 2015, **98**, 51.
32 Q. F. Zeng, A. R. Oganov, A. O. Lyakhov, C. Xie, X. D. Zhang, J. Zhang, Q. Zhu, B. Wei, I. Grigorenko, L. Zhang and L. Cheng, *Acta Crystallogr., Sect. C: Cryst. Struct. Commun.*, 2014, **70**, 76.
33 G. H. Johannesson, T. Bligaard, A. V. Ruban, H. L. Skriver, K. W. Jacobsen and J. K. Norskov, *Phys. Rev. Lett.*, 2002, **88**, 255506.
34 S. Schönborn, S. Goedecker, S. Roy and A. R. Oganov, *J. Chem. Phys.*, 2009, **130**, 144108.
35 J. Wang, S. Deng, Z. Liu and Z. Liu, *Nat. Sci. Rev.*, 2015, **2**, 22.
36 X. Y. Luo, J. Yang, H. Liu, X. Wu, Y. Wang, Y. Ma, S. Wei, X. Gong and H. Xiang, *J. Am. Chem. Soc.*, 2011, **133**, 16285.
37 Y. Wang, M. Miao, J. Lv, L. Zhu, K. Yin, H. Liu and Y. Ma, *J. Chem. Phys.*, 2012, **137**, 224108.
38 Q. Zhu, L. Li, A. R. Oganov and P. B. Allen, *Phys. Rev. B: Condens. Matter Mater. Phys.*, 2013, **87**, 195317.
39 G. Qian, R. M. Martin and D. J. Chadi, *Phys. Rev. B: Condens. Matter Mater. Phys.*, 1988, **38**, 7649.
40 C. P. Brock and J. D. Dunitz, *Chem. Mater.*, 1994, **6**, 1118.
41 Q. Zhu, V. Sharma, A. R. Oganov and R. Ramprasad, *J. Chem. Phys.*, 2014, **141**, 154102.
42 G. Kresse and J. Furthmuller, *Phys. Rev. B: Condens. Matter Mater. Phys.*, 1996, **54**, 11169.
43 J. P. Perdew, K. Burke and M. Ernzerhof, *Phys. Rev. Lett.*, 1996, **77**, 3865.
44 P. E. Blochl, *Phys. Rev. B: Condens. Matter Mater. Phys.*, 1994, **50**, 17953.
45 S. Grimme, *J. Comput. Chem.*, 2006, **27**, 1787.
46 J. Klimes and A. Michaelides, *J. Chem. Phys.*, 2012, **137**, 120901.
47 A. Tkatchenko and M. Scheffler, *Phys. Rev. Lett.*, 2009, **102**, 073005.
48 J. Klimes, D. R. Bowler and A. Michaelides, *J. Phys: Condens. Matter*, 2010, **22**, 022201.
49 T. Bucko, D. Tunega, J. G. Angyan and J. Hafner, *J. Phys. Chem. A*, 2011, **115**, 10097.
50 Q. Zhu, D. Y. Jung, A. R. Oganov, C. W. Glass, C. Gatti and A. O. Lyakhov, *Nat. Chem.*, 2013, **5**, 61.
51 W. Zhang, A. R. Oganov, A. F. Gonchanov, Q. Zhu, S. E. Boulfelfel, A. O. Lyakhov, E. Stavrou, M. Somayazulu, V. B. Prakapenka and Z. Konopkova, *Science*, 2013, **342**, 1502.
52 Q. Zhu, A. R. Oganov and A. O. Lyahkov, *Phys. Chem. Chem. Phys.*, 2013, **15**, 7696.

53 G. R. Qian, A. O. Lyakhov, Q. Zhu, A. R. Oganov and X. Dong, *Sci. Rep.*, 2014, **4**, 5606.
54 Y. Liu, A. R. Oganov, S. W. Wang, Q. Zhu, X. Dong and G. Kresse, *Sci. Rep.*, 2014, **4**, 5606.
55 C. Hu, A. R. Oganov, Q. Zhu, G. R. Qian, G. Frapper, A. O. Lyakhov and H. Y. Zhou, *Phys. Rev. Lett.*, 2013, **110**, 165504.
56 M. Miao, *Nat. Chem.*, 2013, **5**, 846.
57 Q. Zeng, J. Peng, A. R. Oganov, Q. Zhu, C. Xie, X. Zhang, D. Dong, L. Zhang and L. Cheng, *Phys. Rev. B: Condens. Matter Mater. Phys.*, 2013, **88**, 214107.
58 X. Yu, G. B. Thompson and C. R. Weinberger, *Acta Mater.*, 2014, **80**, 341.
59 H. Niu, X. Chen, W. Ren, Q. Zhu, A. R. Oganov, D. Li and Y. Li, *Phys. Chem. Chem. Phys.*, 2014, **16**, 15866.
60 X. Cheng, W. Zhang, X.-Q. Chen, H. Niu, P. Liu, K. Du, G. Liu, D. Li, H. Cheng, H. Ye and Y. Li, *Appl. Phys. Lett.*, 2013, **17**, 171903.
61 Q. Zhu, A. R. Oganov and Q. F. Zeng, *Sci. Rep.*, 2015, **5**, 7875.
62 X. Zhang, Y. Wang, J. Lv, C. Zhu, Q. Li, M. Zhang, Q. Li and Y. Ma, *J. Chem. Phys.*, 2013, **138**, 114101.
63 H. Xiang, B. Huang, E. Kan, S. Wei and X. Gong, *Phys. Rev. Lett.*, 2013, **110**, 118702.
64 V. S. Baturin, S. V. Lepeshkin, N. L. Matsko, A. R. Oganov and Y. A. Uspenskii, *Eur. Phys. Lett.*, 2014, **106**, 37002.
65 Z. A. Piazza, H.-S. Hu, W.-L. Li, Y.-F. Zhao, J. Li and L.-S. Wang, *Nat. Commun.*, 2014, **5**, 3113.
66 H. Tang and S. Ismail-Beigi, *Phys. Rev. Lett.*, 2007, **99**, 115501.
67 N. Gonzalez, A. Sadrzadeh and B. I. Yakobson, *Phys. Rev. Lett.*, 2007, **98**, 166804.
68 S. De, A. Willand, M. Amsler, P. Pochet, L. Genovese and S. Goedecker, *Phys. Rev. Lett.*, 2011, **106**, 225502.
69 X.-F. Zhou, X. Dong, A. R. Oganov, Q. Zhu, Y. Tian and H. T. Wang, *Phys. Rev. Lett.*, 2014, **112**, 085502.
70 B. Aufray, A. Kara, S. B. Vizzini, H. Oughaddou, C. Leandri, B. Ealet and G. L. Lay, *Appl. Phys. Lett.*, 2010, **96**, 183102.
71 S. Cahangirov, M. Topsakal, E. Akturk, H. Sahin and S. Ciraci, *Phys. Rev. Lett.*, 2009, **102**, 236804.
72 J. Ristein, *Surf. Sci.*, 2006, **600**, 3677.
73 K. C. Pandey, *Phys. Rev. lett.*, 1981, **47**, 1913.
74 T. Akiyama, D. Ammi, K. Nakamura and T. Ito, *Phys. Rev. B: Condens. Matter Mater. Phys.*, 2010, **81**, 245317.
75 A. Garcia and J. E. Northrup, *Appl. Phys. Lett.*, 1994, **65**, 708.
76 W. J. Lee and Y. S. Kim, *Phys. Rev. B: Condens. Matter Mater. Phys.*, 2011, **84**, 115318.
77 M. Amsler, S. Botti, M. A. L. Marques and S. Goedecker, *Phys. Rev. Lett.*, 2013, **111**, 136101.
78 C. B. Duke, *Chem. Rev.*, 1996, **96**, 1237.
79 X.-F. Zhou, A. R. Oganov, X. Shao, Q. Zhu and H.-T. Wang, *Phys. Rev. Lett.*, 2014, **113**, 176101.
80 V. Sharma, C. Wang, R. Lorenzini, R. Ma, Q. Zhu, D. W. Sinkovits, G. Pilania, A. R. Oganov, S. Kumar, G. A. Sotzing, S. A. Boggs and R. Ramprasad, *Nat. Commun.*, 2014, **5**, 4845.
81 D. R. Holmes, C. W. Bunn and D. J. Smith, *J. Polym. Sci.*, 1955, **17**, 159.
82 Y. Li and W. A. Goddard, *Macromolecules*, 2002, **35**, 8440.

83 Y. Nishiyama, G. P. Johnson, A. D. French, V. T. Forsyth and P. Langan, *Biomacromolecules*, 2008, **9**, 3133.
84 Y. Nishiyama, J. Sugiyama, H. Chanzy and P. Langan, *J. Am. Chem. Soc.*, 2003, **125**, 14300.
85 J. Yang, W. Hu, D. Usvyat, D. Matthews, M. Schutz and G. K. Chan, *Science*, 2014, **345**, 640.
86 S. Wen and G. J. Beran, *Cryst. Growth Des.*, 2012, **12**, 2169.
87 J. Nyman and G. M. Day, *CrystEngComm*, 2015, **17**, 5154.
88 T.-Q. Yu and M. E. Tuckerman, *Phys. Rev. Lett.*, 2011, **107**, 015701.
89 P. Raiteri, R. Martonak and M. Parrinello, *Angew. Chem., Int. Ed.*, 2005, **44**, 3769.
90 Q. Zhu, A. R. Oganov and A. O. Lyakhov, *CrystEngComm*, 2012, **14**, 3596.
91 G. R. Qian, X. Dong, X. F. Zhou, Y. Tian, A. R. Oganov and H. T. Wang, *Comput. Phys. Commun.*, 2013, **184**, 2111.
92 S. Boulfelfel, A. R. Oganov and S. Leoni, *Sci. Rep.*, 2012, **2**, 471.

Ab initio global optimization of clusters

Jijun Zhao,*[a] Xiaoming Huang,[a] Ruili Shi,[a] Lingli Tang,[b] Yan Su[a] and Linwei Sai[c]
DOI: 10.1039/9781782622703-00249

1 Introduction

Cluster, as the aggregate composed of a few to thousands of atoms or molecules, represents the intermediate state between microscopic atoms or molecules and macroscopic condensed matters.[1–6] Due to quantum size effect, the physical and chemical properties of a cluster may differ from its bulk counterparts significantly. The exotic properties of clusters offer new opportunities as building blocks of nanoscale materials and devices.

For a small or medium-sized cluster with less than 300 atoms, more than half of the atoms locate on the surface. Adding one more atom to the cluster would lead to significant structural reconstruction. As a consequence, the atomic structure of a cluster is usually rather different from a piece of the corresponding bulk solid. Indeed, one of the most fundamental and challenging problems in cluster science is to determine the ground state structure of a cluster. Mathematically, finding global minimum on the multi-dimensional potential energy surface (PES) of a cluster is a well-known nondeterministic polynomial-time hard (NP-hard) problem.[7] As cluster size increases, the number of possible isomers on the PES may increase exponentially with the number of atoms.[8]

In cluster science, clusters are usually generated in gas phase and detected *via* time of flight mass spectroscope and other spectroscopic means, *e.g.*, photoelectron spectrum, infrared spectrum, electron diffraction. However, direct determination of the cluster structure by experiment alone is still rather difficult so far. Alternatively, by combining theoretical simulations and experimental data, a complete description of the geometric structure and the corresponding physical properties of a cluster can be established.

Early theoretical efforts of determining the ground state structure of clusters usually evolve the simulated annealing (SA) technique.[9] In the SA procedure, the system is first heated up to a high temperature and then gradually cooled down *via* molecular dynamics (MD) or Monte Carlo (MC) simulations, which can be incorporated with either empirical potentials or first-principles methods. Even though SA procedure combined with first-principles calculations is a robust method and has been successfully

[a]*Key Laboratory of Materials Modification by Laser, Ion and Electron Beams, Dalian University of Technology, Dalian 116024, China. E-mail: zhaojj@dlut.edu.cn*
[b]*College of Science, Dalian Nationalities University, Dalian 116000, China*
[c]*Department of Mathematics and Physics, Hohai University, Changzhou 213022, China*

employed to a series of clusters, such as Cd_nTe_n (n = 1–14),[10] Li_n (n = 20, 30, 40, 50),[11] P_{2m+1}^+ (m = 1–12),[12] B_{80},[13] and $B_{101-103}$,[14] it cannot guarantee the global minimum of the PES unless the annealing time is sufficiently long.

In the past decade, there has been significant progress in the global optimization of cluster structures using *ab initio* methods combined with more sophisticated search algorithms, such as genetic algorithm, basin hopping, minimum hopping, topological methods, particle swarm optimization, and tabu search. In this chapter, we will give an overview of this rapidly growing field, focusing on the methodology of genetic algorithm, basin hopping, and serval other methods and their applications to clusters. Finally, we give a summary and perspective of all these *ab initio* global optimization methods.

2 Genetic algorithm

2.1 General concept of genetic algorithm

Genetic algorithm (GA) is a kind of widely used method to determine the global minimum structure of a cluster. It can sample the potential energy surface efficiently and hop from one region of the PES to another region rather easily. It is inspired by Darwinian evolution theory that only the fittest individuals can survive. The basic philosophy of GA is to mimic the natural selection and evolution processes in nature.[15–17] The essential idea of GA procedure is to allow a population of a number of individual candidates to evolve under a given selection rule that maximizes the fitness function.

The GA program was first developed by Holland[16] and further generalized by Goldberg.[17] Since 1993, there have been several pioneering works of utilizing GA to determine the cluster structures.[17–24] Even in its early stage, GA demonstrated an impressive efficiency in searching the global minima of a variety of systems ranging from noble gas clusters to semiconductor clusters. An extraordinary example is the direct location of the buckyball structure of C_{60} starting from a random configuration by Deaven and Ho.[22] Motivated by the success of these pioneering works, there have been increasing interests in the applications of GA in the cluster global optimization. In particular, along with the rapid speeding-up of the computer hardware, different groups in the world have developed various GA codes for cluster optimization independently.[22,25–34] In the following subsection, we will describe the detailed implement of GA for cluster optimization. Next are some examples for the practical applications of GA for different kinds of clusters, including elemental clusters and alloy clusters.

GA is based on a population of individuals. In the present situation, each individual represents a cluster structure. The evolutionary principles are adopted to identify the optimal cluster structure.[25,30,31,35] Then evolutionary operations such as mating, mutation and 'natural selection' are used. The fittest members of a population survive and pass their genetic information onto subsequent generations (offspring) to produce descendants of improved quality.

2.2 Implementation of genetic algorithm for clusters

The flow chart of GA for cluster optimization is shown in Fig. 1. Initially, the original population contains a number of cluster structures (N_p) generated by random. Usually, 16 to 64 clusters are adopted to form a population. Larger population should be used if the cluster size is larger or the structure is very complicated. It is preferable to generate a much larger number of the initial population members and discard most of the high-energy (less favorable) members *via* natural selection. After the structures of all initial individuals are optimized, the genetic operators are applied. The two major operators are mating and mutation.

One individual in this population is randomly selected as a father with a probability related to its total energies. Optionally, the roulette technique (a roulette wheel with slot widths corresponding to the fitness values of the cluster) is used and the lower energy structures have more chances to be selected. Then a random number from 0 to 1 is used to decide whether a mother is needed or not. If the random number is bigger than the mutation rate, a mating operation is employed. We should choose another cluster from the population randomly as a

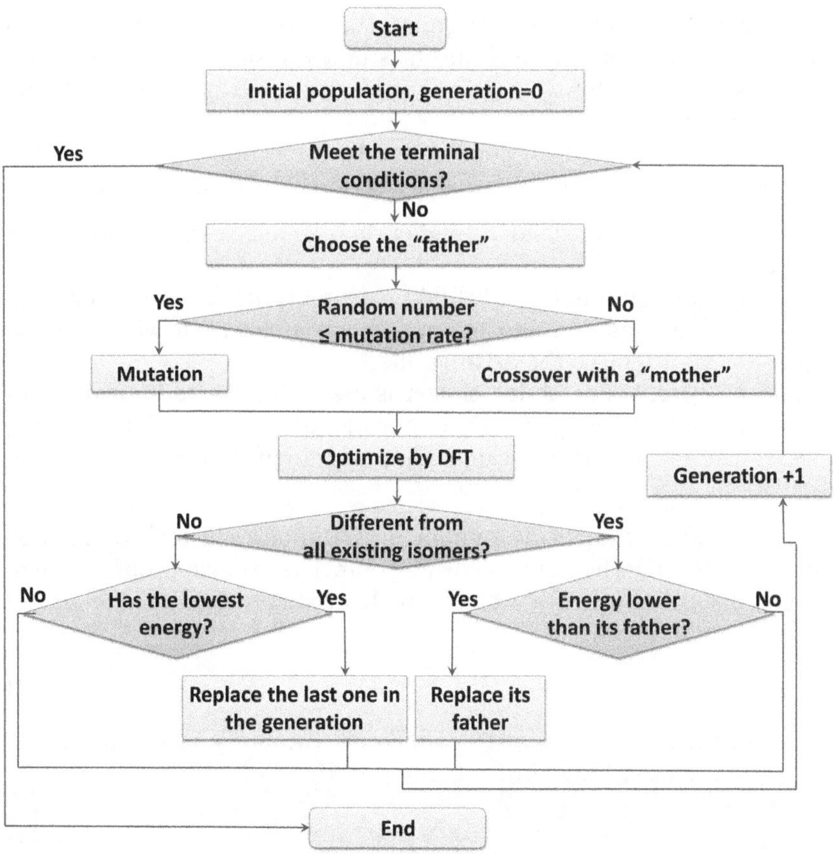

Fig. 1 Flow chart of genetic algorithm combined with DFT calculations for cluster optimization.

"mother" to produce a child cluster *via* a "cut and splice" crossover operation.[25] It is robust but efficient since we can combine the advantages of the father and mother together. Otherwise, the child cluster directly inherits from the father by mutation operation described below.

A mutation operation is chosen to modify the child from single-parent inheriting. In practice, genetic algorithm can converge very rapidly to the minimum of one region of the configuration space, which may not be the global minimum. The advantage of mutation is to avoid a population being trapped in the local minima of the PES. We have adopted a few mutation operations.

1. Give each atom of a cluster a small random displacement less than 10% of the average bond length. This helps the cluster to find its local minimum on the PES nearby.
2. Move a randomly chosen atom to the neighborhood of another random atom.
3. Exchange the atomic type of a pair of different types of atoms for alloy clusters. The influence of exchanging atomic type for the total energy of a cluster could be tremendous.

After mating or mutation procedure, the offspring cluster is then relaxed using DFT optimization. In our studies, DMol^3 [26,27] or VASP program[28] is employed for optimizing the cluster structure and calculating the total energy.

To ensure the diversity of population, the "selection" operation in the GA search is a crucial step. Several selection criteria can be used:

1. If the energy of a "child" is higher than all the existing "parents", it will be thrown away.
2. If the configuration of a "child" is isomorphic to an existing individual in the population and its total energy is lower, it will enter the population and replace the existing one.
3. If the energy of a "child" cluster is lower than some individuals in the population and its geometry is not identical to any of the existing ones, it will replace the worst individual with highest total energy in the population.

Here we adopt the pseudo rotational inertia of a cluster[29] defined by eqn (1) to identify whether two cluster geometries are isomorphic, where r_i is the distance of the ith atom from cluster mass center and m_i is the mass of the ith atom:

$$I = \sum m_i r_i^2 \tag{1}$$

For alloy clusters, one has to distinguish the two kinds of elements. Eqn (2) is then used to define a high-order inertia I' as:

$$I' = \sum m_i' r_i^4 \tag{2}$$

By definition, the high-order inertia can help differentiate different type of atoms.

After the "natural selection" process, the total number of individuals in the population retains N_p and the GA iteration continues until the convergence is achieved or the maximum number of iterations is reached. In principle, all kinds of GA methods are similar, with slight difference in the detailed operations.[22,31,32]

2.3 Application of genetic algorithm to elemental clusters

Since 1993, some pioneering GA studies have been conducted to determine the cluster structures.[20,22–24] The efficiency of GA for elemental clusters is rather high.

Small noble gas clusters have been studied as a benchmark for GA optimization in those earlier studies. Zeiri has studied the lowest energy structures of very small Lennard-Jones clusters ($n = 4$–10) using GA and found that GA based method converges to the global minima much more rapidly than some other techniques.[36]

Metal clusters have attracted tremendous attentions due to their interesting electronic properties and potential applications. Many empirical and semiempirical potentials or models have been used for metal clusters[37–41] and the related works have been reviewed.[4] Compared to the empirical or semiempirical simulations, *ab initio* calculations based on density functional theory (DFT) or Hartree–Fock (HF) theory are more accurate for describing the potential energy surface. In recent years, direct combination of DFT calculations and GA have become computationally affordable, allowing a more reliable search of any elemental clusters without requiring any empirical potential parameters.

Sodium is a simple metal with one 3s valence electron. As predicted by the shell model,[1] a metal cluster with closed electronic shell should possess 8, 20, 40, 58... valence electrons. Therefore, Na_n^+ clusters with closed electronic shells at $n = 9$, 21, 41, and 59 were studied by Zhao's group using GA combined with DFT.[42] Several open-shell Na_n^+ clusters at $n = 15$, 26, 31, 36, and 50 whose sizes lie between the magic numbers were also investigated. The lowest energy structures of these Na_n^+ clusters are displayed in Fig. 2. The simulated photoabsorption spectra of Na_n^+ clusters from time-dependent DFT (TDDFT)[83] simulations shown in Fig. 3 agree excellently with experimental data, confirming the reliability of the theoretical approaches.

Na_9^+ cluster adopts a tricapped trigonal prism configuration with D_{3h} symmetry. Na_{15}^+ can be obtained by adding two sodium atoms (highlighted in Fig. 2) on the nearby triangle faces of a 13-atom icosahedron. The lowest-energy structure of Na_{21}^+ cluster can be regarded as a double icosahedron of Na_{19} with two additional Na atoms capped on the waist with C_{2v} symmetry. Na_{26}^+ is composed of two icosahedra sharing three Na atoms on the middle triangle face remaining three Na atoms locating on the middle region. Structure from empirical simulations using MM and Gupta potentials for the neutral Na_{26} formed by four interpenetrating icosahedra with a high T_d symmetry structure is less energetically favorable by 0.058 eV according to the DFT calculations.[84] The most stable configuration of Na_{31}^+ cluster consists of two

Table 1 A list of GA-DFT studies of different cluster systems.

Types of clusters	Systems	Size	Authors and Ref.
Elemental clusters	Na_n^+	$n=9–59$	Huang *et al.*[42]
	Al_n^+, Al_n^-	$n \leq 34$	Drebov and Ahlrichs[43]
	P_n	$n=20$	Nava and Ahlrichs[44]
	K_n	$n \leq 20$	Abdalla *et al.*[45]
	Ca_n	$n=2–22$	Liang *et al.*[46]
	Cu_n	$n=9$	Assadollahzadeh *et al.*[47]
	Ga_n	$n=20–40$	Sai *et al.*[48]
	Ga_n	$n \leq 25$	Drebov *et al.*[49]
	As_n	$n=20$	Nava and Ahlrichs[44]
	Cs_n	$n=2–20, \infty$	Assadollahzadeh *et al.*[50]
	Au_n	$n=2–20$	Assadollahzadeh and Schwerdtfeger[51]
	Au_n	$n=10, 20$	Shayeghi *et al.*[52]
	Pb_n^+, Pb_n^-	$n=4–15$	Kelting *et al.*[53]
	Bi_n^+	$n=4–14$	Kelting *et al.*[54]
Alloy clusters	Si_4Li_n	$n=1–7$	Kanters and Donald[55]
	Na_nSi_m	$1 \leq n \leq 3$, $1 \leq m \leq 11$	Sai *et al.*[29]
	$Pt_{x-y}M_y$, (M = Ti, V)	$x=2–6$, $y=1, 2$	Jennings and Johnston[56]
	$V_3Si_n^-$,	$n=3–14$	
	$V_xSi_{12}^-$	$x=1–3$	Huang *et al.*[57,58]
	Cu_nAg_{8-n},	$1<n<8$	Heard and Johnston[59]
	Cu_nAu_{8-n}	$1<n<8$	
	Au_mAg_n	$5 \leq m+n \leq 12$	Hong *et al.*[60]
	$Au_{8-m}Ag_m$	$1<m<8$	Heiles *et al.*[61]
	Pt_nSn_n	$n=1–10$	Huang *et al.*[62]
	$Pt_{3m}Sn_m$	$m=1–5$	
	$PtHf_{12}$, $(LaPb_7Bi_7)^{4-}$		Weigend[63]
	$Sn_{9-n}Bi_n$	$0 \leq n \leq 4$	Heiles *et al.*[64]
	$Sn_{m-n}Bi_n$	$m=5–13$, $n=1–2$	Heiles *et al.*[65]
	$A@B_{12}@A_{20}$, A = Sn, Pb; B = Mg, Zn, Cd, Mn		Huang *et al.*[66]
Metal oxide clusters	$(H_2O)_3$		Kanters and Donald[55]
	$(Li_2O)_n$	$n=1–8$	Yuan and Cheng[67]
	$(BeO)_n$	$n=2–24$	Ren *et al.*[68]
	$(B_2O_3)_n$	$n=1–6$	Li and Cheng[69]
	$(MgO)_n$, $(MgO)_n^+$	$n=2–7$	Kwapien *et al.*[70]
	$(MgO)_n$	$n=3–16$	Haertelt *et al.*[71]
	$(TiO_2)_n$	$n \leq 20$	Bhattacharya *et al.*[72]
	$Ce_2O_2^+$, $Ce_3O_4^+$, $(CeO_2)_mCeO^+$	$m=0–4$	Burow *et al.*[73]
	$(WO_3)_n$	$2 \leq n \leq 12$	Sai *et al.*[74]
	$(La_2O_3)_n$	$n=1–6$	Ding *et al.*[75]
Other mixed systems	$H(H_2O)_n$	$n=1–4$	Alexandrova *et al.*[76]
	$(AlH_3)_n$	$n=1–8$	Kiran *et al.*[77]
	$(LiF)_nF^-$	$n=1–30$	Fernandez-Lima *et al.*[78]
	$(ZnF2)_n$	$n=1–3$	Kanters and Donald[55]
	CB^{7-}		Wang *et al.*[79]
	$B_nAl_{6-n}^{2-}$	$n=0–6$	Huynh and Alexandrova[80]
	$[Ag_7(SR)_4]^-$, $[Ag_7(DMSA)_4]^-$		Xiang *et al.*[81]
	M_8 on $TiO_2(110)$ M = Ru, Rh, Pd, Ag, Pt, Au		Vilhelmsen and Hammer[82]

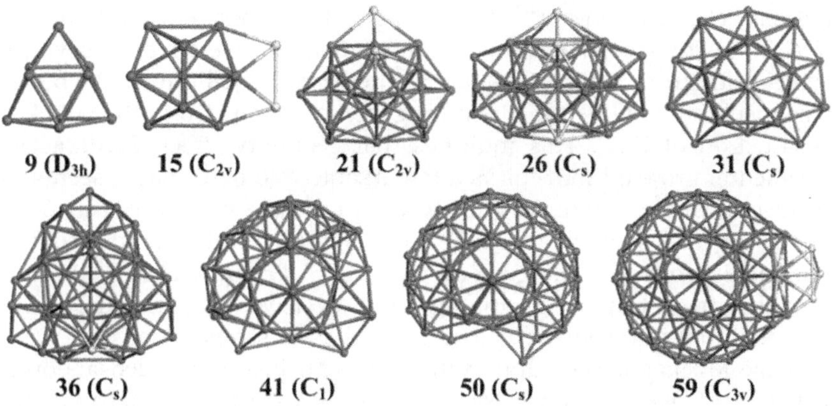

Fig. 2 Lowest-energy structures of the Na_n^+ clusters ($n = 9$, 15, 21, 26, 31, 36, 41, 50, and 59). The symmetry for each cluster is given in the parenthesis. The additional atoms that are not belonging to the major building units of icosahedron or double icosahedron are highlighted in yellow (light) for the guide of the eyes. Reproduced with permission from ref. 42. Copyright (2013) Springer-Verlag GmbH.

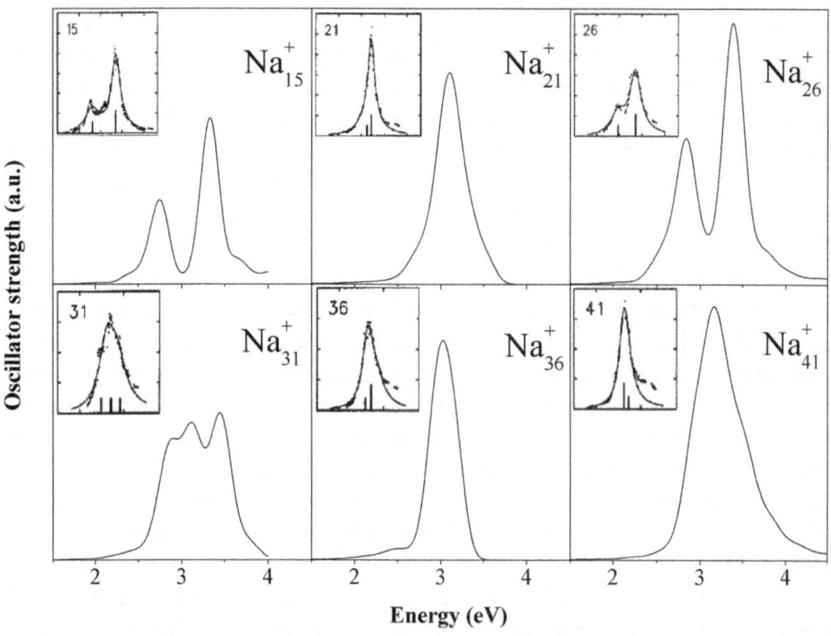

Fig. 3 Photoabsorption spectra of the Na_n^+ clusters ($n = 15$, 21, 26, 31, 36, and 41) from TDDFT simulations. The corresponding experimental data are shown in insets for comparison. Reproduced with permission from ref. 42. Copyright (2013) Springer-Verlag GmbH.

interpenetrating double icosahedra which form an approximate angle of 47° with one additional Na atom capped on the middle region. Na_{36}^+ cluster geometry can be achieved by adding one atom on a C_{3v} symmetric structure of Na_{35}.[85] It prevails the spherical disordered structures with high compactness predicted from Gupta and MM potentials by 0.24 eV and 0.30 eV, respectively.[84] Interestingly, the lowest-energy structure of Na_{41}^+ with closed electronic shell (40 valence electrons) deviates significantly from a perfect sphere as predicted by jellium model but is a semisphere obtained by removing some atoms from the Mackay icosahedron of Na_{55}. This indicates that geometry effect (*i.e.*, Mackay icosahedral growth motif) plays a remarkable role here. Na_{41}^+ seems to be the critical size where the growth pattern of Na_n^+ cluster transits from polyicosahedral sequence to Mackay icosahedral motif. According to the DFT calculations, the energy of present Na_{41}^+ structure is 0.05 eV lower than that from Gupta potential and 0.09 eV lower than that from the MM potential,[84] respectively. The ground state structure of Na_{50}^+ follows the Mackay icosahedral motif and is an incomplete icosahedron with lack of five Na atoms. Na_{59}^+ is also based on the Mackay icosahedron of Na_{55} and the four extra Na atoms are added on a triangle face forming a tetrahedron themselves. Overall, the ground-state structures of Na_n^+ clusters obtained by GA-DFT prevail most of previously reported ones for neutral Na_n clusters by empirical potentials and there is a transition of growth pattern from "polyicosahedral" sequence to the Mackay icosahedral motif at around $n = 40$.

There have been some other GA studies of alkali metal clusters. For example, Reyes-Nava *et al.*[86] performed a global GA optimization of sodium cluster modeled by Gupta potentials. A global optimization of Na, K, Rb, and Cs clusters using both genetic algorithm and basin-hopping with the Gupta potential has been performed by Lai *et al.* up to 56 atoms.[87] The clusters are also found to adopt the icosahedron-based growth sequence in general. By using the density functional tight binding method (DFTB) combined with genetic algorithm, Abdalla and co-workers studied the potassium clusters with up to 20 atoms.[45] K_n clusters showed an even–odd oscillatory behavior and more pronounced peaks of relative stability were found for $n = 8$, 18, and 20, in agreement with the spherical jellium model and the measured mass spectra of potassium cluster beams. Schwerdtfeger *et al.* have carried out a systematic search for the global minimum structures of neutral cesium clusters using GA with DFT method in a scalar-relativistic pseudopotential formalism.[50] They found that the transition from 2D to 3D structures is ambiguous and no unique growth pattern could be identified.

Calcium is a widely distributed divalent metal element on the earth and is very important for biological science. Liang *et al.* have performed DFT based GA search for the ground state configurations of neutral and anionic Ca_n clusters with $n = 2$–22.[46] Independent GA-DFT search with 2000 GA iterations were conducted for each cluster size to ensure global minimum on the PES of Ca_n clusters. Except for the size ranging from 13 to 17, neutral and anionic Ca_n clusters shared the same ground state structures, as shown in Fig. 4. In brief, for the clusters with $n \geq 11$ the ground state structures are with pentagonal bipyramids as skeleton or expand to hexagonal pyramid with $n = 16$–22. The vertical detachment energies and adiabatic detachment energies of anionic Ca_n clusters were compared well with the experimental data.[88]

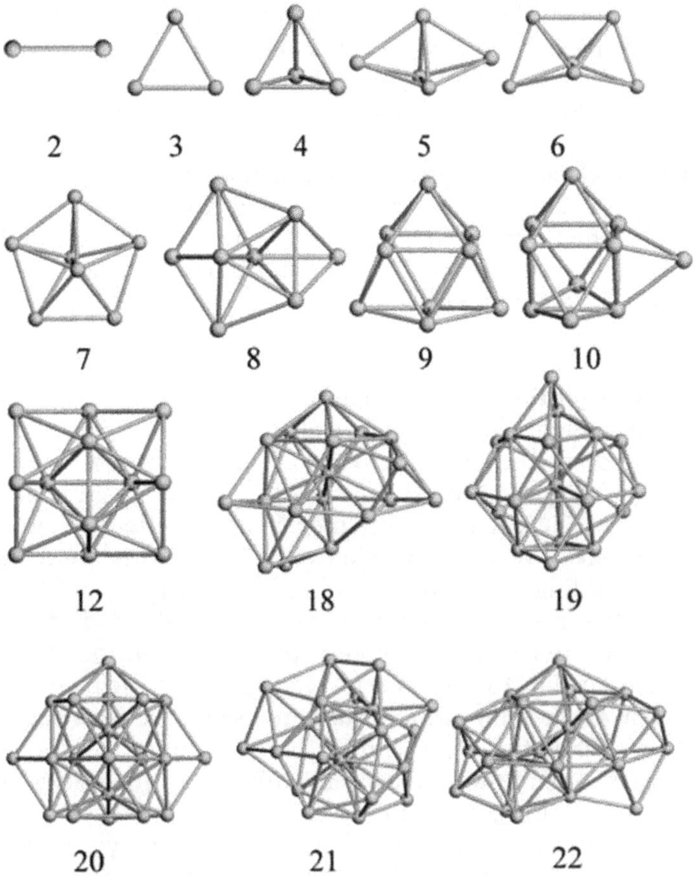

Fig. 4 Ground state structures of neutral and anionic Ca_n clusters with $n = 2–10$, 12, 18–22. Reproduced with permission from ref. 46. Copyright (2015) Elsevier Ltd.

Gallium is a trivalent metal element with unusual structural and bonding characteristics, whose solid state is known to exhibit at least nine polymorphs and only the orthorhombic α phase is stable. First-principles calculations revealed that the α-Ga solid is indeed a metallic molecular crystal with coexistence of metallic and covalent characters.[89] As the embryo of the bulk solid, the Ga clusters have also received considerable attentions. In this regard, Sai *et al.* carried out unbiased GA search combined with DFT to determine the most stable structures of medium-sized Ga_n clusters ($n = 20–40$).[48] Compared to those simple metal clusters, the structures of medium-sized gallium clusters are more complicated and less compact. Most gallium clusters exhibit no symmetry or low symmetry, whereas Ga_{23} (D_{3h}), Ga_{25} (D_{3h}) and Ga_{36} (D_{2d}) have relatively high symmetry. Starting from Ga_{31}, core–shell structures with four to six interior atoms become the dominant structural growth motif. From Ga_{32} to Ga_{39}, the cluster structures are all based on the high-symmetry Ga_{36} by removing or moving several atoms. Ga_{40} initiates a new stage of growth sequence with the number of core atoms increasing to six. Among all Ga_n clusters studied, Ga_{36} stands out as a magic cluster with high symmetry, large binding energy, and large HOMO-LUMO gap, which can be partially related to the electron shell picture. Meanwhile, Drebov *et al.*[49] have also performed GA-DFT study of small Ga_n clusters up to $n = 25$. Particularly stable clusters for Ga_n were found for $n = 7$, 14, and 20.

Besides the clusters discussed above, many other kinds of elemental clusters such as aluminum clusters,[43] P_{20} and As_{20},[44] Cu_9,[47] Au_n ($n = 2$–20),[51] lead clusters[53] and bismuth cluster,[54] have been investigated using GA-based *ab initio* simulations. In Table 1, we present a list of the published works for different clusters.

2.4 Genetic algorithm for mixed clusters

Except for elemental clusters, genetic algorithm has also been successfully employed to study mixed clusters. It is known that small silver clusters prefer 3D structures while small gold clusters tend to adopt 2D configurations. Thus, it would be interesting to explore binary Au–Ag clusters. The effects of size and composition (Au/Ag ratio) on the atomic structures, coordination numbers and binding energies and formation energies of Au–Ag binary clusters are systematically studied by Hong *et al.*[60] As shown in Fig. 5, we can see a split line that divides the 2D and 3D structures of all Au–Ag clusters. Generally, for each size, the 2D structures of Au–Ag alloy clusters share the same skeleton of the pure gold clusters, while the 3D structures share the same skeleton of pure silver clusters. For the binary clusters with 7, 11, and 12 atoms, the transition Au : Ag ratio is close to 1, which means that the influences of Au and Ag atoms are nearly the same. As for the binary systems with 8, 9 and 10 atoms, the 2D–3D transition occurs at Au_5Ag_3, Au_5Ag_4 and Au_6Ag_4, respectively. The results provide valuable insight that the 2D structures result from the s-d hybridization and the relativistic effects of Au atoms and the 3D structures are due to the s valence orbitals in Ag atoms. The Ag atoms sometime play a more important role than Au atoms in determining the most stable configuration in these clusters. Johnston *et al.*[61] also investigated the 8-atom Au–Ag alloy clusters. A transition from 2D to 3D structures is predicted between Au_6Ag_2 and Au_5Ag_3, in line with Hong's result. In addition, they have studied the 8-atom Cu–Ag and Cu–Au clusters.[59] The transition between 3D and 2D structures observed for copper–gold clusters is sharper and appears earlier in the doping series than is known for gold–silver particles.

Small Au–Sc alloy clusters have been studied by Ge *et al.*[90] They found that Au_4Sc_2 cluster may be a fairly good catalyst with higher stability and smaller HOMO-LUMO gap. An unbiased search for all Au_nSc_m ($n + m = 6$) clusters using GA incorporated with all-electron relativistic DFT method is carried out to guarantee the global minima of PES.

GA-DFT method has also been used to explore the low-energy structures of Pt_nSn_n ($n = 1$–10) and $Pt_{3m}Sn_m$ ($m = 1$–5) clusters by Huang *et al.*[62] and $Sn_{9-n}Bi_n$ clusters with $n = 0$–4 and $Sn_{m-n}Bi_n$ clusters with $m = 5$–13, $n = 1$–2 by Heiles *et al.*,[64,65] respectively. In the case of Pt–Sn clusters, Pt and Sn prefer to mix with each other and Sn atoms on the surface tend to

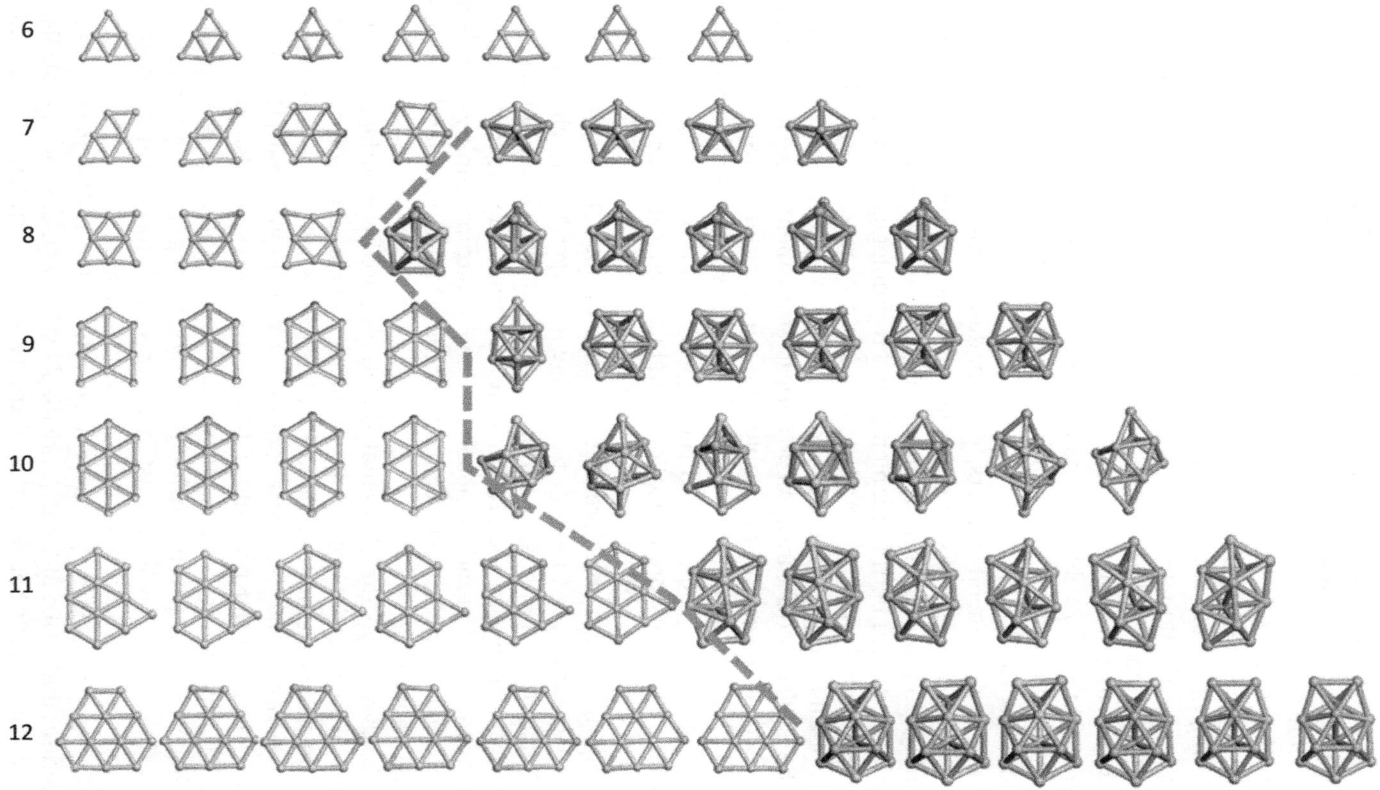

Fig. 5 The lowest-energy structures and some metastable isomers of Au_mAg_n ($6 \leq m+n \leq 12$) clusters.[60] Left side: Au; right side: Ag.

segregate due to the energetically favorable Pt–Sn bonds as well as the larger atomic radius of Sn atoms.[62] This makes one or two Pt atoms available for reaction. When the total number of atoms exceeded four, 3D cluster structures emerge. Pt_2Sn_2 quadrilaterals as basic structural units of PtSn solid can be found in many cluster isomers. For $Pt_{3m}Sn_m$ clusters with more Pt atoms, the Sn atoms are well separated without forming any Sn–Sn bond. As the cluster grows bigger, there is a clear tendency of layered structures consisting of Pt–Sn alternative triangular lattices, resembling the Pt_3Sn solid.[62] For Sn–Bi clusters, experimental and theoretical investigation revealed that structures without Bi–Bi bonds are lower in energy.[64,65] Bi-doped Sn clusters show structural rearrangement compared to pure Sn clusters; for instance, all the Bi-doped $Sn_{9-n}Bi_n$ clusters ($n=0$–4) clusters manifested tricapped trigonal prisms configurations except for Sn_9, which has a bicapped pentagonal bipyramid structure.[64]

Compared to metal clusters, semiconductor clusters might possess more complicated PESs due to the intrinsic nature of the covalent bonding. Sodium adsorption on silicon surfaces has received considerable interest from both fundamental and technological points of view. It has been found that the Na adsorption not only modifies the electronic and optical properties of the Si surface, but also results in reconstruction of the surface structures. As a critical assessment, GA-DFT method was used to determine the stable configurations of Na_mSi_n ($1 \leq n \leq 11$, $1 \leq m \leq 3$) clusters by Sai *et al.*[29] The lowest-energy configurations of neutral and anionic clusters of Na_nSi_m were obtained. For neutral Na–Si clusters, Na atoms prefer to be separated from each other by the Si skeleton while in anionic clusters Na atoms stay together to form Na–Na bonds. The cluster structures are sensitive to both Na/Si stoichiometry and charge state for small clusters; while for larger Na_nSi_m clusters with $m>7$, addition of one Na atom or one extra charge would not result in significant change in the ground state structure. The experimental size dependence of ionization potentials and electron affinities are well reproduced by theoretical calculations, confirming the reliability of the GA-DFT approach.

It is known that the elemental silicon clusters have surface dangling bonds, which renders them chemically reactive and therefore not suitable for nanoscale building blocks. It has been shown both theoretically and experimentally that incorporation of transition metal (TM) atoms into a silicon cluster not only stabilizes the cluster but also brings into peculiar physical properties. However, little attention has been paid to doping of multi-TM atoms inside Si clusters. Huang *et al.*[57] systematically studied the V_3Si_n ($n=3$–14) clusters using GA-DFT method and simulated the anion photoelectron spectroscopy, the vertical detachment energies (VDEs) and adiabatic detachment energies (ADEs). The lowest-energy structures and ADEs of V_3Si_{3-14} are displayed in Fig. 6, showing excellent agreement of ADEs between theory and experiment. Interestingly, V_3Si_{12} carries a sizeable magnetic moment of 4 μ_B and exhibits ferrimagnetic behavior, *i.e.*, +2.4 μ_B and −0.6 μ_B on the surface and the central V atoms, respectively.[58]

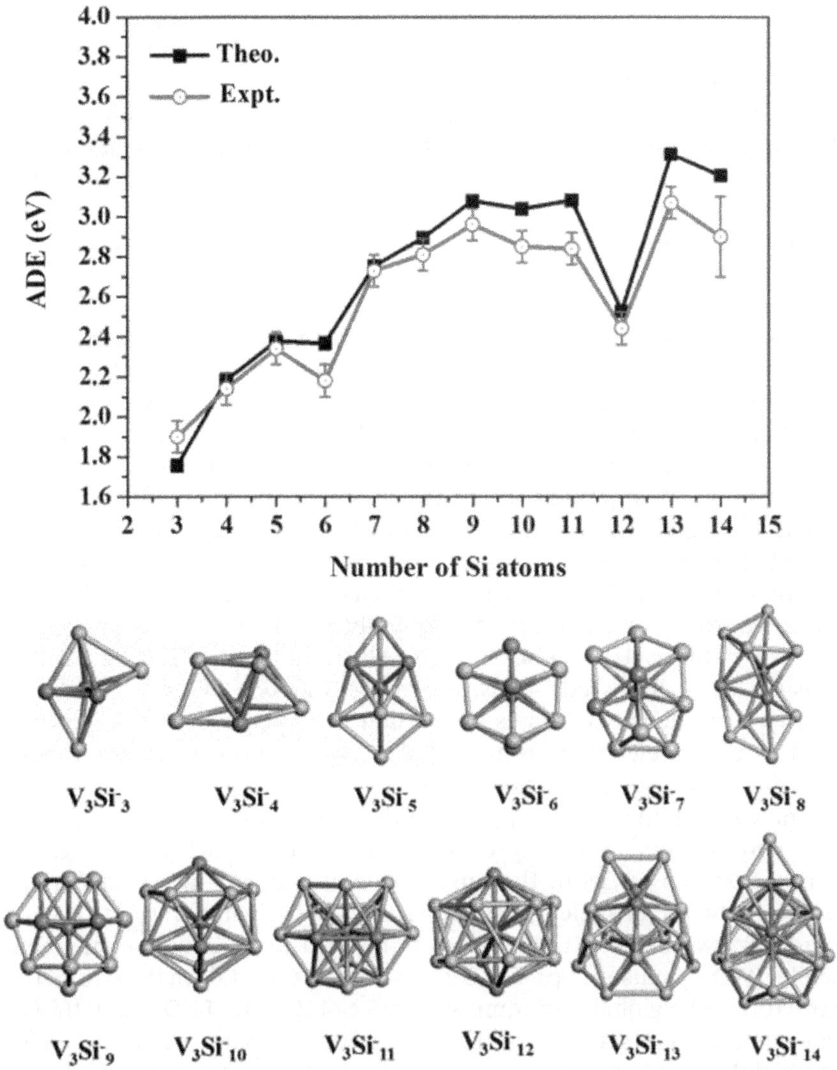

Fig. 6 Adiabatic detachment energies (ADEs) (upper panel) and lowest-energy structures of $V_3Si_n^-$ ($n = 3$–14) clusters. Reproduced with permission from ref. 57. Copyright (2015) American Chemical Society.

The existence of more complicated structures, the onion-like three-shell icosahedral "matryoshka" cluster, is confirmed by GA-DFT method. This kind of 33-atom clusters can be separated into three shells with chemical formula of $A@B_{12}@A_{20}$ (A = Sn, Pb; B = Mg, Zn, Cd, Mn). Zhao's group[66] has studied the HOMO-LUMO gaps, formation energies, electron counting rule and magnetic properties of these cluster. The high stability of these "matryoshka" clusters is attributed to suitable matching of atomic sizes as well as the high I_h symmetry and consequently the splitting of superatom orbitals of high angular momentum. Especially, two magnetic "matryoshka" clusters, *i.e.*, $Sn@Mn_{12}@Sn_{20}$ and $Pb@Mn_{12}@Pb_{20}$, are proposed, which combine a large magnetic moment of 28 μ_B, a moderate HOMO-LUMO gap, and weak inter-cluster interaction energy, making them ideal building blocks in novel magnetic materials and devices.

Apart from the above discussed alloy clusters, metallic oxide clusters have also been studied by GA-DFT. As a prototype of simple metal oxides, MgO solid has a NaCl-type structure and possesses an inherent structural stability. The structures of cationic $(MgO)_n^+$ ($n=2$–7) clusters and neutral $(MgO)_n$ clusters were determined by combining GA-DFT calculation with experiments.[70,71]

Figure 7 shows a comparison between the experimental infrared multiple photon dissociation (IR-MPD) spectra and the calculated linear IR absorption spectra for the gas-phase $(MgO)_n^+$ clusters.[70] The results demonstrate that small neutral and cationic MgO clusters display unusual structural diversity and flexibility. The structures of cationic clusters differ significantly from their neutral counterparts.

Using GA-DFT approach, lowest-energy structures of $(WO_3)_n$ ($2 \leq n \leq 12$) clusters are globally searched by Sai *et al.*,[74] as displayed in Fig. 8. It is found that small $(WO_3)_n$ clusters with $n=3$ or 4 adopt ring-like configurations with W-O alternating arrangement. Starting from $(WO_3)_8$, the tungsten oxide clusters transform to symmetric spherical-like cages. The relative stability, HOMO-LUMO gap, electronic states of these $(WO_3)_n$ clusters were discussed. Analysis of wavefunctions of frontier orbitals and electron density of states shows that the valence bands are dominated by the 2p electrons from O and the conduction bands are mainly contributed by the 5d states from W.

The GA-DFT method is applied to the global optimization of $(Li_2O)_n$ ($n=1$–8) clusters by Cheng *et al.*[67] The planar structures are global minima up to $n=2$, while the cluster structures are all three dimensional at $n>3$. The metal skeleton is similar to $(WO_3)_n$ clusters for $n<3$. When cluster grows bigger, $(Li_2O)_n$ clusters tend to form bulk-like fragment while $(WO_3)_n$ clusters prefer the less compact and high symmetric structures. The global minimum structures of $(Li_2O)_2$, $(Li_2O)_6$, and $(Li_2O)_7$ clusters are more stable than the other sizes. The same group have also studied $(BeO)_n$ clusters ($n=2$–12, 16, 20, and 24).[68] The global minimum structures of $(BeO)_n$ clusters are rings up to $n=5$, double rings at $n=6$ and 7 and cages for $n \geq 8$. In addition, small $(B_2O_3)_n$ ($n=1$–6) clusters are investigated and the cage structure begins at $n=4$.[69]

A kind of more complicated metal oxide clusters, cerium oxide clusters are studied by Sierka's group.[73] $Ce_2O_2^+$, $Ce_3O_4^+$, and $(CeO_2)_m CeO^+$ ($m=0$–4) are unambiguously determined by a combination of GA-DFT method and infrared vibrational predissociation spectroscopy of the cluster-rare gas atom complexes. The most stable structures of even small gas-phase aggregates of cerium oxide with 2–5 cerium atoms show structural motifs reminiscent of the bulk ceria. This is different from main group and transition metal oxide clusters, which often display structural features that are distinctly different from the bulk structure.

Gradient-embedded genetic algorithm (GEGA) program[30,35] is used to search for the global minimum of CB_7^-.[79] Using the same technique,

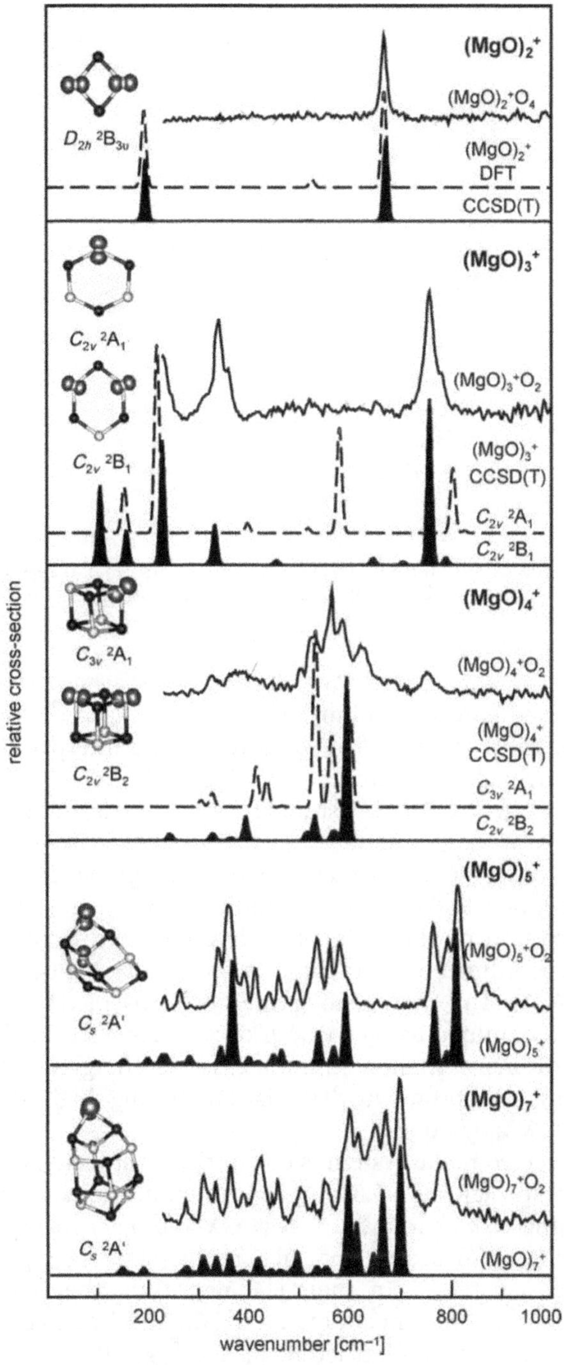

Fig. 7 Comparison of the experimental IR-MPD and calculated linear IR absorption spectra for the most stable $(MgO)_n^+$ gas–phase clusters with $n = 2–5$ and 7 along with their geometrical structures. The calculated CCSD(T) and the DFT (B3LYP) spectra (convoluted with Gaussian functions) are shown for $n = 2–4$ and $n = 2$, 5, and 7, respectively. (Mg: black spheres, O: white spheres, isosurface of spin density: gray). Reproduced with permission from ref. 70. Copyright (2011) Wiley-VCH Verlag GmbH & Co. KGaA, Weinheim.

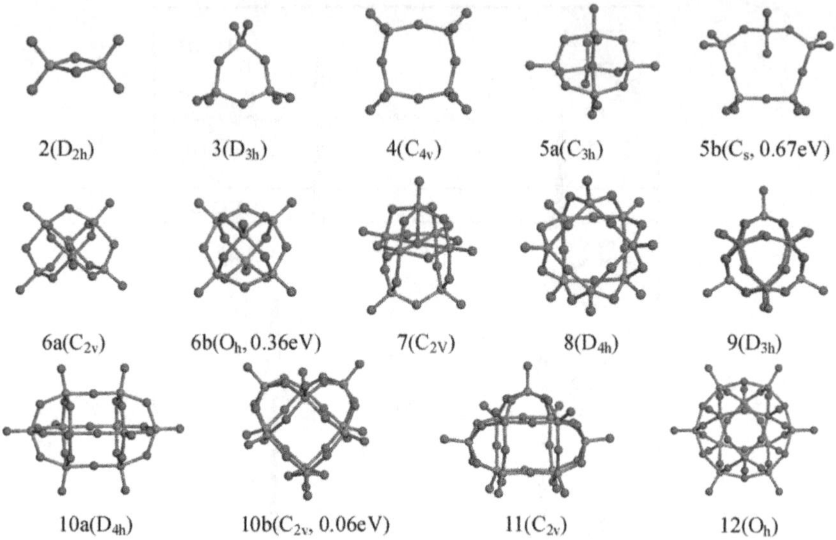

Fig. 8 Lowest-energy structures of $(WO_3)_n$ clusters ($2 \leq n \leq 12$) and selected metastable isomers (labeled as 5b, 6b, 10b). Reproduced with permission from ref. 74. Copyright (2012) Elsevier Ltd.

a series of mixed boron–aluminum cluster ions, $B_nAl_{6-n}^{2-}$ ($n = 0$–6), and their lithium salts are investigated to reveal the interplay between covalent and strongly delocalized bonding.[80] Light element clusters, $(LiF)_nLi^+$ series ($n = 0$–7), are studied by Fernandez-Lima *et al.* both theoretically and experimentally.[78] Hydrogenous mixed clusters like $(H_2O)_nH$[76] or $(AlH_3)_n$[77] are also considered. The global minimum structures of $(AlH_3)_n$ ($n = 1$–8) clusters are studied by Kiran and co-workers and the growth pattern of alanes for $n \geq 4$ is dominated by structures containing hexa-coordinated Al atoms. It is found that Al_6H_{18} is an embryo crystal of γ-AlH_3. The most stable configurations of $(H_2O)_nH$ ($n = 1$–4) clusters are determined using GEGA program. This is a challenging task since there are many shallow minima close in energy while the global minima always are prone to dissociation, having high structural degeneracy with weakly bound.

Ligand-protected metal system is one of the most complicated systems in fundamental science and technological applications. As the first application, Xiang *et al.* proposed a new GA approach incorporated with DFT methodology to search for the ground state of $[Ag_7(SR)_4]^-$ and $[Ag_7(DMSA)_4]^-$ clusters.[81] In their GA algorithm, they simplified the model by assuming only one atom of a ligand attached to the metal core cluster. The lowest-energy structure of the $[Ag_7(SR)_4]^-$ cluster can be viewed as the addition of two –RS–Ag–RS– staples to a D_{2h} planar Ag_5 cluster, the lowest-energy structure of $[Ag_7(DMSA)_4]^-$ had eight Ag–S bonds with its X-ray diffraction pattern in line with experiment. This demonstrated that –RS–Ag–RS– can be a stable motif in thiolate-protected Ag clusters, and GA can be extended to predict the structures of ligand-protected metal clusters.

3 Basin hopping

3.1 Introduction

Basin hopping (BH), another popular approach for global optimization of clusters, is essentially Monte-Carlo (MC) minimization based on the Metropolis algorithm. As one of the most reliable tools to search the ground state structures of chemical and biological systems, this algorithm was first proposed by Li *et al.*[91] for solving the multiple-minima problem in protein folding. Subsequently, Wales and Doye[92] utilized BH to search the PES of Lennard-Jones (LJ) clusters up to 110 atoms and confirmed that BH is an efficient method to find the global minimum of LJ clusters. Later, many other variants of BH[93–98] have been developed to further improve its efficiency and BH algorithm has been implemented with many quantum chemical codes (*e.g.*, DMol^3 program, Q-Chem package[99]) for structural optimization.

3.2 Theory of basin hopping

Simply speaking, BH algorithm is an unbiased global optimization technique combining the Metropolis MC sampling algorithm and a gradient-based local search method. The energy landscape is sampled by the MC simulation to find the energy basin, which is local minimum surrounded by all correlative configurations that could relax into this minimum using downhill relaxations. This transformation of configuration space makes all the isomers associate with the local minimum obtained by a direct optimization, and the transition state (TS) regions can be removed effectively from the problem. Figure 9 is a schematic diagram of the one-dimensional energy landscape searched by the BH algorithm. The PES is transformed into a collection of interpenetrating staircases (black dashed line in Fig. 9) *via* the equation: $\hat{E}(X) = \min\{E(X)\}$, where PES is a function of the generalized coordinate (X), and min

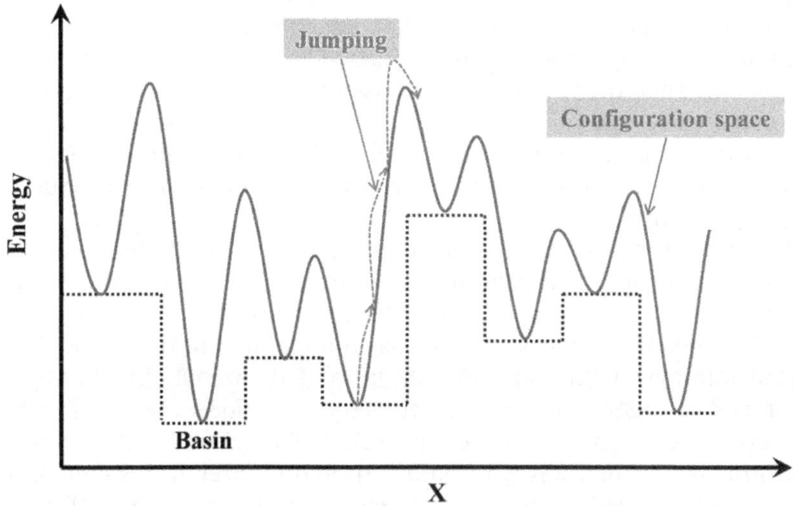

Fig. 9 Schematic diagram of the PES searched by basin hopping algorithm.

denotes the local minimum associated with X. The veritable configuration space is also shown in Fig. 9 for the blue solid line. Hence, the above transformation does not change not only the local minima and the global minimum, but also the relative energies of those minima in the energy landscape.

In the BH procedure, a randomly generated conformer of a given structure is used as the starting configuration. Using a gradient-driven descent routine, the starting structure is relaxed to the closest local minimum. Then, the PES is mapped onto a set of plateaus (basins) where the potential is constant. This modified PES contains less local minima and barriers but keeps the global minimum. A Markovian process is introduced to randomly deform this local minimum configuration, which must ensure the cluster structure not to vaporize or break. After the mutation operations, the newly found structure will be also relaxed to the vicinity of a local minimum. If the local minimum is a new structure, which means the change is sufficient to overcome the nearby TS, the algorithm enables hopping from one minimum to the next on PES. As shown is Fig. 9, BH algorithm hop from one basin to the near one around the starting structure on the PES.

For each BH iteration, the energy difference ΔE between the initial and final isomers is compared using the Metropolis MC criterion[18] to evaluate the newly found minimum structure. If $\Delta E \leq 0$, or for $\Delta E > 0$ if $\exp[-\Delta E/(k_BT)]$ (k_B is the Boltzmann constant and T is the simulation temperature) is larger than a random number between 0 and 1, the newly found minimum is accepted. Thus in BH algorithm, temperature as a crucial parameter must be carefully chosen and adjusted, which will affect the tradeoff between the acceptance ratio and the sampling efficiency.[100]

However, if the basins are surrounded by high TS barriers, normal moves cannot let the system escape from the basins, then the BH searching is trapped. Many methods[95,96,101–103] have been exploited to overcome this problem, and the most famous one is jumping[95] (red dashed line in Fig. 9). In practice, when the structure does not update for a number of steps, the system is judged to be restrained in a local minimum. Then the temperature is raised to ∞, and the jumping moves are executed several times to propel the system to escape from the local minimum. The structure is changed without local minimization in jumping moves, and the changes are always accepted. The introduction of jumping improves the efficiency of BH algorithm dramatically.

Except for the jumping moves, other variants of BH strategy enjoy the favor of scientists for determination of the global minima. White and Mayne used two variants of BH methods, significant structures basin hopping (SSBH) and raw structures basin hopping (RSBH), to search the global minima of $(LJ)_n$ ($n = 19$, 30, 38) and $(benzene)_n$ ($n = 6$, 10, 14) clusters.[104] In SSBH, the update geometry is the local minimum, while the update configuration is left in RSBH. They found that the SSBH performs better than RSBH in both $(LJ)_n$ and $(benzene)_n$ systems. Leary and Doye developed monotonic sequence basin-hopping (MSBH)[93,94] for exploring the energy landscape of $(LJ)_{98}$ cluster. The difference between original BH method and MSBH is that MSBH only accepts downhill moves. MSBH is a multi-start method that samples funnel bottoms rather than local minima due to the fact that the process restarts from a new random configuration when it arrives at a local minimum. Extremal optimization based basin hopping (EOBH) is proposed by Iwamatsu[105] to cross the high TS barriers by accepting every move without Metropolis criterion. This method can effectively explore all the low-energy structures, but gives up the elaboration of local search. All together, the above methods sufficiently demonstrate that BH is one of the most efficient algorithms to search for the lowest-energy structures of atomic and molecular clusters.

In addition to a suitable iterative method, an accurate description of PES is critical for BH global optimization research. Though many studies[106,107] have confirmed that empirical force fields combined with appropriate search algorithms can successfully characterize the global minimum energy structures of many systems, for a given system, the reliable force field is not always available. Alternatively, direct combination of BH with *ab initio* methods (*i.e.*, *ab initio* BH) is a very powerful tool to explore diverse atomic structures and electronic structures of atomic and molecular clusters, without reference to any empirical force fields. The pioneer of *ab initio* BH is that Calvo *et al.*[108] who used a simple quantal Hückel-type model and a density-functional-theory (DFT) based orbital-free model to identify the ionic structure and lowest-energy minima of cationic Na_n^+ clusters with $n = 4–40$. Subsequently, a series of *ab initio* BH studies have been conducted to explore the stable configurations, electronic structures and some chemical reaction paths of a variety of clusters, as summarized in Table 2.

3.3 Basin hopping for atomic clusters

Like GA, the first-principle based BH algorithm has been widely applied to various atomic clusters, including nonmetallic clusters[113,116–118] and metallic clusters.[122,124] For example, Yoo *et al.*[118] performed a biased BH search coupled with DFT to search for the lowest-energy geometric structures of medium-sized Si_{27-39} clusters. Shao *et al.*[122] utilized DFT in conjunction with BH to study the structures of medium sized gold clusters Au_n^- with $n = 27–35$. All these attempts show that first-principle based BH is a powerful tool to search the global minimum of atomic clusters.

Boron, which sits adjacent to carbon in the periodic table of elements, has attracted considerable amount of interest in structural prediction of clusters and nanostructures. Boron cannot form a honeycomb hexagonal framework as graphene due to its electron deficiency; instead the 2D boron sheet prefers triangle lattice with appropriately distributed hexagonal holes. Piazza *et al.* reported the first experimental evidence of a quasi-planar boron cluster.[110] They searched the low-lying isomers of B_{36}^- cluster using a constrained BH code, and found that the B_{36}^- cluster is quasi-planar with a central hexagonal hole (Fig. 10a). Meanwhile, the stable neutral B_{36} is found to have high C_{6v} hexagonal symmetry

Table 2 A list of published works on BH-DFT simulation of different cluster systems.

Types	Systems	Sizes	Authors and ref.
Atomic clusters	B_{35}^-, B_{35}, $C_{22}H_{12}$		Li *et al.*[109]
	B_{36}^-, B_{36}		Piazza *et al.*[110]
	B_{39}^-		Chen *et al.*[111]
	B_{40}^-, B_{40}		Zhai *et al.*[112]
	C_n	$3 \leq n \leq 24$	Yen and Lai[113]
	CTi_7^{2+}		Gao *et al.*[114]
	Na_n	$n = 10–20$	Aguado *et al.*[115]
	Si_n	$n = 12–22$	Yoo and Zeng[116]
	Si_n^-	$20 \leq n \leq 45$	Bai *et al.*[117]
	Si_n	$n = 27–39$	Yoo *et al.*[118]
	Si_n	$n = 7, 10$	Gehrke and Reuter[119]
	Si_{12}^{2-}, Ge_{12}^{2-}, Sn_{12}^{2-}		Shao *et al.*[120]
	Co_n^+, $Co_n^+Ar_m$	$n = 4–8$	Gehrke *et al.*[121]
	Cu_7		Gehrke and Reuter[119]
	Au_n^-	$n = 27–35$	Shao *et al.*[122]
	Au_{20}		Aprà *et al.*[123]
	Au_{40}		Jiang and Walter[124]
	$Au_{17}M$		Tang *et al.*[125]
	M = Cu, Ag, Li, Na, K		
	$Au_{23}S_{11}^-$, $Au_{25}S_{12}^-$, $Au_{27}S_{13}^-$		Jiang *et al.*[126]
	MAu_n^-	$n = 5–8$	Pal *et al.*[127]
	M = Si, Ge, Sn		
	MAu_{16}^-		Wang *et al.*[128]
	M = Fe, Co, Ni		
Molecular clusters	$(H_2O)_n$	$n = 4–11$	Do and Beseley[98]
	$(CH_3OH)_m$	$m = 4–11$	
	$(H_2O)_n(CH_3OH)_m$	$4 \leq m + n \leq 7$	
	$H^+(H_2O)_n$	$n \leq 9$	
	$(H_2O)_n$	$n = 4–10$	Liu *et al.*[129]
	$C_2O_4^{2-}(H_2O)_m$	$m = 1–6$	
	$NO_3^-(H_2O)_h$	$h = 1–7$	
	$CH_3NH_2(H_2O)_n$	$n = 1–7$	Lv *et al.*[130]
	$NH_4^+(H_2O)_n$	$n = 1–10$	Pei *et al.*[131]
	$Cl_2(H_2O)_n$	$n = 1–4$	Jiang *et al.*[100]
	$M(H_2O)_n^+$, $M_2(H_2O)$	$n = 1–2$	Wadey and Besley[132]
	M = PH_3, H_2S, HCl		
	Na^+He_n	$n = 2–24$	Issaoui *et al.*[133]
	Rh_8^+		Harding *et al.*[134]
	Rh_n^+	$n = 6–12$	Harding *et al.*[135]
Complex systems	$(CO)_nAu_7^-$	$n = 1–4$	Pal *et al.*[136]
	$O_2Au_n^-$	$n = 2, 4, 6$	Pal *et al.*[137]
	Ag_n on MgO(100)	$n = 2–10$	Barcaro and Fortunelli[138]
	Ag_n on MgO(100)	$n = 2–4, 6, 8, 10$	Barcaro *et al.*[139]
	$PdAg_n$ on MgO(100)	$n = 1–8$	Barcaro and Fortunelli[140]
	M_2Ag_n, MAg_7, MAg_8 on MgO(100)	$n = 6–7$	Barcaro and Fortunelli[141]
	M = Cu, Au, Pd		
	Au_n on MgO(100)	$n = 4, 6, 8, 10$	Barcaro and Fortunelli[142]
	Ag_3, $(AgAu)_3$ on MgO(100)		Negreiros *et al.*[143]

(Fig. 10b). Afterwards, a similar global minimum search for B_{35}^- was performed, and the results showed that the ground state of B_{35}^- is a quasi-planar structure with a double-hexagonal vacancy.[109]

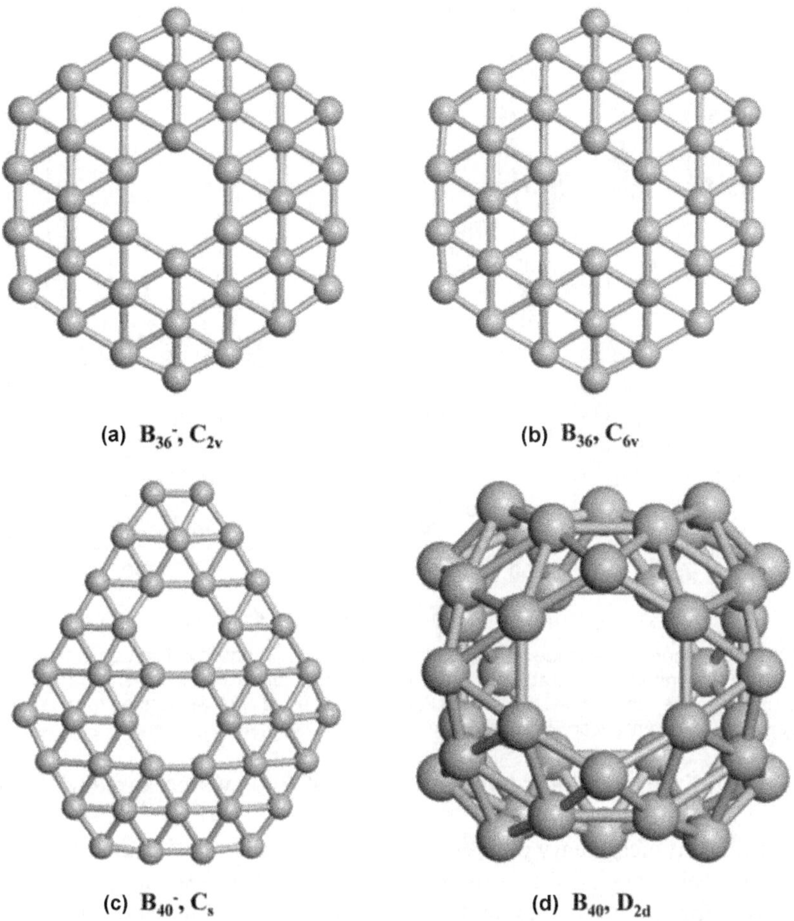

Fig. 10 Lowest-energy geometries of B_{36}^{-}, B_{36}, B_{40}^{-} and B_{40} clusters. (a) and (b) are the most stable structures of anion and neutral B_{36}, (c) and (d) are the ground state of B_{40}^{-} and B_{40} clusters. Reproduced with permission from ref. 110, 112. Copyright (2014) Nature Publishing Group.

After the C_{60} fullerene structure was discovered, many experiments attempted to observe the all-boron fullerene but failed. In contrast, there have been many theoretical investigations on all-boron fullerenes. Recently, Zhai *et al.* conducted a combined experimental and theoretical exploration of all-boron fullerene B_{40} and B_{40}^{-} clusters.[112] They produced B_{40}^{-} clusters in a laser vaporization supersonic source and characterized B_{40}^{-} using photoelectron spectroscopy. They also performed unbiased global-minimum searches for B_{40}^{-} and B_{40} by using the stochastic surface walking[144] and basin hopping algorithms. The lowest-energy structures of B_{40}^{-} and B_{40} from global optimizations are shown in Fig. 10. A quasi-planar configuration with two adjacent hexagonal holes of B_{40}^{-} is more stable than its fullerene conformer. This is in accordance with the structure of B_{36}^{-} found by Piazza *et al.*[110] For neutral B_{40} clusters, the fullerene-like cage is calculated to be the most stable structure. They also calculated the ADE and VDE of B_{40}^{-} isomers using TDDFT to confirm the

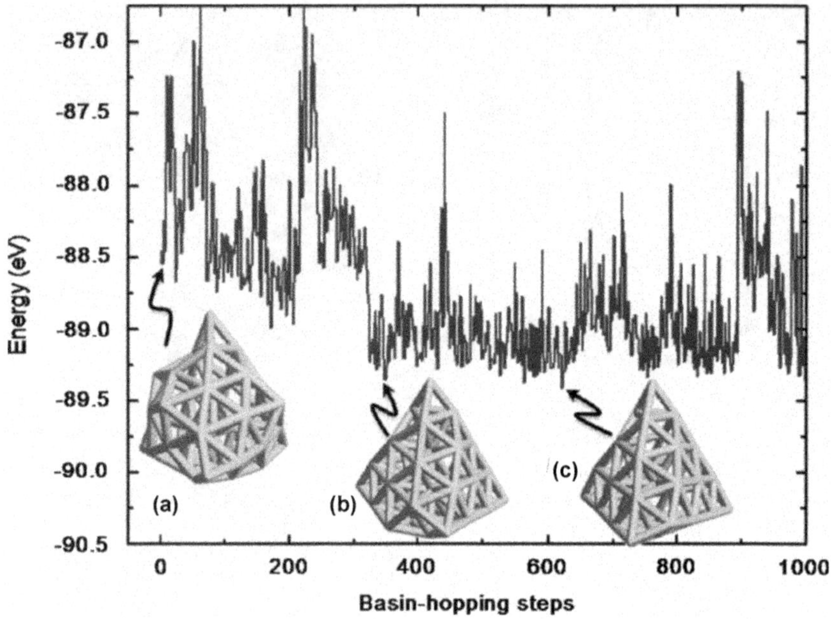

Fig. 11 The energy landscape of Au_{40} obtained from a DFT-based BH search at GGA-PBE level. (a) is the initial configuration of Au_{40} built on Au_{38}, (b) is the lowest-energy structure of Au_{40}, (c) is the metastable structure of Au_{40}. Reproduced with permission from ref. 124. Copyright (2011) American Physical Society.

global minimum and low-lying structures of B_{40}^-, which agree well with the experimental results.

Gold is a typical heavy element with pronounced relativistic effect, and gold nanoclusters exhibit unusual catalytic capabilities. Au_{40} is a magic number for gold clusters with tetrahedral symmetrical configuration, which shares the same symmetry with famous tetrahedral Au_{20} cluster.[145] Jiang and Walter[124] explored the energy landscape of Au_{40} using DFT-based BH method, as shown in Fig. 11. The original configuration of Au_{40} cluster (Fig. 11(a)) was built on the putative global minimum of Au_{38} by manually adding two Au atoms. The BH procedure contains more than 1000 steps of geometry optimization at GGA-PBE level. The ground state configuration is given in Fig. 11(b), and the isomer in Fig. 11(c) has C_3 symmetry and is only 0.15 eV higher in energy than the ground state. The two stable isomers are both chiral, twisted pyramid structures. The Au_{40} is a manifestation of the enhanced stability due to the tetrahedral symmetry as well as close-shell electronic structure, similar to Au_{20}.[123,145] This work indicates that the electronic structure and the complex energy landscape of such heavy element clusters can be accurately described by *ab initio* BH approach instead of empirical potentials.

X-doped Au_n clusters (X = transition metal,[128] semiconductor atoms,[127] sulfur) show distinctive structural, electronic and magnetic properties. Wang *et al.* made a comprehensive comparison between DFT-based BH structure searches and experimental trapped ion electron diffraction and photo electron spectroscopic data on transition metal (Fe, Co, and Ni) doped Au_{16}^- cluster.[128] They found agreement between theoretical and

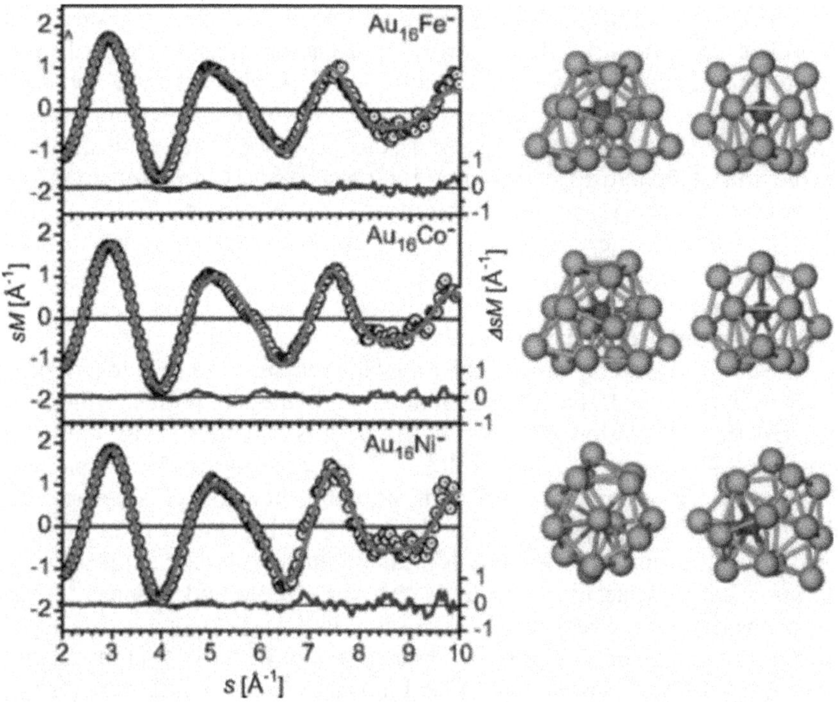

Fig. 12 The structures with two views rotated by 90° (right) and trapped ion electron diffraction (left) for MAu_{16}^- (M = Fe, Co, Ni). The lower traces in each frame show the residuals. Reproduced with permission from ref. 128. Copyright (2009) American Physical Society.

experimental results. For instance, the simulated lowest-energy structures of MAu_{16}^- with M = Fe, Co and Ni fitted the trapped ion electron diffraction very well (Fig. 12). Pal *et al.* made a similar study on the Si-, Ge-, Sn-doped anion gold clusters MAu_n^- with $n = 5$–8.[127] They employed the BH technique coupled with the DFT for global optimization at PBE level, and used well-resolved photoelectron spectra to identify the low-lying structures of MAu_{5-8}^-. Not only the ground state structures of MAu_{5-8}^- clusters were obtained, but also the growth tendency of larger sized MAu_n^- clusters were predicted.

In addition to the atomic clusters by heavy element, the structures of transition metal clusters and semiconductor clusters are studied extensively because they are significant for nanomaterials and microelectronics. Gehrke and co-workers explored the energy landscape of Co_n^+ and $Co_n^+Ar_m$ with $n = 4$–8 through *ab initio* BH method.[121] The spin degree of freedom was added to the local relaxation of BH iterations. They also measured the far-infrared multiple photon dissociation spectra of Co_n^+ and $Co_n^+Ar_m$ clusters to confirm the geometric and electronic global minima of first-principle based BH algorithm, and only a few sizes agreed well with experimental results. This means the multiple close-lying spin states of transition metal clusters require a proper treatment using a multiple determinant wavefunction-based method.[146]

Silicon is the typical semiconductor element, and Si clusters have attracted abundant attention since the 1980s. Determination of most stable structures of Si_n clusters is crucial for understanding their physical and chemical properties. Combining with DFT calculations, the global minimum geometries of small and medium sized Si_n clusters were obtained by Yoo and Zeng using unbiased BH algorithm with $n=12$–16 and biased one for $n=17$–22.[116] The BH search started with several arbitrarily selected initial structures containing 1000 MC steps. Compared to structures reported previously,[147–153] the most stable structures of Si_{12-15} and Si_{19-21} were reproduced and some new configurations were found at sizes of 16, 17, 18, and 22.

3.4 Basin hopping for molecular clusters

Unlike atomic clusters, most of the molecular clusters are held together by weak van der Waals (vdW) interactions or hydrogen bonds, which restricted the application of *ab initio* BH because standard DFT methods cannot describe vdW forces well.[154,155] By accounting for dispersion interactions in a semiempirical manner, some new dispersion-corrected functionals make it possible to combine DTF with BH algorithm to solve the global minima problem of molecular systems. A pioneer application of *ab initio* BH to molecular clusters is the work by Do and Beseley.[98] They used basin hopping algorithm combined with DFT to search the equilibrium structures of neutral water clusters $(H_2O)_n$ with $n=4$–11, methanol clusters $(CH_3OH)_m$ with $m=4$–7, water–methanol complex clusters $(H_2O)_n(CH_3OH)_m$ with $4 \leq n+m \leq 7$, and protonated water clusters $H^+(H_2O)_n$ with $n \leq 9$ at B3LYP-D/6-31+G* level (for which, however, hydrogen bonding instead of vdW interactions are important). Their approach could reproduce the lowest-energy structures from previously theoretical calculations[156–159] and experimental efforts.[160,161] Moreover, they found a new global minimum for $H^+(H_2O)_7$ cluster, whose binding energy is 0.515 kcal mol^{-1} lower than Hodges and Wales's[156] at the same level of theory. Their work clearly demonstrates that *ab initio* BH studies of molecular clusters without empirical parameters are feasible.

Later, *ab initio* BH algorithm has been employed for searching the ground state of a variety of molecular clusters. Water is one of the most significant molecules in the universe and has pivotal meaning in the formation of ice, clouds and biological systems. The exploration of ground state structures of water clusters helps explain the unique behavior of water, such as the solvent effect. Liu *et al.* have combined two different sampling techniques with BH algorithm to explore the PES of $(H_2O)_{4-10}$ clusters at MP2/aug-cc-pVDZ level.[129] One sampling method is random sampling, in which the maximum value of each atom or molecule is constant, and another is compressed sampling with the size of atomic or molecular movement on the surface larger than that of inside ones. For each cluster size, they performed eight independent BH searches consisting of 600 steps at 1000 K. The low-lying configurations of $(H_2O)_{4-10}$ clusters found by BH agree with previous studies.[98,158,162,163] Based on the idea that the structure of M_{n+1} cluster usually comes from one of the low-lying isomers of M_n, they speculated three evolutionary routes for the BH results from $(H_2O)_4$ to $(H_2O)_{10}$, and the three routes are shown in Fig. 13. Route one shows that $n=6$ is the boundary of 2D to 3D

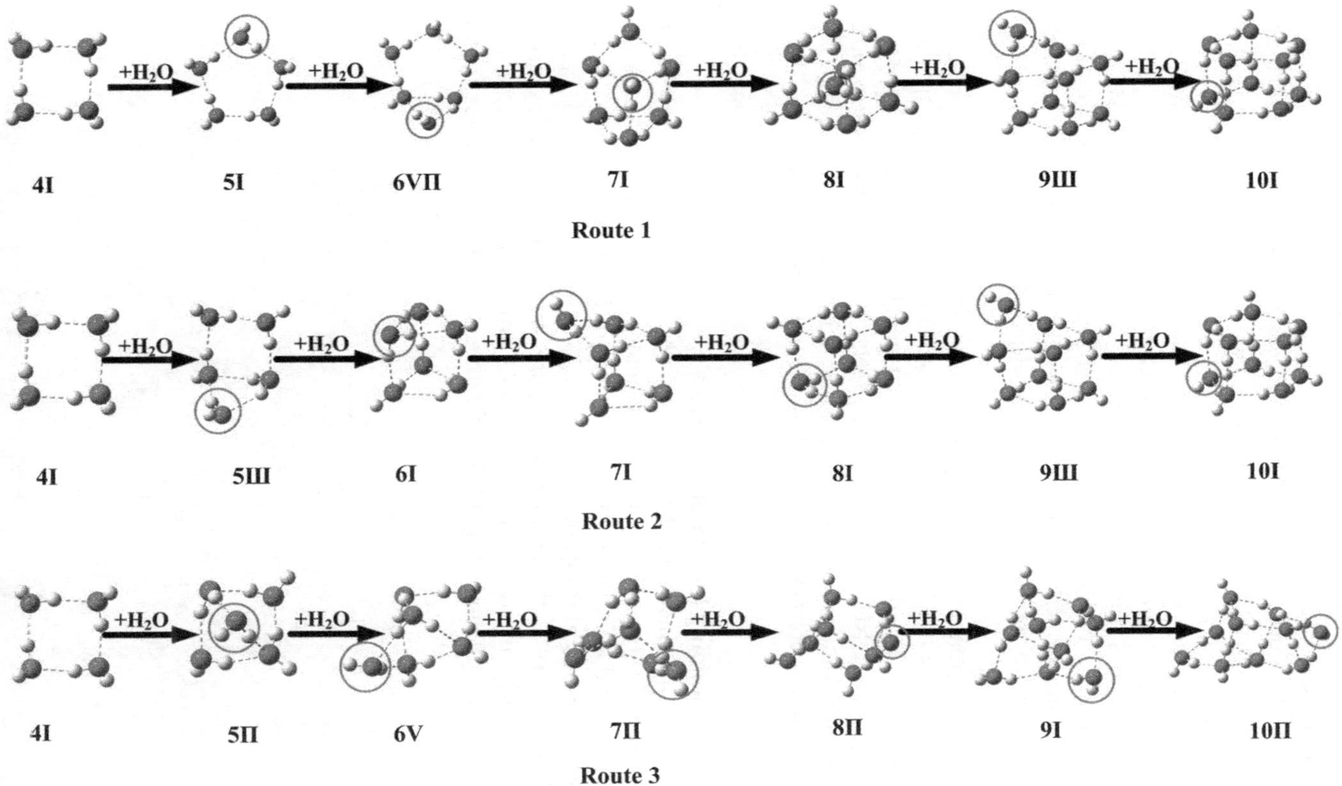

Fig. 13 Schematic diagram of three evolutionary routes of neutral water clusters. The water molecule in red loop can be regarded as the extra one added to front water clusters. Reproduced with permission from ref. 129. Copyright (2014) American Chemical Society.

structures, while route two and route three yields the boundary at $n = 5$. The isomers in the three routes are not all lowest-energy structures of each cluster size, which give the growth information that the lowest-energy structure for each cluster size may result from the cluster isomers of neighbor size.

Jiang *et al.*[100] used *ab initio* BH to successfully predict the global and local minima of $Cl^-(H_2O)_n$ with $n = 1$–4. Their BH algorithm was coupled with DFT calculations implemented in the $DMol^3$ program. For different cluster sizes, they conducted different numbers of independent BH searches ranging from 2 to 5. Every search had 1000 MC steps at 2000 K with the initial structure being generated randomly. At each MC step, the maximum translational distance was 2 Å and the largest rotational displacement was $\pi/2$ for all molecules. For $Cl^-(H_2O)_{1-4}$ clusters, the structures are surface salvation shapes, which is in line with the work of Gora *et al.*[164]

In addition, many groups employed *ab initio* BH to neutral molecular systems such as $CH_3NH_2(H_2O)_{1-7}$,[130] anion clusters ($C_2O_4^{2-}(H_2O)_{1-6}$ and $NO_3^-(H_2O)_{1-7}$[129]), and cation systems such as $PH_3^+(H_2O)_{1-2}$, $H_2S^+(H_2O)_{1-2}$, $HCl^+(H_2O)_{1-2}$,[132] $NH_4^+(H_2O)_{1-10}$,[131] and Na^+He_{2-24}.[133] All these studies demonstrated that *ab initio* based BH can efficiently determine the possible low-lying cluster configurations of different molecular clusters.

3.5 Basin hopping for complex systems

Besides the aforementioned progresses, *ab initio* BH can be applied to many other systems, like ligand-protected clusters, cluster–molecule complexes, and even some chemical reaction. Figure 14 gives the lowest-energy structures of $(CO)_nAu_7^-$ clusters with $n = 0$–4 from the BH search coupled with DFT.[136] Compared to pure Au_7^- cluster, Au_7^- clusters adsorbed with different numbers of CO units presented diverse configurations, as shown in Fig. 14. The flexible geometrical variation makes

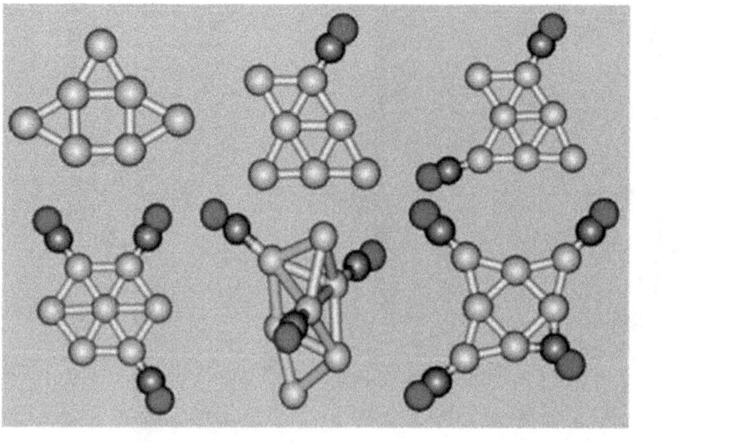

Fig. 14 The lowest-energy structures of Au_7^-, $(CO)Au_7^-$, $(CO)_2Au_7^-$, $(CO)_3Au_7^-$, and $(CO)_4Au_7^-$ optimized using DFT-based BH algorithm. Reproduced with permission from ref. 136. Copyright (2011) American Chemical Society.

Au_7^- cluster to be a unique model to study how small 2D gold clusters change during chemical reactions.

Another brilliant application of *ab initio* BH to complex systems is the series of works by Fortunelli *et al.* for metal clusters adsorbed on the F-center defects MgO(100) surface[138–140,142] and the chemical reaction for propylene partial oxidation by Ag_3 supported on MgO(100)[143] terrace. They used a finite MgO cluster model to represent the F-center defects MgO(100) surface with the atoms of the oxide cluster and the point charges located at the lattice positions of the MgO rock–salt bulk structure. The pure metal or alloy cluster was placed on the F-center of MgO(100) surface, followed by global optimization of the metal cluster by DFT-based BH with the fixed MgO substrate. Later, Fortunelli *et al.*[143] extended DFT-based BH to study the chemical reaction of propylene partial oxidation by Ag_3 supported on F-center MgO using reactive global optimization algorithm based on an eigenvector following internal reaction coordinate method. The DFT-based BH followed a given eigenvector until the TS barrier is overcome instead of changing the configuration randomly. During the reaction path, the O_2 will dissociate from the Ag_3/F/MgO. Then the propene will be adsorbed on the Ag_3O_2/F/MgO system. One reaction path of this process is shown in Fig. 15. With the dissociated O_2 molecule of initial structure, the propylene became epoxide by overcoming 0.2 eV energy barrier. Since the Ag-propylene bond is very weak, the epoxide is desorbed soon, leaving Ag_3O/F/MgO behind.

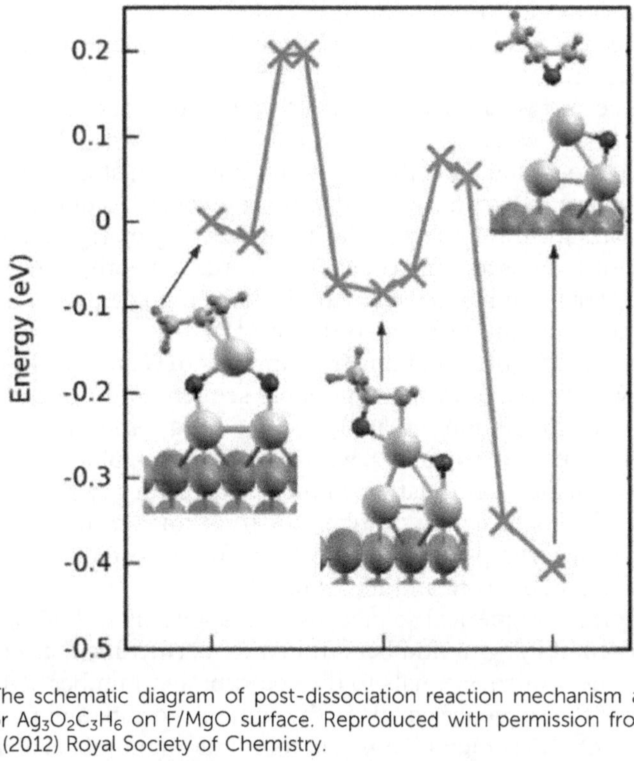

Fig. 15 The schematic diagram of post-dissociation reaction mechanism and energy barriers for $Ag_3O_2C_3H_6$ on F/MgO surface. Reproduced with permission from ref. 143. Copyright (2012) Royal Society of Chemistry.

The above study shows that the combination of *ab initio* method with BH algorithm is not only a powerful method for structure prediction, but also an efficient tool of tracing chemically reaction paths. Nevertheless the application of *ab initio* BH to configuration search of 2D and one-dimensional systems is still limited.

4 Other methods

In addition to the two major approaches described above, there are several other methods or variants, either for generally purpose or designed for specific systems. Here we give an introduction of topological methods, particle swarm optimization and tabu search which have been combined with first-principles calculations recently for global optimization of clusters.

4.1 Topological methods

Topological methods search the ground state structure of a cluster by exploring possible topological connections, which is different from conventional methods on the basis of 3D coordinates of clusters. In this regard, spiral algorithm[165] using the topological relationship was proposed to generate carbon fullerenes. This method has been used widely to study the carbon fullerenes and other clusters with cage configurations.[166–171] Recently, Zhao's group has developed a novel topological method[172] which is appropriate for binary clusters with ionic character and well-defined bonding rules.

The essential idea of the topological methods is to represent topological connection of each cluster conformer by a numerical sequence. As an example, Tang *et al.* described in detail how to establish the correspondence between $(TiO_2)_n$ cluster structures and numerical sequences.[172] First, we assign an integral number i (j) ($1 \leq i \leq 2n$, $1 \leq j \leq n$) to every O (Ti) atom, and label the corresponding O (Ti) atom as O(i) (Ti(j)). Second, for each O atom, we derive its topological sequence by combining the assigned numbers of the Ti atoms (*i.e.* j of Ti(j)) bonded with this O atom in an ascending order of j. Third, the entire sequence of a $(TiO_2)_n$ cluster is obtained by combining the topological sequences of each O(i) atom, in an ascending order of i.[172]

In principle, assignment of integral numbers to O and Ti atoms can be arbitrary, which results in many possible sequences for one equivalent topological configuration of $(TiO_2)_n$ cluster. In order to reduce the number of redundant sequences, we always assign smaller numbers to O atoms with smaller coordination number. Hence, for two O atoms O(i) and O(j), if the coordination number of O(i) is smaller than that of O(j), then $i < j$. As an example, Fig. 16 illustrates the correspondence between a $(TiO_2)_3$ cluster structure and its sequence.[172]

The generated numerical sequences corresponding to $(TiO_2)_n$ clusters are composed of integral numbers from 1 to n. The time that i ($1 \leq i \leq n$) appears in the sequence equals to the coordination number of Ti(i), and the length of the sequence equals to the sum of coordination numbers of all O atoms (or all Ti atoms). Thus, the coordination numbers of all O and

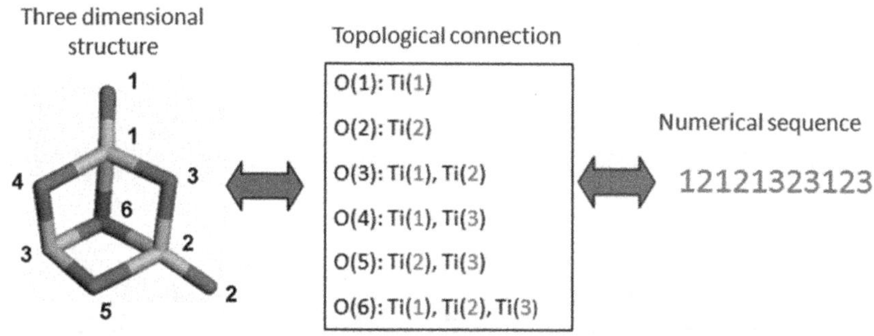

Fig. 16 The correspondence between a $(TiO_2)_3$ cluster structure and its numerical sequence. Reproduced with permission from ref. 172. Copyright (2012) Wiley-VCH Verlag GmbH & Co. KGaA, Weinheim.

Ti atoms have to be determined before sequence generation. Additionally, different sequences may represent one equivalent topological structure due to different ways of encoding and ordering atoms. Therefore, the redundant sequences have to be removed to make sure one topological structure is represented by only one sequence. For a given sequence, all its redundant sequences can be generated by switching the order of O atoms with same coordination numbers (sometimes, the order of bonded Ti atoms changed consequently). They are then removed from the inventory of existing sequences.

Although the procedure of generating numerical sequence described above is for $(TiO_2)_n$ clusters, it can be conveniently applied to any other appropriate binary clusters. Generally, there are four steps to generate the cluster structures using this topological method:

1. Determine the coordination number for each atom in the cluster.
2. Generate numerical sequences. The length and composition of the sequences are determined by the coordination numbers of the atoms in the cluster.
3. Delete redundant sequences to make sure that the correspondence between cluster topological structures and numerical sequences is one to one.
4. Generate the 3D cluster structures from numerical sequences according to the correspondence between them as illustrated in Fig. 16. Some empirical force fields might be utilized to realize this purpose.

Finally, the lowest-energy structures of a given cluster are obtained by optimizing all of the generated structures from the topological method. In principle, this method can explore all the possible isomers of a cluster if the consideration for the possibilities of atomic coordination numbers is comprehensive. Therefore, compared with other global search algorithms, this method makes us more confident to guarantee that the true ground state structures have been identified.

This topological method was first applied to search for the lowest-energy structures of the $(TiO_2)_n$ ($n=1$–6) clusters.[172] Large amounts of isomers were generated, and some new low-energy metastable structures were obtained in addition to the known ground state structures.[172]

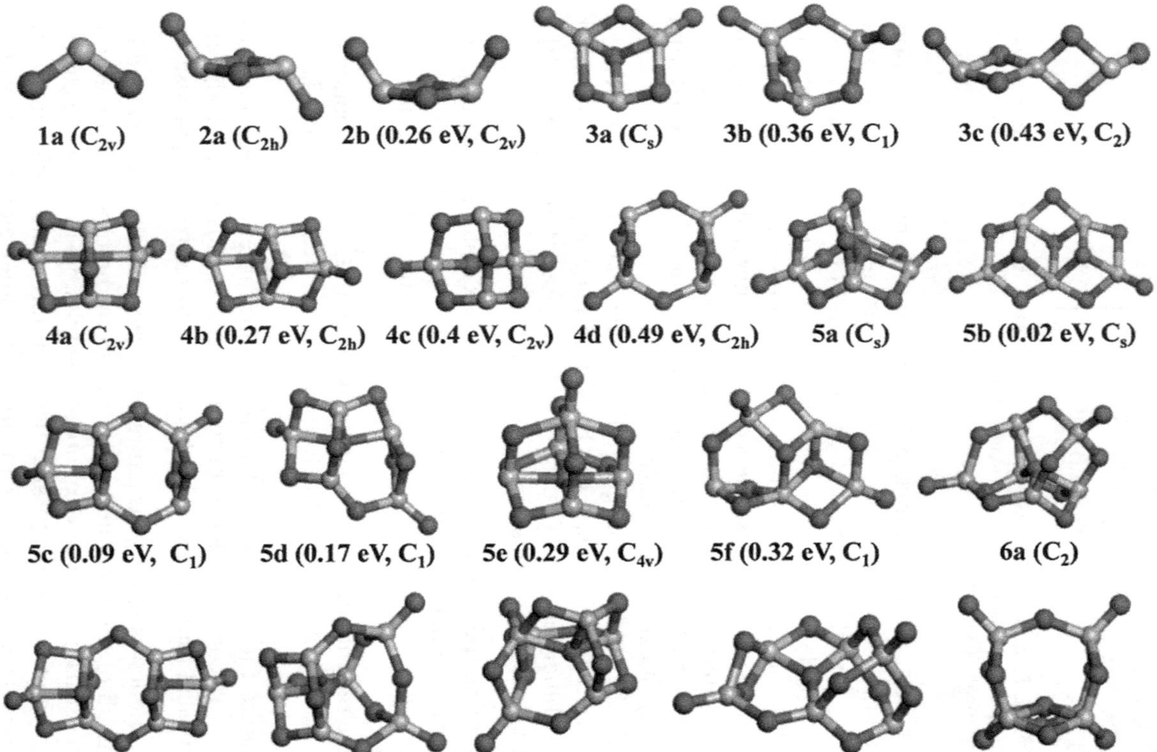

Fig. 17 The ground state structures and some metastable low-energy isomers of $(TiO_2)_n$ ($n = 1$–6) clusters. The point group symmetries and the relative energies (ΔE) with regard to the ground states are given in the parenthesis. Reproduced with permission from ref. 172. Copyright (2012) Wiley-VCH Verlag GmbH & Co. KGaA, Weinheim.

Figure 17 gives the ground state structures and some metastable low-energy isomers of the $(TiO_2)_n$ ($n = 1$–6) clusters. The C_s isomer 5b which is only 0.02 eV less stable than the ground state structure 5a, the C_1 isomer 5c and 5f with $\Delta E = 0.09$ eV and $\Delta E = 0.32$ eV respectively, the C_{2h} isomer 6b ($\Delta E = 0.52$ eV) and the C_s isomer 6d ($\Delta E = 0.72$ eV) are the new structures that have not been reported before by the other methods. Figure 18 depicts the structures, the simulated and experimentally measured photoelectron spectra[173] of the $(TiO_2)_n^-$ ($n = 1$–6) clusters. Clearly, the characteristic features and locations of the peaks in the simulated photoelectron spectra agree very well with the experimental curves.[173]

Later, this method was employed to search for the lowest-energy structures of the ionic $(MgO)_n$ ($n = 2$–7) clusters.[174] Similar to the case of the $(TiO_2)_n$ clusters, the structures of $(MgO)_n$ clusters reported previously were successfully reproduced and some new isomers such as 5b, 7g and 7h (Fig. 19) were obtained. From the acquired isomers, the growth patterns of MgO clusters were observed: the expansion of the 2D ring-like structures, such as the 2a, 3a, 4b and 5e isomers; the combination of basic units like quadrangle and hexagon, for instance, 4a which is a stacking of two quadrangles, 4c constructed by two paralleled quadrangles, and 6a formed by two hexagons; the addition of atoms to small clusters as demonstrated by the growth from 6a to 7b or 7d and from 6b to 7c, 7g or 7h.

The disadvantage of this topological method is that it can be applied only to small sized clusters with dozens of atoms. For larger clusters, this method will generate tremendous amount of isomers that is beyond the current capacity of computer power. At this time, proposing some empirical topological indices which can be used to predict the stabilities of isomers in advance may be helpful to reduce the number of the isomers.

4.2 Particle swarm optimization

Particle swarm optimization (PSO) method[175] is a global optimization strategy inspired by the social behavior of fish schooling or birds flocking to solve multi-agent optimization problems. The system is initialized with a population of random solutions, each solution can be called a particle. Every particle in the population has a velocity to determine its distance and orientation of evolution. The behavior (velocity) of each particle is influenced by the personal best result during itself preceding search and the best result seen by the population or the local best result of partial best particles. PSO method has memory, in which the information of best solution will be kept in all particles. This makes all variables change at the same time, which means PSO method can search the configuration space quickly. In the PSO procedure, the velocity of each particle is limited to a maximum value V_{max}, which is a switch to control the algorithm to be a global exploration (high V_{max}) or a local exploration (low V_{max}).

For cluster science, PSO was developed to identify the global minimum of LJ clusters.[176,177] In the PSO procedure, clusters in the population keep track in the information of best result and itself, which are weighted by

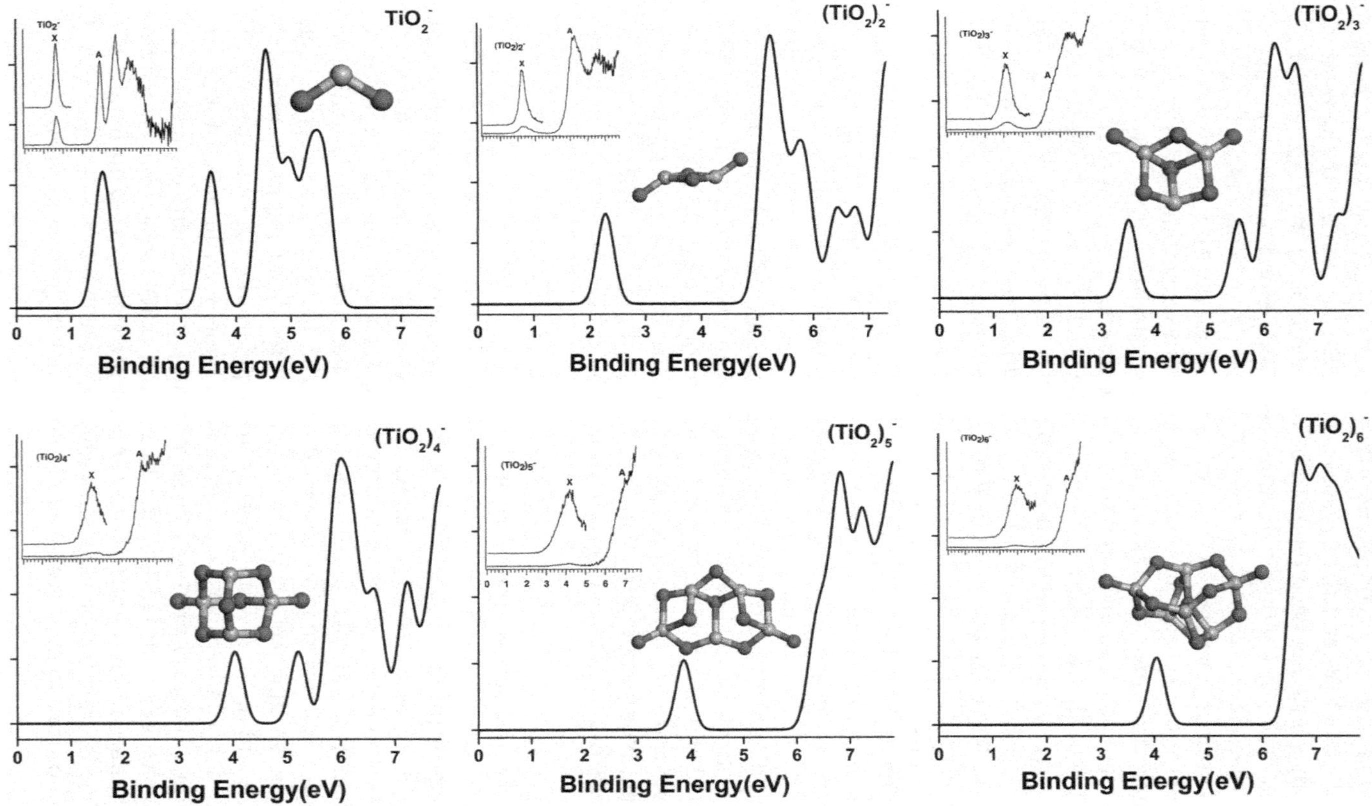

Fig. 18 Atomic structures, simulated and experimentally measured photoelectron spectra of $(TiO_2)_n^-$ ($n = 1$–6) clusters. Reproduced with permission from ref. 172. Copyright (2012) Wiley-VCH Verlag GmbH & Co. KGaA, Weinheim.

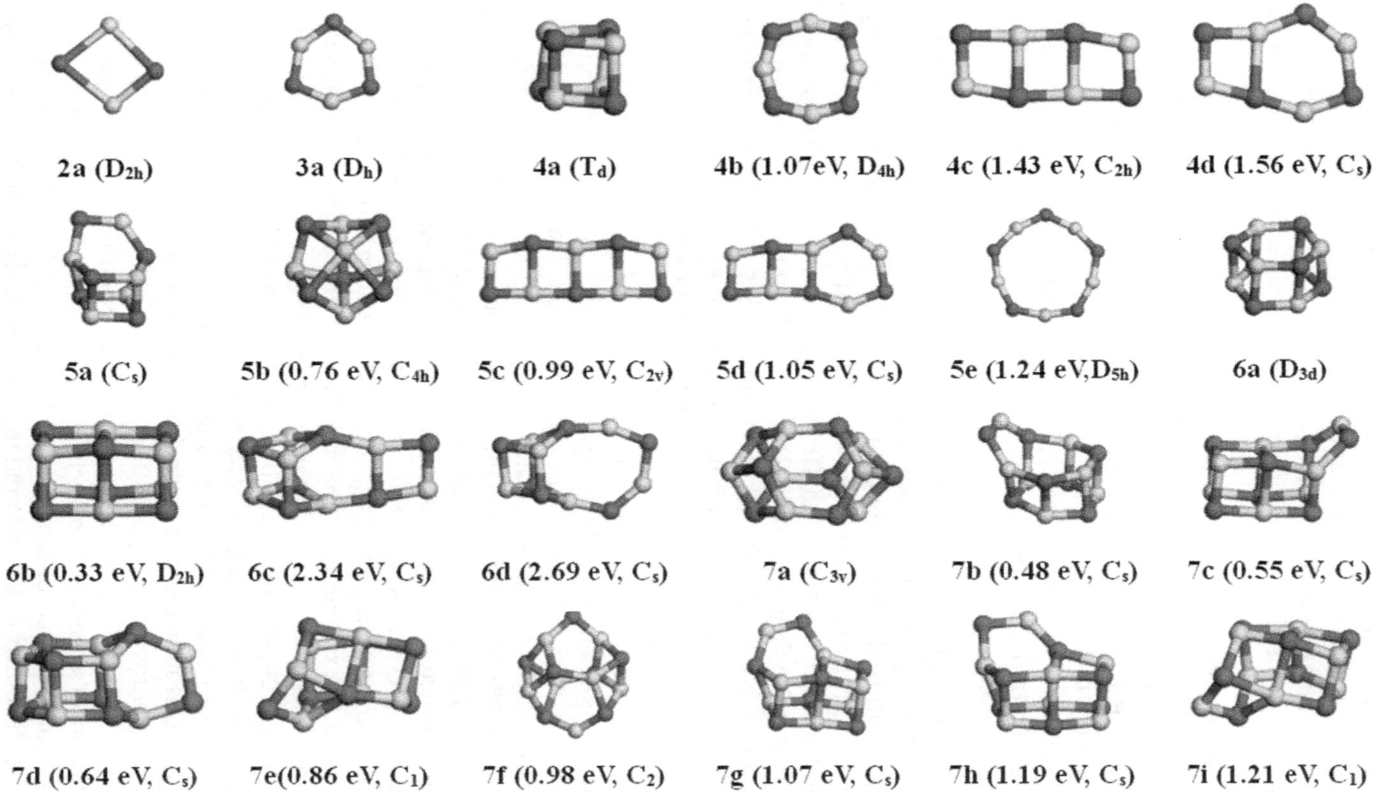

Fig. 19 The ground state structures and some metastable low-energy isomers of $(MgO)_n$ ($n = 2$–7) clusters. The point group symmetries and the relative energies (ΔE) with regard to the ground states are given in the parenthesis. Reproduced with permission from ref. 174. Copyright (2012) Elsevier Ltd.

the energy, geometrical parameters or connectivity descriptors. Five steps to search the PES of clusters by the PSO method are listed below:

1. Initialize the PSO algorithm by generating original population randomly with distance constraints among atoms.
2. A distance matrix of the new generated structure is calculated, which can be used to judge the similarity of new structures with all the previous ones, therefore, eliminate the similar configurations.
3. Geometry optimizations are carried out after all the structures of each population have been generated, which will drive the isomers to the local minima.
4. Like BH, Metropolis criterion is employed to accept or reject the optimized structures.
5. Then PSO operations are applied to produce new structures for the next generation, and the iterative loop is continued until the stop condition is reached.

As the first application of the swarm-intelligence based PSO technique for structure prediction of small clusters or isolated molecules, Call *et al.*[176] utilized PSO method to successfully find the lowest-energy structures of the LJ_{26} cluster, anionic silicon hydride $Si_2H_5^-$ cluster, and triply hydrated hydroxide ion $OH^-(H_2O)_3$ clusters. They demonstrated that this technology required relatively small population sizes and converged fast as compared to GA algorithm. However, their PSO algorithm did not contain the local optimizations, which are critical for structure prediction applications. Thus, PSO is limited to small cluster sizes and failed for larger systems.

In 2010, Ma's group developed a crystal structure analysis by particle swarm optimization methodology, namely the CALYPSO program,[178] which can be implemented in larger systems like crystals and two-dimensional layers systems. Later, Lv *et al.*[177] extended CALYPSO to search the PES of cluster and predicted the ground state of LJ clusters with sizes up to 150 atoms as well as the medium sized Li_n with $n = 20, 40, 58$. Original structure was generated randomly with group point symmetries, which can prevent the appearance of liquid-like clusters effectively. After the new structure is generated, the bond characterization matrix will be calculated to examine the similarity between the newly found structure and previous ones. Ground state configurations of medium sized Li_n clusters ($n = 20, 40, 58$) found by CALYPSO algorithm are verified to be comparable or better than the structures by others.[11,85,179–181] Their results demonstrated that PSO is successful in structure prediction, and sometimes outperforms the other global optimization methods.

Subsequently, many applications of this PSO variant to the cluster systems[177,182–184] emerged. An excellent example is the structure prediction of cage-like boron clusters. B_{80} cluster was proposed to be isoelectronic to that of famous C_{60} fullerene structure.[185] However, the core–shell configuration is found to be more stable than the hollow cage structure for B_{80}[13,186,187] and other large B_n clusters with $n > 68$.[14] CALYPSO combined with *ab initio* calculations find an energetically stable B_{38} fullerene analogue (Fig. 20(a)), which is a hollow cage with

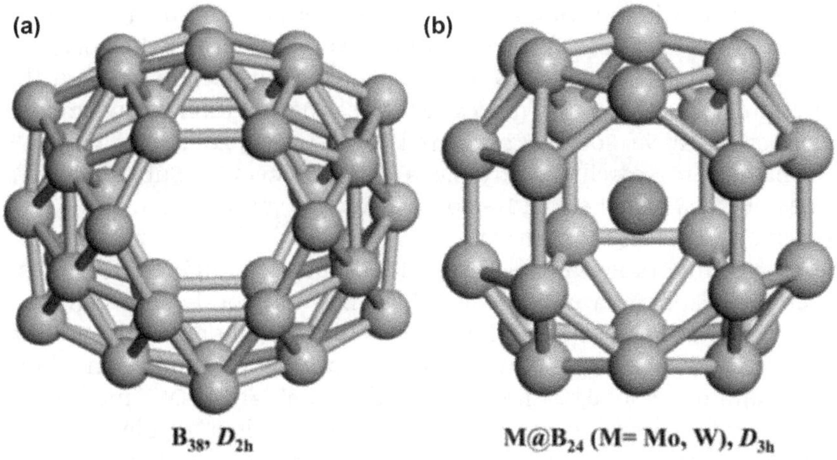

Fig. 20 The cage structures of B_{38} and M@B_{24} (M = Mo, W) clusters found by CALYPSO search. Reproduced with permission from ref. 182, 184. Copyright (2014, 2015) Royal Society of Chemistry.

high symmetry (D_{2h}).[182] Moreover, the large space among the cage-like structure of boron clusters calls for the search of transition metal encapsulated boron cages. Therewith, the same group of authors confirmed the stabilization of transition metal doping (Ti, Zr, Hf, Cr, Mo, W, Fe, Ru and Os) of fullerene-like B_{24} cages *via* first-principle CALYPSO structure prediction.[184] Among the nine TM elements, Mo and W encapsulated B_{24} cages represented highest D_{3h} symmetry, whose configuration is shown in Fig. 20(b). This study indicated that boron clusters with appropriate transition metal element encapsulation will reduce the critical size for the formation of cage configurations.

A similar swarm intelligence global optimization method is the artificial bee colony (ABC) algorithm, which has been proposed by Karaboga and Basturk to optimize multivariable functions.[188] It is based on the intelligent behavior of honey bee swarm, and has recently been applied for global optimization of LJ clusters.[189] ABC was argued to be more efficient than the other global optimization methods due to its ability of escaping from the local minimum.

4.3 Tabu search method

Like particle swarm optimization algorithm, tabu search[190] strategy also contains information about the previous results to avoid exploring the same trial solution more than once. During the structure searching process, the structures found during the search are marked by several descriptors, including the energy at the minimum of the basin, the vibrational frequencies or some geometrical characteristics. In the following iterations, tabu search algorithm can avoid generating structures, which have the same descriptors as previously visited minima. Thus this strategy can efficiently improve the speed of global optimization methods and prevent being trapped in local minimum.

Tabu search in descriptor space (TSDS) method[191] also can generate isomers with favorable geometric structures and interpret the energy of the isomers based on various descriptors. The connection between the descriptors and the energy makes TSDS less computational consumptive and more effective. Fournier *et al.* used TSDS strategy combined with DFT calculations to search the structures of Li clusters.[180] Subsequently a series of TSDS searches for 13-atom clusters of various main-group and transition metal clusters have been conducted.[192–194] The results showed that the configurations of DFT-based TSDS approach differed from those of empirical potential prediction, which had an icosahedron configuration for 13-atom size. Nevertheless, it is difficult to predict the ground state of transition metal clusters due to not only the limitation of the global optimization algorithm but also the reliability of DFT methods.

Another famous tabu search approach is the minima hopping (MH) algorithm, which can be used in conjunction with DFT-based strategy.[195] MH is a very fast global search approach that can find the global minimum of very complex condensed matter systems. This method depends not on the thermodynamic principles but strongly on the type of moves that are used to hop into new local minima. This makes minima hopping efficient and avoiding going to the local minima time after time. The MH method includes two parts, an inner part and an outer part. The inner part jumps into the local minimum of another basin. The most important issue is to have an appropriate kinetic energy that is distributed by the Boltzmann velocity of the atoms. The outer part will accept or reject this new local minimum and the acceptance/rejection is done by simple threshold. In order to guarantee the global minimum, it starts to explore higher energy after the system has explored the low energy configurations regions. In contrast to BH, MH is not a Monte Carlo method which has nothing to do with the temperature, making it to climb out of a "wrong" funnel much faster. It also introduces a feedback mechanism that has a physical counterpart.

Using this efficient MH method combined with DFT calculation, Goedecker *et al.* showed that the energy landscape of boron clusters is glasslike, in contrast to carbon and boron nitride clusters. For the interesting B_{80}, the configurational density of states is shown in Fig. 21. The majority of B_{80} clusters are dome type and the energies of dome-type and fullerene-type structures overlap. The energy differences between the low energy isomers are very small, and the lowest-energy icosahedron-centered core-shell structure found by Goedecker *et al.* is considerably lower in energy than the B_{80} structures proposed by Zhao *et al.*[13] Chen *et al.*[111] and Li *et al.*[109]

Heidari *et al.* have successfully extended MH approach to perform global optimization of small to medium-sized Mg_{10-56} clusters at the DFT level.[196] Their MH iterations started with an educated guess structure. This algorithm controlled the kinetic energy of the molecular dynamics to hop through different energy barriers with low beginning kinetic energy. Analysis shows that the shapes of the clusters vary in a cyclic way from spherical to nonspherical and back again to spherical, which is agreement with the results by others.[197,198] Thus, MH is an outstanding methodology that is able to make accurate predictions on cluster structures.

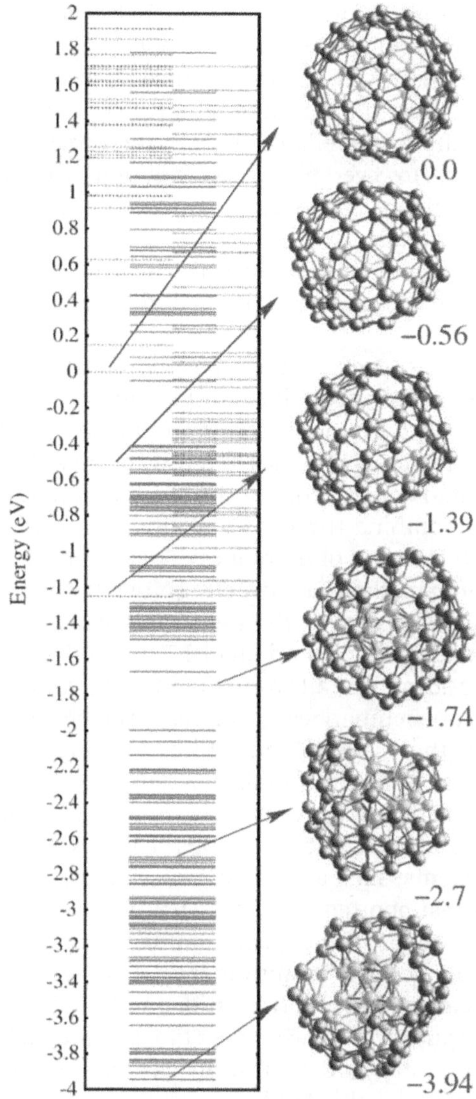

Fig. 21 The configurational density of states of B_{80} clusters. The atoms belonging to the interior B_{12} icosahedra are highlighted in yellow. Reproduced with permission from ref. 186. Copyright (2011) American Physical Society.

5 Summary and perspective

In this chapter, we have presented an overview of the current progresses on global optimization of cluster structures incorporated *ab initio* methods. A number of global search algorithms have been introduced, including genetic algorithm, basin hopping, topological methods, particle swarm optimization, tabu search, and minima hopping. All these methods show remarkable performance in finding the ground state structures of various clusters with different types of chemical bonding, from metallic to covalent, and to van der Waals. The essential idea of this kind of global search methods is to avoid being trapped in local minimum and to explore the entire region of potential energy surface as far as possible. Since the cluster PES is described by first-principles methods (mostly DFT), the global minimum structures come out of the unbiased searches are usually more reliable than those structures from previous empirical searches. However, as a tradeoff of high accuracy, applications of these *ab initio* global optimization methods are limited by the tremendous computational costs. Within the current capability of computer power, the upper limit of cluster size is about 50–60 atoms, depending on the intrinsic nature of chemical bonding and consequent the complexity of the PES.

Generally speaking, each strategy has its own advantages and disadvantages in structure prediction. To guarantee the true global minimum of the PES of a given cluster, it would be desirable to perform several independent global searches and utilize two or more optimization approaches simultaneously. For the future methodology development, one might try to combine the merits of several algorithms in order to overcome the shortcoming of every individual method.

Although *ab initio* methods are usually quite reliable, one still has to be very cautious when dealing with transition metal clusters, molecular clusters, and so on, since conventional DFT methods still contain certain approximations and might fail in distinguishing various structural isomers. Sometimes, benchmark calculations of the performance of various choices of DFT methods are needed before initiating the global search of PES, which consists of a large number of DFT calculations.

In addition to the bare atomic and molecular clusters in the gas phase, the *ab initio* global optimization methods have been recently applied to other relevant systems, such as, ligand-protected metal clusters, reaction processes, surface-supported clusters. Definitely, these pioneer works have extended the scope of the global optimization methods and would provide vital atomistic insights into many other systems and processes. With further developments of optimization methodologies along with the improving of computer speed, we expect the *ab initio* global optimization methods become a standard routine in cluster science and make significant achievements in this field.

Acknowledgements

This work was supported by the National Natural Science Foundation of China (11134005, 11304030).

References

1 W. A. de Heer, *Rev. Mod. Phys.*, 1993, **65**, 611.
2 R. Berry and H. Haberland, *Clusters of Atoms and Molecules I*, Springer-Verlag, Berlin, 1994.
3 J. A. Alonso, *Chem. Rev.*, 2000, **100**, 637.
4 F. Baletto and R. Ferrando, *Rev. Mod. Phys.*, 2005, **77**, 371.
5 R. Ferrando, J. Jellinek and R. L. Johnston, *Chem. Rev.*, 2008, **108**, 845.

6 P. Jena and A. W. Castleman Jr, *Naoclusters: A Bridge Across Disciplines*, Elsevier, Oxford, 2010.
7 L. T. Wille and J. Vennik, *J. Phys. A: Math. Gen.*, 1985, **18**, L419.
8 M. Hoare, *Adv. Chem. Phys.*, 1979, **40**, 49.
9 S. Kirkpatrick, C. D. Gelatt and M. P. Vecchi, *Science*, 1983, **220**, 671.
10 J. Wang, L. Ma, J. Zhao and K. A. Jackson, *J. Chem. Phys.*, 2009, **130**, 214307.
11 Z. Guo, B. Lu, X. Jiang, J. Zhao and R.-H. Xie, *Phys. E*, 2010, **42**, 1755.
12 T. Xue, J. Luo, S. Shen, F. Li and J. Zhao, *Chem. Phys. Lett.*, 2010, **485**, 26.
13 J. Zhao, L. Wang, F. Li and Z. Chen, *J. Phys. Chem. A*, 2010, **114**, 9969.
14 F. Li, P. Jin, D.-e. Jiang, L. Wang, S. B. Zhang, J. Zhao and Z. Chen, *J. Chem. Phys.*, 2012, **136**, 074302.
15 L. Davis, *Genetic Algorithm and Simulated Annealing*, Pitman, London, 1987.
16 J. H. Holland, *Adaption in Natural and Artificial Systems*, Univerisy of Michigan Press, Ann Arbor, 1975.
17 D. E. Goldberg, *Genetic Algorithms in Search, Optimization and Machine Learning*, Addison-Wesley Longman Publishing Co. Inc., Boston, MA, USA, 1989.
18 N. Metropolis, A. W. Rosenbluth, M. N. Rosenbluth, A. H. Teller and E. Teller, *J. Chem. Phys.*, 1953, **21**, 1087.
19 B. Hartke, *J. Phys. Chem.*, 1993, **97**, 9973.
20 Y. Xiao and D. E. Williams, *Chem. Phys. Lett.*, 1993, **215**, 17.
21 H. M. Cartwright, *Applications of Artificial Intelligence in Chemistry*, Oxford University Press, UK, 1993.
22 D. M. Deaven and K. M. Ho, *Phys. Rev. Lett.*, 1995, **75**, 288.
23 B. Hartke, *Chem. Phys. Lett.*, 1995, **240**, 560.
24 Y. Zeiri, *Phys. Rev. E: Stat. Phys., Plasmas, Fluids, Relat. Interdiscip. Top.*, 1995, **51**, R2769.
25 J. Zhao and R. H. Xie, *J. Comput. Theor. Nanosci.*, 2004, **1**, 117.
26 B. Delley, *J. Chem. Phys.*, 1990, **92**, 508.
27 B. Delley, *J. Chem. Phys.*, 2000, **113**, 7756.
28 G. Kresse and J. Furthmüller, *Phys. Rev. B*, 1996, **54**, 11169.
29 L. Sai, L. Tang, J. Zhao, J. Wang and V. Kumar, *J. Chem. Phys.*, 2011, **135**, 184305.
30 A. N. Alexandrova, A. I. Boldyrev, Y. J. Fu, X. Yang, X. B. Wang and L. S. Wang, *J. Chem. Phys.*, 2004, **121**, 5709.
31 R. L. Johnston, *Dalton Trans.*, 2003, **22**, 4193.
32 S. Heiles and R. L. Johnston, *Int. J. Quantum Chem.*, 2013, **113**, 2091.
33 S. Wu, M. Ji, C.-Z. Wang, M. C. Nguyen, X. Zhao, K. Umemoto, R. Wentzcovitch and K.-M. Ho, *J. Phys.: Condens. Matter*, 2014, **26**, 035402.
34 M. Sierka, *Prog. Surf. Sci.*, 2010, **85**, 398.
35 A. N. Alexandrova and A. I. Boldyrev, *J. Chem. Theory Comput.*, 2005, **1**, 566.
36 Y. Zeiri, *Comput. Phys. Commun.*, 1997, **103**, 28.
37 J. Wang, G. Wang and J. Zhao, *J. Phys.: Condens. Matter*, 2001, **13**, L753.
38 J. Zhao, *Phys. Rev. A: At., Mol., Opt. Phys.*, 2001, **64**, 043204.
39 J. Zhao, Y. Luo and G. Wang, *Eur. Phys. J. D*, 2001, **14**, 309.
40 J. Wang, G. Wang and J. Zhao, *Phys. Rev. B: Condens. Matter Mater. Phys.*, 2002, **66**, 035418.
41 J. Wang, G. Wang and J. Zhao, *Phys. Rev. A: At., Mol., Opt. Phys.*, 2003, **68**, 013201.
42 X. Huang, L. Sai, X. Jiang and J. Zhao, *Eur. Phys. J. D*, 2013, **67**, 43.
43 N. Drebov and R. Ahlrichs, *J. Chem. Phys.*, 2010, **132**, 164703.
44 P. Nava and R. Ahlrichs, *Chemistry*, 2008, **14**, 4039.
45 S. Abdalla, M. Springborg and Y. Dong, *Surf. Sci.*, 2013, **608**, 255.

46 X. Liang, X. Huang, Y. Su and J. Zhao, *Chem. Phys. Lett.*, 2015, **634**, 255.
47 B. Assadollahzadeh, P. R. Bunker and P. Schwerdtfeger, *Chem. Phys. Lett.*, 2008, **451**, 262.
48 L. Sai, J. Zhao, X. Huang and J. Wang, *J. Nanosci. Nanotechnol.*, 2012, **12**, 132.
49 N. Drebov, F. Weigend and R. Ahlrichs, *J. Chem. Phys.*, 2011, **135**, 044314.
50 B. Assadollahzadeh, C. Thierfelder and P. Schwerdtfeger, *Phys. Rev. B: Condens. Matter Mater. Phys.*, 2008, **78**, 245423.
51 B. Assadollahzadeh and P. Schwerdtfeger, *J. Chem. Phys.*, 2009, **131**, 064306.
52 A. Shayeghi, D. Gotz, J. B. A. Davis, R. Schafer and R. L. Johnston, *Phys. Chem. Chem. Phys.*, 2015, **17**, 2104.
53 R. Kelting, R. Otterstätter, P. Weis, N. Drebov, R. Ahlrichs and M. M. Kappes, *J. Chem. Phys.*, 2011, **134**, 024311.
54 R. Kelting, A. Baldes, U. Schwarz, T. Rapps, D. Schooss, P. Weis, C. Neiss, F. Weigend and M. M. Kappes, *J. Chem. Phys.*, 2012, **136**, 154309.
55 R. P. F. Kanters and K. J. Donald, *J. Chem. Theory Comput.*, 2014, **10**, 5729.
56 P. C. Jennings and R. L. Johnston, *J. Comput. Theory Chem.*, 2013, **1021**, 91.
57 X. Huang, S.-J. Lu, X. Liang, Y. Su, L. Sai, Z.-G. Zhang, J. Zhao, H.-G. Xu and W. Zheng, *J. Phys. Chem. C*, 2015, **119**, 10987.
58 X. Huang, H.-G. Xu, S. Lu, Y. Su, R. B. King, J. Zhao and W. Zheng, *Nanoscale*, 2014, **6**, 14617.
59 C. J. Heard and R. L. Johnston, *Eur. Phys. J. D*, 2013, **67**, 37.
60 L. Hong, H. Wang, J. Cheng, X. Huang, L. Sai and J. Zhao, *J. Comput. Theory Chem.*, 2012, **993**, 36.
61 S. Heiles, A. J. Logsdail, R. Schafer and R. L. Johnston, *Nanoscale*, 2012, **4**, 1109.
62 X. Huang, Y. Su, L. Sai, J. Zhao and V. Kumar, *J. Cluster Sci.*, 2015, **26**, 389.
63 F. Weigend, *J. Chem. Phys.*, 2014, **141**, 134103.
64 S. Heiles, K. Hofmann, R. L. Johnston and R. Schäfer, *ChemPlusChem*, 2012, **77**, 532.
65 S. Heiles, R. L. Johnston and R. Schäfer, *J. Phys. Chem. A*, 2012, **116**, 7756.
66 X. Huang, J. Zhao, Y. Su, Z. Chen and R. B. King, *Sci. Rep.*, 2014, **4**, 6915.
67 Y. Yuan and L. Cheng, *Int. J. Quantum Chem.*, 2013, **113**, 1264.
68 L. Ren, L. Cheng, Y. Feng and X. Wang, *J. Chem. Phys.*, 2012, **137**, 014309.
69 L. Li and L. Cheng, *J. Chem. Phys.*, 2013, **138**, 094312.
70 K. Kwapien, M. Sierka, J. Döbler, J. Sauer, M. Haertelt, A. Fielicke and G. Meijer, *Angew. Chem.*, 2011, **50**, 1716.
71 M. Haertelt, A. Fielicke, G. Meijer, K. Kwapien, M. Sierka and J. Sauer, *Phys. Chem. Chem. Phys.*, 2012, **14**, 2849.
72 S. Bhattacharya, B. H. Sonin, C. J. Jumonville, L. M. Ghiringhelli and N. Marom, *Phys. Lett. B*, 2015, **91**, 241115.
73 A. M. Burow, T. Wende, M. Sierka, R. Włodarczyk, J. Sauer, P. Claes, L. Jiang, G. Meijer, P. Lievens and K. R. Asmis, *Phys. Chem. Chem. Phys.*, 2011, **13**, 19393.
74 L. Sai, L. Tang, X. Huang, G. Chen, J. Zhao and J. Wang, *Chem. Phys. Lett.*, 2012, **544**, 7.
75 X.-L. Ding, Z.-Y. Li, J.-H. Meng, Y.-X. Zhao and S.-G. He, *J. Chem. Phys.*, 2012, **137**, 214311.
76 A. N. Alexandrova, *J. Phys. Chem. A*, 2010, **114**, 12591.
77 B. Kiran, A. K. Kandalam, J. Xu, Y. H. Ding, M. Sierka, K. H. Bowen and H. Schnöckel, *J. Chem. Phys.*, 2012, **137**, 134303.
78 F. A. Fernandez-Lima, O. P. V. Neto, A. S. Pimentel, M. A. C. Pacheco, C. R. Ponciano, M. A. C. Nascimento and E. F. D. Silveira, *J. Phys. Chem. A*, 2009, **113**, 1813.

79 L. M. Wang, W. Huang, B. B. Averkiev, A. I. Boldyrev and L. S. Wang, *Angew. Chem., Int. Ed.*, 2007, **46**, 4550.
80 M. T. Huynh and A. N. Alexandrova, *J. Phys. Chem. Lett.*, 2011, **2**, 2046.
81 H. Xiang, S.-H. Wei and X. Gong, *J. Am. Chem. Soc.*, 2010, **132**, 7355.
82 L. B. Vilhelmsen and B. Hammer, *J. Chem. Phys.*, 2014, **141**, 044711.
83 R. Bauernschmitt and R. Ahlrichs, *Chem. Phys. Lett.*, 1996, **256**, 454.
84 E. G. Noya, J. P. K. Doye, D. J. Wales and A. Aguado, *Eur. Phys. J. D*, 2007, **43**, 57.
85 A. Aguado and O. Kostko, *J. Chem. Phys.*, 2011, **134**, 164304.
86 J. A. Reyes-Nava, I. L. Garzón, M. R. Beltrán and K. Michaelian, *Rev. Mex. Fís.*, 2002, **48**, 450.
87 S. K. Lai, P. J. Hsu, K. L. Wu, W. K. Liu and M. Iwamatsu, *J. Chem. Phys.*, 2002, **117**, 10715.
88 W. Zheng, *Negative ion photoelectron spectroscopy of metal clusters, metal-organic clusters, metal oxides, and metal-doped silicon clusters*, 2005.
89 X. G. Gong, G. L. Chiarotti, M. Parrinello and E. Tosatti, *Phys. Rev. B: Condens. Matter Mater. Phys.*, 1991, **43**, 14277.
90 G. X. Ge, H. X. Yan, Q. Jing, X. M. Huang, J. G. Wan and G. H. Wang, *Eur. Phys. J. D*, 2013, **67**, 361.
91 Z. Li and H. A. Scheraga, *Proc. Natl. Acad. Sci.*, 1987, **84**, 6611.
92 D. J. Wales and J. P. K. Doye, *J. Phys. Chem. A*, 1997, **101**, 5111.
93 R. Leary, *J. Global Optim.*, 2000, **18**, 367.
94 R. H. Leary and J. P. K. Doye, *Phys. Rev. E: Stat. Phys., Plasmas, Fluids, Relat. Interdiscip. Top.*, 1999, **60**, R6320.
95 M. Iwamatsu and Y. Okabe, *Chem. Phys. Lett.*, 2004, **399**, 396.
96 L. Zhan, J. Z. Y. Chen and W.-K. Liu, *Phys. Rev. E: Stat., Nonlinear, Soft Matter Phys.*, 2006, **73**, 015701.
97 S. Shanker and P. Bandyopadhyay, *J. Phys. Chem. A*, 2011, **115**, 11866.
98 H. Do and N. A. Besley, *J. Chem. Phys.*, 2012, **137**, 134106.
99 Y. Shao, L. F. Molnar, Y. Jung, J. Kussmann, C. Ochsenfeld, S. T. Brown, A. T. B. Gilbert, L. V. Slipchenko, S. V. Levchenko, D. P. O'Neill, R. A. DiStasio Jr, R. C. Lochan, T. Wang, G. J. O. Beran, N. A. Besley, J. M. Herbert, C. Yeh Lin, T. Van Voorhis, S. Hung Chien, A. Sodt, R. P. Steele, V. A. Rassolov, P. E. Maslen, P. P. Korambath, R. D. Adamson, B. Austin, J. Baker, E. F. C. Byrd, H. Dachsel, R. J. Doerksen, A. Dreuw, B. D. Dunietz, A. D. Dutoi, T. R. Furlani, S. R. Gwaltney, A. Heyden, S. Hirata, C.-P. Hsu, G. Kedziora, R. Z. Khalliulin, P. Klunzinger, A. M. Lee, M. S. Lee, W. Liang, I. Lotan, N. Nair, B. Peters, E. I. Proynov, P. A. Pieniazek, Y. Min Rhee, J. Ritchie, E. Rosta, C. David Sherrill, A. C. Simmonett, J. E. Subotnik, H. Lee Woodcock III, W. Zhang, A. T. Bell, A. K. Chakraborty, D. M. Chipman, F. J. Keil, A. Warshel, W. J. Hehre, H. F. Schaefer III, J. Kong, A. I. Krylov, P. M. W. Gill and M. Head-Gordon, *Phys. Chem. Chem. Phys.*, 2006, **8**, 3172.
100 S. Jiang, Y.-R. Liu, T. Huang, H. Wen, K.-M. Xu, W.-X. Zhao, W.-J. Zhang and W. Huang, *J. Comput. Chem.*, 2014, **35**, 159.
101 W. Pullan, *J. Comput. Chem.*, 2005, **26**, 899.
102 G. Rossi and R. Ferrando, *Chem. Phys. Lett.*, 2006, **423**, 17.
103 A. Grosso, M. Locatelli and F. Schoen, *Math. Program.*, 2007, **110**, 373.
104 R. P. White and H. R. Mayne, *Chem. Phys. Lett.*, 1998, **289**, 463.
105 M. Iwamatsu, *Co-evolutionary Global Optimization Algorithm*, IEEE, New York, 2002.
106 W. L. Jorgensen and J. Tirado-Rives, *J. Am. Chem. Soc.*, 1988, **110**, 1657.

107 W. L. Jorgensen, D. S. Maxwell and J. Tirado-Rives, *J. Am. Chem. Soc.*, 1996, **118**, 11225.
108 F. Calvo, S. Tran, S. A. Blundell, C. Guet and F. Spiegelmann, *Phys. Rev. B: Condens. Matter Mater. Phys.*, 2000, **62**, 10394.
109 W.-L. Li, Q. Chen, W.-J. Tian, H. Bai, Y.-F. Zhao, H.-S. Hu, J. Li, H.-J. Zhai, S.-D. Li and L.-S. Wang, *J. Am. Chem. Soc.*, 2014, **136**, 12257.
110 Z. A. Piazza, H.-S. Hu, W.-L. Li, Y.-F. Zhao, J. Li and L.-S. Wang, *Nat. Commun.*, 2014, **5**, 3113.
111 Q. Chen, W.-L. Li, Y.-F. Zhao, S.-Y. Zhang, H.-S. Hu, H. Bai, H.-R. Li, W.-J. Tian, H.-G. Lu, H.-J. Zhai, S.-D. Li, J. Li and L.-S. Wang, *ACS Nano*, 2015, **9**, 754.
112 H. J. Zhai, Y. F. Zhao, W. L. Li, Q. Chen, H. Bai, H. S. Hu, Z. A. Piazza, W. J. Tian, H. G. Lu, Y. B. Wu, Y. W. Mu, G. F. Wei, Z. P. Liu, J. Li, S. D. Li and L. S. Wang, *Nat. Chem.*, 2014, **6**, 727.
113 T. W. Yen and S. K. Lai, *J. Chem. Phys.*, 2015, **142**, 084313.
114 Y. Gao, N. Shao, R. Zhou, G. Zhang and X. C. Zeng, *J. Phys. Chem. Lett.*, 2012, **3**, 2264.
115 A. Aguado, A. Vega and L. C. Balbás, *Phys. Rev. B: Condens. Matter Mater. Phys.*, 2011, **84**, 165450.
116 S. Yoo and X. C. Zeng, *Angew. Chem., Int. Ed.*, 2005, **44**, 1491.
117 J. Bai, L.-F. Cui, J. Wang, S. Yoo, X. Li, J. Jellinek, C. Koehler, T. Frauenheim, L.-S. Wang and X. C. Zeng, *J. Phys. Chem. A*, 2006, **110**, 908.
118 S. Yoo, J. Zhao, J. Wang and X. C. Zeng, *J. Am. Chem. Soc.*, 2004, **126**, 13845.
119 R. Gehrke and K. Reuter, *Phys. Rev. B: Condens. Matter Mater. Phys.*, 2009, **79**, 085412.
120 N. Shao, S. Bulusu and X. C. Zeng, *J. Chem. Phys.*, 2008, **128**, 154326.
121 R. Gehrke, P. Gruene, A. Fielicke, G. Meijer and K. Reuter, *J. Chem. Phys.*, 2009, **130**, 034306.
122 N. Shao, W. Huang, Y. Gao, L.-M. Wang, X. Li, L.-S. Wang and X. C. Zeng, *J. Am. Chem. Soc.*, 2010, **132**, 6596.
123 E. Aprà, R. Ferrando and A. Fortunelli, *Phys. Rev. B: Condens. Matter Mater. Phys.*, 2006, **73**, 205414.
124 D.-e. Jiang and M. Walter, *Phys. Rev. B: Condens. Matter Mater. Phys.*, 2011, **84**, 193402.
125 C. Tang, W. Zhu, K. Zhang, X. He and F. Zhu, *J. Comput. Theory Chem.*, 2014, **1049**, 62.
126 D. E. Jiang, M. Walter and S. Dai, *Chem. – Eur. J.*, 2010, **16**, 4999.
127 R. Pal, L.-M. Wang, W. Huang, L.-S. Wang and X. C. Zeng, *J. Am. Chem. Soc.*, 2009, **131**, 3396.
128 L.-M. Wang, J. Bai, A. Lechtken, W. Huang, D. Schooss, M. M. Kappes, X. C. Zeng and L.-S. Wang, *Phys. Rev. B: Condens. Matter Mater. Phys.*, 2009, **79**, 033413.
129 Y.-R. Liu, H. Wen, T. Huang, X.-X. Lin, Y.-B. Gai, C.-J. Hu, W.-J. Zhang and W. Huang, *J. Phys. Chem. A*, 2014, **118**, 508.
130 S. S. Lv, Y. R. Liu, T. Huang, Y. J. Feng, S. Jiang and W. Huang, *J. Phys. Chem. A*, 2015, **119**, 3770.
131 S.-T. Pei, S. Jiang, Y.-R. Liu, T. Huang, K.-M. Xu, H. Wen, Y.-P. Zhu and W. Huang, *J. Phys. Chem. A*, 2015, **119**, 3035.
132 J. D. Wadey and N. A. Besley, *Chem. Phys. Lett.*, 2014, **601**, 110.
133 N. Issaoui, K. Abdessalem, H. Ghalla, S. J. Yaghmour, F. Calvo and B. Oujia, *J. Chem. Phys.*, 2014, **141**, 174316.

134 D. J. Harding, T. R. Walsh, S. M. Hamilton, W. S. Hopkins, S. R. Mackenzie, P. Gruene, M. Haertelt, G. Meijer and A. Fielicke, *J. Chem. Phys.*, 2010, **132**, 011101.
135 D. J. Harding, P. Gruene, M. Haertelt, G. Meijer, A. Fielicke, S. M. Hamilton, W. S. Hopkins, S. R. Mackenzie, S. P. Neville and T. R. Walsh, *J. Chem. Phys.*, 2010, **133**, 214304.
136 R. Pal, W. Huang, Y.-L. Wang, H.-S. Hu, S. Bulusu, X.-G. Xiong, J. Li, L.-S. Wang and X. C. Zeng, *J. Phys. Chem. Lett.*, 2011, **2**, 2288.
137 R. Pal, L.-M. Wang, Y. Pei, L.-S. Wang and X. C. Zeng, *J. Am. Chem. Soc.*, 2012, **134**, 9438.
138 G. Barcaro and A. Fortunelli, *Phys. Rev. B: Condens. Matter Mater. Phys.*, 2007, **76**, 165412.
139 G. Barcaro, E. Aprà and A. Fortunelli, *Chem. – Eur. J.*, 2007, **13**, 6408.
140 G. Barcaro and A. Fortunelli, *J. Phys. Chem. C*, 2007, **111**, 11384.
141 G. Barcaro and A. Fortunelli, *Faraday Discuss.*, 2008, **138**, 37.
142 G. Barcaro and A. Fortunelli, *Chem. Phys. Lett.*, 2008, **457**, 143.
143 F. R. Negreiros, E. Apra, G. Barcaro, L. Sementa, S. Vajda and A. Fortunelli, *Nanoscale*, 2012, **4**, 1208.
144 C. Shang and Z.-P. Liu, *J. Chem. Theory Comput.*, 2013, **9**, 1838.
145 J. Li, X. Li, H.-J. Zhai and L.-S. Wang, *Science*, 2003, **299**, 864.
146 A. Szabo and N. S. Ostlund, *Modern Quantum Chemistry: Introduction to Advanced Electronic Structure Theory*, Dover Publications, 2012.
147 K.-M. Ho, A. A. Shvartsburg, B. Pan, Z.-Y. Lu, C.-Z. Wang, J. G. Wacker, J. L. Fye and M. F. Jarrold, *Nature*, 1998, **392**, 582.
148 B. Liu, Z.-Y. Lu, B. Pan, C.-Z. Wang, K.-M. Ho, A. A. Shvartsburg and M. F. Jarrold, *J. Chem. Phys.*, 1998, **109**, 9401.
149 I. Rata, A. A. Shvartsburg, M. Horoi, T. Frauenheim, K. W. M. Siu and K. A. Jackson, *Phys. Rev. Lett.*, 2000, **85**, 546.
150 B.-X. Li, P.-L. Cao and S.-C. Zhan, *Phys. Lett. A*, 2003, **316**, 252.
151 S. Yoo, X. C. Zeng, X. Zhu and J. Bai, *J. Am. Chem. Soc.*, 2003, **125**, 13318.
152 A. Tekin and B. Hartke, *Phys. Chem. Chem. Phys.*, 2004, **6**, 503.
153 K. A. Jackson, M. Horoi, I. Chaudhuri, T. Frauenheim and A. A. Shvartsburg, *Phys. Rev. Lett.*, 2004, **93**, 013401.
154 S. Grimme, J. Antony, T. Schwabe and C. Muck-Lichtenfeld, *Org. Biomol. Chem.*, 2007, **5**, 741.
155 T. Schwabe and S. Grimme, *Acc. Chem. Res.*, 2008, **41**, 569.
156 M. P. Hodges and D. J. Wales, *Chem. Phys. Lett.*, 2000, **324**, 279.
157 D. J. Wales and M. P. Hodges, *Chem. Phys. Lett.*, 1998, **286**, 65.
158 F. Li, L. Wang, J. Zhao, J.-H. Xie, K. Riley and Z. Chen, *Theor. Chem. Acc.*, 2011, **130**, 341.
159 O. Matsuoka, E. Clementi and M. Yoshimine, *J. Chem. Phys.*, 1976, **64**, 1351.
160 C. J. Gruenloh, J. R. Carney, C. A. Arrington, T. S. Zwier, S. Y. Fredericks and K. D. Jordan, *Science*, 1997, **276**, 1678.
161 U. Buck, I. Ettischer, M. Melzer, V. Buch and J. Sadlej, *Phys. Rev. Lett.*, 1998, **80**, 2578.
162 R. M. Shields, B. Temelso, K. A. Archer, T. E. Morrell and G. C. Shields, *J. Phys. Chem. A*, 2010, **114**, 11725.
163 C. Pérez, M. T. Muckle, D. P. Zaleski, N. A. Seifert, B. Temelso, G. C. Shields, Z. Kisiel and B. H. Pate, *Science*, 2012, **336**, 897.
164 R. W. Gora, S. Roszak and J. Leszczynski, *Chem. Phys. Lett.*, 2000, **325**, 7.

165 P. Fowler and D. E. Manolopoulos, *An Atlas of Fullerenes*, Oxford University Press, New York, 1995.
166 P. W. Fowler, T. Heine, D. E. Manolopoulos, D. Mitchell, G. Orlandi, R. Schmidt, G. Seifert and F. Zerbetto, *J. Phys. Chem.*, 1996, **100**, 6984.
167 E. Hernández, P. Ordejón and H. Terrones, *Phys. Rev. B: Condens. Matter Mater. Phys.*, 2001, **63**, 193403.
168 E. Albertazzi, C. Domene, P. W. Fowler, T. Heine, G. Seifert, C. Van Alsenoy and F. Zerbetto, *Phys. Chem. Chem. Phys.*, 1999, **1**, 2913.
169 Y. Achiba, P. W. Fowler, D. Mitchell and F. Zerbetto, *J. Phys. Chem. A*, 1998, **102**, 6835.
170 L. Tang, L. Sai, J. Zhao and R. Qiu, *J. Comput. Theory Chem.*, 2011, **969**, 35.
171 L. Tang, Y. Su, Y. Liu, J. Zhao and R. Qiu, *J. Chem. Phys.*, 2012, **136**, 224508.
172 L. Tang, L. Sai, J. Zhao and R. Qiu, *J. Comput. Chem.*, 2012, **33**, 163.
173 H.-J. Zhai and L.-S. Wang, *J. Am. Chem. Soc.*, 2007, **129**, 3022.
174 L. Hong, H. Wang, J. Cheng, L. Tang and J. Zhao, *J. Comput. Theory Chem.*, 2012, **980**, 62.
175 R. Eberhart and J. Kennedy, *A New Optimizer Using Particle Swarm Theory*, IEEE, New York, 1995.
176 S. T. Call, D. Y. Zubarev and A. I. Boldyrev, *J. Comput. Chem.*, 2007, **28**, 1177.
177 J. Lv, Y. Wang, L. Zhu and Y. Ma, *J. Chem. Phys.*, 2012, **137**, 084104.
178 Y. Wang, J. Lv, L. Zhu and Y. Ma, *Phys. Rev. B: Condens. Matter Mater. Phys.*, 2010, **82**, 094116.
179 G. Gardet, F. Rogemond and H. Chermette, *J. Chem. Phys.*, 1996, **105**, 9933.
180 R. Fournier, J. B. Y. Cheng and A. Wong, *J. Chem. Phys.*, 2003, **119**, 9444.
181 J. Centeno and P. Fuentealba, *Int. J. Quantum Chem.*, 2011, **111**, 1419.
182 J. Lv, Y. Wang, L. Zhu and Y. Ma, *Nanoscale*, 2014, **6**, 11692.
183 M. Ju, J. Lv, X.-Y. Kuang, L.-P. Ding, C. Lu, J.-J. Wang, Y.-Y. Jin and G. Maroulis, *RSC Adv.*, 2015, **5**, 6560.
184 J. Lv, Y. Wang, L. Zhang, H. Lin, J. Zhao and Y. Ma, *Nanoscale*, 2015, **7**, 10482.
185 N. G. Szwacki, A. Sadrzadeh and B. I. Yakobson, *Phys. Rev. Lett.*, 2008, **100**, 159901.
186 S. De, A. Willand, M. Amsler, P. Pochet, L. Genovese and S. Goedecker, *Phys. Rev. Lett.*, 2011, **106**, 225502.
187 H. Li, N. Shao, B. Shang, L.-F. Yuan, J. Yang and X. C. Zeng, *Chem. Commun.*, 2010, **46**, 3878.
188 D. Karaboga and B. Basturk, *J. Global Optim.*, 2007, **39**, 459.
189 C. Wehmeyer, G. Falk von Rudorff, S. Wolf, G. Kabbe, D. Schärf, T. D. Kühne and D. Sebastiani, *J. Chem. Phys.*, 2012, **137**, 194110.
190 F. Glover, *ORSA J. Comput.*, 1989, **1**, 190.
191 J. Cheng and R. Fournier, *Theor. Chem. Acc.*, 2004, **112**, 7.
192 Y. Sun, M. Zhang and R. Fournier, *Phys. Rev. B: Condens. Matter Mater. Phys.*, 2008, **77**, 075435.
193 M. Zhang and R. Fournier, *Phys. Rev. A: At., Mol., Opt. Phys.*, 2009, **79**, 043203.
194 Y. Sun, R. Fournier and M. Zhang, *Phys. Rev. A: At., Mol., Opt. Phys.*, 2009, **79**, 043202.
195 S. Goedecker, *J. Chem. Phys.*, 2004, **120**, 9911.
196 I. Heidari, S. De, S. M. Ghazi, S. Goedecker and D. G. Kanhere, *J. Phys. Chem. A*, 2011, **115**, 12307.
197 T. Diederich, T. Döppner, J. Braune, J. Tiggesbäumker and K.-H. Meiwes-Broer, *Phys. Rev. Lett.*, 2001, **86**, 4807.
198 S. M. Ghazi, S. Zorriasatein and D. G. Kanhere, *J. Phys. Chem. A*, 2009, **113**, 2659.

Nitrogen- and phosphine-binding ligands in interaction with gold atoms, clusters, nanoparticles and surfaces

Doreen Mollenhauer

DOI: 10.1039/9781782622703-00293

1 Introduction

Gold clusters and gold nanoparticles have been shown to have very interesting and promising electronic, optical and catalytic properties.[1–3] These properties are very useful for several applications in diverse fields as for instance in biomedicine (*e.g.* sensing and labelling), nanoelectronics, photonics, sensing and catalysis.[1–4] Whereas the term cluster describes a particle with a well-defined number of atoms which usually is smaller than 3 nm in size, the term nanoparticle covers all particle sizes, but a particle larger than 1 nm in size up to 100 nm is meant. The number of atoms plays no crucial role for the nanoparticle, but instead the nanoparticle shape, for example visualised by a Wulff construction which determines the crystal shape in the thermodynamic equilibrium. The crystal shape is constructed in such a way that the free energy of the surface is minimised by having a constant volume.

Small clusters exhibit different properties compared to their larger counterparts. Clusters sized below 5 nm (and in particular below 3 nm) have been found to catalyse various reactions.[3,5] Thereby gold cluster are of special importance on the one hand for low-temperature catalytic reactions, *e.g.* the CO oxidation at room temperature.[6,7] On the other hand, gold cluster reveal a high selectivity of reactions such as oxidation of benzyl alcohol to benzaldehyde or epoxidation of styrene.[8–13] Furthermore, absorption spectra of small clusters exhibit multiple peaks in contrast to larger nanoparticles.[14] Due to the discrete electronic transition which can be assigned to a specified character, one can distinguish between different cluster sizes.[15] The molecular electronic structure and presence of a HOMO-LUMO gap lead to unique luminescence and two-photon absorption properties.[16–22] This makes small gold clusters interesting for the assembly of nano-architectures with specific optical properties. Also, small gold clusters have been used in intercluster compounds with fullerides or polyoxometalates, which are interesting building blocks for future optic or electronic devises and model systems to understand self-assembly at the nanoscale.[23–25]

Larger gold nanoparticles (>3 nm) are interesting for bio- and nanomedicine. For example, gold nanoparticles have been used to tag proteins or bio-conjugates for electron spectroscopy or to treat rheumatoid arthritis.[26] Bronchial asthma or malaria are also diseases to be treated

Justus-Liebig-Universität Gießen, Physikalisch-Chemisches Institut, Heinrich-Buff-Ring 58, Giessen, Germany. E-mail: doreen.mollenhauer@phys.chemie.uni-giessen.de

potentially by gold nanoparticles.[26] Highly promising are the applications of gold nanoparticles in photodynamic therapy, biosynthesis or as sensing elements for DNA, sugars or bio-conjugates such as peptides, lipids, enzymes, drugs and viruses.[4,27] Also, they are discussed as high potential treatment against cancer, for example for anti-cancer drug delivery or targeting cells.[28,29] Regarding optical properties, nanoparticles of the size 5–20 nm reveal a single, sharp plasmon resonance peak around 530 nm. This is the result of catched free and mobile electrons in the nanoparticle (quantum-size effect) which show a collective oscillator frequency of the plasmon resonance.[1] On the basis of reflection of the surface plasmon resonance, a deeply red colour results. The size-dependent applications of functionalised gold nanoparticles in catalysis and biomedicine reveal the strong variation in reactivity along different nanoparticles sizes.

Gold clusters and nanoparticles are stabilised by ligands in solution. Typical ligands for small clusters include thiolate and phosphine groups as well as halides.[30,8,31,32] Larger gold nanoparticles usually are stabilised by thiolates, citrates and polymers, but also by pyridine derivatives, amines and bioconjugates.[1] Examples of ligand-passivated gold clusters and nanoparticles can be found in Fig. 1a and b. The common synthesis combines a gold salt and the ligands with a strong reducing agent (usually $NaBH_4$) in chloroform or dichloromethane. Thereby, it is possible to control and isolate particles of a particular size and certain stoichiometry.[14,15,33] Due to the high surface area of clusters and nanoparticles, the influence of the protecting ligands to the properties of the nanosystems is crucial. They influence the electronic, optical and catalytic properties, but also the solubility or the aggregation behaviour. Thus, the understanding of the ligand–particle interaction at the interfaces is essential for controlling or fine-tuning various properties of nanosystems. Besides nanomaterials, some of these ligands have also importance in gold complex chemistry. Thereby, gold complexes can offer the first step on the understanding of the ligand–metal bond that is important for ligand–particle interactions. Furthermore, there is high interest of ligand–surface interaction in the formation of monolayers on gold for the use in electronic and optoelectronic devices.[34–36] The obtained information about the molecule–surface interaction regarding

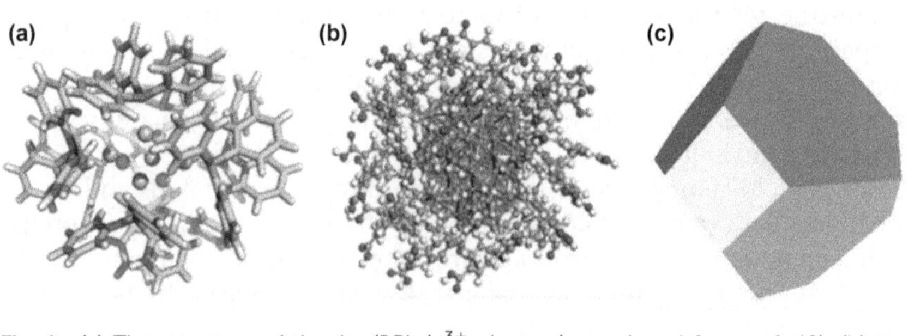

Fig. 1 (a) The structure of the $Au_9(PPh_3)_8{}^{3+}$ cluster (reproduced from ref. 42), (b) the Au_{102}(p-MBA)$_{44}$ cluster (reproduced from ref. 43) and (c) Wulff construction of a nanoparticle in the size range of 3 to 100 nm based on surface energies of ref. 44 are shown.

binding modes and coverage effects are helpful for understanding the interaction to gold nanoparticle surfaces as well.

One special focus in experimental and theoretical work lies on gold–sulphur nanosystems. The gold–sulphur interaction is the most studied organo/gold system.[37–39] The binding between thiolates and the gold system is strong in nature and in the binding energy range of the gold–gold bond. One issue in the use of thiolates is the difficulty to control monolayer composition at the nanoparticle. Also, irreversible functionalization is possible, which complicates the use for further applications.[40] In contrast, weakly binding ligands have the advantage of not necessarily functionalising the nanoparticles, so that the ligand shell or monolayer at surfaces can be removed. Thus, it can serve as a gold nanoparticle precursor and recycling strategies might be developed.[40,41] In this review, the focus will be limited to weak-binding ligands,[†] which undergo nitrogen–gold or phosphine–gold bindings. Therefore, for each case one example ligand has been chosen. Thus, on the one hand the phosphine ligand PR_3 with P in sp^3 hybridization is used, and on the other hand pyridine NC_5H_5 with N in sp^2 hybridization.

Phosphine ligands are used for gold clusters with special relevance in catalysis, in intercluster compounds or due to their optical properties. Several X-ray structures of the phosphine-passivated gold clusters are present, whereas the series of crystal structures has started with the early work of Malatesta in 1969.[45] Mingos stated already in 1984 that the metal clusters may have alternative structures and are separated just by small energy differences.[46] Thus, several small gold clusters can be generated. Special stability is noticed for magic numbers of clusters up to 150 atoms and explained by shell closing of delocalised electrons assigned to angular momentum orbitals over the entire cluster system. The concept is called super-atomic complex. Here, mainly thiolate-protecting gold clusters have been assessed, but also a few phosphine-stabilised or mixed phosphine-halide stabilised ones. However, to cluster stability contribute on the one hand the geometric structure, and on the other hand the electronic structure.

Actually, pyridine derivatives are used to stabilize gold nanoparticles with sizes larger than 4 nm.[40,41,47–49] Based on several experimental studies the orientation of the molecule at the surface is perpendicular *via* the nitrogen *e.g.* for pyridine and 4-dimethylaminopyridine or flat for 4-methoxypyridine. This demonstrates the interesting feature of pyridine and its derivatives which can bind to surfaces generally *via* two different binding modes. In both cases the binding is assumed to be weak physisorption, for pyridine on *e.g.* Au(111) the ΔG was estimated to -10 kcal mol^{-1}.[50] Thus, the ligand can be easily removed from surfaces and nanoparticles. Pyridine and pyridine derivatives are studied on surfaces and nanoparticles as they are interesting molecular building blocks for organo-electronics or host-guest networks.[34,35] Furthermore, the building block pyridine has relevance as model system for medical

[†]Ligands which are discussed to bind weak to larger gold nanoparticle systems.

applications.[51] It is the most common heterocyclic framework of pharmaceutics.[51] Also, one can find the motive in many enzymes of living organism or vitamins.[51]

In addition to experimental techniques a great insight into binding modes and fundamental questions of ligands interacting with gold complexes, clusters, nanoparticles and surfaces have been allowed by electronic structure calculations. Reviews of theoretical studies in the field of gold complexes, clusters, nanoparticles and surfaces are for example ref. 52–59. Furthermore, there are reviews in the field of small passivated gold clusters[30] and for the gold–sulphur interface.[37] Especially density functional theory (DFT) or related to optical properties time-dependent DFT (TD-DFT) has been employed. The theoretical methods allow insight into geometric and electronic structure as well as charge distribution or binding nature and strength. The general understanding of the origin of stability of clusters and nanoparticles and binding modes between the ligand and the gold system as well as for example growth mechanism achieve a much deeper view inside the chemistry and physics of the nanosystems. Also, experimental data such as STM images or absorption spectra can be simulated. This allows a better interpretation and understanding of experimental work and opens up design and tuning possibilities.

This chapter focuses on phosphine and pyridine derivative-gold systems and their related investigation in recent years. Furthermore, studies will be included which have led to the present understanding. The viewpoint of this review is on theoretical work (quantum mechanics) which has been done in the field. Thereby, we will discuss theoretical work on phosphine and pyridine derivatives in complexes to gain better understanding of binding situation and origin of optical properties. Theoretical studies of ligand-stabilised clusters are considered and compared to related experiments. Here, of special importance is the understanding of the influence of the ligand on the cluster and how this can be modified by ligand variation. Further, the interpretation of stability of geometric, electronic or superatom complex structure will be discussed in context. Ligand-passivated gold nanoparticles and surfaces will be analysed regarding binding modes and relevant properties by electronic structure calculations. Relevant experiments will be briefly discussed if necessary because the interplay between experiments and theory gains deepest insight into the phosphor- and nitrogen-based gold interfaces. The related methology of the computations will be discussed, too. It is known that the size of particles in field of nanotechnology determines the properties and can vary much over certain particle sizes. Following up, the investigations of differently sized gold systems (atoms to surfaces) will be looked at in context. Similarities and differences of the considered systems will be pointed out in a size dependent manner.

2 Computational approaches

Up to now, electronic structure calculations have been used to describe relatively small clusters or relatively large nanoparticles surrounded by a ligand shell. The reason therefore lies in the difficulty to describe the size-dependency of such ligand-surrounded metallic nanoparticle systems with current methods. On the one hand, quantum chemical calculations using the overall cluster-ligand system have been performed for small systems. Thereby, electronic structure methods such as DFT support the understanding of the structure, reactivity and properties of gold clusters coated with phosphine or thiol ligands.[8,30,31,42,60] On the other hand, quantum chemical methods with periodic-boundary constraints are used to simulate surface facets of larger nanoparticles.[37,41,61–64] The shape of gold nanoparticles can be estimated for the region of 3 to 100 nm by help of the free energies of different surfaces by a Wulff construction (see Fig. 1c). As the large nanoparticles exhibit surface facets with the Au(111) surface as the most stable and representative one, the molecule–surface interaction can serve as a model for calculations to understand ligand stabilisation of larger nanoparticles.

For the treatment of gold atoms, one has to consider that relativistic effects are of high importance, because the electrons which are close to the nucleus can have velocities close to speed of light.[65] Due to the relativistic perturbation near the nucleus, the inner core shells contract (direct relativistic effect) which lead to a stronger screening of the nucleus. The screening causes a reduction of Z_{eff} (effective nuclear charge) which induces an energetic destabilization and expansion of all orbitals (indirect relativistic effect). Due to this interplay between direct and indirect relativistic effects s orbitals get contracted and d/f orbitals expanded.[66] Gold is the element which exhibit strongest relativistic effects up to $Z = 100$ and the physical and chemical properties of gold systems are determined by relativity. Following, a reliable description of the relativistic effects in the quantum chemical calculations is crucial.

Calculations of valence properties of molecular systems which contain gold atoms are most commonly done by replacing core orbitals by pseudopotentials (PP) and effective core potential (ECP). The term ECP is used more often regarding molecular systems and PP regarding the solid state, but the basic idea is the same. Applying ECPs/PPs leads to large computational savings. Often employed in finite electronic structure calculations to treat relativistic effects of gold atoms are the scalar-relativistic ECPs of Stuttgart group and Stuttgart/Cologne group.[67–69] The number of core electrons for each ECP is 60 electrons and the semi-local pseudopotentials are adjusted for the reproduction of the atomic valence–energy spectra (multi-fit). The reference data are quasi-relativistic or fully relativistic ones. Calculations with ECPs which contain 78 electrons in core have been shown to give no reliable results.[70] The ECP is mostly combined with the corresponding Dunning type or Pople basis sets for describing the valence electrons of the gold atom. An often used combination of an ECP and basis set is called LANL2DZ.[71] However, other approaches which are sometimes applied in order to describe relativity of gold atoms are the Douglas–Kroll–Hess (DKH) theory or zeroth-order regular approximation (ZORA) equations.[72–77] Periodic approaches most commonly use scalar relativistic pseudo potentials

(68 electron core), such as ultra-soft potentials or projected augmented wave generated potentials combined with plane wave basis sets.[78–81]

Most investigations which employ wave-function-based methods such as MP2, MP2 + SQM, SCS-MP2 or CCSD(T) study very small systems and are benchmark studies.[70,82–84] Thereby, the wave-function-based methods are used to determine the performance of DFT and to find DFT functionals which can accurately describe the system under consideration. The MP2 calculations have been used for small clusters, CCSD(T) method up to the gold dimer. Benchmark calculations have been done for the weakly binding ligands phosphine and pyridine in interaction to gold systems, because the DFT functional needs to describe a metallic as well as molecular system and in addition the weak interaction between the systems. Various DFT functionals have been used for finite systems ranging from LDA, GGA, meta-GGA, hybrid and double-hybrid functionals to long-range corrected and van der Waals exchange-correlation functionals. The periodic calculations are mostly limited to LDA and GGA functionals. In order to study optical properties of gold ions and clusters interacting with ligands, usually TD-DFT calculations are performed.

For the description of weak interactions such as pyridine and phosphine gold interactions, it is important to consider that standard DFT fails to correctly account for dispersion effects in the medium and long range. In order to account for dispersion effects, dispersion-correction approaches or improved functionals have to be employed. The DFT-D approach adds an empirical correction term on the basis of distance-dependent interatomic potential functions to the DFT energy (D2 and D3 version).[85,86] Within the D2 version the pairwise damped C_6R^{-6} potentials are of the order 6, the dispersion coefficients are especially determined up to the third main group, the ones for transition metals are poorly determined. Tonigold *et al.* determined the C_6 coefficients for gold interactions by two different approaches, namely first based on atomic properties by using metal–ion containing complexes and second by using a hybrid QM:QM approach of small metal clusters which was suggested by Tuma and Sauer.[87,88] The extended D3 correction up to element 94 by Grimme is constituted by the two-body energy or two- and three-body energy terms with potentials of orders 6 or higher (8, 10). The non-additive three-body energy E(3) as a correction for interactions between three ground-state atoms plays a role especially in larger systems. For dispersion-dominated interaction to gold surfaces, the dispersion correction determines nearly the total value of, *e.g.*, the interaction energy. It has been shown that the C_6 coefficients based on atomic properties by Tonigold within the D2 correction as well as the D3 correction by Grimme lead to reliable results regarding aromatic molecule-Au(111) interactions.[88]

Modified functionals, which are generated to account for dispersion effects, have been also utilised for gold systems. Double hybrid functionals, which include MP2 equations to calculate a part of the correlation energy, have been shown to give accurate results for gold phosphine complexes and dimers, if additionally a D3 correction is applied.[60] Regarding the use of hybrid functionals in combination with the D2/D3 correction for gold complexes, Koppen *et al.* claimed that there is no guarantee that for donor–acceptor interactions the HOMO(D) and LUMO(A) gap will be correctly described and range-separated functionals reveal a general better route.[89] Thus, the authors tuned the monomer properties of range-separated functionals in a two-parameter process to find an optimal functional for acceptor and donor; they called it LRC-ωPBEα.[89] Benchmark calculations on pyridine-Au_4 systems and others have been determined to be in good agreement to CCSD(T) reference values.[89]

It has been demonstrated that the highly parameterized meta-GGA functional M06-L is able to treat weak dispersive effects compared to standard DFT for several systems.[90] Furthermore, van der Waals exchange-correlation functionals, namely the vdW-DF2 and vdW-DF1 and KBM which has been developed by Klimes *et al.* in combination with localised basis sets on the basis on short-ranged pseudo-atomic orbitals have been employed to the ligand–gold interaction and obtained reasonable adsorption structure and energies.[91]

3 Ligand binding in gold complexes

3.1 Phosphine gold fragments and complexes

The chemistry of gold at the nano-scale looks much different to the chemically inert gold bulk and is more complex. Whereas gold is known in oxidation states of −I to +V, the common ones are +I (linear coordination), +III (quadratic planar) and +V (octahedral).[37] However, in order to understand the gold–phosphine interaction related to ligand-stabilised gold complexes and clusters, first of all the basic interaction within the neutral fragment $AuPR_3$ is considered more precise. In methological terms, the inclusion of electron correlation has been found for the $AuPH_3$ system to be essential to describe the binding energy, otherwise the binding is repulsive at HF-level of theory (see Fig. 2a).[60,92,93] This indicates a weak dispersion–dominated interaction.

Benchmark studies using CCSD(T) at the basis set limit have been performed for the smallest phosphine PH_3 in order to determine a DFT functional which describes the weak interaction sufficient. One study determines PBE0 and TPSSh (error smaller than 9%) among others with the best match to the benchmark value.[93] Another investigation has observed best agreement regarding bond length for PBE-D3 and BP86-D3.[60] The binding energy was descripted within a variance of $+9/-8$ kcal mol^{-1} considering all tested density functionals. Some functionals give results within 1 kcal mol^{-1} variance such as PBE0, B3P86, B2PLYP-D and PBErev-D3. In accordance with the study of ref. 93 PBE0 without dispersion correction yields the best agreement to CCSD(T). For a few DFT functionals the dispersion correction improves the result, for others an overestimation of binding energies results for the small system (see Fig. 2b).

Analyzing the binding nature in the $AuPH_3$ fragment, one-component (scalar) CCSD(T) calculations at basis set limit revealed a relatively strong binding energy (15.9 kcal mol^{-1}) compared to the corresponding complexes containing copper or silver (7.9 or 2.7 kcal mol^{-1}, respectively).[92]

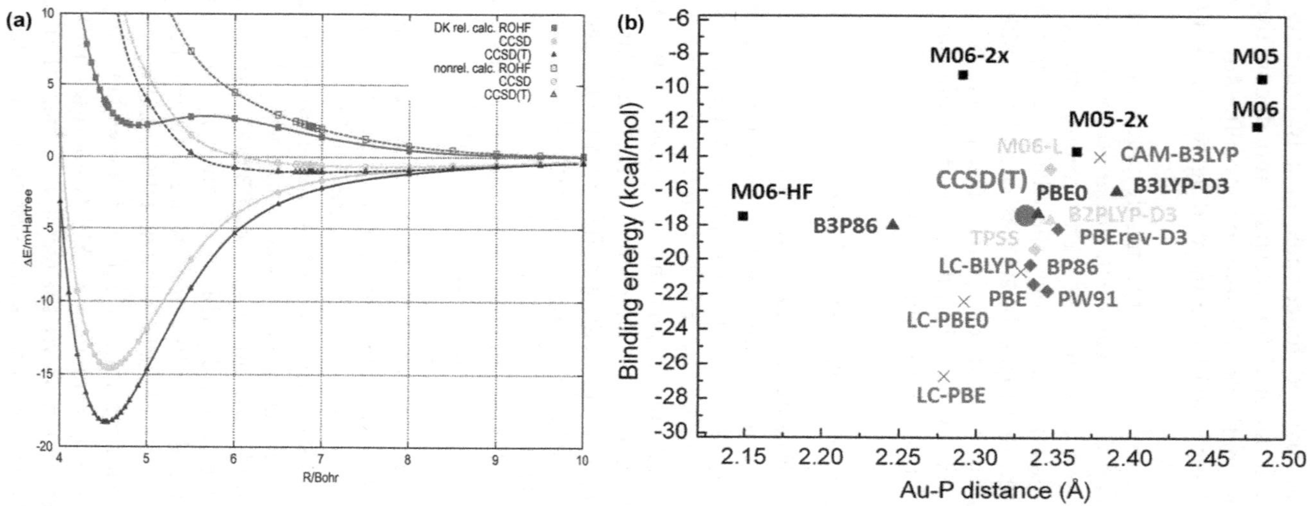

Fig. 2 (a) Potential energy curves are shown at non-relativistic and DKH relativistic BSSE corrected ROHF, CCSD and CCSD(T) level of theory for the $AuPH_3$ complex over the Au–P distance, reproduced from ref. 92; (b) Binding energy and bond distances of the $AuPH_3$ complex at different level of theories and basis set limit (ECP60MDF), reproduced from ref. 60.

The reason has been shown to have its origin especially in relativistic effects, see potential energy surface in Fig. 2a. Neglecting these effects results in a very weak van der Waals complex (around 1 kcal mol^{-1}). The relativistic effects enhance the electron affinity of the metal, which leads to a bond with the main contribution in a charge-transfer effect from the phosphorous lone-pair to the metal atom. Based on NBO analysis the charge transfer mechanism has been described by a n_P (lone pair orbital of phosphorous)→Au 6s donation of the lone pair donor ligand and metal combined with an Au 6s→n_P* back-donation.[92] Minor contribution has been attributed to π back-donation from d orbitals of the metal atom and p orbitals of phosphorous. The calculated charge transfer is strongly dependent on the used method, so Mulliken, NPA and Bader charge analysis give values between 0.1 to 0.4 e, but reveal always the charge transfer from the ligand to the metal.[60] However, also the relativistic shrinking of the orbital sizes leads to an improved overlap between the metal and the ligand orbitals.

According to the molecular-orbital diagram of the $AuPH_3$ fragment Au 5 and 6s orbitals hybridize, and a fully occupied orbital between this hybridised orbital and n_P can be found (HOMO-7, see Fig. 3), whereas the corresponding antibonding orbital (single occupied molecular orbital) is singly occupied.[60] The HOMO-1 mostly represents the Au 5d and 6s hybridised orbital with a small bonding contribution to n_P. This gives evidence of a half bond, which agrees well with the calculated Wiberg bond index (quantum chemical estimate of the bond order).[60] Also, some interaction between 5d gold states and n_P can be attributed to small distribution of back donation.

Within a series of different ligands in interaction to a gold atom, the complexes containing PX_3 ligands (X = H, F, Cl or Me) result in the strongest interaction followed by SX_2, NX_3 and OX_2 ligands.[93] The low ionization potential of phosphorous has been described as being responsible for this effect. The binding energies of the different $AuPX_3$ complexes follow the order $AuPMe_3 \gg AuPH_3 > AuPCl_3 > AuPF_3$. Thus, whereas the donating methyl group enhances the interaction energy, the accepting fluorinating ion leads to much lower interaction strength.[93] Considering PPh_3, which is a widely used ligand for catalysis, the binding energy calculated at DFT level of theory is along $PMe_3 > PPh_3 > PH_3$ due to higher charge-transfer effects for the Me group. Anyway this outcome changes to $PPh_3 > PMe_3 > PH_3$ by applying dispersion correction, which is more important for the aryl phosphine then the alkyl phosphine. Accordingly, the dispersion correction becomes more and more important for larger phosphine ligands.

Regarding the binding nature and geometric as well as electronic structure and charge distribution, the PH_3 ligand can sufficiently serve to describe the experimentally more important $AuPPh_3$ system as computational less demanding model system, whereas the PMe_3 ligand yields better agreement regarding the binding energy.[60] Similarly, Roesch *et al.* found that in the $MeAuPR_3$ series (R = H, Me, Ph) the PH_3 ligand gives already realistic structures, but for energy values and dipole moment a PMe_3 ligand has to be used to model the larger PPh_3 ligand.[94]

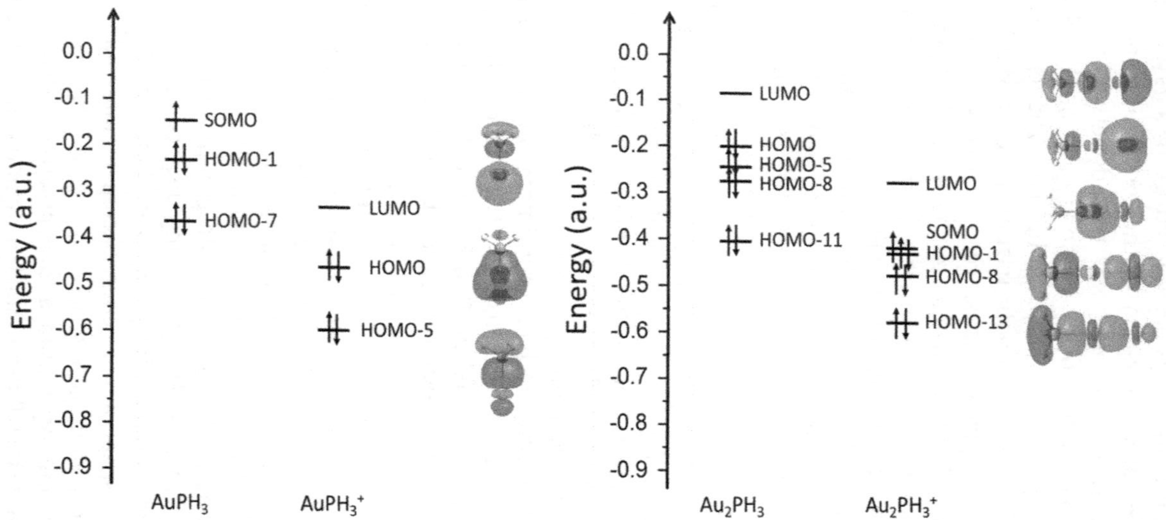

Fig. 3 Molecular orbital diagrams of $AuPH_3$, $AuPH_3^+$, Au_2PH_3 and $Au_2PH_3^+$ as well as the visualization of the most important orbitals ordered along the energy of appearance are shown.

DFT investigations of several $OCAuPR_3$ complexes show that for the gold complexes, the expected correlation between the Tolman electronic parameter (TEP) and the charge transfer cannot be found.[95] The TEP is assigned to the change of the vibrational stretching mode of the carbonyl group depending on the phosphine ligand (or in general other ligands) in the complex. The parameter serves to measure the electron donating or withdrawing property of a ligand. If just the charge transfer from the metal to ligand back-donation is considered, the correlation to TEP is obtained.[95] It is likely that the charge transfer induced by relativistic effects distorts the correlation in comparison to for example nickel complexes.

The charged fragment $[AuPR_3]^+$ has relevance as "terminal ligand" in several clusters such as $X(AuPR_3)_n^{m+}$ with X = C, N, O, ... or $[MPR_3]_n^+$ with $n = 1–4$.[96–100] Considering the charged $[AuPR_3]^+$ fragment compared to the corresponding neutral system, the binding energy is increased by a factor of 4–5 considering CCSD(T) or DFT-D3 results.[60] The increase for the CCSD(T) result is just originated by the HF part. The charge transfer effect enhances up to 1 *e*, depending on the method. Again, relativistic effects have been determined by MP2 and DFT calculations as the most important feature resulting binding strength along the order (quasi relativistic)-$[AuPR_3]^+ > [CuPR_3]^+ > [AgPR_3]^+ >$ (non-relativistic)-$[AuPR_3]^+$.[101] Analysis of the MO diagram reveals a bond order of about 1, because the antibonding SOMO of the corresponding neutral complex is not occupied. This is in agreement to the calculated Wiberg bond index. The HOMO-LUMO gap enhances by 1 eV to 3 eV compared to the neutral complex and functional of choice. Along different substituents, the binding energy decreases similarly as for the neutral complex by the sequence of $PPh_3 > PMe_3 > PH_3$. Interestingly, this is the case for applying dispersion correction or not. The correction is not as important for the strong bond as for the weaker bound neutral system. As for the larger R (ligand), the bond strength increases, and as a consequence the electrophilic power decreases.[101]

Going away from the single-ligand system, the two coordination case has been determined to be preferred for the model systems $[Au(PH_3)_n]^+$ with $n = 1–4$ and the complexes $[AuCl(PH_3)_n]$ with $n = 1–3$.[99] Regarding the $[Au(PR_3)_2]^+$ system it has been demonstrated experimentally and explained theoretically that the reactivity of the cationic gold complex for the oxidation addition of aryl iodine can be manipulated by ligand design.[102] The reaction using mono-coordinating phosphine ligands leads just to a short living $(R_3P)_2Au(PhI)^+$ system. Instead, by using a bidendate phosphine ligand the reactivity is enhanced and the oxidative addition can take place.[102] DFT calculations revealed the difference in the binding modes: the PhI can better bind in a square planar-like geometry (bidendate phosphine ligand) to the gold system than in the linear arrangement (monodendate phosphine ligand) (see Fig. 4a). In addition larger activation barriers as well as less energetically preference for monodendate phosphine compared to bidendate phosphine ligands have been determined as explanation (see Fig. 4b).[102]

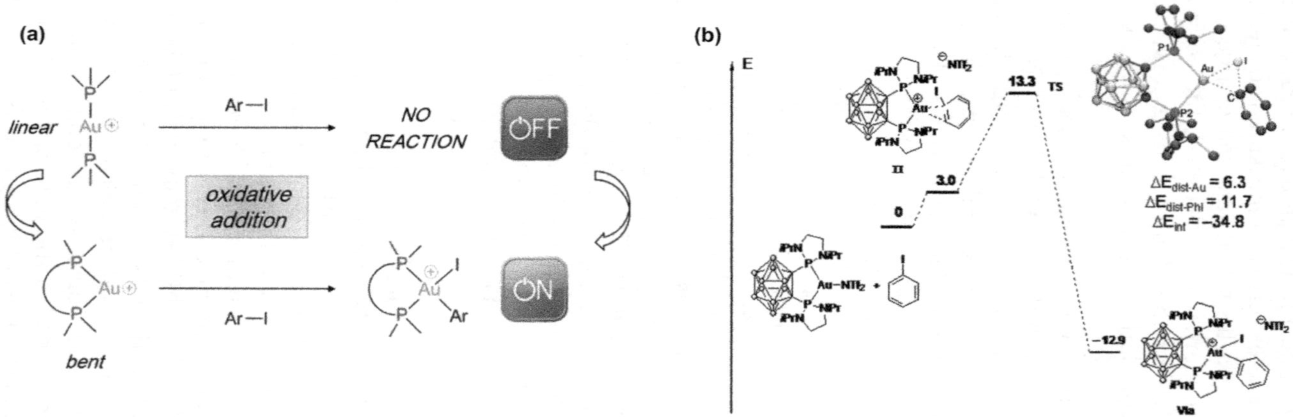

Fig. 4 (a) Mono- and bidendate phosphine Au(I) complexes in the oxidative addition of aryl iodides; (b) Energy profile of the oxidative addition of phenyl iodide to gold. Reprinted with permission from ref. 102. Copyright 2014 American Chemical Society.

Next to catalysis, Au(I) complexes have been shown to be interesting for luminescent effects. The origin of luminescence is generally assigned to charge transfer from the ligand to the gold system, transitions centred at the gold system or aurophilic interactions.[103–105] DFT and TD-DFT calculations support the experiments by understanding the interplay between geometric and electronic structure that is important for the origin of the luminescence effects. Considering Au(I) phosphine complexes, differences in the appearance of luminescence have been found that depend on the number of coordinated phosphine ligands.[106] Whereas Au-based luminescence is found for $Au(PR_3)_3^+$, the effect does only appear for $Au(PR_3)_2^+$ when Au…Au interactions take place, and have not been found for $Au(PR_3)_4^+$ complexes at all.[106]

In order to understand the origin of luminescence of three-coordinated luminescent Au(I) phosphine complexes a model study has been done using DFT and the smallest Au(I) phosphine complex, namely $Au(PH_3)_3^+$.[106] DFT calculations of the singlet ground state reveal the electronic structure with degenerated e′ orbitals as HOMO, which have Au $5d_{xy}$, $5d_{x2-y2}$ character and an antibonding Au–P contribution (see Fig. 5a). The LUMO in contrast shows bonding character between Au p_z and n_P orbitals. Experimentally visible luminescence with large Stokes shifts has been explained by the Au–P bond shortening induced by the excitation from an antibonding HOMO to a bonding LUMO.[107] However, extended studies including the triplet state reveal for the lowest triplet state a T-shaped $Au(PH_3)_3^+$ structure instead of the predicted trigonal planar one (D_{3h} symmetry), see Fig. 5a.[106] The structure change has been allocated to a Jahn–Teller effect. The decreased singlet–triplet gap between the ground and the triplet T-shaped state explains the experimentally obtained long phosphorescence wavelength.[106] The gap does not change noticeable when the Au–P distance is varied, so the first explanation does not hold. QM:MM calculations confirm the effect for larger ligands such as PMe_3, PPh_3 and $PPhCy_2$.[106] Furthermore, the evidence of obtaining even higher emission energies opens up a tuning possibility of the wavelength.

Two-coordinated Au(I) complexes such as [AuCl{$Ph_2P(CNS)(C_6F_4)$}] and [$Au(C_6F_5)${$Ph_2P(CNS)(C_6F_4)$}] with the polyfluorinated benzothiazolephosphine ligand, show phosphorescence effects in the solid state (see Fig. 5b).[108] DFT and TD-DFT calculations have identified the emission of the free phosphine ligand ($n_P \rightarrow \pi^*$) to be different to the Au(I) complex ($\pi \rightarrow \pi^*$(heterocyclic ring)) due to the formation of a gold phosphine bond. In contrast, the spectra of two-coordinated di-gold complexes with ethynyl pyridine derivatives (X) as linker unit [$PR_3Au(C{=}C)X(C{=}C)AuPR_3$] have been obtained very similar for the free ligands and di-complexes with the origin in $\pi \rightarrow \pi^*$ ligand-centred transitions (assigned by DFT calculations).[109] Interestingly, photoluminescence properties in the solid state show also the tetrahedral-coordinated mono- and dinuclear Au(I) complexes $Au(diphos)_2PF_6$ and $Au_2(tetraphos)_2(PF_6)_2$.[110] By the help of DFT and TD-DFT calculations, the emission has been assigned to ligand-to-metal charge-transfer nature [π^*(Ph)→Au(d)-P(p)] and phenyl stacking [π^*(Ph–Ph)→Au(d)-P(p)]. This agrees well with experimental data and the fact that

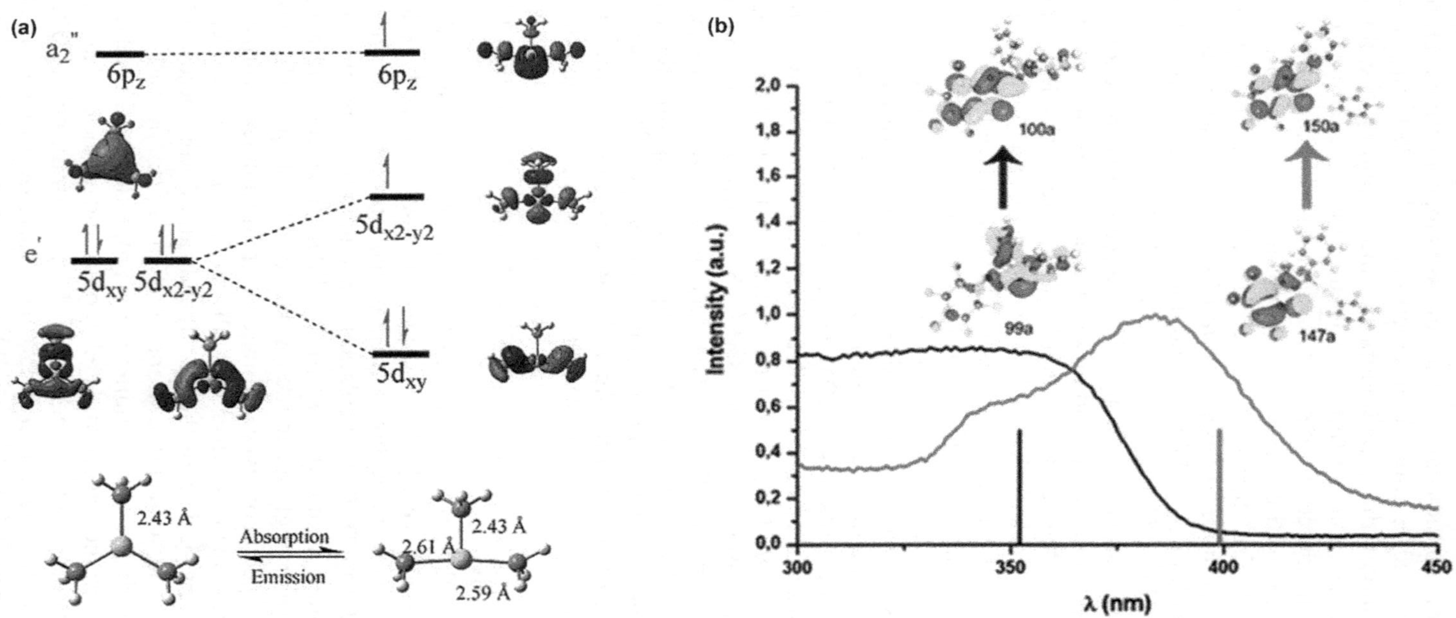

Fig. 5 (a) The optimized singlet ground state structure of the $[Au(PH_3)_3]^+$ model (bottom, left) and the lowest triplet state (bottom, right) are shown as well as the corresponding molecular orbital diagrams (top). Reprinted with permission from ref. 106. Copyright 2003 American Chemical Society. (b) Experimental spectra of two-coordinated Au(I) complexes and the first triplet excitation for model systems are given. Reproduced from ref. 108 with permission from The Royal Society of Chemistry.

lifetimes of the excited states have been found to depend on phenyl-phenyl pairing in an inter- and intramolecular way.[110]

Strong luminescence properties due to aurophilic interactions exhibit tri- and tetranuclear Au(I) complexes. Thereby, the design of the phosphine ligand can be used to determine the geometric structure for the multinuclear Au(I) complexes. Thus, trinuclear Au(I) complexes with tris-indole and triphenylphosphine ligands lead to metallodendrimer-like complexes, whereby the use of triphosphine results in a metallo-cryptands-complex (see Fig. 6a).[111] TD-DFT calculations reveal the lowest energy absorption, which appears to be a ligand-to-metal charge-transfer effect with N(p, tertiary amine)→Au transitions in the solid state. Furthermore, tetranuclear Au(I) complexes constructed by tetraphosphine ligands have been observed to be aligned in a syn or anti arrangement depending on the two ligands' relation to the metal chain.[112] Following a linear ordered gold string, a zigzag gold chain or a bent gold core with intercorporated anion to the bent pocket has been analysed. The emission energy based on HOMO and LUMO analysis of DFT calculations has the main contribution on the tetranuclear Au(I) arrangement. However, the counter ions have been observed as highly important, as strong luminescence effects have been found just in the cases of large anions, which do not coordinate to the complex.[112]

Tong *et al.* also have found that the appearance of high emission or low emission bands for trinuclear Au(I) complexes $[Au_3(dcmp)_2]X_3$ (dcmp - bis(dicyclohexylphosphinomethyl)-cyclohexylphosphine) (HOMO and LUMO orbitals are plotted in Fig. 6b) is depending on the counter ion X and whether this has close contact or no contact to the complex.[113] DFT calculations assign the emission bands to the three-coordinate Au(I) chain and show that the spectroscopic properties are affected by aurophilic interactions (Au–Au distances of about 3.0 to 3.1 Å). DFT calculations of two- to four-nuclear coordinated model complexes present evidence that both absorption and emission energies are inversely proportional to the number of Au(I) atoms if gold-anion contacts are not close.[113]

3.2 Pyridine derivatives gold fragments and complexes

With its aromatic, planar ring structure, its dipole moment of 2.2 D (higher electronegativity of nitrogen) and its free electron pair at the nitrogen atom, pyridine constitutes an interesting system with generally two possible binding modes to surfaces, namely the N-binding and the π-binding mode. To gain a basic understanding of these two binding modes, the pyridine gold complex has been considered as a model system (see Fig. 7). A benchmark study regarding the accuracy of DFT and DFT-D3 approaches of bond lengths and interaction energies for both binding modes (to treat both modes in the same accuracy) reveals PBE0-D3 in best agreement to CCSD(T) (ref. 114). However, different DFT functionals give a good description of the N-binding situation. Even if it has been discussed that the N-binding mode gets overestimated by DFT calculations because photoelectron spectra studies have found the reversed order for the HOMO and HOMO-1 of the free pyridine molecule

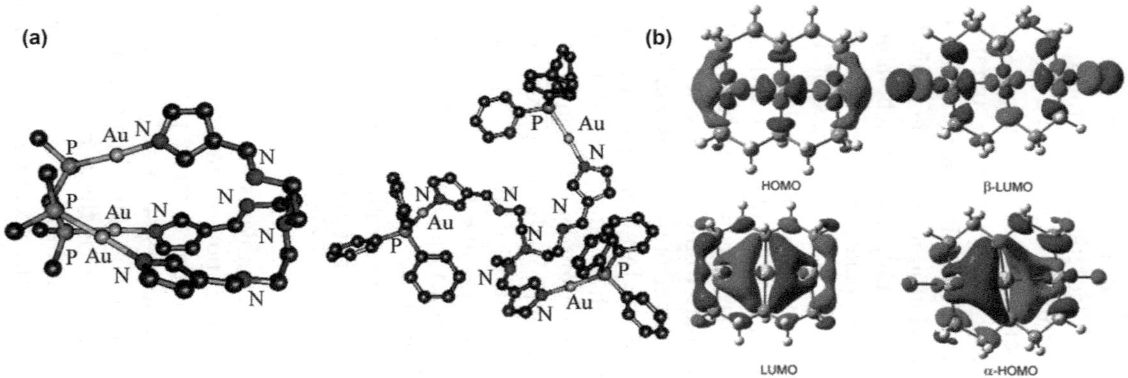

Fig. 6 (a) Optimized structures of a trinuclear Au(I) complex with tris-indole and triphenylphosphine (right) or triphosphine (left) ligands. Reprinted with permission from ref. 111. Copyright 2008 American Chemical Society. (b) HOMO and LUMO orbitals of the $[Au_3(H_2PCH_2PHCH_2PH_2)_2]^{3+}$ (left) and $[Au_3(dcmp)_2Cl_2]^{+}$ (right) complexes which have been used to explain the origin of the luminescence effects, reproduced from ref. 113.

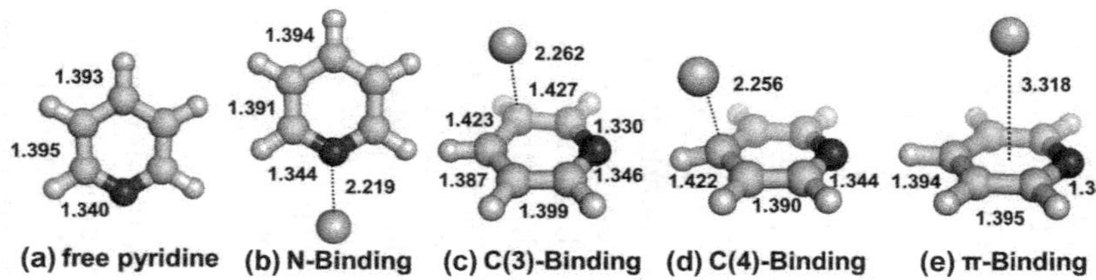

Fig. 7 Pyridine (a) and four different binding possibilities within the pyridine gold complex, namely *via* N (b), C(3) (c), C(4) (d) and π (e), optimized in structure at DFT level of theory, reproduced from ref. 114.

(n_N and π-orbital) compared to DFT level.[88,115,116] Similar to the phosphine gold complex, applying a dispersion correction (D3) has improved the accuracy regarding the interaction energy for the N-binding situation mainly for hybrid and double hybrid functionals. The π-binding situation, in which the gold is directly placed above the centre of the pyridine ring, is even more difficult to describe by standard DFT calculations. For this configuration the dispersion correction is absolutely necessary for GGA as well as hybrid functionals, because the bare DFT value is much too low – except for LDA – or even no binding is received.[114]

For the N-binding AuPy fragment, binding energies in the order of 12 kcal mol^{-1} have been obtained.[114] The analysis of the binding situation takes place by considering the molecular orbitals of the complex (see Fig. 8).[84] The singly occupied HOMO is a hybridization of the gold 6s orbital and some small contribution of the nitrogen lone pair (n_N) in an antibonding constitution. Also the HOMO-1 has *anti*-bonding character between the lone pair and the gold orbital. The corresponding bonding orbital can be detected at lower energy (HOMO-8). Thus, the main contribution of binding can be assigned to an n_N interaction to the gold atom. The occupation of bonding and *anti*-bonding orbitals destabilizes the Au–N bond and leads to a binding situation with no or very small covalent binding contribution. Also, there has been found a small perturbation between the π-orbitals and the gold d-states. This minor contribution is discussed as back-donation of metal d orbitals to the ring π-system. Charge-transfer effects have been determined of importance for the N-binding pyridine gold system, too. The binding energy of the π-binding pyridine gold complex is 1/4 of the N-binding constitution. The interaction is a bare vdW interaction which is also indicated by the MO-diagram (see Fig. 8).[114] There is just a weak mixing of the gold d

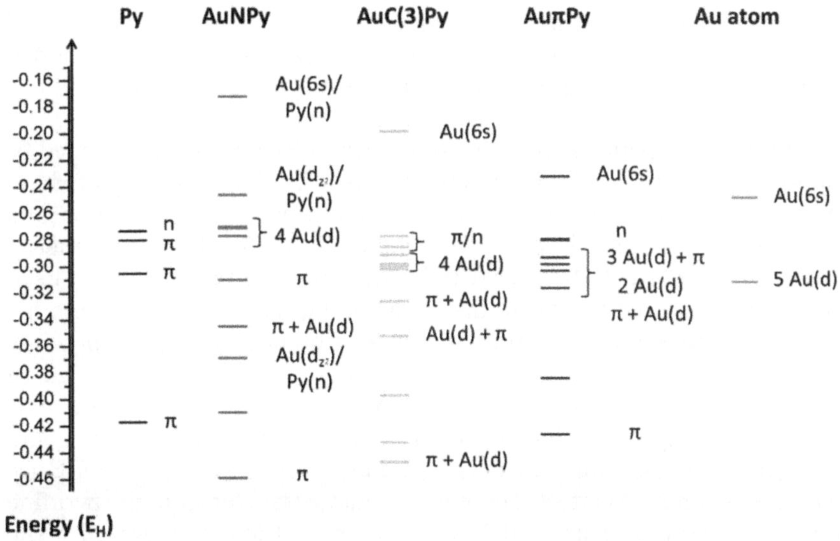

Fig. 8 Molecular orbital diagrams of pyridine and the gold atom as well as the N-binding (AuNPy), C(3)-binding (AuC(3)Py) and π-binding (AuπPy) gold complexes.[84]

orbitals and the pyridine π orbitals, the MO-diagram remains nearly as sum of the ones of free pyridine and the gold atom.

Further binding modes have been determined by calculating potential energy surfaces of the gold atom moved above the pyridine ring (Fig. 9).[114] Approaching the gold atom to the pyridine ring gives at largest distance (around 3.5 Å) a central vdW binding followed by atom-centred binding situations to nitrogen, carbon(3) and carbon(5) atoms at closer distances. Structure optimizations followed by frequency calculations at DFT level of theory exhibit the N-, C(3)- and π-binding pyridine gold complexes as minimum structures (see Fig. 7). In addition a C(4)-binding complex has been found. For carbon-binding complexes, the carbon atom moved very slightly out of the ring plane. The interaction energy and MO-diagram of the carbon(3)-binding pyridine gold complex reveal averaged values and a split up between the N-binding and π-binding ones (see Fig. 8).

Considering relativistic effects of the element gold, DFT-D3 calculations of the pyridine copper, silver and gold complexes *via* N-atom display the decrease of the interaction energy along Cu > Au > Ag, whereby copper and gold show similar values.[84] The same trend has been observed for the corresponding C(3)-binding complexes. Viewing the influence of the charge to the pyridine gold complex, the positive charged N-binding pyridine gold complex exhibits a 6–8 times stronger binding energy than the neutral one using CCSD(T)//MP2 or DFT level of theory.[117] Photodissociation studies by laser vaporization in combination with supersonic molecular beams of the one times positively charged complex estimate the upper bond strength limit for the charged complex to around 60 kcal mol^{-1} which is in agreement with theoretical findings.[117] Thus, the bond strength indicates a covalent binding in contrast to the weakly binding neutral complex. In contrast to the neutral complex, the gold atom binds to the carbon(2)/(3) or carbon(3)/(4) in a bridging configuration in the charged complex.[117] Binding energies of 2/3 of the N-binding complex are present for the bridged carbon-bonded complexes. Considering a negative charge, no minimum with the ring framework has been found by DFT calculations.[84]

The aromatic pyridine building block can be electronically changed by donor and acceptor substituents which open up the possibility of changing or controlling the binding to surfaces. In this view, the pyridine gold complex has been investigated for simple donor substituents, namely –CH_3, –OH, –OCH_3, –NH_2 and $N(CH_3)_2$, whereby the substitution has been placed at position 4.[114] Due to the pushing of electronic density from the substituent to the pyridine ring, the interaction energy between pyridine and the gold atom enhances by 21% and 78% for the N-Binding and C-binding configurations, respectively. Thus, the enhancement is relatively stronger for the carbon-binding complex. All structural and energetic changes along increasing donor strength are in accordance with the +I and +M effect. Thereby the charge transfer from the pyridine derivative to the gold atom slightly increases (0.31 to 0.35 *e*) as well as the total dipole moment.

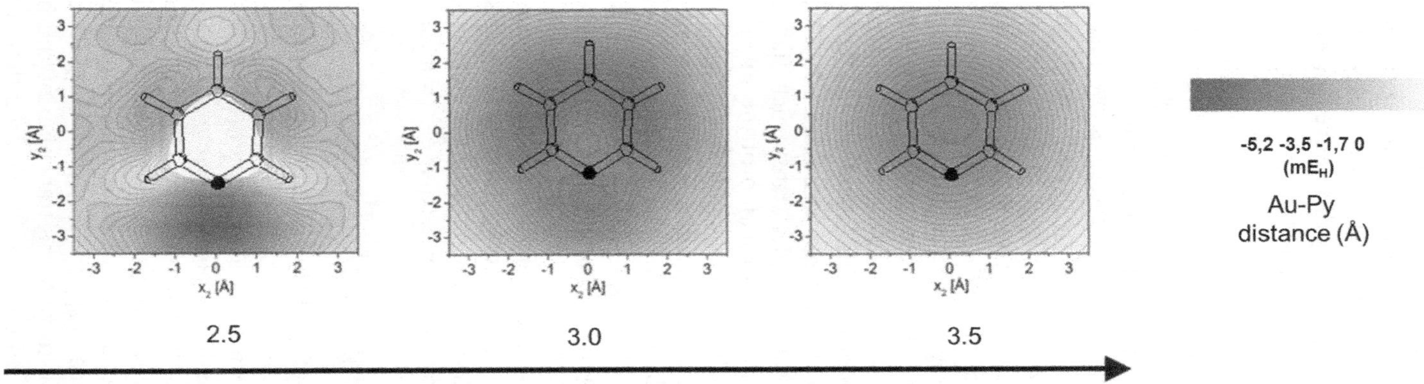

Fig. 9 Potential energy surfaces for the gold atom moved above the pyridine ring at different AuPy distances at CCSD(T) level of theory. Reproduced from ref. 114.

Considering pyridine derivatives as ligands in Au(III) $AuCl_3Py$ complexes, the reactivity for the formation of $AuCl_4^-$ has been determined to follow the order pyridine > pyridine substituted in position 4 > pyridine substituted in position 3, which has been supported by frontier orbital analysis at DFT level of theory.[118] The minor reactivity of the $AuCl_3$(Py-3R) has its reason in the bigger energy difference between the LUMO and the interacting Cl^- 3s orbital compared to $AuCl_3$(Py-4R). Also, a linear relationship between the Mulliken charge of the AuX_3 fragment (in the AuX_3L complex, X = CN, Cl and Br, L = different N-heterocycles) and the proton affinity has been discovered.[118–120] The proton affinity of the N-heterocycles (free molecules) compared to the AuX_3L complex is lowered by about 25 kcal mol^{-1} due to bond formation.

Planar pyridine Au(I) alkyl complexes show just thermodynamically stability for electron-rich pyridine ligands such as 4-dimethylaminopyridine (DMAP) or 4-aminopyridine (AP).[121] DFT calculations give no indication for aurophilic interaction of the neutral, linear and two coordinative complexes which form dimer structures in the solid state (X-Ray).[121] Gold complexes containing pyridine and aryl or alkynyl ligands also have been shown to be interesting for their optical properties *e.g.* for organic light emitting diode applications.[122–125] For example, Au(III) diaryl complexes with 2-phenylpyridine coordinating ligands reveal long-life emission in solution at room temperature and TD-DFT calculations devote the emission origin from a triplet state which has mainly intra-ligand character and some small perturbation of the metal center.[126] Also, the absorption bands of alkynyl Au(II) with 6-phenyl-2,2′-bipyridine ligands exhibit metal-perturbed intra-ligand $\pi \rightarrow \pi^*$ character.[127]

In the context of larger complexes, metal-organic trimetallic macromolecular complexes of the type $[Au(\mu\text{-}L)]_3$ with pyridine-based ligands have been assessed by DFT computations regarding π-acidity and π-basicity which has been determined to be higher than for the corresponding organic molecules.[128] As the strength of the π-acidity and π-basicity can be systematically tuned by the variation of the ligand and substituents on the ligands, theoretical studies can be used to design novel $[Au(\mu\text{-}L)]_3$ systems, which are highly interesting in molecular electronics.

4 Bonding in small gold clusters

4.1 General aspects of gold clusters and phosphine surrounded gold clusters

There have been performed various theoretical studies at different levels of theory regarding small neutral and anionic bare gold clusters and their geometric and electronic structure, ref. 129 and 130 are two examples, and ref. 59 a review. Calculations of neutral bare gold clusters showed the interesting fact that global minima for small cluster sizes appear as planar structures. This has been explained by relativistic effects. Several studies have followed the task to find the transition of the induced planar 2D structures to 3D ones. Different studies reveal different results and cluster sizes from 7 to 15 have been given as transitions from 2D to 3D (one example is shown in Fig. 10).[52,59,129–132] Thus, several results regarding

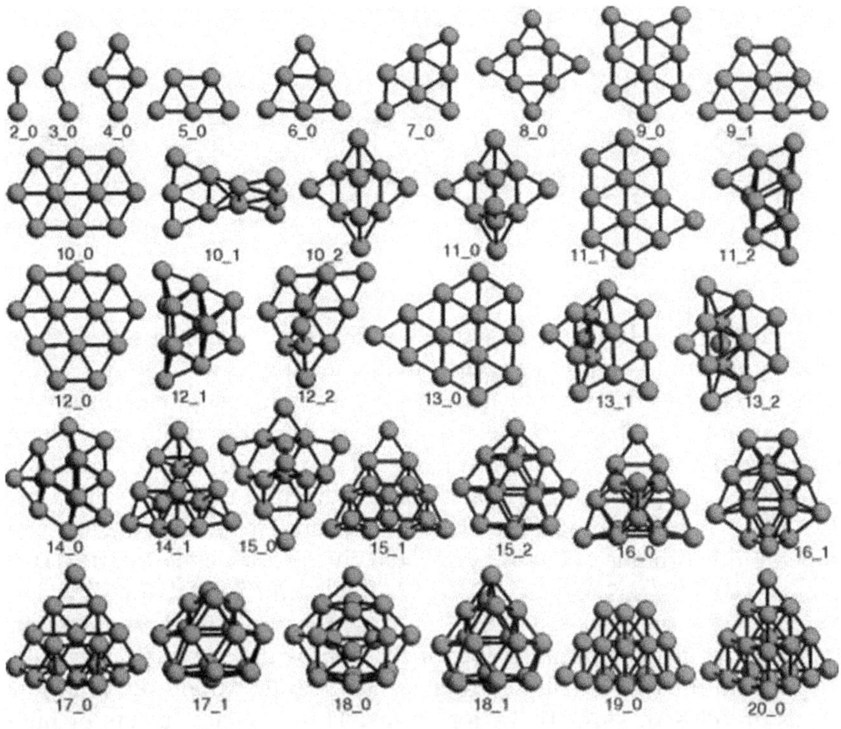

Fig. 10 Predicted global minima and lowest-energy isomer structures of Au_{2-20} which are ordered regarding increasing cluster size and energy. The notation n_m means the mth energetic isomer from the cluster which contains n atoms, reprinted from ref. 130.

structure and stability are obtained on certain level of theories. One reason is the flat potential energy surface of the different cluster structures which shows the need of accurate and high level theoretical calculations.

Regarding small phosphine-stabilised gold clusters, several X-ray structures have been characterised. A summary of gold clusters up to 11 atoms is given in ref. 30. Also, larger cluster compounds have been identified such as the Schmidt Au_{55} cluster: $Au_{55}(PPh_3)_{12}Cl_6$, $Au_{46}(PPh_3)_{12}Cl_6$ and $Au_{67}(PPh_3)_{14}Cl_8$ or $Au_{101}(PPh_3)_{21}Cl_5$ and $Au_{99}(PPh_3)_{24}Cl_2$.[133–135] The larger clusters have not yet been determined in composition and structure and are debated.[135] Furthermore, the larger compounds are covered by a mixed phosphine/halide ligand shell whereas the amount of halide is smaller. Based on the X-ray structures of the small clusters it is known, that phosphine ligands bind only to one gold atom in the cluster in an on-top configuration. Following, the gold phosphine complexes are already a very good model system for a basic understanding of binding nature and property origin. However, phosphine ligands are generally discussed to bind weakly to gold clusters, affect the electronic structure hardly and fulfil mainly steric stabilization. The overall stability of ligand-stabilised clusters is attributed at the one hand to geometric factors, which means for example a highly symmetric gold core structure, and on the other hand to electronic factors. The HOMO/LUMO gap has been often used to describe the cluster stability: if it is high, the cluster is more stable than if it is a small gap.

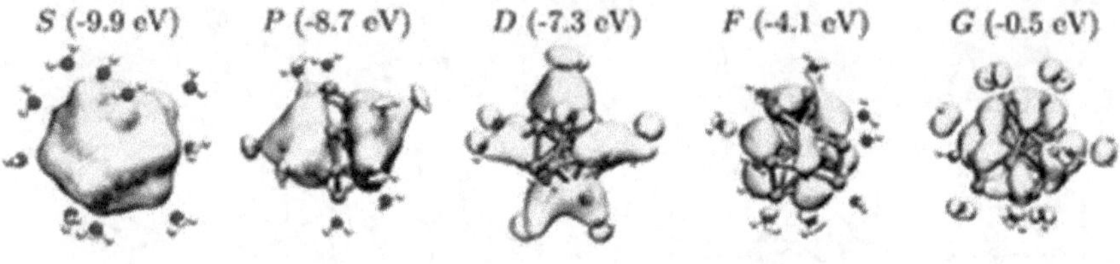

Fig. 11 The isosurfaces of electron charge density, which resemble atomic-orbital like topologies, are shown for the Ga_{23} core of a metalloid cluster (the particular energy bands also are given), reproduced from ref. 139.

Regarding the electronic-structure stabilization, several cluster sizes up to 150 gold atoms can be associated to magic numbers.[135] These clusters have an enhanced stability which is associated to the global electronic structure of the delocalised electrons. The delocalised electrons have their origin in the 6s atomic orbital of gold and are discussed to generate delocalised "superatomic orbitals" which cover the entire metallic cluster.[43,136–138] As for atoms there exists an appropriate aufbau rule for such superatomic orbitals of ligand-passivated clusters, which is $1S^2/1P^6/1D^{10}/2S,^2\ 1F^{14}/2P^6\ 1G^{18}/,\ 2D,^{10}\ 3S,^2\ 1H^{22}/\ldots$ (superatomic orbitals are shown in Fig. 11). The symbols S, P, D, F, ... describe the angular-momentum followed by the maximum number of electrons which can fill the orbitals. The superatoms are attributed to a special stability when the electronic shells are closed, similarly as for atoms. Thus, magic clusters of high stability are associated with electron numbers (closed electron shells) of $n^* = 2, 8, 34, 58, 92, 138, \ldots$.

The concept served for passivated gold clusters whereby this has been assessed for thiolate-surrounded gold clusters such as the $Au_{102}(p\text{-MBA})_{44}$ cluster.[43] Thiolates or halides (X) withdraw electrons or localise them into a covalent bond, thus they influence the number of delocalised electrons of the metallic core. Phosphine ligands (weak Lewis base, L) are discussed to weakly coordinate to the cluster and do not influence the number of delocalised electrons. As a result for the complex $(L_sAu_NX_M)^z$ the number of delocalised electrons has been defined by $n^* = N\nu_a - M - z$ related to the shell closing rule as described above. The formula contains N for the number of gold atoms which is multiplied by the atomic valence ν_a which is 1 (electron from gold 6s orbital). M is the number of withdrawing ligands X and z the charge of the ligand-stabilised cluster. The calculation of the global electronic structure is usually rationalised by projecting the Kohn–Sham orbitals onto spherical harmonic functions within a defined sphere with the origin in the centre of mass.[139] Otherwise some authors just analyse the electronic structure and consider the higher and highest occupied molecular orbitals and lowest unoccupied molecular orbitals.

Methodological, the phosphine gold dimer reveals a variance of DFT results within $+/-10$ kcal mol^{-1} compared to CCSD(T); accurate results can be obtained by PBE0-D3, B2PLYP-D3 and PBErev-D3.[60] Even by having a closed-shell system similar trends as for the pyridine gold complex have been found. The interaction energy (DFT with dispersion correction) is increasing along $PPh_3 > PMe_3 > PH_3$. However, MP2 results and experimental results for small gold cluster dimer systems reveal that DFT gives too small and DFT-D2 too large aurophilic interactions.[140] The use of DFT and DFT-D2 does not just lead to different energies, in cluster growth studies also different product structures for gold-trimer systems have been observed.[140] Difficulties in the methology occur by increasing the gold core. Whereas gold-gold distances are well described by LDA in tetranuclear phosphine-covered gold clusters, PBE-D3, PBE0-D3 or B2PLYP-D3 give an excellent agreement for gold–phosphine interaction energies in hexanuclear clusters compared to SCS-MP2 results (see Fig. 12).[60,141] Thereby, the importance of dispersion correction increases with the volume of the ligands. In order to consider ligand–ligand interactions which also can contribute to hold the system together, the application of dispersion correction is crucial (see Fig. 12). Nevertheless, long-range corrected functionals have not been recommended for structure optimizations of phosphine gold clusters because of the poor agreement of theoretical and experimental bond-length values.[141] On the other side, the ONIUM approach has been used for triphenylphosphine

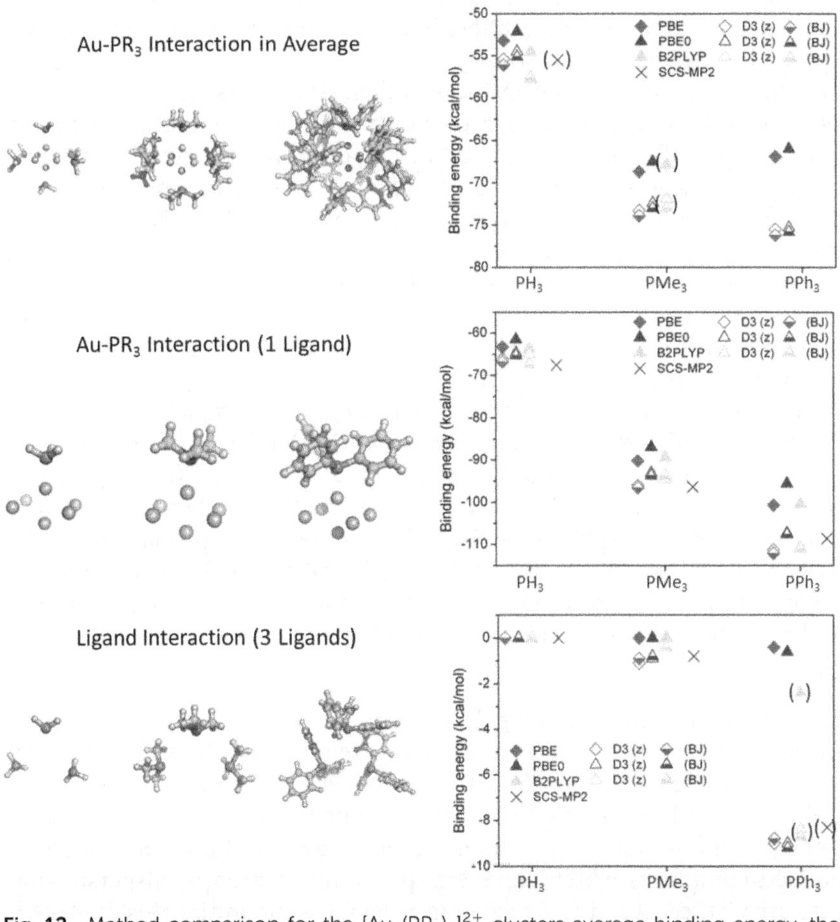

Fig. 12 Method comparison for the $[Au_6(PR_3)_6]^{2+}$ clusters average binding energy, the $[Au_6(PR_3)]^{2+}$ binding energy as well as ligand–ligand interaction energy, reproduced from ref. 60.

ligands and results in very good agreement to the experimental structure.[141] Excitation spectra profiles as well as relative intensities of transitions strong in nature are qualitatively not changed by applying different methods and basis set.[141]

4.2 Phosphine gold dimers

Compared to the gold phosphine complex (one ligand), the gold dimer with one PR_3 ligand (R = H, Me, Ph) exhibits a shorter Au–P distance and an increase of the interaction energy by a factor of two.[60] The analysis of the molecular orbital diagram (see Fig. 3) and the Wiberg bond index evidences a single bond (0.8 to 0.9) instead of a half bond which has been found for the complex. The charge transfer effect is similar in size as for the complex, the gold atom to which PR_3 is bound shows a less negative (or slightly positive) charge compared to the other gold atom. The gold dimer has a bond length similar to the free dimer (2.47 Å).[142] Applying a positive charge to the phosphine gold dimer exhibit increased interaction energy values by a factor of two.[60] This is similar for adding a charge to the phosphine gold complex. Interestingly, the bond index is the same for the neutral and charged phosphine gold dimers. But the transferred charge has increased, and values between 0.4 to 0.9 e have been found depending on the method (NPA, Mulliken, Bader) as well as the phosphine rest. Thus, the bigger part of the charge is located at the phosphine ligand, especially when PMe_3 and PPh_3 ligands are present. This leads to a slight elongation of the Au–Au bond length. The gold dimer covered by two PPh_3 ligands has been suggested, but not been found experimentally.[52] In a theoretical MP2 study for the related $Au_2(PH_3)_2$ model a linear P–Au–Au–P configuration has been indicated.[143]

Gold hydride dimers $(AuPR_3)_2H^+$ have been generated by electrospray ionization with a mixture of $AuPR_3Cl$ complexes. Corresponding DFT calculations (model system: R = H) have revealed a ligand migration during the fragmentation.[144] Furthermore, a symmetrical triangular structure for the ground state and a PH_3-bridged transition state has been determined. Related to this context the cationic $Au_2(DPPP)_2Cl^+$ has been found as partial decomposition product of melanoma cytotoxic chlorotriphenylphosphine-1,3-bis(diphenylphosphino)-propanegold(I) by 31P-NMR, MS and X-ray measurements.[145] DFT calculations of the obtained dimer reveal a triangular coordination sphere in contrast to the T-shaped geometry of related compounds.

4.3 Phosphine surrounded gold clusters

In order to understand cluster growth mechanisms, the experimentally known complex $AuClPR_3$ (R = H, Me, and Ph) with linear coordination geometry has been applied as starting point for a DFT investigation.[140] The attempt to force two complexes to reaction does not work and aligned complexes which show aurophilic interaction of dispersive nature are yielded. A reducing agent is crucial for the initial cluster growth step which generates an anionic complex $[AuClPR_3]^-$. Two anionic complexes $[AuClPR_3]^-$ lead to a reaction with different dimer products which depend on the chosen starting orientation of the reactants. In each case, dissociation processes lead to a stable product, and two Cl^- or one Cl^- and one PR_3 or two PR_3 leave the system.[140] The chloride-ion dissociation is thermodynamically more favorable. The phosphine dissociation is less favorable the bulkier the rest R.

For the generation of gold trimer clusters, at least one anionic complex has to be present for a reaction. The trimer structures tend to be in a triangular shape mostly having a negative charge.[140] Fig. 13 presents trimer gold clusters surrounded by ligands. The formation of tetramers leads to metal cores of diamond, tetrahedral, Y-shaped as well as linear wires.[140] The total cluster charges range from +1 to −2. The tetrahedrally-shaped clusters have positive charges, whereas the Y-shaped ones have negative ones. This shows the cluster-geometry dependence on the charge. The tetrahedral structure $Au_4Cl_2(PH_3)_4$ which is similar to the experimentally determined and characterised (X-ray) $Au_4(\mu\text{-}I)_2(PPh_3)_4$ has been found by the theoretical cluster-growth approach to be the energetically lowest cluster among the others.[140] However, the preferred chlorine dissociation explains the resulting cluster form of $Au_x(PR_3)_yCl_z$ with $x > y > z$.

Small gold phosphine cluster systems have been shown to be interesting for optical properties.[58,141,146–150] In this context the smallest polyhedral gold phosphine cluster systems are $Au_4(PR_3)_4{}^{2+}$ and $Au_4(\mu\text{-}I)_2(PR_3)_4$ with R = H, Me and Ph (X-ray structure with R = Ph). Interaction energies have been obtained for the charged cluster to be stronger (factor 2) than for the one where iodine is involved that screens charge. For example, the interaction energy for R = H is decreased from 55–60 kcal mol^{-1} to 20–29 kcal mol^{-1} depending on the level of theory. Solvent effects yield an energy reduction of 40% for the dicationic cluster. The solvent seems to stabilize the positive core, indeed for the $Au_4(\mu\text{-}I)_2(PR_3)_4$ cluster the solvent effect does nearly not change the result. The molecular-orbital diagram reveals several s-like orbitals for the high occupied orbitals whereas low unoccupied orbitals show p-like character. The authors conclude a special stability due to the presence of a superatomic closed shell structure. The peak of the experimental excitation spectra at 2.94 eV has been assigned to a Au(5d) → π^* (Ph) excited state. Accordingly, the exchange of PPh_3 ligands by PH_3 or PMe_3 as model ligands cannot be done to calculate electronic transitions.

Small Au_6 cluster such as $[Au_6(Ph_2P\text{-}o\text{-tolyl})_6]^{2+}$ and $[Au_6(PPh_3)_6]^{2+}$ have been used as catalysts supported for example on silica or titania.[8,151] Absorption spectra of the bare Au_6 and $Au_6{}^{2+}$ cluster calculated by a highly correlated configuration interaction (CI) approach emphasized in comparison to experimental UV-Vis spectra that the gold cluster surrounded by the ligand shell has a neutral charge.[151] The charge of +2 has been determined to be delocalised on the phosphine ligands. The fact has been used to explain that the phosphine-surrounded gold clusters adsorb to the electron-rich TiO_2 and the repulsive interaction between the clusters could be responsible for hindering the aggregation. Anyway, the phosphine ligands of the supported gold cluster are removed before or during the reaction.[2,8,31,152,153] The partial removal of the

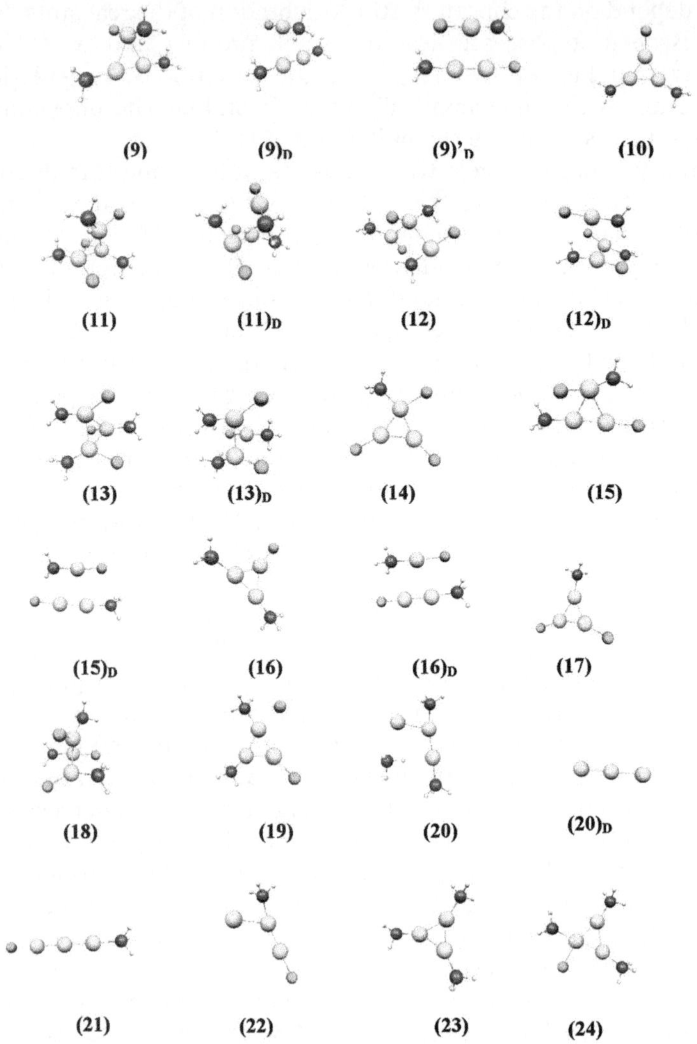

Fig. 13 Trimer ligand-surrounded cluster structures which have been obtained from the cluster growth study (DFT) and the reaction of a monomer and dimer. D indicates that dispersion correction has been included (the numbering is not relevant for this review). Reprinted with permission from ref. 140. Copyright 2011 American Chemical Society.

ligands could avoid cluster aggregation.[31] Electronic structure calculations have suggested a sequentially ligand removal process for which it gets harder and harder to remove the ligands. Thereby, the interaction energies have been obtained to increase from 38 kcal mol^{-1} for the first ligand up to 125 kcal mol^{-1} for the last ligand.[8] The cluster rearranges thereby dramatically (see Fig. 14). Reduction of the cluster change reduces the interaction energy and helps significantly to remove the ligands.[154] The studies show that an understanding of the influence of the ligands to the cluster is crucial, the cluster rearrangement as well as the influence by the ligand shell can change the catalytic properties next to the influence by the support.

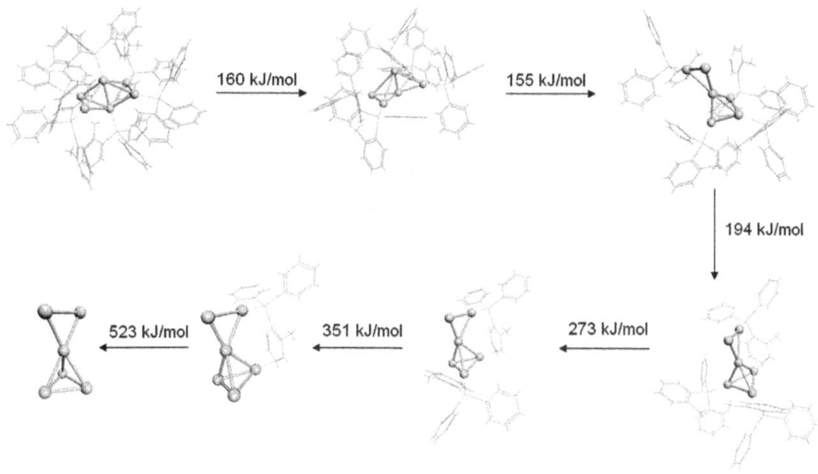

Fig. 14 Minimum energy path calculated by DFT for the removal of ligands from the $[Au_6(Ph_2P\text{-}o\text{-}tolyl)_6](NO_3)_2$ system. The required energies are given for each step. Reproduced from ref. 8.

Comparing charged ligand-surrounded gold clusters (Au_n with $n = 6$–9) electronic structure with the corresponding bare clusters one, it can be shown that remarkable changes can take place in contrast to general assumptions.[159] Regarding the superatomic-complex concept, an electronic-shell closure has been determined for the $Au_7(PPH_3)_7{}^+$ cluster and a nearly electronic-shell closure for $Au_8(PPh_3)_8{}^{2+}$ and $Au_9(PPh3)_8{}^{3+}$. The number n^* is not determined to be equal to a magic number. Interestingly, all three clusters have been used as intercluster compounds consisting of fullerides or polyoxometalates and the gold clusters.[23,25] For the hypothetic octahedral $Au_6X_6{}^{2+}$ cluster a partial occupancy of the HOMO, which is a triply degenerate p-state, is found. Following the system is compressed along one 4-fold axis which reduces the cluster symmetry from O_h to D_{2h} or D_{4h}.[141] The distortion of the structure has been discussed in relation to Jahn–Teller effect.[141,155]

Au_{11} and Au_{13} phosphine-halide passivated clusters are characterised by X-ray diffraction.[156,157] The $Au_{11}(PR_3)_7Cl_3$ with an approximated C_{3v} gold-core structure has been determined within the superatomic-complex concept with a closed s and p shell ($n^* = 8$) as well (see Fig. 15). The phosphine ligands have been determined to act as weak surfactants which do not modify the cluster core or the electronic structure.[43] A similar structure, namely $Au_{11}(dpb)_4Cl_2{}^+$ (C_2 symmetry), has been determined to exhibit chirality. The achiral metal cores has been induced chiroptical properties by a chiral field which has been arisen by the bidentate phosphine ligands.[158] The phosphine gold interaction has been determined to be charge dependent for the Au_{11} cluster.[159] A single phosphine binds weak to a neutral gold cluster, whereas a rather strong binding is found to tricationic gold cluster.[159] In the latter case, the structural and electronic properties of the gold cluster get affected by the phosphine ligands. The Au_{11} cluster can bind up to 8 phosphines, thereby solvation effects (water) enhances the ligand-to-metal charge-transfer

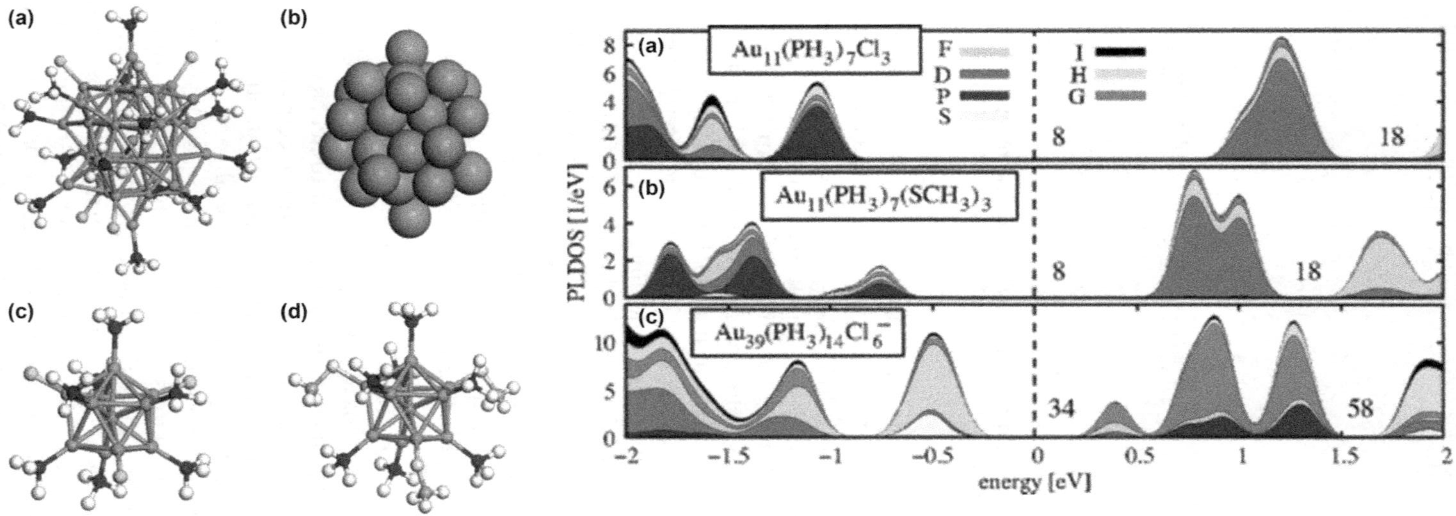

Fig. 15 Structures of phosphine (and chloride or thiolate) protected Au_{39} and Au_{11} clusters. Left: the $Au_{39}Cl_6(PH_3)_{14}^-$ cluster (a), Au_{39} core (b), $Au_{11}Cl_3(PH_3)_7$ cluster (c) and $Au_{11}(SMe)_3(PH_3)_7$ cluster (d) structures are shown. Right: the angular-momentum-projected local density of electron states (PLDOS) for the gold core of the following protected clusters is shown (a) $Au_{11}(PH_3)_7Cl_3$. (b) $Au_{11}(PH_3)_7(SMe)_3$. (c) $Au_{39}(PH_3)_{14}Cl_6^-$. Reproduced from ref. 43.

effect for the neutral species, whereas these disfavour the trication compared to gas phase.[159]

The experimentally characterised icosahedral gold cluster $[Au_{13}(PMePh)_{10}Cl_2]^{3+}$ has been theoretically determined to be particularly stable by the separating of P and D states within the superatomic-complex concept ($n^* = 8$) and by a large HOMO-LUMO gap.[43,160] Considering differently charged states (0, +3 and +5) of the Au_{13} cluster structure surrounded by the phosphine ligands exhibit strong charge dependence and changes between bilayer, icosahedral and flake structures.[160] The origin of cluster stability has been related to charge transfer effects from the ligand shell to the metal core. Thereby, the phosphine-covered Au_{13} cluster with a +5 charge reveals HOMO states as triply P and then five D-LUMO states as lowest unoccupied ones, which fulfills the superatom-shell closure.

The intrinsically chiral $[Au_{20}(PP_3)_4]Cl_4$ cluster (PP_3: tris(2-(diphenylphophino)ethyl)-phosphine) crystal structure has been recently characterised.[161,162] The system has been identified to be stable within the superatom-complex model by an electron configuration of $1S^2/1P^6/1D^6/2S^2$ which leads to a large HOMO-LUMO gap.[161] Optical properties measured by experiments could be well reproduced by TD-DFT calculations and electronic transitions could be assigned to metal-to-ligand charge-transfer effects and core–core effects. An even larger cluster, namely $[Au_{39}(PPh_3)_{14}Cl_6]^-$, has been characterised with a gold core close to D_3 symmetry.[163] The gold core can also be descripted by two hexagonal antiprismas with a central gold atom. Also, for this cluster, electron-shell closure of the global electronic structure can be found by gap closing with F states and the appearance of G states above the gap (see Fig. 15).[43]

Larger clusters, such as the Schmidt $Au_{55}(PPh_3)_{12}Cl_6$ cluster with a size of 1.4 nm, have been mainly assigned by stoichiometry and geometry related ideas to an cubo-octahedral (later icosahedral or decahedral) gold cluster core.[134,164] The assignment has been debated experimentally as well as theoretically.[134,164–169] Gas-phase studies show for example that the Au_{55} cluster has no geometric shell closure.[170] The authors of ref. 135 shed light on the debate by theoretically analyzing several gold-phosphine-halide Au:PR3:X combinations with 55–69 gold atoms. All these clusters could be assigned to a gold core of approximately 1.4 nm. A strong correlation between the delocalised electronic shell closing and the cluster stability has been observed. The 58-delocalised electron-shell closing has been found as driving force behind the cluster energetics, also the HOMO-LUMO gap is increased for such cases.[135]

The most stable cluster has been determined to be $[Au_{69}(PR_3)_{20}Cl_{12}]^-$ which has a truncated decahedral 37-atom gold core which is encapsulated by 20 $AuPR_3$ and 12 AuCl units (see Fig. 16).[135] The C_5 symmetric system can be also considered as a layered structure with a top and bottom layer with 10 $AuPR_3$ propeller-like chiral conformations arranged as a ring and containing a central atom. Other stable structures, but slightly less preferred ones are $Au_{55}(PPh_3)_{12}Cl_6$ and $[Au_{69}(PR_3)_{15}Cl_{12}]^-$ (about 0.2 eV between the most stable structure and the following ones). Whereas the $[Au_{69}(PR_3)_{20}Cl_{12}]^-$ structure reveals shell closing by 58

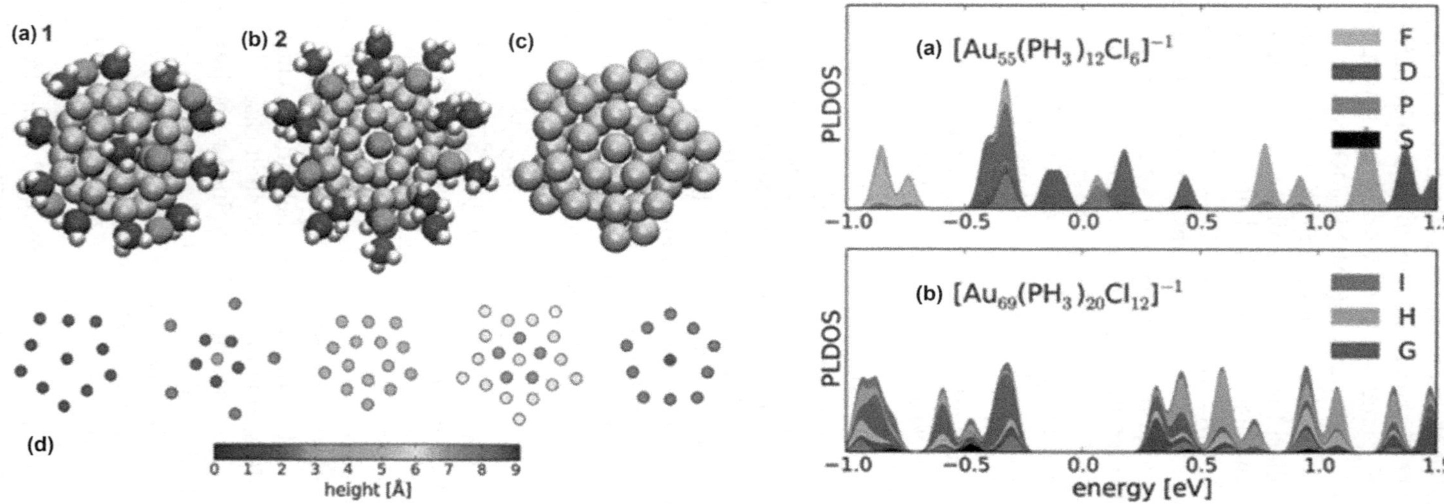

Fig. 16 Left: Structure model of ref. 171 (a) and energetically preferred structure (b) of ref. 135 for the $Au_{55}(PPh_3)_{12}Cl_6$ cluster. The gold core of (b) is shown in (c) and can be split into layers different in height (d). Right: Angular-momentum-projected local density of electron states (PLDOS) around the Fermi energy. Reproduced from ref. 135.

delocalised electrons with large HOMO-LUMO gap, the so far expected $Au_{55}(PPh_3)_{12}Cl_6$ shows a small HOMO-LUMO gap and no shell closure (see Fig. 16).[135] Considering several experiments so far, there are no precise determinations regarding the Au_{55} system, also the Au_{69} could be possible structure.[135]

5 Bonding to larger nanoparticles and surfaces

5.1 Pyridine interaction to surfaces and nanoparticles by periodic approach

The molecule pyridine is studied as a model system on gold surfaces, because it can be seen as a molecular building block for organic electronics or basic component for natural products or drugs. Also, the studies contribute to the understanding of ligand-stabilised nanoparticles. Considering the adsorption of a single pyridine to a gold surface (low-coverage limit), not so much is known experimentally for the gas phase by *e.g.* ultrahigh vacuum (UHV) studies. However, investigations of pyridine to single-crystal surfaces of metals such as Ag, Ni or Cu are done in UHV and reveal a coverage- and temperature-dependent behaviour of pyridine on the surface regarding the orientation (an overview presents Fig. 1 of ref. 172) which can be vertical or parallel.[‡]

The orientation of pyridine adsorbed to a Au(111) surface in solution has been investigated by chronocoulometry, scanning-tunnelling spectroscopy, difference-frequency generation and surface-enhanced infrared absorption.[173–177] For positive and zero potentials, pyridine shows the vertical orientation and is changing to flat orientation by applying a negative potential. The surface concentration implies one pyridine molecule at nine surface gold atoms, and the Gibbs energy has been estimated to be -10 kcal mol^{-1}. Another investigation of pyridine interacting with Au(100) and Au(110) indicates similar results as for the Au(111) surface.[§]

As the Au(111) surface is the most stable one, most investigations for the pyridine–gold interaction have been done for this interface. Considering the different slab models of periodic DFT calculations, several differences can be worked out. First, some studies have used the lattice parameter determined with the GGA functional which is mostly in the order of 4.17 Å, others used the experimental one of 4.08 Å. However, recent investigations indicate the later one in perfect agreement with PBE-D3 level of theory, thus when DFT-D3 is used, the experimental value seems to be fine to use.

‡Anyway, also a partial dehydrogenation has been observed on, *e.g.*, Cu(110) surfaces leading to the formation of α-pyridyl. This effect has not been found for the interaction with gold surfaces.

§Certain theoretical studies relate their results to a preferred parallel adsorption of pyridine to gold(111) based on ref. 177. Therefore one has to note, that a parallel arrangement just has been determined, when a negative potential has been applied, whereas for a zero potential also a perpendicular orientation could be found.

Further, the number of layers is varying from 3 to 5, and the molecules are either placed on one side or both sides to avoid dipol–dipol interactions. Zero to two layers are usually relaxed. Furthermore, the vacuum region is varied from 10 to 25 Å. Just few studies discussed tests regarding these parameters. The arrangement and orientation of pyridine on the gold surface is usually just descripted as vertical and parallel. However, it should be noted that there are several binding mode arrangements, which result in different binding structures and energies. The separation of the molecule from the surface which includes structural changes is called interaction energy.

Different methods, models and technical details lead to different results which is demonstrated on the vertical and flat adsorption modes of pyridine to a Au(111) surface (Table 1). Considering the bare DFT results, variances up to 4 kcal mol^{-1} occur for interaction energies using the same density functional due to different technical parameters and constructed structure models. However, all functionals result in preference of the vertical adsorption in agreement with experimental data due to the lack of standard DFT in treating the medium- and long-rang dispersion effects. The inclusion of medium- and long-range dispersion effects by Grimme's semiempirical approach or modified functionals brings all results closer to the estimated experimental value of the interaction energy. It can clearly be pointed out, that the accuracy of the dispersion correction determines the value of the dispersion–dominated interaction energy. Interaction energies vary up to 12.7 kcal mol^{-1} for the semiempirical D2 or D3 energy-add on and up to 22 kcal mol^{-1} considering all approaches to include dispersion effects. The experimentally determined vertical orientation has not been found by all approaches. The comparison shows the need of interpreting results carefully when applying dispersion correction or using modified functionals. The following variation is much stronger for this example than considering the technical details or differences in model systems. However, the best agreement to experimental data[¶] for the pyridine-Au(111) interaction shows PBE0-D3.[61]

A first computational study of the single pyridine molecule adsorbed to an Au(111) surface by Bilić *et al.* in 2002[178] has investigated intensively different configurations of pyridine on top of the surface. Four vertical and twelve perpendicularly oriented adsorption modes have been considered in top, bridge and 3-fold hollow sites (fcc and hcp) (see Fig. 17). Their DFT computations reveal the vertical on-top orientations as the preferred ones by interaction strength of about 7 kcal mol^{-1}, whereas all other configurations show rather weak binding energies in the range of 1.8 to 2.5 kcal mol^{-1}. DFT results are qualitatively in agreement with experimental results, but reproduce interaction energies and distances not sufficiently. The preference of the vertical orientation has been determined by DFT for several gold surfaces with the strength of

¶The interaction energy in the gas phase has been estimated to be around 22 kcal mol^{-1} based on the Gibbs energy of electrochemical experiments of pyridine to Au(111) in aqueous solution and considering the hydration energy of pyridine, pyridine–pyridine interactions, solvation changes and entropy effects.[174,178]

Table 1 Results of computation of the pyridine–Au(111) interaction in vertical and parallel orientation are given at different levels of theory and adjustment of different technical parameters and models. They are in the order as appearing in literature ranging from 2002 to 2014. The corresponding experimental value for the gas phase has been estimated to be 22 kcal mol^{-1}.[174,178]

Ref.	Method	Lattice par. (A)	Au–N Dist. (A)	Au–N BE (kcal mol^{-1})	Au-flat Dist. (A)	Au-flat BE (kcal mol^{-1})
178[a]	PW91	4.20	2.46	7.3	3.40	2.5
88[b]	PBE	4.17	—	6.5	3.60	1.4
	PBE-D2 (1)	4.17	—	18.2	—	16.4
	PBE-D2 (2)	4.17	—		—	29.1
179[c]	PBE	4.17	2.59	3.8	—	—
61[d]	PBE/-D3	4.08	—	4.9/17.8	—	1.1/17.1
	PW91	4.08	—	5.9	—	1.5
	PBEsol	4.08	—	10.6	—	0.8
	PBE0/-D3	4.08	—	5.0/19.5	—	0.4/18.6
91[e]	vdW1	4.08	2.80	14.8	3.68	14.8
	vdW2	4.08	2.90	10.8	3.93	12.0
	KBM	4.08	2.66	10.4	3.87	10.1
180[f]	PW91	4.17	2.68	3.2	—	—
90[g]	M06-L	4.17	—	7.8	3.2	8.1
	M06-L[a]	4.17	—	7.1	—	7.1
178,174	Exp.	22 kcal mol^{-1}, vertical orientation preferred				

[a] PW91/pw (US-PP), E_{cut} of 350 eV, 5×5×1 k-mesh, 0.2 eV smearing, 3×3 overstructure, 4 layers (top and bottom molecule and relaxed), 10 Å vacuum.
[b] PBE/pw(PAW-PP), D2 (C6 atomic properties, QM:QM approach), E_{cut} of 400 eV, 4×4×1 k-mesh (MP), 0.1 eV smearing, 3×3 overstructure, 5 layers (molecule and 2 layers relaxed), 25 Å vacuum.
[c] PBE/pw(PAW-PP), E_{cut} not given, 9×9×1 k-mesh, smearing not given, 2×2 overstructure, 5 layers (relaxation ns), 25 Å vacuum.
[d] PBE, PW91, PBEsol, PBE0/pw(PAW-PP), D3 for PBE and PBE0, Ecut of 400 eV, 6×6×1 and 12×12×1 k-mesh, 0.2 eV smearing (MP), 3×3 overstructure, 3 layers (molecule relaxed), 19 Å vacuum.
[e] vdW-DF1, vdW-DF2, KBM/localised basis set (long-range PAO, norm-consisting PP), 2×2×1 (MP), 5×5 overstructure, 4 layers (molecule and two layer relaxed), 25 Å vacuum.
[f] PW91, HSE/pw-PAW, E_{cut} not given, 8×5×1 k-mesh, 0.2 eV smearing, P(SQRT3×3) overstructure, 5 layers (molecule and 2 layer relaxed), >20 Å vacuum.
[g] M06-L, M06-L, PBE/pw-PAW, E_{cut} not given, 8×5×1 k-mesh, 0.2 eV smearing, 4×4×1 overstructure, 4 layers (molecule and 2 layer relaxed), >20 Å vacuum.

the interaction energy along Au(111) < Au(100) < Au(110) (3.8 kcal mol^{-1} to 10.1 kcal mol^{-1}).[179] The trend of interaction energies agrees with Gibbs energies of pyridine on Au(hkl) surfaces obtained by chronocoulometry.[174,181,182]

Applying D2 or D3 dispersion correction to DFT values for the pyridine–Au(111) interaction has shown that the dispersion correction is the dominant part of the interaction energy for either the vertical or parallel arrangement on the surface. The dispersion coefficients for the DFT-D2 calculations have been determined by atomic properties or *via* the QM : QM approach.[88] The determination of the C_6 coefficients by atomic properties has given better results for different aromatic molecules on the gold surface compared with experimental studies, whereas the QM : QM approach leads to an overestimation of the binding situation. Including the dispersion correction D2 or D3 changes the

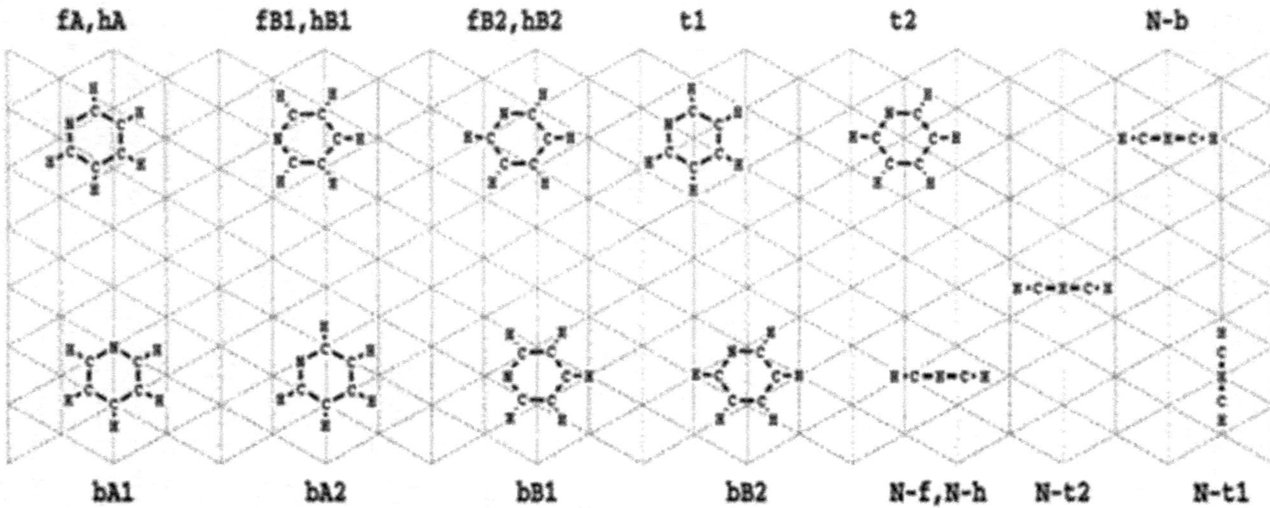

Fig. 17 Initial guess adsorption modes of pyridine on a Au(111) surface reproduced from ref. 178.

interaction energy of the parallel arrangement remarkably from nearly no interaction to 16 up to 29 kcal mol^{-1}.[61,88] The vertical and parallel adsorbed pyridine on gold yields an interaction energy that is very similar to the slight preference of the vertical orientation, which is in agreement with studies in solution.

Highly parameterized meta-GGA functionals have also been used to describe the pyridine-Au(111) binding situation and yield about 7–8 kcal mol^{-1} for vertical as well as parallel arrangements.[90] Thus, the functional obtains improved interaction energies compared to, *e.g.*, the standard functional PBE without any dispersion correction, but the experimentally preferred vertical orientation cannot be reproduced. Better descriptions of dispersion-corrected functionals give van der Waals exchange-correlation functionals in combination with localised basis sets. Interaction energies up to 70% in agreement with experimental results have been obtained.[91] The interaction energies are very similar, but the parallel arrangement is preferred which is not in agreement to the experimental finding. Thus, DFT-D3 approaches give best accuracy regarding pyridine–Au(111) adsorption.

Regarding the variation of the adsorption modes, it has been found that the vertical adsorbed pyridine can be tilted by about 30° without having a notable change in the interaction energy.[178] While no transition going from the vertical orientation to the parallel arrangement by tilting the molecule and considering a variation of the molecule-surface distance at the same time has been found at DFT level of theory, there has been one transition encountered by DFT-D3 calculations. This occurs at an Au–N distance of 2.7 and at a tilting angle of the molecule to the surface plane of about 50°; the interaction-energy change for this transition is just about 2 kcal mol^{-1} (see Fig. 18).[61] Furthermore, the parallel adsorbed pyridine ring can be moved above the surface by energy differences of about 5 kcal mol^{-1}.[178]

Besides the interaction energy, also the molecule–surface distance is strongly changed by inclusion of dispersive effects. Therefore, the dispersion correction as simple energy add-on also changes the electronic structure due to closer molecule–surface distances. Tonigold *et al.* have shown the increase from DFT to DFT-D2 for the small covalent contribution by charge-difference plots.[88] Electronic-structure calculations (LDOS) indicate a hybridization of the p_z-orbital of nitrogen (n_p) and Au d_{z^2}-states for the vertical adsorbed pyridine.[61,178] Furthermore, an interaction with the π-ring and gold-states occurs combined with a slight shift to smaller energies. Differential change-density plots offer just a little net charge transfer for the system.[88] Nevertheless, a charge transfer of 0.1 e would be enough to account for the 1.2 D change in effective dipole moment which has been determined.[178] The parallel adsorbed pyridine ring shows just the interaction of Au d-states and the π states.[61] The binding nature of the pyridine–gold surface interaction agrees well with the one determined for the pyridine–gold complex.[61,114]

Next to the adsorption on ideal surfaces, the adsorption of pyridine on gold ad-atoms on the surface has been investigated as a simple model for studying the interaction energy and nature on edges and corners

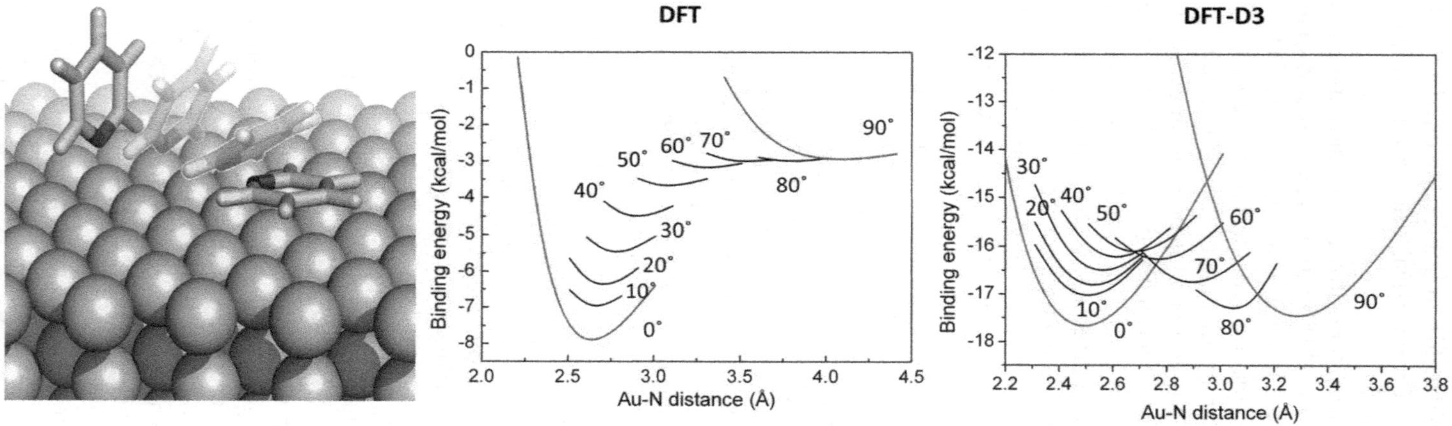

Fig. 18 Potential energy surfaces of tilting pyridine between the perpendicular and parallel adsorption modes on a Au(111) surface. Reproduced with permission from ref. 61. Copyright 2013 American Chemical Society.

(low coordinated gold atoms) of nanoparticles or rough surfaces. A pitting of atoms out of the gold surface is not expected for the weak binding pyridine.[61] The interaction energy of pyridine vertical adsorbed to the gold ad-atom is slightly stronger (about 4 kcal mol^{-1}) than to the ideal Au(111) surface.[61] The bare DFT contribution to the interaction energy has been increased from the surface (30%) to the ad-atom (80%), which indicated a change in binding nature. Electronic-structure analysis by means of LDOS as well as bond analysis for pyridine–gold cluster models reveal further evidence of the change of binding nature from weak dispersive-dominated interaction for the ideal (111) surface to covalent for low coordinated gold atoms on the surface.[61] Similar results have been found for parallel or tilted pyridine molecules in interaction with the ad-atoms (see Fig. 19). However, the extension of low coordinated gold atoms to a gold row on the surface yields similar results for the pyridine adsorption, but with weaker interaction energies. The outcome is generally in agreement with the general assumption of binding modes in gold catalysis.[5]

In regard to understand Raman scattering spectroscopic studies, the *zz*-component of the polarizability of clean and pyridine-modified

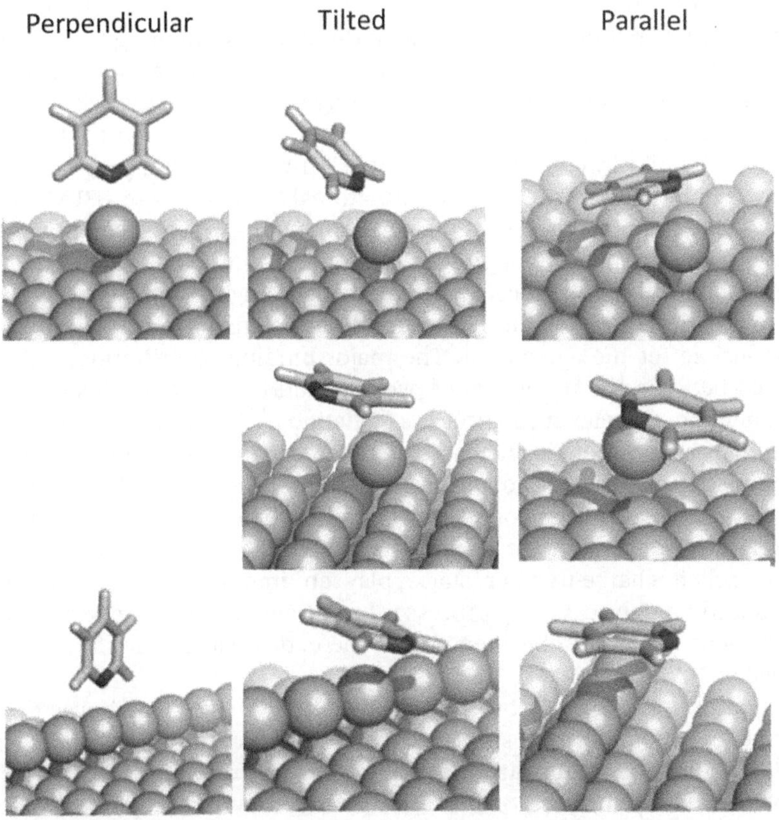

Fig. 19 Preferred perpendicular, tilted, and parallel adsorption structures of pyridine on an Au(111) surface (calculated at DFT-D3 level of theory). Reproduced with permission from ref. 61. Copyright 2013 American Chemical Society.

surfaces as well as the Raman scattering factors for the ring-breathing mode have been calculated by applying a finite field.[179] The results reveal that the covered surfaces get polarizable in the *z*-direction along the same order as the interaction energy for different gold surfaces [Au(111) < Au(100) < Au(110)]. The same holds for the Raman signal strength of the considered ring breathing modes.[183] The proposal of higher enhancement by stronger interaction with the surface would also be of importance for gold-ad-atoms and rough surfaces.

5.2 Pyridine interaction to surfaces by finite cluster approach

Next to periodic pyridine–gold surface calculations also the cluster approach to model the surface has been employed to understand the adsorption modes and the nature of the pyridine–gold interaction. The pyridine–gold cluster model calculations have been specially carried out to interpret and analyse Raman spectroscopy studies or to have a higher variety of methods. The cluster models for the surface are created in different ways. Smallest surface models include just 2 to 4 atoms with the intention to model a part of a rough surface. Different geometries of the clusters are assumed and lead to different ranges of results regarding the binding properties.[184,185] A benchmark study with various density functionals of mainly the pyridine gold dimer/cluster binding *via* nitrogen has been performed against the MP2 approach. Structural values and interaction energies are strongly depending on the method of choice. The binding energies for cluster numbers $n = 2$–4 are very similar in the order of 26 kcal mol^{-1} by choosing for example B3LYP functional (one exception represents one cluster with four gold atoms).[185] Also, these B3LYP functional results are in agreement with experimental surface calculations, but one has to note that the rough surface model is compared to experimental data of an ideal gold surface. The change in binding nature and interaction energy as described for the adsorption to ad-atoms on surfaces might be responsible for the agreement. The major binding contribution is determined between the n_N orbital of pyridine and the metallic cluster in accordance to complex and surface calculations.[185] However, the covalent binding contribution is increased compared to the later ones and the binding nature assigned to a chemisorption related to a surface. A minor contribution of binding comes from weak back-donation from the metal atoms to the antibonding π-orbital of pyridine. TD-DFT calculations reveal that excited charge-transfer states play an important role in surface-enhanced Raman scattering processes for the small cluster models.[185] Best agreement between experimental and theoretical vibrational frequencies reveals B3LYP.

Another approach has been used by an Au_{20} cluster which was determined by photoelectron spectroscopy in combination with relativistic DFT calculations to appear in tetrahedral geometry where the four faces represent (111) surfaces of bulk face-centered cubic (fcc) gold.[186] Two binding models to the Au_{20} cluster have been considered, one where pyridine binds to the centre of a surface facet and one where pyridine binds to the corner of the tetrahedron (see Fig. 20). The binding mode of

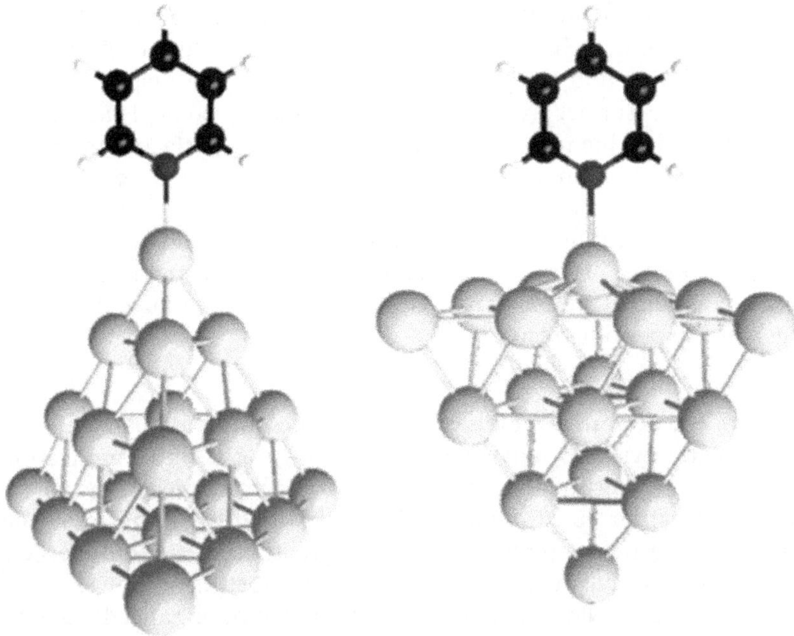

Fig. 20 Two pyridine–Au_{20} cluster models to describe the pyridine–gold surface interaction. Reprinted with permission from ref. 186. Copyright 2006 American Chemical Society.

pyridine to the corner of the tetrahedron results in an about three times larger interaction energy than the one to the surface facet, and SERS enhancement is about one magnitude larger. This indicates the enhanced SERS effect for rough modelled gold surfaces.

A similar study with pyridine modelled on a Au_{20} cluster, constructed like a surface facet, has obtained that the gold–nitrogen (pyridine) distance decreases or increases due to the applied external electric field, and accordingly the charge-transfer effect gets influenced.[187] As a consequence, the energies of the charge-transfer transition are tuned to be or not to be resonant with incident light, which result in selectively enhanced Raman intensities sensitive to the direction of the field. Another approach employed a cluster composed of 28 gold atoms to simulate a gold surface. 4-Cyanopyridine has been adsorbed to a (111) surface facet of the cluster model.[70,82] DFT and MP2 calculations reveal potential-dependent adsorption modes with vertical orientation for negative potential and flat orientation for positive potential. However, this outcome is not in agreement with experimental studies, a reason could be the use of a pseudo potential, which covers a 78 electron core in combination with a small basis set and no applied dispersion correction.

With respect to cluster calculations which model surface facets, recent investigations model the pyridine gold complex over gold clusters different in size up to surfaces.[83,84] It could be shown that the results and accuracies of the DFT approach are strongly depending on the cluster size and corresponding binding nature. Overall, PBE-D3 and PBE0-D3 have been chosen to give the best performance with respect to CCSD(T) or SCS-MP2 outcome at basis set limit. An analysis of the binding nature between pyridine and the gold system shows that the dispersion-dominated pyridine–surface interaction can be reliably modelled *via* a cluster approach, which includes a fragment of the gold surface and, for example, is embedded in a dispersion correction of an extended cluster.[84] Therefore, a relatively small Au_{10} cluster is necessary to obtain results related to the binding properties close to the periodic calculation of Au(111) surfaces.[84] However, if ad-atoms are to be considered the approach fails and large cluster models have to be constructed to be in agreement to periodic results.

5.3 Pyridine derivatives interaction to surfaces and nanoparticles

The aromatic pyridine molecule is interesting for certain applications due to the possibility of changing the electronic properties of the ring by donor or acceptor substituents and thereby modifying the molecule–surface/nanoparticle interaction. Accordingly, the interaction energy of pyridine substituted in position 4 by donor substituents ($-CH_3$, –OH, $-OMe_3$, $-NH_2$ and $-NMe_2$) to a Au(111) surface can be enhanced along the donor strength by 2 to 11% (DFT-D3 level of theory) considering the vertical adsorption mode (see Fig. 21).[61,||] The donor substituent pushes electron density in the pyridine ring and nitrogen can bind a bit more efficient to the surface. The binding of an acceptor substituent such as 4-Nitropyridine yields in contrast an about 5% weaker binding situation. Also, the parallel orientated pyridine derivatives at the Au(111) surface increase in interaction strength, but depend on the nature and number of substituent atoms which interact dispersive with the surface. Enhancement of the interaction energy is reached between 16 to 58%

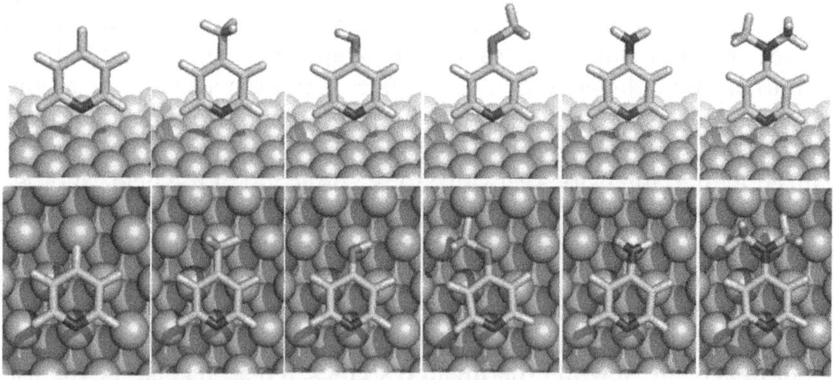

Fig. 21 Perpendicular and parallel adsorption structures of pyridine derivatives (ordered along the increasing donor strength of the substituent) on a Au(111) surface. The preferred orientation is marked in red. Reproduced from ref. 61.

[||] The overall binding situation is very similar to the pyridine complex and the same trend has been observed.

(see Fig. 21). Also the acceptor-substituent (–NO_2) resulted understandably in an enhancement of the binding energy by 19%. Comparing the absolute values of the interaction energies, the enhancement is remarkably stronger for the parallel orientation than for the vertical one, and a change of the preferred orientation occurs from pyridine vertical to substituted pyridine in a parallel arrangement.[61]

Experimental studies by Raman spectroscopy and NMR techniques indicate that 4-methoxypyridine-stabilised gold nanoparticles (about 6 nm) prefer a parallel adsorption mode,[41] which is in agreement to the theoretical finding, not taking into account coverage or solvent effects. More intensive DMAP in interaction to gold nanoparticles sized 5.5 nm have been studied.[40,47] In agreement to the theoretical calculation, no evidence for a covalent bonding between DMAP and the gold nanoparticles has been found and a weak, labile donor-acceptor complex with atoms of the nanoparticle surface is discussed. Surface–plasmon resonance spectroscopy (SPR) which determine the thickness of the DMAP monolayer as well as investigations in acid milieu conclude in contrast to the theoretical calculations a vertical adsorption mode to nanoparticles.[40] Also, surface-enhanced infrared investigations of DMAP adsorbed to gold surfaces which contains polycrystalline and Au(111) regions evidence vertical adsorption modes.[188] However, solvent effects and also coverage effects, which not have been treated within the DFT calculations might have an influence to the adsorption mode. Also, the scaling of the dispersion correction which has been identified to increase too strong along system size for pyridine-graphene system might have an impact.[189] Similar to pyridine, DMAP has been investigated on ad-atoms and an enhancement of the binding strength has been found.[61] Thus, for rough surfaces or edges and corners of nanoparticles (which get more important for smaller nanoparticles with higher edge/facet ratio), the binding and nature might be different than to the surface facets.

A parallel arranged molecule–surface interaction with weak as well as strong binding units, is the 4,4′-dithiopyridine (PySSPy)-gold system which has been studied under UHV conditions and low temperature.[190] The molecule, which has been determined in a boat-like conformation, yields interaction energies between 32 and 43 kcal mol^{-1} depending on the level of theory. DFT-D3 or vdW-DFT reveal a large contribution of dispersion effects, but also a significant covalent contribution is seen by charge-density plots. The Tersoff–Hamann approach has been utilised to simulate STM images and revealed an oblong alignment with the N–N direction in agreement with cigar-like experimental STM images (see Fig. 22). Three molecules resemble to a triskelion with molecules rotated by 60 °C (small energy gain of 0.1 eV per bond) as found theoretically and experimentally.

Zhang *et al.*[191] have investigated the self-assembly of 2,4,6-tris(2-pyridyl)-1,3,5-triazine (TPTZ) on Au(111) in ultrahigh vacuum at low temperature. They compared STM images with their DFT calculations (no dispersion correction) and found agreement that the adsorbed TPTZ undergoes a spontaneous chiral resolution on the surface. Waldmann *et al.*[192] studied the oxidation of a pyridin-based adlayer on Au(111) in

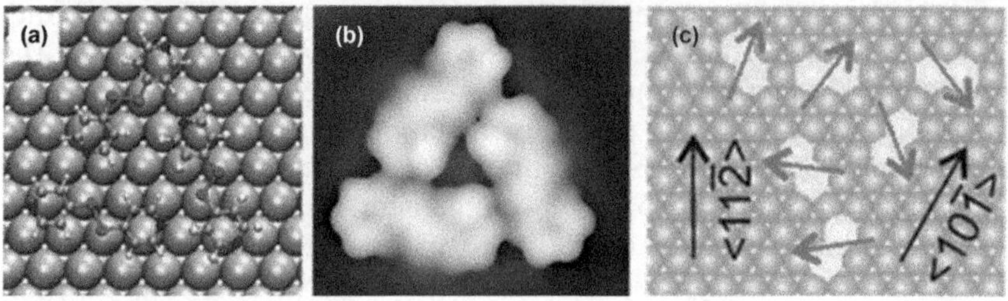

Fig. 22 (a) DFT-optimized triangular arrangement and (b) simulated STM image of PySSPy molecules adsorbed to an Au(111) surface. (c) The adsorbed molecule on the Au(111) surface as a schematic model. Above the light marked surface atoms, the S–S axis of the molecules lies; the arrows indicate the orientation of the N–N-axis. Reprinted with permission from ref. 190. Copyright 2013 American Chemical Society.

comparison to Ag(111) by DFT-D as model system for inorganic/organic contacts. Other heteroaromates are for example included in studies of phenanthrolin and dipyridophenazin,[193] melamine films,[194] Histidine,[195,196] different nucleobases or [3 + 2] cycloaddition reaction between azides and alkynes on Au(111) surfaces.[196–198] Questions remain mostly on the interpretation of STM or STC images, adsorption modes or regio-isomer stability. Whereas early studies did not include dispersion correction, the newer ones showed that the inclusion of dispersive effects is crucial for the accurate description of the systems.

5.4 Pyridine derivatives as wires between gold contacts

Pyridine docking groups have been found to show bonding-induced rearrangements to interfaces, which have the potential to small charge injections that are interesting for molecular electronics.[199–203] Hence, pyridine-based SAM models containing one to three pyridine units as linkers and electron-donating or electron-accepting tailing groups (4th position) have been studied on Au(111) surfaces (herringbone pattern) by DFT calculations.[180] The nearly vertical bound SAM models yield interaction energies between 3.2 to 4.6 kcal mol^{-1}, because no dispersion correction has been applied. An enhancement of the binding strength, except when employing the $-NO_2$ tailing group, has been obtained for adding more than one pyridine unit to the linker.[180] Electronic structure (calculated by PDOS) for the free-standing monolayers are quite complex with close-lying σ and π-states which change by a function of the substituent. Charge-rearrangements along the whole molecular backbone have been associated with local polarization of the SAMs. Very small barriers for electron injections are found for all adsorbed systems.

Built-up, linked pyridine units between two gold contacts have been studied by several authors in regard to understand molecular electronics. Thus, 4,4′-bipyridine has been considered between different electrode models, namely, a dicomplex, trimer units and extended Au_{10} cluster by DFT calculations based on Green's function method.[204] Significant binding has been obtained just for the interaction *via* the nitrogen atom. The structure of the optimised gold–molecule–gold systems (interaction energy, Au–N distance and molecule torsion) strongly varies with gold cluster size and arrangement. The electronic structure of the molecule placed between the electrode models changes from discrete to continuous. The calculated transmission coefficient fits very good to experimental data. Stretching the Au–N bond decreases the transmission, whereas the torsion angle change has just a very small impact to it.

Another study has modelled the contact geometries for the gold electrode as a Au(111) surface with one to nine ad-atoms constructed to a plateau.[206] Always an atop N-binding pyridine molecule has been placed above the central ad-atom. The DFT calculations yield Au–N bond length and interaction energy variations (9 to 18 kcal mol^{-1}) depending on the number of ad-atoms.[206] The latter one decreases with increasing number of ad-atoms and indicates a bonding between van der Waals interactions and a covalent bond. Based on the varying Au–N bond length the derivatives at the inflexion point are used to calculate breaking forces, which range between 1.2 nN to 0.3 nN.[206] Thus, the forces, but also the conductance's depend on the modelled contact geometries.

DFT-D2 calculations in comparison to AFM measurements have been employed to study the interplay between the so far described Au–N binding and vdW interactions between pyridine terminated structures (4,4-bipyridine and the 1,2-bis(4-pyridyl)ethylene) and gold electrodes.[205] For the low-conductance case, modelled by a gold dicomplex, the dispersion contribution is minimal to the bond rupture force. The high conductance of a rough electrode surface is modelled by a Au(111) surface in addition with a ridge of two layers (Fig. 23). The chemical Au–N

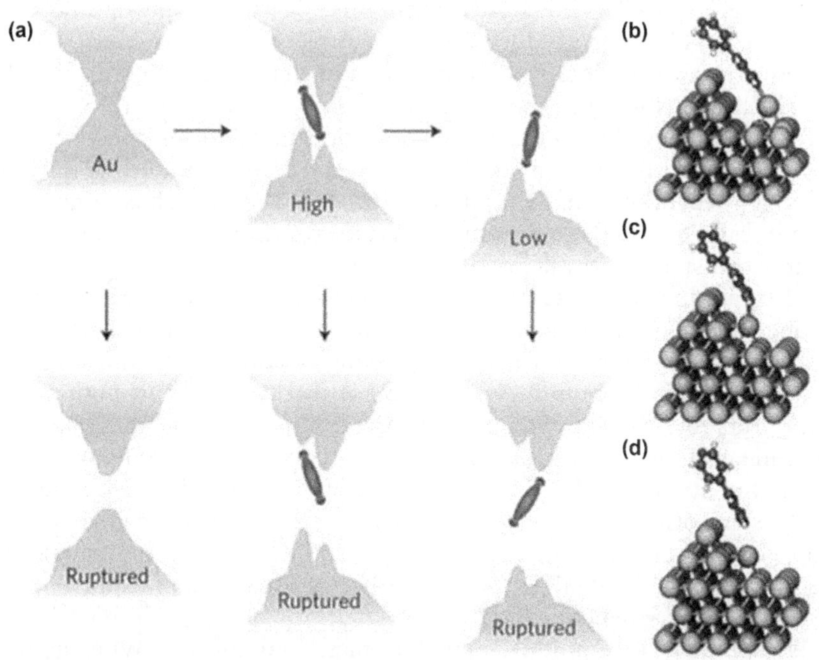

Fig. 23 (a) Structural pathways of experimental observations where high-conductance junctions result in structural rearrangement and a switch to the low-conductance one. (b)–(d) Different model structures for DFT calculations. Reproduced from ref. 205.

binding is calculated to be about 23 kcal mol^{-1}, and an additional contribution of dispersion correction is found with 18 kcal mol^{-1}.[205] They simulated the junction elongation by a vertical movement of the molecule and relaxing the overall structure but freezing the nitrogen atoms. In agreement with the experiment an abrupt structural rearrangement has been found. The simulations indicate that the molecule pulls the adatom to an adjacent hollow site which is accompanied by a force drop. The Au–N bond maintains. Along the elongation, the pyridine ring passes the edge and an additional loss of vdW interaction energy results to the maximum force. Both contributions are necessary to explain the experimental measured forces.[205]

A model study for 4,4-bipyridine gold nanojunctions has been performed by pyrazines which are substituted in position 2 connecting with gold nanoclusters (six atoms build to a tip).[207] DFT has demonstrated that electron-withdrawing substituents lead to reduced rupture forces necessary to disconnect the wire. Modification of the cluster model to gold chains lead to gold chain rupture for chains containing two or more gold atoms in line. This is explained by the fact that rupture forces for breaking the gold–gold bonds are similar in size to the Au–N bond breakage. The increase of interaction energy for pyridine gold dimer systems is in agreement to this.

6 Bonding at different scales

The phosphine and pyridine gold system interaction has been considered over different gold system scales within this chapter. The phosphine and pyridine (N-binding) gold complex (neutral, with one ligand) holds together by a weak binding with the origin in charge transfer and a small covalent contribution. The gold phosphine complex exhibits a bit stronger binding due to the lower ionization potential of phosphorous compared to nitrogen. The binding nature and strength in the complex is strongly depending on the charge state. The charged phosphine and pyridine (N-binding) gold complex is strong in nature by increased charge transfer and covalent contribution.

Also, an increasing of the gold system to the neutral gold dimer changes the interaction strength to phosphine or pyridine. The enhancement can be explained by the missing electron in the antibonding $n_{P/N}$/golds orbital; accordingly an increase in covalent contribution occurs. Thereby, the interaction strength is strongly depending on the alignment of the dimer ligand system, thus maximal values are obtained for an Au–Au–N/P angle of 180°. An angle of 90° results in weak binding in the same order as the complex.[84] This effect is also evidenced for small and neutral gold clusters covered by phosphine, the Au–Au–P angle is aligned to 180° if possible (steric hindrance for bulky phosphine's or cluster geometry dependent).[42]

Phosphine is a common ligand for small gold clusters. Whereas for very small clusters such as Au_6 to Au_9 the interaction energy is in middle interaction strength (about 35 kcal mol^{-1}), values have increased by a factor of two for charged clusters.[42] The increase has its origin in charge-transfer effects. Similar strong interactions have been found for Au_{20} clusters passivated by phosphine ligands. However, in the superatomic concept the phosphine cluster interaction (also within mixed ligand shells) has been described by a weak interaction, for which the cluster's geometric and electronic structure does hardly get affected. It has been discussed that phosphine ligands next to thiol and halide ligands just sterically fulfil the ligand shell. Considering the weak, dispersion-dominated phosphine gold surface interaction (which has not been calculated, but experimentally investigated), the later statement is reasonable for larger clusters and nanoparticles.[208]

Also, a weak and dispersion-dominated interaction has been estimated for the pyridine–gold surface interaction. In contrast to phosphine, two binding modes (vertical and parallel) similar in binding strength result for the aromatic ring system. The N-binding situation holds together by interaction between n_N and Au states, whereby just a very small covalent contribution as well as charge transfer is obtained. The π-binding has its origin in dispersive π-states and Au d-states interaction. The adsorption of pyridine to low-coordinated gold atoms on the surface (*via* nitrogen or carbon atoms) leads to a change in binding nature, the covalent contribution is increasing. However, binding strength and nature is strongly depending from the geometric arrangement of the ad-atoms (*e.g.* number of ad-atoms as well as arrangement). This has been also demonstrated for the modelling of molecular wires between gold contacts. The binding nature and strength is highly depending from the used structure of the modelled contact.

In the context of ligand-stabilised clusters and nanoparticles, the ligand-gold interaction can be concluded to be strongly size, charge and shape dependent. Considering the size effect, the ligand interaction to small clusters can be considered to be strong and covalent in nature, whereas large nanoparticles should show weak interaction to surface facets and intermediate interaction to edges and corners. In today's quantum-chemical modelling mainly small ligand-covered cluster systems have been investigated by finite computational studies, and large ligand-passivated nanoparticles have been modelled as ligand–surface interaction by periodic boundary conditions. The reason lies in the difficulty of the description of the size-dependency of ligand-surrounded metallic nanoparticles with current methods. Size-dependent investigations of nanoparticle systems to date have been mostly employed for uncoated systems.[62,209–212] For example DFT, force field approaches, QM : MM or statistical methods have been used over a range of sizes. The understanding of ligand-stabilised nanoparticles is experimentally highly important in order to control nano-system properties, thus, strategies should be developed to model the size-dependency of ligand-surrounded clusters and nanoparticles over a large scale.

7 Conclusions

The chapter focused on the weak binding phosphine and pyridine ligands in interaction with gold clusters and nanoparticles. In order to learn about the binding nature and properties also corresponding complex and surface studies have been analysed. On the one hand, finite electronic-structure methods are employed for complexes, small clusters or surface models in literature, on the other hand calculations with periodic boundary conditions for ligand–surface (nanoparticle) interactions can be found. The differently used quantum-chemical approaches for finite or periodic calculations have been discussed. Whereas electron correlation methods are rarely used for benchmarking, the DFT approach is the method of choice. It has been pointed out that the inclusion of dispersion correction to DFT is crucial in order to describe reliable the ligand–gold surface interaction but also complexes or clusters. Comparing different results of different dispersion corrected DFT approaches a large variation of interaction energies has been obtained. Thus, accurate corrections are necessary to describe dispersion-dominated ligand-gold system interactions.

Electronic structure calculations have been widely employed to understand phosphine covered gold clusters. The binding nature is covalent and for charged systems very strong due to additional charge-transfer effects. The basic phosphine–gold interaction can be well understood from neutral and charged gold phosphine complexes. For the initial cluster growth starting from $AuClPR_3$ salt the calculations have indicated the necessity of a reducing agent to initiate the formation of the first gold–gold bond. The thermodynamically preferred leaving group is chlorine compared to phosphine, and accordingly the cluster form $Au_x(PR_3)_yCl_z$ with $x>y>z$ can be explained. The origin of optical properties could be mainly assigned to gold-to-ligand transitions. Some phosphine-covered gold clusters can be assigned by superatomic-complex concepts to special stability in agreement to experimental studies.

The pyridine surface interaction has been investigated by DFT and DFT-D2/D3 methods to understand the binding situation, ligand-nanoparticle stabilisation, vibrational or electronic properties. The binding can take place in a weakly chemisorbed N-binding or *via* a dispersion-dominated π-binding situation. Adsorption to modelled edges, corners or ad-atomic constructed tips increases the covalent contribution and thus changes the binding nature. Also, therefore, the interaction in complexes and dimer systems improves the basic understanding of the interface interaction. The theoretical modelling has been used to determine the orientation and binding strength of pyridine derivatives on gold surfaces or molecular wires have been studied to one or two gold contacts in order to interpret experimental data.

Overall the review shows a strong size, charge and shape dependency of the phosphine and pyridine ligands binding to gold clusters and nanoparticles surfaces and binding-strength ranges from weak to strong, binding nature from dispersive to covalent. In order to model ligand-stabilised clusters and nanoparticles for nanotechnology reliably, theoretical models should be constructed carefully regarding their structure, and methods have to be employed which can cover weak as well as strong interactions. Future approaches to describe accurately the size-depended ligand-passivated clusters and nanoparticles over a larger size will be important in order to understand and interpret experimental data and optimize relevant properties.

References

1 M.-C. Daniel and D. Astruc, Gold Nanoparticles: Assembly, Supramolecular Chemistry, Quantum-Size-Related Properties, and Applications toward Biology, Catalysis, and Nanotechnology, *Chem. Rev.*, 2004, **104**, 293–346.

2 Y. Zhu, H. Qian and R. Jin, Catalysis opportunities of atomically precise gold nanoclusters, *J. Mater. Chem.*, 2011, **21**, 6793–6799.

3 R. Meyer, C. Lemire, S. K. Shaikhutdinov and H. J. Freund, Surface chemistry of catalysis by gold, *Gold Bull.*, 2004, **37**, 72–124.

4 W. Jahn, Review: Chemical Aspects of the Use of Gold Clusters in Structural Biology, *J. Struct. Biol.*, 1999, **127**, 106–112.

5 B. Hvolbæk, T. V. W. Janssens, B. S. Clausen, H. Falsig, C. H. Christensen and J. K. Nørskov, Catalytic activity of Au nanoparticles, *Nano Today*, 2007, **2**, 14–18.

6 M. Haruta, Gold as a novel catalyst in the 21st century: Preparation, working mechanism and applications, *Gold Bull.*, 2004, **37**, 27–36.

7 M. Haruta, Nanoparticulate gold catalysts for low temperature CO oxidation, *J. New Mater. Electrochem. Syst.*, 2004, **7**, 163–172.

8 J. Kilmartin, R. Sarip, R. Grau-Crespo, D. Di Tommaso, G. Hogarth, C. Prestipino and G. Sankar, Following the Creation of Active Gold Nanocatalysts from Phosphine-Stabilized Molecular Clusters, *ACS Catal.*, 2012, **2**, 957–963.

9 V. R. Choudhary, A. Dhar, P. Jana, R. Jha and B. S. Uphade, A green process for chlorine-free benzaldehyde from the solvent-free oxidation of benzyl alcohol with molecular oxygen over a supported nano-size gold catalyst, *Green Chem.*, 2005, **7**, 768–770.

10 V. Choudhary and D. Dumbre, Supported Nano-Gold Catalysts for Epoxidation of Styrene and Oxidation of Benzyl Alcohol to Benzaldehyde, *Top. Catal.*, 2009, **52**, 1677–1687.

11 N. Dimitratos, J. A. Lopez-Sanchez, D. Morgan, A. Carley, L. Prati and G. J. Hutchings, Solvent free liquid phase oxidation of benzyl alcohol using Au supported catalysts prepared using a sol immobilization technique, *Catal. Today*, 2007, **122**, 317–324.

12 D. Enache, D. Knight and G. Hutchings, Solvent-free Oxidation of Primary Alcohols to Aldehydes using Supported Gold Catalysts, *Catal. Lett.*, 2005, **103**, 43–52.

13 P. Haider, B. Kimmerle, F. Krumeich, W. Kleist, J.-D. Grunwaldt and A. Baiker, Gold-Catalyzed Aerobic Oxidation of Benzyl Alcohol: Effect of Gold Particle Size on Activity and Selectivity in Different Solvents, *Catal. Lett.*, 2008, **125**, 169–176.

14 T. G. Schaaff, M. N. Shafigullin, J. T. Khoury, I. Vezmar, R. L. Whetten, W. G. Cullen, P. N. First, C. Gutiérrez-Wing, J. Ascensio and M. J. Jose-Yacamán, Isolation of Smaller Nanocrystal Au Molecules: Robust Quantum Effects in Optical Spectra, *J. Phys. Chem. B*, 1997, **101**, 7885–7891.

15 R. B. Wyrwas, M. M. Alvarez, J. T. Khoury, R. C. Price, T. G. Schaaff and R. L. Whetten, The colours of nanometric gold, *Eur. Phys. J. D*, 2007, **43**, 91–95.

16 S. Chen, R. S. Ingram, M. J. Hostetler, J. J. Pietron, R. W. Murray, T. G. Schaaff, J. T. Khoury, M. M. Alvarez and R. L. Whetten, Gold

Nanoelectrodes of Varied Size: Transition to Molecule-Like Charging, *Science*, 1998, **280**, 2098–2101.
17 D. Lee, R. L. Donkers, J. M. DeSimone and R. W. Murray, Voltammetry and Electron-Transfer Dynamics in a Molecular Melt of a 1.2 nm Metal Quantum Dot, *J. Am. Chem. Soc.*, 2003, **125**, 1182–1183.
18 R. S. Ingram, M. J. Hostetler, R. W. Murray, T. G. Schaaff, J. T. Khoury, R. L. Whetten, T. P. Bigioni, D. K. Guthrie and P. N. First, 28 kDa Alkanethiolate-Protected Au Clusters Give Analogous Solution Electrochemistry and STM Coulomb Staircases, *J. Am. Chem. Soc.*, 1997, **119**, 9279–9280.
19 T. P. Bigioni, R. L. Whetten and Ö. Dag, Near-Infrared Luminescence from Small Gold Nanocrystals, *J. Phys. Chem. B*, 2000, **104**, 6983–6986.
20 G. Wang, T. Huang, R. W. Murray, L. Menard and R. G. Nuzzo, Near-IR Luminescence of Monolayer-Protected Metal Clusters, *J. Am. Chem. Soc.*, 2005, **127**, 812–813.
21 E. S. Shibu, M. A. H. Muhammed, T. Tsukuda and T. Pradeep, Ligand Exchange of Au25SG18 Leading to Functionalized Gold Clusters: Spectroscopy, Kinetics, and Luminescence, *J. Phys. Chem. C*, 2008, **112**, 12168–12176.
22 G. Ramakrishna, O. Varnavski, J. Kim, D. Lee and T. Goodson, Quantum-Sized Gold Clusters as Efficient Two-Photon Absorbers, *J. Am. Chem. Soc.*, 2008, **130**, 5032–5033.
23 M. Schulz-Dobrick and M. Jansen, Intercluster Compounds Consisting of Gold Clusters and Fullerides: [Au7(PPh3)7]C60·THF and [Au8(PPh3)8](C60)2, *Angew. Chem., Int. Ed.*, 2008, **47**, 2256–2259.
24 M. Schulz-Dobrick and M. Jansen, Structure-Directing Effects in the Supramolecular Intercluster Compound [Au9(PPh3)8]2[V10O28H3]2: Long-Range versus Short-Range Bonding Interactions, *Inorg. Chem.*, 2007, **46**, 4380–4382.
25 M. Schulz-Dobrick and M. Jansen, Characterization of Gold Clusters by Crystallization with Polyoxometalates: the Intercluster Compounds [Au9(dpph)4][Mo8O26], [Au9(dpph)4][PW12O40] and [Au11(PPh3)8Cl2]2[W6O19], *Z. Anorg. Allg. Chem.*, 2007, **633**, 2326–2331.
26 C. F. Shaw, *III In Gold: Progress in Chemistry, Biochemistry and Technology*, 1999.
27 E. C. Dreaden, A. M. Alkilany, X. Huang, C. J. Murphy and M. A. El-Sayed, The golden age: gold nanoparticles for biomedicine, *Chem. Soc. Rev.*, 2012, **41**, 2740–2779.
28 S. D. Brown, P. Nativo, J.-A. Smith, D. Stirling, P. R. Edwards, B. Venugopal, D. J. Flint, J. A. Plumb, D. Graham and N. J. Wheate, Gold Nanoparticles for the Improved Anticancer Drug Delivery of the Active Component of Oxaliplatin, *J. Am. Chem. Soc.*, 2010, **132**, 4678–4684.
29 B. Kang, M. A. Mackey and M. A. El-Sayed, Nuclear Targeting of Gold Nanoparticles in Cancer Cells Induces DNA Damage, Causing Cytokinesis Arrest and Apoptosis, *J. Am. Chem. Soc.*, 2010, **132**, 1517–1519.
30 C. M. Aikens, Modelling small gold and silver nanoparticles with electronic structure methods, *Mol. Simul.*, 2012, **38**, 607–614.
31 D. P. Anderson, R. H. Adnan, J. F. Alvino, O. Shipper, B. Donoeva, J.-Y. Ruzicka, H. Al Qahtani, H. H. Harris, B. Cowie, J. B. Aitken, V. B. Golovko, G. F. Metha and G. G. Andersson, Chemically synthesised atomically precise gold clusters deposited and activated on titania. Part II, *Phys. Chem. Chem. Phys.*, 2013, **15**, 14806–14813.
32 P. Maity, S. Xie, M. Yamauchi and T. Tsukuda, Stabilized gold clusters: from isolation toward controlled synthesis, *Nanoscale*, 2012, **4**, 4027–4037.
33 Y. Negishi, K. Nobusada and T. Tsukuda, Glutathione-Protected Gold Clusters Revisited: Bridging the Gap between Gold(I) – Thiolate Complexes

and Thiolate-Protected Gold Nanocrystals, *J. Am. Chem. Soc.*, 2005, **127**, 5261–5270.
34 M. A. Reed, C. Zhou, C. J. Muller, T. P. Burgin and J. M. Tour, Conductance of a Molecular Junction, *Science*, 1997, **278**, 252–254.
35 J. Chen, M. A. Reed, A. M. Rawlett and J. M. Tour, Large On-Off Ratios and Negative Differential Resistance in a Molecular Electronic Device, *Science*, 1999, **286**, 1550–1552.
36 J. M. Seminario, C. E. De La Cruz and P. A. Derosa, A Theoretical Analysis of Metal – Molecule Contacts, *J. Am. Chem. Soc.*, 2001, **123**, 5616–5617.
37 H. Hakkinen, The gold-sulfur interface at the nanoscale, *Nat. Chem.*, 2012, **4**, 443–455.
38 D. Leff, L. Brandt and J. R. Heath, Synthesis and Characterization of Hydrophobic, Organically-Soluble Gold Nanocrystals Functionalized with Primary Amines, *Langmuir*, 1996, **12**, 4723.
39 C. D. Bain, J. Evall and G. M. Whitesides, Formation of Monolayers by the Coadsorption of Thiols on Gold: Variation in the Head Group, Tail Group, and Solvent, *J. Am. Chem. Soc.*, 1989, **111**, 7155–7164.
40 V. J. Gandubert and R. B. Lennox, Assessment of 4-(Dimethylamino)pyridine as a Capping Agent for Gold Nanoparticles, *Langmuir*, 2005, **21**, 6532–6539.
41 H. Lange, J. Maultzsch, W. Meng, D. Mollenhauer, B. Paulus, N. Peica, S. Schlecht and C. Thomsen, Adsorption Behavior of 4-Methoxypyridine on Gold Nanoparticles, *Langmuir*, 2011, **27**, 7258–7264.
42 D. Mollenhauer, N. Gaston, unpublished.
43 M. Walter, J. Akola, O. Lopez-Acevedo, P. D. Jadzinsky, G. Calero, C. J. Ackerson, R. L. Whetten, H. Grönbeck and H. Häkkinen, A unified view of ligand-protected gold clusters as superatom complexes, *Proc. Natl. Acad. Sci.*, 2008, **105**, 9157–9162.
44 A. Barnard, X. M. Lin and L. A. Curtiss, *J. Phys. Chem. B*, 2005, **109**, 24465.
45 M. McPartlin, R. Mason and L. Malatesta, Novel cluster complexes of gold(0)-gold(I), *J. Chem. Soc. D*, 1969, 334.
46 D. M. P. Mingos, *Polyhedron*, 1986, **3**, 1289–1297.
47 D. I. Gittins and F. Caruso, Spontaneous Phase Transfer of Nanoparticulate Metals from Organic to Aqueous Media, *Angew. Chem., Int. Ed.*, 2001, **40**, 3001–3004.
48 D. I. Gittins, A. S. Susha, B. Schoeler and F. Caruso, Dense Nanoparticulate Thin Films via Gold Nanoparticle Self-Assembly, *Adv. Mater.*, 2002, **14**, 508–512.
49 T. Biver, A. Corti, N. Eltugral, E. Lorenzini, M. Masini, A. Paolicchi, A. Pucci, G. Ruggeri, F. Secco and M. Venturini, Analysis of 4-dimethylaminopyridine (DMAP)-gold nanoparticles behaviour in solution and of their interaction with calf thymus DNA and living cells, *J. Nanopart. Res.*, 2012, **14**, 1–12.
50 J. Lipkowski and L. Stolberg, *In Adsorption of Molecules at Metal Electrodes*, ed. J.Lipkowski and P. N.Ross, VCH, New York, 1992, p. 171.
51 G. D. Henry, De novo synthesis of substituted pyridines, *Tetrahedron*, 2004, **60**, 6043–6061.
52 P. Pyykkö, Theoretical Chemistry of Gold, *Angew. Chem., Int. Ed.*, 2004, **43**, 4412–4456.
53 F. Mohr, *Gold Chemistry: Applications and Future Directions in the Life Sciences*, WILEY-VCH Verlag GmbH & Co. KgaA, Weinheim, 2009.
54 P. Pykkö, *Relativistic Theory of Atoms and Molecules*, Bd III; Springer, Berlin, 2000.
55 P. Schwerdtfeger, Relativistic effects in properties of gold, *Heteroat. Chem.*, 2002, **13**, 578–584.

56 P. Schwerdtfeger, Nanostrukturen von Gold – von kleinen Clustern zu niederdimensionalen Anordnungen, *Angew. Chem.*, 2003, **115**, 1936–1939.
57 P. Pyykkö, Theoretical chemistry of gold. II, *Inorg. Chim. Acta*, 2005, **358**, 4113–4130.
58 P. Pyykko, Theoretical chemistry of gold. III, *Chem. Soc. Rev.*, 2008, **37**, 1967–1997.
59 H. Hakkinen, Atomic and electronic structure of gold clusters: understanding flakes, cages and superatoms from simple concepts, *Chem. Soc. Rev.*, 2008, **37**, 1847–1859.
60 D. Mollenhauer and N. Gaston, A balanced procedure for the treatment of cluster–ligand interactions on gold phosphine systems in catalysis, *J. Comput. Chem.*, 2014, **35**, 986–997.
61 D. Mollenhauer, N. Gaston, E. Voloshina and B. Paulus, Interaction of Pyridine Derivatives with a Gold (111) Surface as a Model for Adsorption to Large Nanoparticles, *J. Phys. Chem. C*, 2013, **117**, 4470–4479.
62 V. Tripkovic, I. Cerri, T. Bligaard, J. Rossmeisl, V. Tripkovic, I. Cerri, T. Bligaard and J. Rossmeisl, The Influence of Particle Shape and Size on the Activity of Platinum Nanoparticles for Oxygen Reduction Reaction: A Density Functional Theory Study, *Catal. Lett.*, 2014, **144**, 380–388.
63 J. V. Maya Girón, E. Zelaya, A. Rubert, G. Benítez, P. Carro, R. C. Salvarezza and M. E. Vela, Surface Chemistry of 4-Mercaptobenzoic Acid Self-Assembled on Ag(111) and Ag Nanoparticles, *J. Phys. Chem. C*, 2013, **117**, 24967–24974.
64 E. Fertitta, E. Voloshina and B. Paulus, Adsorption of multivalent alkylthiols on Au(111) surface: Insights from DFT, *J. Comput. Chem.*, 2014, **35**, 204–213.
65 J. P. Desclaux, Relativistic Dirac-Fock expectation values for atoms with Z = 1 to Z = 120, *At. Data Nucl. Data Tables*, 1973, **12**, 311–406.
66 *Gold Chemistry: Applications and Future Directions in the Life Sciences*, ed. F. Mohr, WILEY-VCH Verlag GmbH & Co. KGaA, Weinheim, 2009.
67 D. Figgen, G. Rauhut, M. Dolg and H. Stoll, Energy-consistent pseudopotentials for group 11 and 12 atoms: adjustment to multi-configuration Dirac–Hartree–Fock data, *Chem. Phys.*, 2005, **311**, 227–244.
68 D. Andrae, U. Häußermann, M. Dolg, H. Stoll and H. Preuß, Energy-adjustedab initio pseudopotentials for the second and third row transition elements, *Theor. Chim. Acta*, 1990, **77**, 123–141.
69 P. Schwerdtfeger, M. Dolg, W. H. E. Schwarz, G. A. Bowmaker and P. D. W. Boyd, Relativistic effects in gold chemistry. I. Diatomic gold compounds, *J. Chem. Phys.*, 1989, **91**, 1762–1774.
70 M. Tadjeddine and J. P. Flament, Ab initio interpretation of sum frequency generation spectra of 4-cyanopyridine adsorbed on Au(111), *Chem. Phys.*, 2001, **265**, 27–46.
71 P. J. Hay and W. R. Wadt, Ab initio effective core potentials for molecular calculations. Potentials for K to Au including the outermost core orbitals, *J. Chem. Phys.*, 1985, **82**, 299–310.
72 M. Reiher, Douglas–Kroll–Hess Theory: a relativistic electrons-only theory for chemistry, *Theor. Chem. Acc.*, 2006, **116**, 241–252.
73 M. Reiher and A. Wolf, Exact decoupling of the Dirac Hamiltonian. II. The generalized Douglas–Kroll–Hess transformation up to arbitrary order, *J. Chem. Phys.*, 2004, **121**, 10945–10956.
74 M. Reiher and A. Wolf, Exact decoupling of the Dirac Hamiltonian. I. General theory, *J. Chem. Phys.*, 2004, **121**, 2037–2047.
75 E. V. Lenthe, E. J. Baerends and J. G. Snijders, Relativistic regular two-component Hamiltonians, *J. Chem. Phys.*, 1993, **99**, 4597–4610.
76 E. van Lenthe, E. J. Baerends and J. G. Snijders, Relativistic total energy using regular approximations, *J. Chem. Phys.*, 1994, **101**, 9783–9792.

77 E. Van Lenthe, A. Ehlers and E.-J. Baerends, Geometry optimizations in the zero order regular approximation for relativistic effects, *J. Chem. Phys.*, 1999, **110**, 8943–8953.
78 D. Vanderbilt, Soft self-consistent pseudopotentials in a generalized eigenvalue formalism, *Phys. Rev. B: Condens. Matter Mater. Phys.*, 1990, **41**, 7892–7895.
79 G. Kresse and J. Hafner, Norm-conserving and ultrasoft pseudopotentials for first-row and transition elements, *J. Phys.: Condens. Matter*, 1994, **6**, 8245.
80 P. E. Blöchl, Projector augmented-wave method, *Phys. Rev. B: Condens. Matter Mater. Phys.*, 1994, **50**, 17953–17979.
81 G. Kresse and D. Joubert, From ultrasoft pseudopotentials to the projector augmented-wave method, *Phys. Rev. B: Condens. Matter Mater. Phys.*, 1999, **59**, 1758–1775.
82 O. Pluchery, M. Tadjeddine, J.-P. Flament and A. Tadjeddine, Adsorption of 4-cyanopyridine on Au(111): ab initio calculations and SFG measurements, *Phys. Chem. Chem. Phys.*, 2001, **3**, 3343–3350.
83 D. Mollenhauer, B. Paulus, unpublished.
84 D. Mollenhauer, Dissertation 2011, Freie Universität Berlin, Quantenchemische Untersuchungen zur Wechselwirkung von Pyridinderivaten mit Goldnanopartikeln.
85 S. Grimme, Semiempirical GGA-type density functional constructed with a long-range dispersion correction, *J. Comput. Chem.*, 2006, **27**, 1787–1799.
86 S. Grimme, J. Antony, S. Ehrlich and H. Krieg, A consistent and accurate ab initio parametrization of density functional dispersion correction (DFT-D) for the 94 elements H-Pu, *J. Chem. Phys.*, 2010, **132**, 154104.
87 C. Tuma and J. Sauer, Treating dispersion effects in extended systems by hybrid MP2:DFT calculations-protonation of isobutene in zeolite ferrierite, *Phys. Chem. Chem. Phys.*, 2006, **8**, 3955–3965.
88 K. Tonigold and A. Groß, Adsorption of small aromatic molecules on the (111) surfaces of noble metals: A density functional theory study with semiempirical corrections for dispersion effects, *J. Chem. Phys.*, 2010, 132.
89 J. V. Koppen, M. Hapka, M. Modrzejewski, M. M. Szczęśniak and G. Chałasiński, Density functional theory approach to gold-ligand interactions: Separating true effects from artifacts, *J. Chem. Phys.*, 2014, **140**, 244313.
90 L. Ferrighi, G. K. H. Madsen and B. Hammer, Self-consistent meta-generalized gradient approximation study of adsorption of aromatic molecules on noble metal surfaces, *J. Chem. Phys.*, 2011, **135**, 084704.
91 L. Buimaga-Iarinca and C. Morari, Adsorption of small aromatic molecules on gold: a DFT localized basis set study including van der Waals effects, *Theor. Chem. Acc.*, 2014, **133**, 1–11.
92 J. Granatier, M. Urban and A. J. Sadlej, 'Relativistic bond' in complexes of Cu, Ag, and Au elements with phosphine, *Chem. Phys. Lett.*, 2010, **484**, 154–159.
93 L. F. Pašteka, T. Rajský and M. Urban, Toward Understanding the Bonding Character in Complexes of Coinage Metals with Lone-Pair Ligands. CCSD(T) and DFT Computations, *J. Phys. Chem. A*, 2013, **117**, 4472–4485.
94 O. D. Haeberlen, H. Schmidbaur and N. Roesch, Stability of Main-Group Element-Centered Gold Cluster Cations, *J. Am. Chem. Soc.*, 1994, **116**, 8241–8248.
95 G. Ciancaleoni, N. Scafuri, G. Bistoni, A. Macchioni, F. Tarantelli, D. Zuccaccia and L. Belpassi, When the Tolman Electronic Parameter Fails: A Comparative DFT and Charge Displacement Study of [(L)Ni(CO)3]0/– and [(L)Au(CO)]0/+, *Inorg. Chem.*, 2014, **53**, 9907–9916.
96 H. Schmidbaur, S. Hofreiter and M. Paul, Synthesis of the gold analogue of the elusive doubly protonated water molecule, *Nature*, 1995, **377**, 503–504.

97 K. Angermaier and H. Schmidbaur, The supramolecular structures of complex tri[gold(I)]sulfonium cations, *Chem. Ber.*, 1994, **127**, 2387–2391.

98 F. Canales, M. C. Gimeno, P. G. Jones and A. Laguna, Aurophilicity at Sulfur Centers: Synthesis and Structure of the Tetragold(I) Species [(Ph3PAu)4S](CF3SO3)2 · 2CH2Cl2, *Angew. Chem., Int. Ed. Engl.*, 1994, **33**, 769–770.

99 P. Schwerdtfeger, H. L. Hermann and H. Schmidbaur, Stability of the Gold(I) – Phosphine Bond. A Comparison with Other Group 11 Elements, *Inorg. Chem.*, 2003, **42**, 1334–1342.

100 M. Melnínk and R. V. Parish, Classification and analysis of gold compounds on the basis of their x-ray structural and mössbauer spectroscopic data, *Coord. Chem. Rev.*, 1986, **70**, 157–257.

101 D. Burgos, C. Olea-Azar and F. Mendizabal, Theoretical study of the local reactivity of electrophiles of the type MPR 3 + (M = Cu, Ag, Au; R = -H, -Me, -Ph), *J. Mol. Model.*, 2012, **18**, 2021–2029.

102 M. Joost, A. Zeineddine, L. Estévez, S. Mallet – Ladeira, K. Miqueu, A. Amgoune and D. Bourissou, Facile Oxidative Addition of Aryl Iodides to Gold(I) by Ligand Design: Bending Turns on Reactivity, *J. Am. Chem. Soc.*, 2014, **136**, 14654–14657.

103 J. M. Forward, D. Bohmann, J. P. Fackler and R. J. Staples, Luminescence Studies of Gold(I) Thiolate Complexes, *Inorg. Chem.*, 1995, **34**, 6330–6336.

104 L. Gao, D. V. Partyka, J. B. Updegraff, N. Deligonul and T. G. Gray, Synthesis, Structures, and Excited-State Geometries of Alkynylgold(I) Complexes, *Eur. J. Inorg. Chem.*, 2009, (2009), 2711–2719.

105 E. R. T. Tiekink and J.-G. Kang, Luminescence properties of phosphine-gold(I) halides and thiolates, *Coord. Chem. Rev.*, 2009, **253**, 1627–1648.

106 K. A. Barakat, T. R. Cundari and M. A. Omary, Jahn – Teller Distortion in the Phosphorescent Excited State of Three-Coordinate Au(I) Phosphine Complexes, *J. Am. Chem. Soc.*, 2003, **125**, 14228–14229.

107 T. M. McCleskey and H. B. Gray, Emission spectroscopic properties of 1,2-bis(dicyclohexylphosphino)ethane complexes of gold(I), *Inorg. Chem.*, 1992, **31**, 1733–1734.

108 E. J. Fernandez, A. Laguna, J. M. Lopez-de-Luzuriaga, M. Monge, M. Montiel, M. E. Olmos and M. Rodriguez-Castillo, Synthesis, coordination to Au(I) and photophysical properties of a novel polyfluorinated benzothiazolephosphine ligand, *Dalton Trans.*, 2006, 3672–3677.

109 P. Li, B. Ahrens, A. D. Bond, J. E. Davies, O. F. Koentjoro, P. R. Raithby and S. J. Teat, A new series of dinuclear Au(I) complexes linked by diethynylpyridine groups, *Dalton Trans.*, 2008, 1635–1646.

110 C. D. Delfs, H. J. Kitto, R. Stranger, G. F. Swiegers, S. B. Wild, A. C. Willis and G. J. Wilson, Photoluminescence Properties of Four-Coordinate Gold(I) – Phosphine Complexes of the Types [Au(diphos)2]PF6 and [Au2(tetraphos)2](PF6)2, *Inorg. Chem.*, 2003, **42**, 4469–4478.

111 L. Rodríguez, C. Lodeiro, J. C. Lima and R. Crehuet, Neutral Gold(I) Metallosupramolecular Compounds: Synthesis and Characterization, Photophysical Properties, and Density Functional Theory Studies, *Inorg. Chem.*, 2008, **47**, 4952–4962.

112 T. Tanase, R. Otaki, T. Nishida, H. Takenaka, Y. Takemura, B. Kure, T. Nakajima, Y. Kitagawa and T. Tsubomura, Strongly Luminous Tetranuclear Gold(I) Complexes Supported by Tetraphosphine Ligands, meso- or rac-Bis[(diphenylphosphinomethyl)phenylphosphino]methane, *Chem. – Eur. J.*, 2014, **20**, 1577–1596.

113 G. S. M. Tong, S. C. F. Kui, H.-Y. Chao, N. Zhu and C.-M. Che, The 3[ndσ*(n + 1)pσ] Emissions of Linear Silver(I) and Gold(I) Chains with Bridging Phosphine Ligan, *Chem. – Eur. J.*, 2009, **15**, 10777–10789.
114 D. Mollenhauer, J. Flob, H.-U. Reissig, E. Voloshina and B. Paulus, Accurate quantum-chemical description of gold complexes with pyridine and its derivatives, *J. Comput. Chem.*, 2011, **32**, 1839–1845.
115 M. Sabaye Moghaddam, A. D. O. Bawagan, K. H. Tan and W. von Niessen, Study of the electronic structure of pyridine by synchrotron photoelectron spectroscopy, *Chem. Phys.*, 1996, **207**, 19–29.
116 S. Tixier, W. A. Shapley, Y. Zheng, D. P. Chong, C. E. Brion, Z. Shi and S. Wolfe, Electron densities for the outer valence orbitals of pyridine: comparison of EMS measurements with near Hartree–Fock limit and density functional theory calculations, *Chem. Phys.*, 2001, **270**, 263–276.
117 H.-C. Hsu, F.-W. Lin, C.-C. Lai, P.-H. Su and C.-S. Yeh, Photodissociation and theoretical studies of the Au + -(C5H5N) complex, *New J. Chem.*, 2002, **26**, 481–484.
118 M. Bortoluzzi, G. Paolucci, G. Annibale and B. Pitteri, A new π-acidity scale for several N-donor heterocycles as ligands in neutral gold(III) complexes, *Polyhedron*, 2009, **28**, 1079–1084.
119 M. Bortoluzzi, G. Paolucci and B. Pitteri, The roles of ligands proton affinity, π-back donation and metal fragment hardness on the Au–N bond in N-donor heterocycles gold(III) complexes, *Polyhedron*, 2010, **29**, 767–772.
120 M. Bortoluzzi and B. Pitteri, Synthesis of neutral gold(III) pyrimidine-complexes and theoretical studies on the proton affinity of the coordinated ligands, *Polyhedron*, 2010, **29**, 1833–1836.
121 K. J. Kilpin, R. Horvath, G. B. Jameson, S. G. Telfer, K. C. Gordon and J. D. Crowley, Pyridyl Gold(I) Alkynyls: A Synthetic, Structural, Spectroscopic, and Computational Study, *Organometallics*, 2010, **29**, 6186–6195.
122 K. M.-C. Wong, X. Zhu, L.-L. Hung, N. Zhu, V. W.-W. Yam and H.-S. Kwok, A novel class of phosphorescent gold(III) alkynyl-based organic light-emitting devices with tunable colour, *Chem. Commun.*, 2005, 2906–2908.
123 A. Vogler and H. Kunkely, Electronic spectra and photoreactivity of cyclopentadienyl complexes, *Coord. Chem. Rev.*, 2001, **211**, 223–233.
124 V. W.-W. Yam, S. W.-K. Choi, T.-F. Lai and W.-K. Lee, Syntheses, crystal structures and photophysics of organogold(III) diimine complexes, *J. Chem. Soc., Dalton Trans.*, 1993, 1001–1002.
125 C.-W. Chan, W.-T. Wong and C.-M. Che, Gold(III) Photooxidants. Photophysical, Photochemical Properties, and Crystal Structure of a Luminescent Cyclometalated Gold(III) Complex of 2,9-Diphenyl-1,10-Phenanthroline, *Inorg. Chem.*, 1994, **33**, 1266–1272.
126 J. A. Garg, O. Blacque, T. Fox and K. Venkatesan, Stable and Tunable Phosphorescent Neutral Cyclometalated Au(III) Diaryl Complexes, *Inorg. Chem.*, 2010, **49**, 11463–11472.
127 V. K.-M. Au, W. H. Lam, W.-T. Wong and V. W.-W. Yam, Luminescent Cyclometalated Alkynylgold(III) Complexes with 6-Phenyl-2,2′-Bipyridine Derivatives: Synthesis, Characterization, Electrochemistry, Photophysics, and Computational Studi, *Inorg. Chem.*, 2012, **51**, 7537–7545.
128 S. M. Tekarli, T. R. Cundari and M. A. Omary, Rational Design of Macrometallocyclic Trinuclear Complexes with Superior π-Acidity and π-Basicity, *J. Am. Chem. Soc.*, 2008, **130**, 1669–1675.
129 G. Zanti and D. Peeters, DFT Study of Bimetallic Palladium – Gold Clusters PdnAum of Low Nuclearities ($n + m \leq 14$), *J. Phys. Chem. A*, 2010, **114**, 10345–10356.

130 B. Assadollahzadeh and P. Schwerdtfeger, A systematic search for minimum structures of small gold clusters Aun(n = 2–20) and their electronic properties, *J. Chem. Phys.*, 2009, **131**, 064306.

131 Y. Dong, M. Springborg and I. Warnke, Structural and thermodynamic properties of Au2–20 clusters, *Theor. Chem. Acc.*, 2011, **130**, 1001–1008.

132 S. Bulusu and X. C. Zeng, Structures and relative stability of neutral gold clusters: Aun(n = 15–19), *J. Chem. Phys.*, 2006, **125**, 154303.

133 Gold clusters with the molecular formula [X(AuPH3)n]m+ with X = C,N,O,P,S and s n = 1–6 have been also studied, but will not be discussed within this review.

134 G. Schmid, R. Pfeil, R. Boese, F. Bandermann, S. Meyer, G. H. M. Calis and J. W. A. van der Velden, Au55[P(C6H5)3]12CI6 — ein Goldcluster ungewöhnlicher Größe, *Chem. Ber.*, 1981, **114**, 3634–3642.

135 M. Walter, M. Moseler, R. L. Whetten and H. Hakkinen, A 58-electron superatom-complex model for the magic phosphine-protected gold clusters (Schmid-gold, Nanogold[registered sign]) of 1.4-nm dimension, *Chem. Sci.*, 2011, **2**, 1583–1587.

136 S. N. Khanna and P. Jena, Assembling crystals from clusters, *Phys. Rev. Lett.*, 1992, **69**, 1664–1667.

137 B. Yoon, P. Koskinen, B. Huber, O. Kostko, B. von Issendorff, H. Häkkinen, M. Moseler and U. Landman, Size-Dependent Structural Evolution and Chemical Reactivity of Gold Clusters, *ChemPhysChem*, 2007, **8**, 157–161.

138 M. Walter and H. Hakkinen, A hollow tetrahedral cage of hexadecagold dianion provides a robust backbone for a tuneable sub-nanometer oxidation and reduction agent via endohedral doping, *Phys. Chem. Chem. Phys.*, 2006, **8**, 5407–5411.

139 D. Schebarchov and N. Gaston, Throwing jellium at gallium-a systematic superatom analysis of metalloid gallium clusters, *Phys. Chem. Chem. Phys.*, 2011, **13**, 21109–21115.

140 E. B. Guidez, A. Hadley and C. M. Aikens, Initial Growth Mechanisms of Gold-Phosphine Clusters, *J. Phys. Chem. C*, 2011, **115**, 6305–6316.

141 S. A. Ivanov, I. Arachchige and C. M. Aikens, Density Functional Analysis of Geometries and Electronic Structures of Gold-Phosphine Clusters. The Case of Au4(PR3)42+ and Au4(μ2-I)2(PR3)4, *J. Phys. Chem. A*, 2011, **115**, 8017–8031.

142 K. P. Huber and G. Herzberg, *Constants of Diatomic Molecules, Molecular Spectra and Molecular Structure*, Van Nostrand, New York, 1979, vol. 4.

143 P. Schwerdtfeger and P. D. W. Boyd, Role of phosphine ligands in gold cluster chemistry. Relativistic SCF calculations on Au2 and Au2(PH3)2, *Inorg. Chem.*, 1992, **31**, 327–329.

144 G. N. Khairallah, R. A. J. O'Hair and M. I. Bruce, Gas-phase synthesis and reactivity of binuclear gold hydride cations, (R3PAu)2H+ (R = Me and Ph), *Dalton Trans.*, 2006, 3699–3707.

145 F. Caruso, C. Pettinari, F. Paduano, R. Villa, F. Marchetti, E. Monti and M. Rossi, Chemical Behavior and in Vitro Activity of Mixed Phosphine Gold(I) Compounds on Melanoma Cell Lines, *J. Med. Chem.*, 2008, **51**, 1584–1591.

146 B. A. Collings, K. Athanassenas, D. Lacombe, D. M. Rayner and P. A. Hackett, Optical absorption spectra of Au7, Au9, Au11, and Au13, and their cations: Gold clusters with 6, 7, 8, 9, 10, 11, 12, and 13 s-electrons, *J. Chem. Phys.*, 1994, **101**, 3506–3513.

147 W. Harbich, S. Fedrigo, J. Buttet and D. M. Lindsay, Optical spectroscopy on size selected gold clusters deposited in rare gas matrices, *ZPhys-e.D: At., Mol. Clusters*, 1991, **19**, 157–159.

148 B. Palpant, B. Prével, J. Lermé, E. Cottancin, M. Pellarin, M. Treilleux, A. Perez, J. L. Vialle and M. Broyer, Optical properties of gold clusters in the size range 2–4 nm, *Phys. Rev. B: Condens. Matter Mater. Phys.*, 1998, **57**, 1963–1970.
149 J. P. Wilcoxon, J. E. Martin, F. Parsapour, B. Wiedenman and D. F. Kelley, Photoluminescence from nanosize gold clusters, *J. Chem. Phys.*, 1998, **108**, 9137–9143.
150 S. Link, A. Beeby, S. FitzGerald, M. A. El-Sayed, T. G. Schaaff and R. L. Whetten, Visible to Infrared Luminescence from a 28-Atom Gold Cluster, *J. Phys. Chem. B*, 2002, **106**, 3410–3415.
151 M. A. Omary, M. A. Rawashdeh-Omary, C. C. Chusuei, J. P. Fackler and P. S. Bagus, Electronic structure studies of six-atom gold clusters, *J. Chem. Phys.*, 2001, **114**, 10695–10701.
152 G. G. Andersson, V. B. Golovko, J. F. Alvino, T. Bennett, O. Wrede, S. M. Mejia, H. S. Al Qahtani, R. Adnan, N. Gunby, D. P. Anderson and G. F. Metha, Phosphine-stabilised Au9 clusters interacting with titania and silica surfaces: The first evidence for the density of states signature of the support-immobilised cluster, *J. Chem. Phys.*, 2014, 141.
153 J. F. Alvino, T. Bennett, D. Anderson, B. Donoeva, D. Ovoshchnikov, R. H. Adnan, D. Appadoo, V. Golovko, G. Andersson and G. F. Metha, Far-infrared absorption spectra of synthetically-prepared, ligated metal clusters with Au6, Au8, Au9 and Au6Pd metal cores, *RSC Adv.*, 2013, **3**, 22140–22149.
154 D. Mollenhauer and N. Gaston, Tunable electronic structure of phosphine passivated gold clusters (Au6–9): Understanding the size, shape, and charge dependence, in preparation.
155 D. M. P. Mingos, Molecular-orbital calculations on cluster compounds of gold, *J. Chem. Soc., Dalton Trans.*, 1976, 1163–1169.
156 D. M. P. Mingos, Gold - a flexible friend in cluster chemistry, *J. Chem. Soc., Dalton Trans.*, 1996, 561–566.
157 C. E. Briant, B. R. C. Tobald, J. W. White, L. K. Bell and D. M. P. Mingos, Synthesis and x-ray structural characterization of the centered icosahedral gold cluster compound [Au13(PMe2Ph)10Cl2](PF6)3; the realization of a theoretical prediction, *Chem. Commun.*, 1981, 201–202.
158 M. R. Provorse and C. M. Aikens, Origin of Intense Chiroptical Effects in Undecagold Subnanometer Particles, *J. Am. Chem. Soc.*, 2010, **132**, 1302–1310.
159 F. Dufour, B. Fresch, O. Durupthy, C. Chaneac and F. Remacle, Ligand and Solvation Effects on the Structural and Electronic Properties of Small Gold Clusters, *J. Phys. Chem. C*, 2014, **118**, 4362–4376.
160 B. Fresch, E. Hanozin, F. Dufour and F. Remacle, Interplay of structural and electronic stabilizing factors in neutral and cationic phosphine protected Au13 clusters, *Eur. Phys. J. D*, 2012, **66**, 1–9.
161 S. Knoppe, L. Lehtovaara and H. Häkkinen, Electronic Structure and Optical Properties of the Intrinsically Chiral 16-Electron Superatom Complex [Au20(PP3)4]4 +, *J. Phys. Chem. A*, 2014, **118**, 4214–4221.
162 X.-K. Wan, S.-F. Yuan, Z.-W. Lin and Q.-M. Wang, A Chiral Gold Nanocluster Au20 Protected by Tetradentate Phosphine Ligands, *Angew. Chem., Int. Ed.*, 2014, **53**, 2923–2926.
163 B. K. Teo, X. Shi and H. Zhang, Pure gold cluster of 1:9:9:1:9:9:1 layered structure: A novel 39-metal-atom cluster [(Ph3P)14Au39Cl6]Cl2 with an interstitial gold atom in a hexagonal antiprismatic cage, *J. Am. Chem. Soc.*, 1992, **114**, 2743–2745.
164 G. Schmid, The relevance of shape and size of Au55 clusters, *Chem. Soc. Rev.*, 2008, **37**, 1909–1930.

165 J. P. Fackler, C. J. McNeal, L. H. Pignolet and R. E. P. Winpenny, Californium-252 plasma desorption mass spectrometry as a tool for studying very large clusters; evidence for vertex-sharing icosahedra as components of Au67(PPh3)14Cl8, *J. Am. Chem. Soc.*, 1989, **111**, 6434–6435.
166 E. Gutiérrez, R. D. Powell, F. R. Furuya, J. F. Hainfeld, T. G. Schaaff, M. N. Shafigullin, P. W. Stephens and R. L. Whetten, Greengold, a giant cluster compound of unusual electronic structure, *Eur. Phys. J. D*, 1999, **9**, 647–651.
167 W. Chen and S. Chen, Oxygen Electroreduction Catalyzed by Gold Nanoclusters: Strong Core Size Effects, *Angew. Chem., Int. Ed.*, 2009, **48**, 4386–4389.
168 W. Vogel, B. Rosner and B. Tesche, Structural investigations of gold (Au55) organometallic complexes by x-ray powder diffraction and transmission electron microscopy, *J. Phys. Chem.*, 1993, **97**, 11611–11616.
169 R. Balasubramanian, R. Guo, A. J. Mills and R. W. Murray, Reaction of Au55(PPh3)12Cl6 with Thiols Yields Thiolate Monolayer Protected Au75 Clusters, *J. Am. Chem. Soc.*, 2005, **127**, 8126–8132.
170 H. Häkkinen, M. Moseler, O. Kostko, N. Morgner, M. A. Hoffmann and B. V. Issendorff, Symmetry and Electronic Structure of Noble-Metal Nanoparticles and the Role of Relativity, *Phys. Rev. Lett.*, 2004, **93**, 093401.
171 Y. Pei, N. Shao, Y. Gao and X. C. Zeng, Investigating Active Site of Gold Nanoparticle Au55(PPh3)12Cl6 in Selective Oxidation, *ACS Nano*, 2010, **4**, 2009–2020.
172 C. M. Mate, G. A. Somorjai, H. W. K. Tom, X. D. Zhu and Y. R. Shen, Vibrational and electronic spectroscopy of pyridine and benzene adsorbed on the Rh(111) crystal surface, *J. Chem. Phys.*, 1988, **88**, 441–450.
173 P. Hébert, A. Le Rille, W. Q. Zheng and A. Tadjeddine, Vibrational spectroscopic study of the adsorption of pyridine at the Au(111)-electrolyte interface by in situ difference frequency generation, *J. Electroanal. Chem.*, 1998, **447**, 5–9.
174 L. Stolberg, S. Morin, J. Lipkowski and D. E. Irish, Adsorption of pyridine at the Au(111)-solution interface, *J. Electroanal. Chem. Interfacial Electrochem.*, 1991, **307**, 241–262.
175 W.-B. Cai, L.-J. Wan, H. Noda, Y. Hibino, K. Ataka and M. Osawa, Orientational Phase Transition in a Pyridine Adlayer on Gold(111) in Aqueous Solution Studied by in Situ Infrared Spectroscopy and Scanning Tunneling Microscopy, *Langmuir*, 1998, **14**, 6992–6998.
176 G. Andreasen, M. E. Vela, R. C. Salvarezza and A. J. Arvia, Dynamics of Pyridine Adsorption on Gold(111) Terraces in Acid Solution from in-Situ Scanning Tunneling Microscopy under Potentiostatic Control, *Langmuir*, 1997, **13**, 6814–6819.
177 A. Tadjeddine, A. L. Rille, O. Pluchery, P. Hébert, W. Q. Zheng and T. Marin, Adsorption of pyridine on gold, studied by difference frequency generation (DFG) using the CLIO-FEL, *Nucl. Instrum. Methods Phys. Res., Sect. A*, 1999, **429**, 481–484.
178 A. Bilić, J. R. Reimers and N. S. Hush, Adsorption of Pyridine on the Gold(111) Surface: Implications for "Alligator Clips" for Molecular Wires, *J. Phys. Chem. B*, 2002, **106**, 6740–6747.
179 J.-F. Li, S.-Y. Ding, Z.-L. Yang, M.-L. Bai, J. R. Anema, X. Wang, A. Wang, D.-Y. Wu, B. Ren, S.-M. Hou, T. Wandlowski and Z.-Q. Tian, Extraordinary Enhancement of Raman Scattering from Pyridine on Single Crystal Au and Pt Electrodes by Shell-Isolated Au Nanoparticles, *J. Am. Chem. Soc.*, 2011, **133**, 15922–15925.
180 Z. Ma, F. Rissner, L. Wang, G. Heimel, Q. Li, Z. Shuai and E. Zojer, Electronic structure of pyridine-based SAMs on flat Au(111) surfaces: extended

charge rearrangements and Fermi level pinning, *Phys. Chem. Chem. Phys.*, 2011, **13**, 9747–9760.
181 L. Stolberg, J. Lipkowski and D. E. Irish, Adsorption of pyridine at the Au (100)-solution interface, *J. Electroanal. Chem. Interfacial Electrochem.*, 1987, **238**, 333–353.
182 J. Lipkowski, L. Stolberg, D. F. Yang, B. Pettinger, S. Mirwald, F. Henglein and D. M. Kolb, Molecular adsorption at metal electrodes, *Electrochim. Acta*, 1994, **39**, 1045–1056.
183 S.-Y. Ding, J. Yi, J.-F. Li and Z.-Q. Tian, A theoretical and experimental approach to shell-isolated nanoparticle-enhanced Raman spectroscopy of single-crystal electrodes, *Surf. Sci.*, 2015, **631**, 73–80.
184 D. Y. Wu, M. Hayashi, C. H. Chang, K. K. Liang and S. H. Lin, Bonding interaction, low-lying states and excited charge-transfer states of pyridine–metal clusters: Pyridine–Mn(M = Cu, Ag, Au; n = 2–4), *J. Chem. Phys.*, 2003, **118**, 4073–4085.
185 D. Y. Wu, M. Hayashi, Y. J. Shiu, K. K. Liang, C. H. Chang, Y. L. Yeh and S. H. Lin, A Quantum Chemical Study of Bonding Interaction, Vibrational Frequencies, Force Constants, and Vibrational Coupling of Pyridine – Mn (M = Cu, Ag, Au; n = 2 – 4), *J. Phys. Chem. A*, 2003, **107**, 9658–9667.
186 C. M. Aikens and G. C. Schatz, TDDFT Studies of Absorption and SERS Spectra of Pyridine Interacting with Au20, *J. Phys. Chem. A*, 2006, **110**, 13317–13324.
187 X. Zhao and M. Chen, DFT study on the influence of electric field on surface-enhanced Raman scattering from pyridine–metal complex, *J. Raman Spectrosc.*, 2014, **45**, 62–67.
188 S. M. Rosendahl, B. R. Danger, J. P. Vivek and I. J. Burgess, Surface Enhanced Infrared Absorption Spectroscopy Studies of DMAP Adsorption on Gold Surfaces, *Langmuir*, 2009, **25**, 2241–2247.
189 D. Mollenhauer, C. Brieger, E. Voloshina and B. Paulus, Performance of Dispersion-Corrected DFT for the Weak Interaction between Aromatic Molecules and Extended Carbon-Based Systems, *J. Phys. Chem. C*, 2015, **119**, 1898–1904.
190 B. Koslowski, A. Tschetschetkin, N. Maurer, P. Ziemann, J. Kučera and A. Groß, 4,4′-Dithiodipyridine on Au(111): A Combined STM, STS, and DFT Stu, *J. Phys. Chem. C*, 2013, **117**, 20060–20067.
191 J. Zhang, B. Li, X. Cui, B. Wang, J. Yang and J. G. Hou, Spontaneous Chiral Resolution in Supramolecular Assembly of 2,4,6-Tris(2-pyridyl)-1,3,5-triazine on Au(111), *J. Am. Chem. Soc.*, 2009, **131**, 5885–5890.
192 T. Waldmann, D. Künzel, H. E. Hoster, A. Groß and R. J. Behm, Oxidation of an Organic Adlayer: A Bird's Eye View, *J. Am. Chem. Soc.*, 2012, **134**, 8817–8822.
193 P. F. Cafe, A. G. Larsen, W. Yang, A. Bilic, I. M. Blake, M. J. Crossley, J. Zhang, H. Wackerbarth, J. Ulstrup and J. R. Reimers, Chemisorbed and Physisorbed Structures for 1,10-Phenanthroline and Dipyrido[3,2-a:2′,3′-c]phenazine on Au(111), *J. Phys. Chem. C*, 2007, **111**, 17285–17296.
194 M. Mura, N. Martsinovich and L. Kantorovich, Theoretical study of melamine superstructures and their interaction with the Au(111) surface, *Nanotechnology*, 2008, **19**, 465704.
195 F. Iori, S. Corni and R. Di Felice, Unraveling the Interaction between Histidine Side Chain and the Au(111) Surface: A DFT Study, *J. Phys. Chem. C*, 2008, **112**, 13540–13545.
196 S. Piana and A. Bilic, The Nature of the Adsorption of Nucleobases on the Gold [111] Surface, *J. Phys. Chem. B*, 2006, **110**, 23467–23471.

197 O. Díaz Arado, H. Mönig, H. Wagner, J.-H. Franke, G. Langewisch, P. A. Held, A. Studer and H. Fuchs, On-Surface Azide–Alkyne Cycloaddition on Au(111), *ACS Nano*, 2013, **7**, 8509–8515.

198 R. E. A. Kelly, W. Xu, M. Lukas, R. Otero, M. Mura, Y.-J. Lee, E. Lægsgaard, I. Stensgaard, L. N. Kantorovich and F. Besenbacher, An Investigation into the Interactions Between Self-Assembled Adenine Molecules and a Au(111) Surface, *Small*, 2008, **4**, 1494–1500.

199 G. Heimel, L. Romaner, E. Zojer and J.-L. Brédas, Toward Control of the Metal – Organic Interfacial Electronic Structure in Molecular Electronics: A First-Principles Study on Self-Assembled Monolayers of π-Conjugated Molecules on Noble Metals, *Nano Lett.*, 2007, **7**, 932–940.

200 M. J. Crossley, P. L. Burn, S. J. Langford and J. K. Prashar, Porphyrins with appended phenanthroline units: a means by which porphyrin [small pi]-systems can be connected to an external redox centre, *J. Chem. Soc., Chem. Commun.*, 1995, 1921–1923.

201 B. Xu and N. J. Tao, Measurement of Single-Molecule Resistance by Repeated Formation of Molecular Junctions, *Science*, 2003, **301**, 1221–1223.

202 X. Li, J. Hihath, F. Chen, T. Masuda, L. Zang and N. Tao, Thermally Activated Electron Transport in Single Redox Molecules, *J. Am. Chem. Soc.*, 2007, **129**, 11535–11542.

203 S. Y. Quek, H. J. Choi, S. G. Louie and J. B. Neaton, Thermopower of Amine – Gold-Linked Aromatic Molecular Junctions from First Principles, *ACS Nano*, 2010, **5**, 551–557.

204 H. Shimin, Z. Jiaxing, L. Rui, N. Jing, H. Rushan, S. Ziyong, Z. Xingyu, X. Zenquan and W. Quande, First-principles calculation of the conductance of a single 4,4 bipyridine molecule, *Nanotechnology*, 2005, **16**, 239.

205 S. V. Aradhya, M. Frei, M. S. Hybertsen and L. Venkataraman, Van der Waals interactions at metal/organic interfaces at the single-molecule level, *Nat. Mater.*, 2012, **11**, 872–876.

206 R. Stadler, K. S. Thygesen and K. W. Jacobsen, Forces and conductances in a single-molecule bipyridine junction, *Phys. Rev. B: Condens. Matter Mater. Phys.*, 2005, **72**, 241401.

207 M. E. Z. Michoff, P. Vélez and E. P. M. Leiva, Substituent Effect on the Mechanical Properties of Au – N Nanojunctions, *J. Phys. Chem. C*, 2009, **113**, 3850–3854.

208 A. D. Jewell, E. C. H. Sykes and G. Kyriakou, Molecular-Scale Surface Chemistry of a Common Metal Nanoparticle Capping Agent: Triphenylphosphine on Au(111), *ACS Nano*, 2012, **6**, 3545–3552.

209 S. H. Brodersen, U. Grønbjerg, B. Hvolbæk and J. Schiøtz, Understanding the catalytic activity of gold nanoparticles through multi-scale simulations, *J. Catal.*, 2011, **284**, 34–41.

210 X. Zhang and G. Lu, Computational Design of Core/Shell Nanoparticles for Oxygen Reduction Reaction, *J. Phys. Chem. Lett.*, 2014, **5**, 292–297.

211 B. Zhao, J. Huang and L. S. Bartell, Molecular dynamics studies of the size and temperature dependence of the kinetics of freezing of Fe nanoparticles, *J. Solid State Chem.*, 2013, **207**, 35–41.

212 J. Meier, J. Schiøtz, P. Liu, J. K. Nørskov and U. Stimming, Nano-scale effects in electrochemistry, *Chem. Phys. Lett.*, 2004, **390**, 440–444.